In this book the theories, techniques and applications of reflection electron microscopy (REM), reflection high-energy electron diffraction (RHEED) and reflection electron energy-loss spectroscopy (REELS) are comprehensively reviewed for the first time.

The book is divided into three parts: diffraction, imaging and spectroscopy. The text is written to combine basic techniques with special applications, theories with experiments, and the basic physics with materials science, so that a full picture of RHEED and REM emerges.

An entirely self-contained study, the book contains much invaluable reference material, including FORTRAN source codes for calculating crystal structures data and electron energy-loss spectra in different scattering geometries. This and many other features make the book an important and timely addition to the materials science literature and an ideal guide for graduate students and scientists working on quantitative surface structure characterizations using reflected electron techniques.

Reflection electron microscopy and spectroscopy for surface analysis

Reflection electron microscopy and spectroscopy for surface analysis

Zhong Lin Wang
Georgia Institute of Technology
Atlanta, Georgia,
USA

CAMBRIDGE UNIVERSITY PRESS
Cambridge, New York, Melbourne, Madrid, Cape Town, Singapore, São Paulo

Cambridge University Press
The Edinburgh Building, Cambridge CB2 2RU, UK

Published in the United States of America by Cambridge University Press, New York

www.cambridge.org
Information on this title: www.cambridge.org/9780521482660

First published 1996
This digitally printed first paperback version 2005

A catalogue record for this publication is available from the British Library

Library of Congress Cataloguing in Publication data

Wang, Zhong Lin.
Reflection electron microscopy and spectroscopy for surface analysis / Zhong Lin Wang.
p. cm.
Includes bibliographical references and index.
ISBN 0 521 48266 6
1. Materials – Microscopy. 2. Surfaces (Technology) – Analysis.
3. Reflection electron microscopy. I. Title.
TA417.23.W36 1996
620′.44–dc20 95-33552 CIP

ISBN-13 978-0-521-48266-0 hardback
ISBN-10 0-521-48266-6 hardback

ISBN-13 978-0-521-01795-4 paperback
ISBN-10 0-521-01795-5 paperback

Contents

Preface

This book was written following my review article 'Electron reflection, diffraction and imaging of bulk crystal surfaces in TEM and STEM', published by *Reports on Progress in Physics* [**56** (1993) 997]. Thanks are due to Dr Simon Capelin, the Editorial Manager of Cambridge University Press, for inviting me to write this book. The book is intended for surface scientists and microscopists who are interested in surface characterizations using reflected electron diffraction and imaging techniques.

Many of the ideas illustrated in the book were collected from my past working experiences with Professor J. M. Cowley, Dr J. Bentley, Professor R. F. Egerton, Dr Ping Lu and Dr J. Liu, to whom I am very grateful. Thanks also go to Professor J. C. H. Spence for his initial suggestions when this book was proposed. I heartily thank Dr Nea Wheeler for her careful and critical reviewing of the manuscript, which significantly improved its quality.

I am grateful to Drs C. C. Ahn, H. Banzhof, E. Bauer, P. A. Crozier, R. Garcia-Molina, J. M. Gibson, H. Homma, T. Hsu, A. Ichimiya, S. Ino, M. Iwatsuki, A. V. Latyshev, G. Lehmpfuhl, J. Liu, L. V. Litvin, H. Marten, G. Meyer-Ehmsen, H. Nakahara, H. Nakayama, N. Osakabe, J. C. H. Spence, Y. Tanishiro, K. Yagi, Y. Yamamoto and N. Yao, for permission to use their micrographs to illustrate the text. Thanks also go to Drs T. Hsu and L. M. Peng, who kindly provided the original REM bibliography text file.

Zhong Lin Wang

School of Materials Science and Engineering
Georgia Institute of Technology, Atlanta, GA, 30332-0245 *USA*.
e-mail: zhong.wang@mse.gatech.edu

Symbols and definitions

Listed below are some of the symbols frequently used in this book. All quantities are defined in SI units.

h	Planck's constant
$\hbar$	$=h/2\pi$
c	The speed of light in a vacuum
m_0	The rest mass of an electron
m_e	The mass of an electron with relativistic correction
e	The absolute charge of an electron
k_B	Boltzmann's constant
U_0	The accelerating voltage of an electron microscope
λ	The electron wavelength in free space
$\boldsymbol{p}$	The momentum of an incident electron
$\boldsymbol{K}_0$	The wavevector of an incident electron beam, $K_0=1/\lambda$
$\boldsymbol{K}$	The wavevector of a diffracted electron beam, $K=1/\lambda$
ω	Frequency
ϑ	The electron scattering semi-angle
f_α^e	The electron scattering factor of the αth atom
f_α^x	The X-ray scattering factor of the αth atom
κ	The κth atom in a crystal
$\sum_\kappa$	Sum over all atoms in crystal
$\sum_\alpha$	Sum over atoms within the unit cell
FT	Fourier transform from real space to reciprocal space
FT^{-1}	Inverse Fourier transform
$\boldsymbol{r}$	$=(x,y,z)$. A real-space vector
$\boldsymbol{b}$	$=(x,y)$. A real-space vector
$\boldsymbol{g}$(or $\boldsymbol{h}$)	A reciprocal-lattice vector
$\boldsymbol{u}$(or $\boldsymbol{\tau}$)	A reciprocal-space vector
$V(\boldsymbol{r})$	The electrostatic potential distribution in a crystal
$V(\boldsymbol{u})$	$=\mathrm{FT}(V(\boldsymbol{r}))$. The kinematic scattering amplitude of the crystal

$V_\kappa(\boldsymbol{r})$	The electrostatic potential of the κth atom
$\rho_\kappa(\boldsymbol{r})$	The electron density distribution of the κth atom
$\boldsymbol{s}$	The scattering vector, $\boldsymbol{s}=\boldsymbol{u}/2$, $s=(\sin\vartheta)/\lambda$
Z	Atomic number
S_p	The shape function of the crystal
V_g	The Fourier coefficient of the crystal potential
$V_\alpha(\boldsymbol{g})$	The Fourier transform of the αth atom in the unit cell
$\exp(-W_\alpha)$	The Debye–Waller factor of the αth atom
Ω	The volume of a unit cell
$\boldsymbol{r}_\alpha$	$=\boldsymbol{r}(\alpha)$. The position of the αth atom within the unit cell
$\boldsymbol{R}_n$	The position vector of the nth unit cell
$\boldsymbol{a}, \boldsymbol{b}, \boldsymbol{c}$	Base vectors of the unit cell
$\boldsymbol{a}^*, \boldsymbol{b}^*, \boldsymbol{c}^*$	Base vectors of the reciprocal lattice vector
θ_g	The Bragg angle
d_g	The interplanar distance
t	Time
$\otimes$	Convolution
$T_{obj}(\boldsymbol{u})$	The transfer function of the objective lens in reciprocal space
$A_{obj}(\boldsymbol{u})$	The shape function of the objective aperture in reciprocal space
C_s	The spherical aberration coefficient of the objective lens
Δf	The defocus of the objective lens
Δf_c	The focus shift introduced by chromatic aberration effects
Δf_s	The Schertzer defocus
C_c	The chromatic aberration coefficient of the objective lens
ΔE	Electron energy loss
$\boldsymbol{A}$ and $\boldsymbol{B}$	Basis vectors of the crystal lattice at the surface plane
$\boldsymbol{A}_s$ and $\boldsymbol{B}_s$	Basis vectors of the surface lattice
$\boldsymbol{A}_s^*$ and $\boldsymbol{B}_s^*$	Reciprocal-lattice vectors of the surface lattice
V_s	The surface potential
N_s	The number of surface unit cells
N_I	The number of surface islands
H	Step height
L	The width of a surface terrace
$\langle\rangle_c$	The configurational average over atom arrangements on a surface
υ	Surface coverage

γ	$=1[1-(v/c)^2]^{1/2}$. The relativistic correction factor
E(or E_0)	$=eU_0[1+eU_0/(2m_0c^2)]$. The energy of an incident electron
w	$=E+m_0c^2$. The total energy of an incident electron
$U(\boldsymbol{r})$	$=(2\gamma m_0 e/h^2)V(\boldsymbol{r})$, modified crystal potential
U_g	The Fourier coefficient of the modified potential U
v	The velocity of an incident electron
$\Psi(\boldsymbol{r})$	The electron wave function
$\Phi(\boldsymbol{r})$	The electron wave function excluding $\exp(2\pi i\boldsymbol{K}\cdot\boldsymbol{r})$ factor, $\Phi(\boldsymbol{r})=\Psi(\boldsymbol{r})\exp(-2\pi i\boldsymbol{K}\cdot\boldsymbol{r})$
$B_i(\boldsymbol{r})$	The ith branch Bloch wave
$\boldsymbol{k}^{(i)}$	The wavevector of the ith Bloch wave
α_i	Superposition coefficients of Bloch waves
$\boldsymbol{r}_\alpha$	The position of an atom in the unit cell
$C_g^{(i)}$	The eigenvector of the ith Bloch wave
v_i	The eigenvalue of the ith Bloch wave
$\boldsymbol{S}_g$	Excitation error
ξ_g	Two-beam extinction distance
$\boldsymbol{R}(\boldsymbol{r})$	The static displacement vector of atoms in an imperfect crystal
$\boldsymbol{b}_B$	Burgers vectors of dislocations
σ	$=\dfrac{\pi e\gamma}{\lambda E}=\dfrac{1}{\hbar v}$
Δz	The thickness of a crystal slice
$\boldsymbol{K}_h$	The component of a wavevector in the $\boldsymbol{b}$ plane
θ	The beam's incident angle with respect to the crystal surface
φ	The beam's deviation angle parallel to the surface with respect to the zone axis
$P(\boldsymbol{b},\Delta z)$	The propagation function of a slice with thickness Δz
Q	The phase grating function of a slice with thickness Δz
d	The width of the incident beam in a perpendicular-to-surface multislice calculation
$G(\boldsymbol{r},\boldsymbol{r}')$	Green's function
$\bar{V}_0$	Average crystal inner potential
ω	Angular frequency
E_k	Electron kinetic energy
n_r	The electron refraction index at the crystal surface
$\boldsymbol{B}$	The incident beam's azimuth
ϑ_E	The characteristic angle of inelastic scattering
F	The foreshortening factor

α_{mis}	The surface mis-cut angle
$\boldsymbol{M}_t$	The rotation matrix
ϕ_p	The phase jump at a surface step
D_p	The electron penetration depth into the surface
X_c	The coherence distance
d_r	Image resolution
α	$=\Theta/2$, semi-angle of the objective aperture
D_f	Depth of field
D_i	Depth of focus
d_0	Thickness of crystal foil
a_n	Crystal states
$\boldsymbol{q}$ (or $\boldsymbol{Q}$)	Change in crystal wavevector
$S(\tau, \tau')$	Mixed dynamic form factor
τ	Reciprocal space vector
$\rho_{n0}(\boldsymbol{u})$	Charge density matrix
Λ	Mean-free-path length of inelastic electron scattering
$\boldsymbol{R}_n$	$=\boldsymbol{R}(n)$, the position of the nth unit cell
$\boldsymbol{u}\binom{n}{\alpha}$	The vibrational displacement of αth atom inside the nth unit cell
$\boldsymbol{r}(\alpha)$	Equilibrium position of the atom in the unit cell
$\boldsymbol{u}_\kappa$	The time-dependent displacement vector of the κth atom
M_α	The mass of the αth atom in the unit cell
ε	The polarization vector of the phonon mode
$a^+\binom{-\boldsymbol{q}}{j}$	Creation operators of a phonon with wavevector $\boldsymbol{q}$ and dispersion surface ω_j
$a\binom{\boldsymbol{q}}{j}$	The annihilation operator of a phonon with wavevector $\boldsymbol{q}$ and dispersion surface ω_j
V_α	The time-dependent potential of the αth atom in the unit cell
$V_{0\alpha}$	The time-averaged atomic potential
ΔV	$=V-V_0$, perturbation of crystal potential due to atomic thermal vibration
N_0	The number of primitive cells in a crystal
n_0	The number of atoms in the primitive cell
ω_i	The phonon frequency
$\langle n_s \rangle$	The average occupation number of phonon state $\|n_s\rangle$
q_m	The radius of the Brillouin zone
V_{BZ}	The volume of the Brillouin zone
T	Temperature
T_D	The Debye temperature

$\overline{a_\kappa^2}$	The mean square vibration amplitude of the κth atom
v_j	The phonon velocity
$S_{\mathrm{TDS}}(\boldsymbol{Q}, \boldsymbol{Q}')$	The scattering function in TDS
$\varepsilon(\omega, \boldsymbol{q})$	The dielectric function of a solid
$\dfrac{\mathrm{d}^2 P}{\mathrm{d}z\,\mathrm{d}\omega}$	The differential excitation probability of valence states
$\varepsilon_{\tau\tau'}(\omega)$	The generalized dielectric function
q_{c}	The cut-off value of a wavevector
ω_{p}	The resonance frequency of the volume plasmon
ω_{s}	The resonance frequency of the surface plasmon
$\bar{m}$	The average number of plasmons excited
$\boldsymbol{E}(\boldsymbol{r}, t)$	The electric field vector
$\boldsymbol{B}(\boldsymbol{r}, t)$	The magnetic field vector
$\boldsymbol{J}(\boldsymbol{r}, t)$	The electron current density
$\rho(\boldsymbol{r}, t)$	The electron charge density
Π	The Hertz vector
$H'(\boldsymbol{r})$	The interaction Hamiltonian
σ_{n0}	The ionization cross-section of the nth state
σ_{t}	The total ionization cross-section
f_{E}	The electron single inelastic scattering function
$J(R, E_0 - \Delta E)$	The electron energy-loss distribution function
m_{v}	The average number of volume plasmons excited
m_{s}	The average number of surface plasmons excited
L_{s}	The average distance that an electron travels along the surface
Λ	The inelastic mean free path length
σ_{I}	The angular integrated ionization cross-section
Θ	Solid angle
$\mathfrak{R}$	The Rydberg energy
a_{B}	The Bohr radius
β	The collection semi-angle of an EELS spectrometer
Δ	The energy width of an integration window
n_{A}	The atom concentration
σ_{eff}	The effective angular integrated ionization cross-section
i_{A}	The channeling current density at atom sites
n_{x}	The X-ray refraction index

θ_c The critical angle for total external X-ray reflection

Sign conventions

Free-space plane wave $\exp(2\pi i \boldsymbol{K}\cdot\boldsymbol{r} - i\,\omega t)$

Fourier transforms

real space to reciprocal space $f(\boldsymbol{u}) = \int d\boldsymbol{r} \exp(-2\pi i \boldsymbol{u}\cdot\boldsymbol{r}) f(\boldsymbol{r}) \equiv \mathrm{FT}[f(\boldsymbol{r})]$,

reciprocal space to real space $f(\boldsymbol{r}) = \int d\boldsymbol{u} \exp(2\pi i \boldsymbol{u}\cdot\boldsymbol{r}) f(\boldsymbol{u}) \equiv \mathrm{FT}^{-1}[f(\boldsymbol{u})]$,

where the limis of integration are $(-\infty,\infty)$ unless otherwise specified.

Introduction

In 1986, E. Ruska was awarded the Nobel Physics Prize for his pioneering work of building the world's first transmission electron microscope (TEM) in the late 1920s. The mechanism of TEM was originally based on the physical principle that a charged particle could be focused by magnetic lenses, so that a 'magnifier' similar to an optic microscope could be built. The discovery of wave properties of electrons really revolutionized people's understanding about the potential applications of a TEM. In the last 60 years TEM has experienced a revolutionary development both in theory and in electron optics, and has become one of the key research tools for materials characterization (Hirsch *et al.*, 1977; Buseck *et al.*, 1989). The point-to-point image resolution currently available in TEM is better than 0.2 nm, which is comparable to the interatomic distances in solids.

High-resolution TEM is one of the key techniques for real-space imaging of defect structures in crystalline materials. Quantitative structure determination is becoming feasible, particularly with the following technical advances. The installation of an energy-filtering system on a TEM has made it possible to form images and diffraction patterns using electrons with different energy losses. Accurate structure analysis is possible using purely elastically scattered electrons, scattering of which can be exactly simulated using the available theories. The traditional method of recording images on film is being replaced by digital imaging with the use of a charge-coupled device (CCD) camera, which has a large dynamical range with single-electron detection sensitivity. Thus, electron diffraction patterns and images can be recorded linearly in intensity, and a quantitative fitting is feasible between an experimentally observed image and a theoretically simulated image. This is the future direction of electron microscopy, which allows quantitative structure determination with an accuracy comparable to that of X-ray diffraction. A modern TEM is a versatile machine, which can not only explore the crystal structure using imaging and diffraction techniques but also can perform high-spatial resolution microanalysis using energy-dispersive X-ray spectroscopy (EDS) and electron energy-loss spectroscopy (EELS). Thus the chemical composition in a region of diameter smaller than a few nanometers can be determined. Therefore, TEM is usually known as high-resolution analytical electron microscopy, which is becoming an indispensable technique for materials research.

A wide variety of diffraction, spectroscopy, and microscopy techniques are now available for the characterization of thin films and surfaces; but only the microscope methods, primarily those using electrons, are able to provide direct real-space information about local inhomogeneities. Accompanying the extended applications in materials science and thin crystal characterizations, TEM has been employed to image the surface structure. There are several techniques, such as weak-beam dark-field and surface profile imaging techniques (Cowley, 1986; Smith, 1987), that have been developed for studying surface structures in TEM. This book is about reflection high-energy electron diffraction (RHEED), reflection electron microscopy (REM), scanning REM (SREM) and the associated analytical techniques for studying bulk crystal surfaces and surfaces deposited with thin films. Emphasis is placed on real-space imaging of surface structures at high resolution. These techniques can be applied to perform *in situ* studies of surfaces prepared in the molecular beam epitaxy (MBE) chamber.

A surface is a special state of condensed matter, and it is the boundary between materials and a vacuum. In the semiconductor device industry, for example, techniques are needed to control surface structures in order to control some specific transport properties. Epitaxial growth of thin films is becoming an indispensable technique for synthesizing new materials, such as superconductor thin films, semiconductor superlattices, metallic superlattices (or multilayers) and diamond films, which have important applications in advanced technologies. Therefore, surface characterization is an essential branch of materials science.

Techniques that have been applied to investigate surface structures are classified into the following categories: surface crystallography, diffraction and imaging, electron spectroscopy, incident ion techniques, desorption spectroscopy, tunneling microscopy, work function techniques, atomic and molecular beam scattering, and vibration spectroscopy. An introduction to these techniques has been given by Woodruff and Delchar (1994). Table 0.1 compares various imaging and diffraction techniques that have been developed for surface studies. Each of these techniques has its unique advantages, and most of the techniques use an electron beam as the probe. As limited by the physical mechanisms and the equipment designs, however, most of these techniques may not be adequate to be applied for imaging *in situ* surface phenomena. In this book, we introduce the reflection high-energy electron diffraction (RHEED) and reflection electron microscopy and spectrometry techniques, which can be applied to *in situ* observations of thin film nucleation and growth.

For surface studies it is rarely satisfactory to use only one technique. Information regarding structure, composition and electronic structure is usually required in order to accurately determine the surface structure. Therefore, imaging techniques are usually applied in conjunction with other techniques that can provide surface-sensitive chemical and electronic structures. The two most commonly used tech-

Table 0.1. *Techniques for imaging surface structures; TEM: transmission electron microscopy; STEM: scanning transmission electron microscopy; REM: reflection electron microscopy; SREM: scanning reflection electron microscopy; LEEM: low-energy electron microscopy; SLEEM: scanning low-energy electron microscopy; SP-LEEM: spin polarized LEEM; SEM: scanning electron microscopy; SEMPA: SEM with polarization analysis; SAM: scanning Auger microscopy; PEEM: photoemission electron microscopy; STM: scanning tunneling microscopy; AFM: atom force microscopy; MFM: magnetic force microscopy; SNFOM: scanning near field optical microscopy; FIM: field ion microscopy; and FEM: field emission microscopy. Diffraction and analytical techniques associated with the above techniques: TED: transmission electron diffraction; EELS: electron energy-loss spectroscopy; RHEED: reflection high-energy electron diffraction; LEED: low-energy electron diffraction; TRAXS: total reflection angle X-ray spectroscopy; AES: Auger electron spectroscopy; UPS: ultraviolet photoelectron spectroscopy: XPS: soft X-ray photoemission electron spectroscopy; and EDS: energy dispersive spectroscopy.*

Technique	Contrast mechanism	Resolution (nm)	Features	Chemical analysis
TEM	Diffraction and phase grating	0.2	Atomic resolution, thin film and fine particles	AES
STEM	Diffraction and phase grating	0.2	Microdiffraction, microanalysis	AES, EELS
REM	Phase and diffraction	0.5	Bulk crystals	TRAXS, EELS, AES, RHEED
SREM	Phase and diffraction	0.5	Bulk crystal, microdiffraction	TRAXS, EELS, AES, RHEED
LEEM	LEED	5	No foreshortening	
SLEEM	LEED			
SP-LEEM	Magnetic force	10	Magnetic domain	
SEM	Secondary electron	1	Topography	EDS, Auger
SEMPA	Spin scattering		Magnetic domain	
SAM	Auger electron	2	Chemical mapping	Auger
PEEM	Photoelectron	10	Work function, XPS, UPS	Energy analysis
STM	Tunneling effect	0.02 (z) 0.1 (x, y)	High resolution	
AFM	Atomic force	0.02 (z) 0.1 (x, y)	High resolution, non-conducting surface	
MFM	Magnetic force		Surface magnetic domain	
SNFOM	Photon		No surface damage	
FIM	Ionization	0.2	High resolution, depth profile	Atom probe mass spectrometer
FEM	Tunneling		Work function	

niques are LEED and AES. LEED provides a simple and convenient characterization of the surface crystallography whereas AES provides some indication of chemical composition. Table 0.2 gives a summary of the diffraction and analytical techniques that have been widely used for surface studies.

0.1 Historical background

The reflection electron imaging technique was first devised by Ruska (1933) shortly after the invention of TEM. This development was initiated in order to exceed the resolution limit of surface imaging by optical microscopes. Reflection electron microscopy has experienced an unsteady development (Fert and Saport, 1952; Menter, 1953; Watanabe, 1957) due to competition from other surface imaging techniques, such as scanning electron microscopy (SEM) and the replica technique for TEM. Reflection electron microscopy was advanced by Halliday and Newman (1960), who used Bragg-reflected beams in reflection high-energy electron diffraction (RHEED) patterns for REM imaging. In the 1970s, Cowley and colleagues (Cowley and Hojlund Nielsen, 1975; Hojlund Nielsen and Cowley, 1976) renewed the interest in REM with an emphasis on diffraction contrast, combining both real- and reciprocal-space analyses. A resolution of about 2 nm was achieved for directions parallel to the surface, exceeding the resolution limit of 10 nm for SEM at that time. Since then, REM has experienced rapid development due to improvement in techniques for preparing atomic flat surfaces and the introduction of ultra-high vacuum (UHV) TEMs. Applications of REM have been expanded to various fields, such as semiconductor surface reconstructions, and metal and ceramic surfaces, by many research groups (Cowley, 1986 and 1987; Bleloch *et al.*, 1987; Yagi, 1987; Hsu *et al.*, 1987; Hsu and Peng, 1987a; Yagi *et al.*, 1992; Latyshev *et al.*, 1992; Claverie *et al.*, 1992; Wang, 1993; Wang and Bentley, 1992; Uchida *et al.*, 1992a, b). In recent years, extensive theoretical calculations have been carried out to understand the basic scattering processes of high-energy (10 keV to 1 MeV) electrons from crystal surfaces in a RHEED geometry. Various other techniques, such as STM and electron holography, have been developed and used in conjunction with REM, to provide comprehensive characterization tools for surface studies. In addition, the application of REM and RHEED for *in situ* examinations of MBE growth has attracted much interest. The development of an energy-filtering system for TEM has important implications for REM and RHEED. Before the invention of this technology it was not possible to perform quantitative surface structure analysis, because only elastically scattering processes can be accurately calculated using the available theories.

Accompanying the rapid experimental progress in REM, analytical techniques, such as reflection electron energy-loss spectroscopy (REELS), have been developed.

Table 0.2. *Diffraction and analytical techniques for surface studies: ● indicates the most direct but not limited application of the technique*

Techniques	Atomic structure	Chemical composition	Electronic structure	Vibrational properties
Low-energy electron diffraction (LEED)	●			●
Reflection high-energy electron diffraction (RHEED)	●			●
Surface X-ray diffraction (SXRD)	●			●
X-ray photoelectron spectroscopy (XPS)	●	●	●	●
Surface extended X-ray absorption fine structure (SEXAFS)	●	●		●
Photoelectron diffraction (PhD)	●			●
Auger electron spectrosocopy (AES)		●	●	
Appearance potential spectroscopy (APS)		●	●	
Ionization loss spectroscopy (ILS)		●	●	
Ultraviolet photoelectron spectroscopy (UPS)	●	●	●	
Inverse photoemission spectroscopy (IPES)			●	
Ion neutralization spectroscopy (INS)			●	
Low-energy ion scattering (LEIS)	●	●		●
High-energy ion scattering (HEIS)	●	●		●
Secondary ion mass spectroscopy (SIMS)		●		
Temperature programmed desorption (TPD)		●	●	
Electron- and proton-stimulated desorption (ESD and PSD)	●	●	●	
ESD ion angular distribution (ESDIAD)	●		●	●
Molecular beam scattering (MBS)	●	●		
High-resolution atom scattering (HRAS)	●	●		
Infrared reflection–absorption spectroscopy (IRAS)	●	●	●	●
High-resolution EELS (HREELS)	●	●	●	●
Reflection electron energy-loss spectroscopy (REELS)		●	●	
Transmission high-energy electron diffraction (THEED)	●			●
Total reflection angle X-ray spectroscopy (TRAXS)		●		

Both low-energy-loss and high-energy-loss (Wang and Bentley, 1992) signals can be applied in determining the electronic and chemical structures of surfaces. Scanning reflection electron microscopy has been developed for imaging *in situ* surface structure evolution during MBE growth. The image resolution has been found to be improved remarkably when a field emission gun (FEG) is used.

0.2 The scope of the book

This book describes the theories, calculations, and RHEED, REM, SREM and REELS experiments using reflected electrons for studying bulk crystal surfaces and surface thin films grown by MBE. The entire text was written to combine basic techniques with special applications, theories with experiments, and the basic physics with materials science, so that a full picture about RHEED and REM is exhibited.

The book was written for graduate students and scientists who are interested in surface characterizations using reflected electron techniques. Surface scientists would find it useful for structure determinations using RHEED, REM and SREM techniques. Electron microscopists can obtain useful theories and references in applying microscopy techniques for surface studies. The book is self-contained and serves as a comprehensive source regarding the use of electron microscopy and spectrometry techniques for surface studies.

Chapter 1 provides some preliminary knowledge regarding the basics of kinematical electron diffraction and imaging theory. This chapter is indispensable for understanding the basics of RHEED. The concepts introduced in this chapter will be applied in describing the scattering of crystal surfaces.

The book is composed of three parts, concerning diffraction, imaging and spectrometry of reflected electrons. Each part is intended to give a full coverage of all the related topics. The entire text is given in a sequence so that readers can easily follow the flow of ideas and materials.

Part A is devoted to RHEED. The basic techniques and interpretations of RHEED patterns are illustrated in Chapter 2. Fundamental characteristics of RHEED and their physical basis are introduced. The two-dimensional description of surface crystallography is given, followed by surface structure determination using RHEED. The RHEED oscillation, a remarkable phenomenon for monitoring layer-by-layer crystal growth in MBE, has been described in detail. Chapter 3 addresses the main theoretical schemes for dynamical electron diffraction in RHEED geometry. Each theory is derived directly from the Schrödinger equation, and its applications in RHEED calculations are illustrated. A comparison is made between the existing theories in order to illustrate their unique advantages for some particular problems. Chapter 4 concentrates on the surface resonance effect in

RHEED. This effect is the most remarkable phenomenon in RHEED, which is responsible for the high surface sensitivity (or monolayer resonance) of RHEED and REM. The experimental conditions under which the resonance is observed, the physics of resonance, and applications of the resonance effect are comprehensively described. The scattering processes of electrons from crystal surfaces under different diffracting conditions are investigated with the assistance of dynamically calculated results. The scattering picture is applied to illustrate many fundamental characteristics of REM and REELS.

Part B consists of Chapters 5–8, in which the experimental techniques, contrast mechanism and applications of REM are given. Detailed experimental procedures for obtaining REM images in TEM are shown. Various effects observed in REM are described in Chapter 5. Chapter 6 is designed to illustrate the imaging theory of REM, and it is the key chapter for image interpretation. The resolution of REM is discussed with consideration of electron optics and scattering geometry. Numerous examples are shown in Chapters 7 and 8 illustrating the applications of REM for studying metal, semiconductor and ceramic surfaces under ultra-high-vacuum (UHV) and non-UHV conditions. The surfaces that have been studied are summarized in tables. It is demonstrated that REM is a powerful technique for *in situ* imaging of surface nucleation and growth of thin films that are of technological importance.

Part C concerns the spectrometry of reflected high-energy electrons. The fundamental inelastic scattering processes in RHEED are introduced, and their applications for surface measurements are shown. The basic physics, related mathematical description, and associated applications of each process are given. Chapter 9 is about phonon scattering in RHEED, in which the kinematical and dynamical diffraction theories of thermal diffusely scattered electrons are described. Applications of valence excitations in RHEED are given in Chapter 10. A complete introduction of the dielectric response theory with and without considering retardation effects is given. Examples are shown to demonstrate their applications for simulating REELS spectra. It is shown that the classical dielectric response theory gives exactly the same result as the quantum mechanical scattering theory under the first-order approximation. The valence-loss spectra are applied to measure some fundamental properties associated with RHEED, such as electron penetration depth into the surface and electron mean traveling distance along the surface. Chapter 11 outlines the applications of high-energy-loss EELS for determining surface chemical structures. It is worth pointing out first that the EELS technique described in this book is not the high-resolution EELS, which uses an incident electron beam energy of a few electron-volts and is sensitive to the vibration properties of surface atoms (Ibach and Mills, 1982). The EELS described here uses electrons of incident energies 30–400 keV. The factors that may affect the accuracy of surface chemical analysis are described in detail. Finally, Chapter 12 is furnished to show the various techniques that are used in conjunction with REM, such as secondary electron imaging, surface holography and STM. Techniques that assist RHEED in solving surface structures are also demonstrated. This chapter may serve to indicate the direction of future development of REM.

To experts in the field, the examples in this book are selected from typical studies representing the applications of REM to a wide choice of materials. In Appendix F, a bibliography of REM, SREM and REELS is given.

CHAPTER 1

Kinematical electron diffraction

In a conventional TEM, electrons are emitted from an electron gun and focused by the condenser lens as a beam, which illuminates the specimen. The electron beam interacts with the specimen and is scattered (or diffracted) by the crystal atoms. Thus, the electron wave at the exit face of the specimen contains information about the potential distribution in the specimen. Since electrons are charged particles, their interaction with a solid is rather strong in comparison with either X-rays or neutrons, so that multiple scattering effects are always present in electron diffraction. This means that electron diffraction must be described by dynamical scattering theory, especially when quantitative structural analysis is necessary. However, many characteristics of electron diffraction can be qualitatively treated based on the kinematical scattering theory, which is actually a single-scattering theory. The purpose of this chapter is to outline some basic concepts of kinematical electron diffraction theory and to introduce the imaging theory of TEM, which will be applied in the future chapters for REM imaging. A systematic kinematical treatment of electron diffraction for perfect and imperfect crystals has been given by Cowley (1981). A complete description of dynamical electron diffraction theories has been given by Wang (1995).

1.1 Electron wavelength

Although TEM was invented based on the optical behavior of charged particles, the diffraction of electrons is purely a result of the wave property of particles. The wavelength λ of an electron is related to its momentum, p, by the de Broglie relation,

$$\lambda = h/p. \tag{1.1}$$

In TEM, if the electron is accelerated to a kinetic energy of eU_0, then relativity theory indicates that the momentum p satisfies

$$eU_0 = (m_0^2c^4 + p^2c^2)^{1/2} - m_0^2c^2, \tag{1.2}$$

where $m_0c^2 = 511.77$ keV is the rest energy of an electron. Substituting the solution of p from (1.2) into (1.1), the electron wavelength is

$$\lambda = \frac{h}{\left[2m_0 e U_0 \left(1 + \frac{eU_0}{2m_0 c^2}\right)\right]^{1/2}}. \tag{1.3}$$

Substituting in the numerical values of the fundamental physical constants yields

$$\lambda = \frac{1.226}{[U_0(1 + 0.9788 \times 10^{-6} U_0)]^{1/2}} \text{ nm}, \tag{1.4}$$

where U_0 is measured in volts. A table of electron wavelengths for various accelerating voltages is given in Appendix A. For 100 kV electrons, $\lambda = 0.0037$ nm. Thus, in comparison with the size of an atom, which is on the order of 0.05 nm, the electron cannot be viewed as a particle but rather as a wave. Since the electron wavelength is small, a significant phase shift can be created if there is a small change in the scattering potential. A slight modification in crystal potential due to structural evolution can dramatically affect the scattering behavior of the electron. Since the atom's size is less than 0.05 nm, a phase variation can be created after the electron interacts with the atom. The change of electron phase is the origin of phase contrast in electron microscopy.

1.2 Plane wave representation of an incident electron

In electron imaging and diffraction, an incident electron is expressed as a plane wave. The electron scattering behavior must be treated using wave mechanics, because the electron wavelength is much smaller than the size of the atom. In free space and for a non-relativistic case, the propagation of an electron satisfies the time-dependent Schrödinger equation

$$-\frac{\hbar^2}{2m_0} \nabla^2 \Psi = i\hbar \frac{\partial \Psi}{\partial t}. \tag{1.5}$$

For an infinitely large free space, the plane wave solution of this equation is

$$\Psi = \exp(2\pi i \boldsymbol{K}_0 \cdot \boldsymbol{r} - i\omega t), \tag{1.6}$$

where $\hbar\omega = h^2 K_0^2 / 2m_0 = E$ is the electron kinetic energy and $K_0 = 1/\lambda$ is the electron wave number. The scattering of the electron by a crystal is treated as the scattering of the plane wave by the potential field of the solid. Therefore, electron diffraction by a specimen becomes a scattering process of the plane wave by the electrostatic potential field of the specimen. The interaction between the incident electron and the crystal is characterized by the crystal potential. Thus, the incident electron 'sees' the potential distribution in the crystal. This is different from X-ray diffraction in which

X-rays 'see' charge distribution in the crystal (due to nuclei and crystal electrons). The potential is related to the charge density by Poisson's equation $\varepsilon_0 \nabla^2 V = -\rho$. In general, the effective distribution range of the potential function V is significantly larger than that of the charge density function ρ, meaning that the effective size of an atom 'seen' by electrons is larger than that 'seen' by X-rays.

The time-dependent factor $\exp(i\omega t)$ is a common factor in the electron wave function. This factor is dropped for convenience of discussion and the final result is unaffected. In this book $\exp(2\pi i \boldsymbol{K}_0 \cdot \boldsymbol{r})$ is taken as a plane wave with wave vector $\boldsymbol{K}_0$, which characterizes the momentum of the incident electron.

In TEM, the spherical wave emitted by the electron source is equivalent to a plane wave when it falls on the specimen, because the effective distance between the source and the specimen is infinity. Also the distance between the image plane and the specimen is infinity. Thus, electron diffraction is a scattering process in which a plane wave is scattered by the crystal, and the observing point is at infinity.

1.3 The Born approximation and single-atom scattering

The simplest case in electron diffraction is that of scattering by a single atom. In this case, the electron is scattered due to its interaction with the electrostatic potential of the atom. For a plane wave, if the scattering potential $V(\boldsymbol{r})$ is weak and it is distributed in a small region, then the wave observed at a large distance from the scattering zone is

$$\Psi \approx \exp(2\pi i \boldsymbol{K}_0 \cdot \boldsymbol{r}) + f(\vartheta) \frac{\exp(2\pi i r)}{r}, \tag{1.7}$$

where $f(\vartheta)$ indicates the amplitude of the scattered wave as a function of the scattering semi-angle. The amplitude with which an incident plane wave $\exp(2\pi i \boldsymbol{K}_0 \cdot \boldsymbol{r})$ is scattered to an exit plane wave $\exp(2\pi i \boldsymbol{K} \cdot \boldsymbol{r})$ is calculated based on the first Born approximation,

$$f(\boldsymbol{u}) = -\frac{m_0}{2\pi\hbar^2} \int_{-\infty}^{\infty} \mathrm{d}\boldsymbol{r} \exp(-2\pi i \boldsymbol{u} \cdot \boldsymbol{r})\, V(\boldsymbol{r}), \tag{1.8}$$

where $h\boldsymbol{u} = h(\boldsymbol{K} - \boldsymbol{K}_0)$ is the momentum transfer of the incident electron. The Born approximation assumes single scattering; that is, the electron is scattered only once. This assumption is the basis of kinematical scattering theory. Equation (1.8) explicitly means that the scattering amplitude under the single-scattering approximation is proportional to the Fourier transform of the scattering object potential. If the atomic potential is spherical symmetric, i.e., $V(\boldsymbol{r}) = V(r)$, then Eq. (1.8) becomes

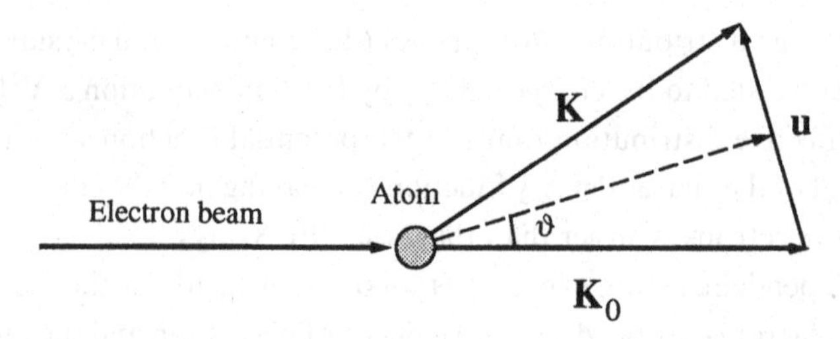

Figure 1.1 Scattering of a plane wave by a single atom under the Born approximation.

$$f(\vartheta) = -\frac{2m_0}{\hbar^2} \int_0^\infty \mathrm{d}r \, \frac{\sin(2\pi ur)}{2\pi ur} V(r)\, r^2, \tag{1.9}$$

where 2ϑ is the scattering angle as shown in Figure 1.1, and $\boldsymbol{u} = \boldsymbol{K} - \boldsymbol{K}_0$ with $u = 2K_0 \sin\vartheta$. The elastic scattering obeys the law of conservation of energy, or equivalently $|\boldsymbol{K}|^2 = |\boldsymbol{K}_0|^2$.

From Eq. (1.8), the scattering power of the atom is determined by the Fourier transform (FT) of its electrostatic potential, thus the electron scattering factor of the atom is defined as

$$f^{\mathrm{e}}(\boldsymbol{u}) = \int_{-\infty}^{\infty} \mathrm{d}\boldsymbol{r} \exp(-4\pi \mathrm{i} \boldsymbol{s} \cdot \boldsymbol{r})\, V(\boldsymbol{r}), \tag{1.10}$$

where the scattering vector is defined as $\boldsymbol{s} = \boldsymbol{u}/2$, with $|\boldsymbol{s}| = \sin\vartheta/\lambda$ and $\boldsymbol{u}$ a reciprocal space vector. The electron scattering factor is usually given as a function of $\boldsymbol{s}$. For convenience in mathematical expressions in this book, we use $\boldsymbol{u}$ as the variable of $f^{\mathrm{e}}(\boldsymbol{u})$. The electron scattering factor defined in Eq. (1.10) is a quantity that characterizes the scattering power of an atom and is independent of accelerating voltage.

1.4 The Fourier transform

Following the introduction of the electron scattering factor in Eq. (1.10), the Fourier transform is the most useful mathematical technique in electron diffraction, which correlates the electron waves in reciprocal space (or the diffraction plane) with those in real space (or the image plane) based on Abbe's imaging theory (to be given in Section 1.9). In this book, the form of the Fourier transform is defined as

$$\Psi(\boldsymbol{u}) = \mathrm{FT}[\Psi(\boldsymbol{b})] = \int_{-\infty}^{\infty} \mathrm{d}\boldsymbol{b} \exp(-2\pi \mathrm{i} \boldsymbol{u} \cdot \boldsymbol{b})\, \Psi(\boldsymbol{b}), \tag{1.11a}$$

and the inverse Fourier transform is

$$\Psi(\boldsymbol{b}) = \mathrm{FT}^{-1}[\Psi(\boldsymbol{u})] = \int_{-\infty}^{\infty} \mathrm{d}\boldsymbol{u} \exp(2\pi \mathrm{i}\boldsymbol{u} \cdot \boldsymbol{b})\, \Psi(\boldsymbol{u}), \tag{1.11b}$$

The Fourier transformation is an important mathematical calculation in describing the imaging and diffraction theories and its initial introduction and many features have been given by Cowley (1981). Two useful relations in imaging theory are

$$\delta(\boldsymbol{u} - \boldsymbol{K}_b) = \int_{-\infty}^{\infty} \mathrm{d}\boldsymbol{b} \exp(-2\pi \mathrm{i}\boldsymbol{u} \cdot \boldsymbol{b}) \exp(2\pi \mathrm{i}\boldsymbol{K}_b \cdot \boldsymbol{b}), \tag{1.12a}$$

$$\int_{-\infty}^{\infty} \mathrm{d}\boldsymbol{u}\, \delta(\boldsymbol{u} - \boldsymbol{K}_b)\, \psi(\boldsymbol{u}) = \psi(\boldsymbol{K}_b). \tag{1.12b}$$

1.5 The scattering factor and the charge density function

The electron scattering factor is usually difficult to measure experimentally because electron scattering in crystals is a multiple-scattering process. X-ray diffraction, on the other hand, is usually treated kinematically for thin crystals, and quantitative data analysis is possible using kinematical scattering theory. For this reason, X-ray diffraction is one of the most accurate methods for structural determination. The interaction of X-rays (or other electromagnetic waves) with a crystal is determined by the charge distribution, which can be derived from the potential function using Poisson's equation. For an atom of atomic number Z, the electrostatic potential function (V_κ) and electron charge density function (ρ_κ) are related by

$$\nabla^2 V_\kappa(\boldsymbol{r}) = -\frac{e}{\varepsilon_0} [Z\delta(\boldsymbol{r}) - \rho_\kappa(\boldsymbol{r})], \tag{1.13}$$

where $\delta(\boldsymbol{r})$ is the Dirac delta function indicating the position of the nucleus. Taking a Fourier transform of Eq. (1.13), using Eq. (1.10) and the X-ray scattering factor defined by

$$f_\kappa^{\mathrm{x}}(s) = \int \mathrm{d}\boldsymbol{r} \exp(-4\pi \mathrm{i}\boldsymbol{s} \cdot \boldsymbol{r})\, \rho_\kappa(\boldsymbol{r}), \tag{1.14}$$

we have

$$(4\pi \mathrm{i}s)^2 f_\kappa^{\mathrm{e}}(\boldsymbol{u}) = -\frac{e}{\varepsilon_0} [Z - f_\kappa^{\mathrm{x}}(s)]$$

or

$$f_\kappa^e(\boldsymbol{u}) = \frac{e}{16\pi^2\varepsilon_0}\frac{[Z - f_\kappa^x(s)]}{s^2}. \tag{1.15}$$

This is known as Mott's formula, based on which the electron scattering factor can be directly calculated from the experimentally measured X-ray scattering factor. When interacting with crystals, X-rays directly 'see' the charge distribution, and electrons 'see' the potential.

For general purposes, f^x can be written in analytical form in terms of the Gaussian functions (Doyle and Turner, 1968)

$$f^x(s) = \sum_{i=1}^{4} [a_i \exp(-b_i s^2) + c_i], \tag{1.16}$$

where the fitting parameters have been determined for most of the elements in neutral and ionized states and the results are listed in Appendix B. For high-energy electrons, the correction introduced by the exchange effect of the valence electrons on the atomic scattering factor is negligible, so that the scattering factor defined in Eq. (1.16) is independent of the incident energy of the electrons. This approximation may not hold if the energy of the incident electron is lower than a few kilo-electron-volts (Qian *et al.*, 1993), for which the exchange effect is significant. For atoms with spherical symmetry, the atomic scattering factor is a real function. The atomic scattering factor may be a complex function if the atomic potential is not absolutely spherically symmetric due to charge redistribution or exchange resulting from solid bonding. Thus the imaginary atomic scattering factor is a measurement of solid bonding.

The asymptotic behavior of the scattering factors for large and small values of s must be considered (Peng and Cowley, 1988a). The electron scattering factor must converge to the Rutherford scattering factor at high angles. The convergence of f^e at $s = 0$ requires that $f^x(0) = Z$ according to the Mott formula, or equivalently

$$f^x(0) = Z = \sum_{i=1}^{4} (a_i + c_i).$$

Figure 1.2(a) shows a comparison of the electron scattering factors of silicon and gold, and Figure 1.2(b) shows the corresponding atomic potentials. It is apparent that the scattering power of Au is much greater than that of Si. The height of the potential peak is a direct measurement of the high-angle scattering power of the atom. Thus, the specimen thickness less than which the kinematic scattering theory holds would be much smaller for Au than for Si. The full width at half maximum of the potential peak is about 0.2 Å.

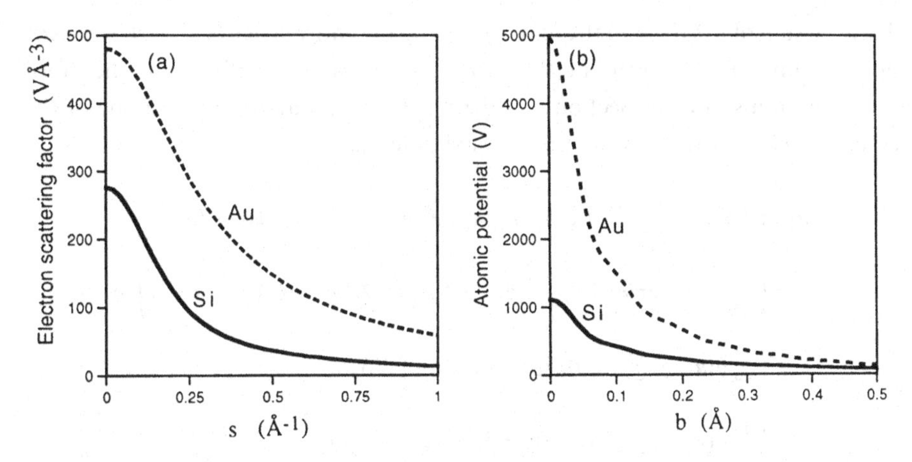

Figure 1.2 *A comparison of (a) electron scattering factors and (b) atomic potentials of gold and silicon atoms, showing the increase of atomic scattering power with increasing atomic number, where* $s = (\sin\theta)/\lambda$ *and* $b = (x^2 + y^2)^{\frac{1}{2}}$.

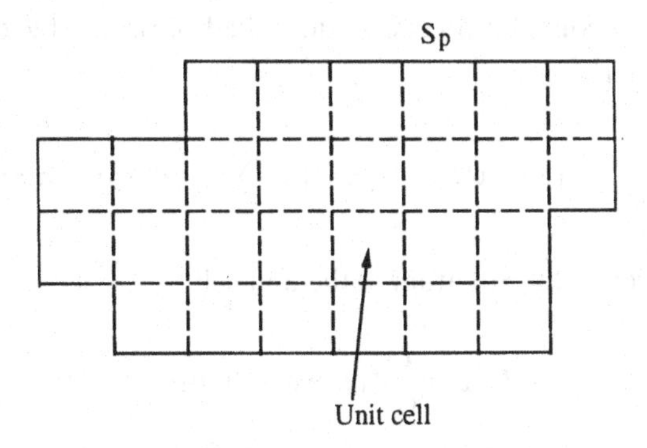

Figure 1.3 *A finite-size periodically structured crystal defined by a shape function* S_p. *The blocks enclosed by the dashed lines represent the distribution of crystal unit cells.*

1.6 Single-scattering theory

If a crystal is very thin and if the crystal potential is weak, then electron scattering can be treated kinematically, i.e., the electron is scattered only once and there are no multiple-scattering effects. We first consider a three-dimensional periodic crystal of finite size. The shape of this crystal is described by a shape function $S_p(\mathbf{r})$, as schematically illustrated in Figure 1.3. Under the rigid body approximation, the potential distribution in real space $\mathbf{r}$ is written as a superposition of the atomic potentials

$$V(\mathbf{r}) = \left(\sum_n \sum_\alpha V_\alpha(\mathbf{r} - \mathbf{R}_n - \mathbf{r}_\alpha) \right) S_p(\mathbf{r}), \tag{1.17}$$

where the sum of n is over all the unit cells filling the entire space (i.e., $-\infty < n < \infty$) and the sum of α is over all the atoms within a unit cell. From the Born approximation shown in Section 1.3, the kinematical scattering amplitude of the crystal is a Fourier transform of the crystal potential,

$$
\begin{aligned}
V(\boldsymbol{u}) = \mathrm{FT}[V(\boldsymbol{r})] &= \left(\sum_n \sum_\alpha \int \mathrm{d}\boldsymbol{r}\, V_\alpha(\boldsymbol{r} - \boldsymbol{R}_n - \boldsymbol{r}_\alpha) \exp(-2\pi i \boldsymbol{u}\cdot\boldsymbol{r}) \right) \otimes S_p(\boldsymbol{u}) \\
&= \left(\sum_n \exp(-2\pi i \boldsymbol{u}\cdot\boldsymbol{R}_n) \sum_\alpha \int \mathrm{d}\boldsymbol{r}\, V_\alpha(\boldsymbol{r}) \exp(-2\pi i \boldsymbol{u}\cdot\boldsymbol{r}) \exp(-2\pi i \boldsymbol{u}\cdot\boldsymbol{r}_\alpha) \right) \otimes S_p(\boldsymbol{u}) \\
&= \left(\sum_g \delta(\boldsymbol{u}-\boldsymbol{g}) \sum_\alpha V_\alpha(\boldsymbol{u}) \exp(-2\pi i \boldsymbol{u}\cdot\boldsymbol{r}_\alpha) \right) \otimes S_p(\boldsymbol{u}) \\
&= \left(\sum_g \delta(\boldsymbol{u}-\boldsymbol{g}) \sum_\alpha V_\alpha(\boldsymbol{g}) \exp(-2\pi i \boldsymbol{g}\cdot\boldsymbol{r}_\alpha) \right) \otimes S_p(\boldsymbol{u}) \\
&= \left(\sum_g \delta(\boldsymbol{u}-\boldsymbol{g})\, V_g \right) \otimes S_p(\boldsymbol{u}),
\end{aligned} \tag{1.18}
$$

where $\otimes$ indicates a convolution calculation; the Fourier coefficient of the crystal potential is defined as

$$
V_g = \frac{1}{\Omega} \sum_\alpha f_\alpha^e(g) \langle \exp[2\pi i \boldsymbol{g}\cdot(\boldsymbol{r}_\alpha + \boldsymbol{u}_\alpha)] \rangle = \sum_\alpha V_\alpha(\boldsymbol{g}) \exp(-2\pi i \boldsymbol{g}\cdot\boldsymbol{r}_\alpha); \tag{1.19a}
$$

the scattering power of the αth atom in the unit cell is

$$
V_\alpha(\boldsymbol{g}) = \frac{1}{\Omega} f_\alpha^e(\boldsymbol{g}) \exp(-W_\alpha(\boldsymbol{g})), \tag{1.19b}
$$

where Ω is the volume of the unit cell; and $\exp(-W_\alpha(\boldsymbol{g})) = \exp(-2\pi^2 \langle A_\alpha^2 \rangle g^2)$ is the Debye–Waller (DW) factor, which is introduced in order to take into account the thermal vibration of the atom near its equilibrium position. This factor will be included in all elastic scattering calculations in order to characterize the weakening of the scattering power of an atom due to its thermal vibration. In deriving Eq. (1.18) an identity

$$
\sum_n \exp(-2\pi i \boldsymbol{u}\cdot\boldsymbol{R}_n) = \sum_g \delta(\boldsymbol{u}-\boldsymbol{g}) \tag{1.20}
$$

was used, where $\boldsymbol{g}$ is defined as a reciprocal lattice vector satisfying

$$
\boldsymbol{g}\cdot\boldsymbol{R}_n = \boldsymbol{g}\cdot(n_1\boldsymbol{a} + n_2\boldsymbol{b} + n_3\boldsymbol{c}) = \text{integer}, \tag{1.21}
$$

n_1, n_2 and n_3 are integers; $\boldsymbol{a}$, $\boldsymbol{b}$ and $\boldsymbol{c}$ are the basis vectors of the unit cell. The Dirac delta function, $\delta(\boldsymbol{u}-\boldsymbol{g})$, allows only those reflections that satisfy Bragg's law in reciprocal space $\boldsymbol{u}$. It is very important to point out that Eq. (1.20) is a key equation for converting a summation over unit cells in real space into a summation of reciprocal lattice vectors.

The intensity of each Bragg reflection is scaled by $|V_g|^2$; the conditions under which $V_g = 0$ for some $\boldsymbol{g}$ values give the extinction rules for forbidden reflections under the kinematical scattering approximation.

For body-centered-cubic (b.c.c.) metals there are identical atoms at (000) and $(\frac{1}{2}\frac{1}{2}\frac{1}{2})$ in the unit cell, the diffraction intensity is

$$|V_g|^2 = |V_\alpha(\boldsymbol{g})|^2\{1 + \cos[\pi(h+k+l)]\}^2. \tag{1.22}$$

Thus, $|V_g|^2 = 0$ if $(h+k+l)$ is odd.

For face-centered-cubic (f.c.c.) metals with identical atoms in the unit cell, the diffraction intensity is

$$\begin{aligned}|V_g|^2 = |V_\alpha(\boldsymbol{g})|^2 \{&\{1 + \cos[\pi(h+k)] + \cos[\pi(k+l)] + cos[\pi(h+l)]\}^2 \\ &+ \{\sin[\pi(h+k)] + \sin[\pi(k+l)] + \sin[\pi(h+l)]\}^2\}.\end{aligned} \tag{1.23}$$

Thus, $|V_g|^2 = 0$ if h, k and l are mixed odd and even.

The forbidden reflections, however, may appear in the diffraction pattern if the multiple-scattering effect is dominant (Gjønnes and Moodie, 1965).

Appendix C gives the kinematically calculated low-index zone-axis electron diffraction patterns of b.c.c., f.c.c., diamond and closely packed hexagonal structures. These patterns are useful for indexing RHEED patterns.

The sharp Bragg reflections are smeared out due to the convolution of the crystal shape factor S_p. For a near circular shape factor (Figure 1.4(a)), its Fourier transform is an oscillating, damped ring-shape function (Figure 1.4(b)). The Fourier transform of a narrow slit function (Figure 1.4(c)) results in streaking in the direction perpendicular to the slit (Figure 1.4(c)). This streaking effect occurs in RHEED due to small electron penetration depth into the surface, and the streaks should be perpendicular to the surface. The full width at half maximum of the Fourier transformed shape factor $S_p(\boldsymbol{u})$ is inversely proportional to the size of the crystal. In general, the shape factor smearing effect becomes significant in electron diffraction when the crystal size is less than a few nanometers.

To illustrate the nature of the terms appearing in Eq. (1.18), Figure 1.5(a) shows an electron diffraction pattern of Si oriented along [110]. The position of each Bragg reflection is at a reciprocal lattice point $\boldsymbol{g}$, and the intensity of this reflection is scaled (in kinematical scattering theory) according to $|V_g|^2$. The intensities at non-Bragg reflection positions are simply zero, as characterized by the $\delta(\boldsymbol{u}-\boldsymbol{g})$ function. The diffraction spots that are absent in the diffraction patterns are those whose structure factors vanish (i.e., forbidden reflections). The (001) reflection of Si is such an example. Figure 1.5(b) is a diffraction pattern of $SrTiO_3$ oriented along [001]. The structure factor of $SrTiO_3$ is

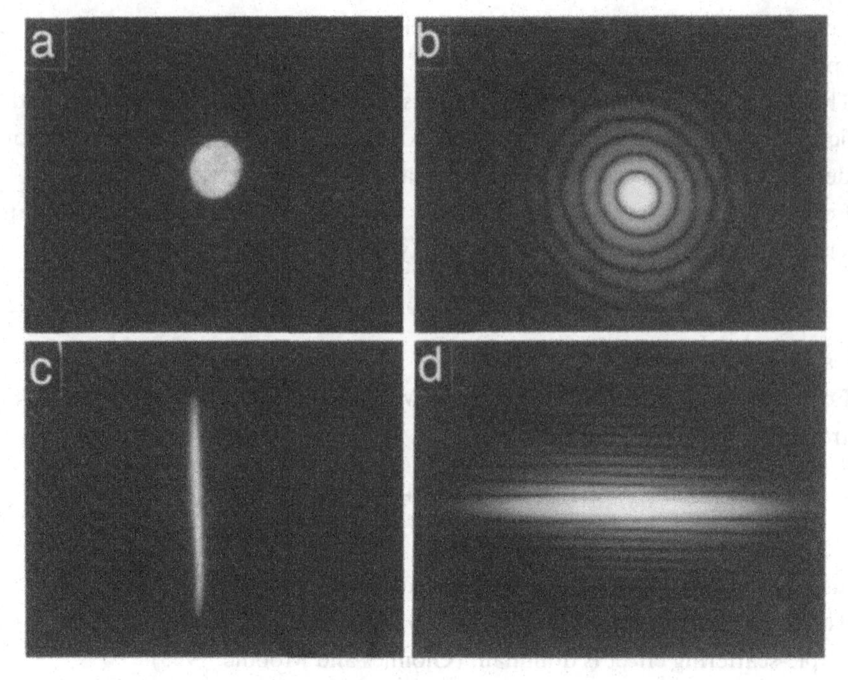

Figure 1.4 *Real-space shape factor functions (a, c) and the corresponding Fourier transform* ($\mathrm{FT}(S_p)$) *(b, d) respectively.*

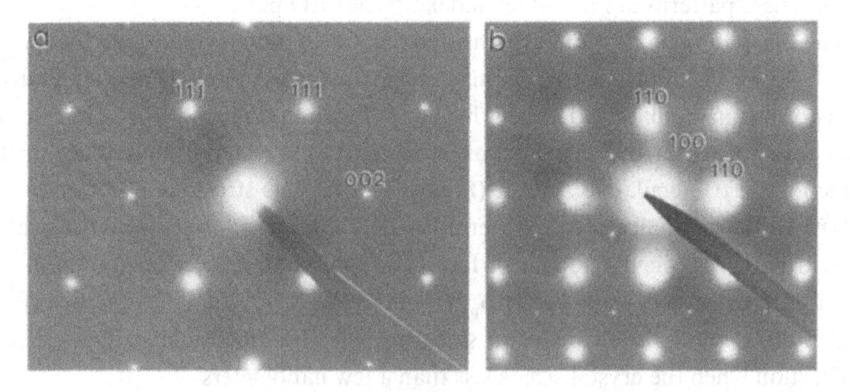

Figure 1.5 *Transmission electron diffraction patterns recorded at 100 kV from (a) Si [110] and (b) $SrTiO_3$ [001]. The central transmitted (000) beam was blocked.*

$$V_g = V_{Sr}(\boldsymbol{g}) + V_{Ti}(\boldsymbol{g})\exp[i\pi(h+k+l)] + V_O(\boldsymbol{g})\,[\exp(i\pi h) + \exp(i\pi k) + \exp(i\pi l)]. \quad (1.24)$$

For the (100) reflection, $V_{(100)} = V_{Sr}(\boldsymbol{g}) - V_{Ti}(\boldsymbol{g}) + V_O(\boldsymbol{g})$; for the (110) reflection, $V_{(110)} = V_{Sr}(\boldsymbol{g}) + V_{Ti}(\boldsymbol{g}) - V_O(\boldsymbol{g})$. Thus, $|V_{(110)}|^2 \gg |V_{(100)}|^2$, which means that the intensity of the (110) reflection is much greater than that of (100), as observed in Figure 1.5(b). The position of Bragg reflected beams can be accurately predicted by the kinematical scattering theory.

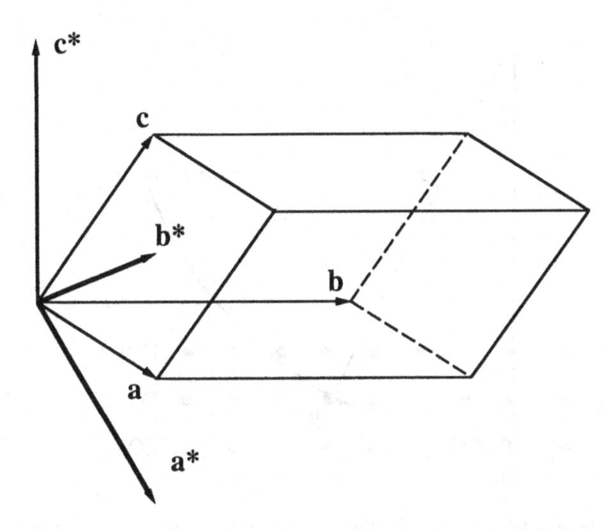

Figure 1.6 Relationships between the lattice vectors of a crystal unit cell and the corresponding reciprocal lattice vectors.

1.7 Reciprocal space and the reciprocal-lattice vector

As shown in Eq. (1.21), the reciprocal lattice is defined by a vector $\boldsymbol{g} = l\boldsymbol{a}^* + m\boldsymbol{b}^* + n\boldsymbol{c}^*$, where $(l\ m\ n)$ are the Miller indices of the $\boldsymbol{g}$ reflection of a crystal plane $(l\ m\ n)$ and satisfy

$$(l\boldsymbol{a}^* + m\boldsymbol{b}^* + n\boldsymbol{c}^*) \cdot (n_1\boldsymbol{a} + n_2\boldsymbol{b} + n_3\boldsymbol{c}) = \text{integer}. \tag{1.25}$$

This equation can be satisfied by choosing the reciprocal-lattice vectors as

$$\boldsymbol{a}^* = \frac{\boldsymbol{b} \times \boldsymbol{c}}{\Omega}, \qquad \boldsymbol{b}^* = \frac{\boldsymbol{c} \times \boldsymbol{a}}{\Omega}, \qquad \boldsymbol{c}^* = \frac{\boldsymbol{a} \times \boldsymbol{b}}{\Omega}, \tag{1.26}$$

where $\Omega = (\boldsymbol{a} \times \boldsymbol{b}) \cdot \boldsymbol{c}$ is the volume of the unit cell. It can be easily proven that

$$\boldsymbol{a}^* \cdot \boldsymbol{b} = \boldsymbol{a}^* \cdot \boldsymbol{c} = \boldsymbol{b}^* \cdot \boldsymbol{a} = \cdots = 0, \tag{1.27a}$$

$$\boldsymbol{a}^* \cdot \boldsymbol{a} = \boldsymbol{b}^* \cdot \boldsymbol{b} = \boldsymbol{c}^* \cdot \boldsymbol{c} = 1. \tag{1.27b}$$

The space defined by the basis vectors $(\boldsymbol{a}^*, \boldsymbol{b}^*, \boldsymbol{c}^*)$ is called reciprocal space (or momentum space). A point in reciprocal space specifies electrons traveling in the same direction in real space. A high-resolution lattice image in real space is the interference result of electron beams located at different points in reciprocal space. Therefore, the angular distribution of scattered electrons is more conveniently described in reciprocal space. Figure 1.6 schematically shows a crystal unit cell and the corresponding reciprocal-lattice vectors.

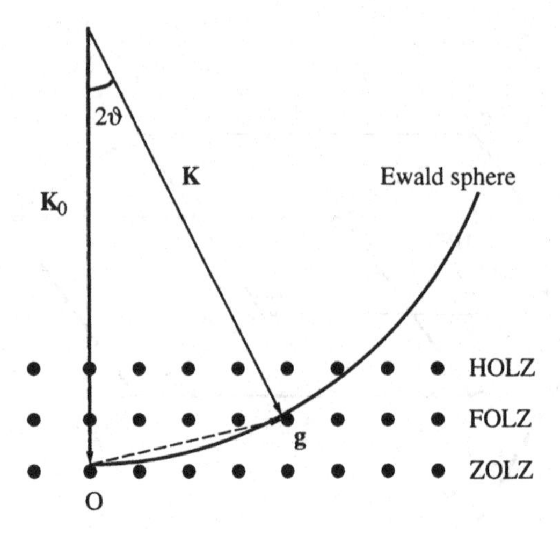

Figure 1.7 Construction of the Ewald sphere in reciprocal space, illustrating the electron diffracting condition. The Bragg reflections belonging to different Laue zones are indicated by dark spots.

1.8 Bragg's law and the Ewald sphere

As shown in Eq. (1.18), the diffraction of a perfect crystal forms a set of sharp peaks distributed at the reciprocal-lattice sites defined by $\boldsymbol{g} = l\boldsymbol{a}^* + m\boldsymbol{b}^* + n\boldsymbol{c}^*$. A strong reflection peak will be generated if the change of electron wavevector $\boldsymbol{u} = \boldsymbol{K} - \boldsymbol{K}_0$ equals the reciprocal-space vector $\boldsymbol{g}$,

$$\boldsymbol{u} = l\boldsymbol{a}^* + m\boldsymbol{b}^* + n\boldsymbol{c}^*. \tag{1.28}$$

This is exactly the result of conservation of momentum in elastic electron scattering. The meaning of Eq. (1.28) can be clearly stated using the Ewald sphere construction in reciprocal space, as shown in Figure 1.7. For elastic scattering, conservation of energy requires $K = K_0 = 1/\lambda$, thus a sphere of radius K is drawn in reciprocal space with wavevector of the incident electron beam ending at the origin. The wavevector of the diffracted beam is drawn from the center of the sphere, and the diffracted intensity at $\boldsymbol{u}$ is $|V(\boldsymbol{u})|^2$. The possible beams that will be generated in a diffraction pattern are the reciprocal lattice vectors intersecting the surface of the Ewald sphere. Those beams that do not fall on the sphere surface will not be excited (for an infinitely large perfect crystal). The circles made by the intersections of the Ewald sphere surface with the planes of reciprocal lattice vectors are called the Laue circles. The reflections are thus classified as the zeroth-order Laue zone (ZOLZ), the first-order Laue zone (FOLZ) and higher-order Laue zones (HOLZs). The ZOLZ is usually defined as the plane of reciprocal lattice points that passes through the origin. A HOLZ is any other reciprocal-lattice plane parallel to the plane of the ZOLZ, not passing through the origin, as illustrated in Figure 1.7. The intensities of the HOLZ reflections are sensitive to the electron scattering along the beam

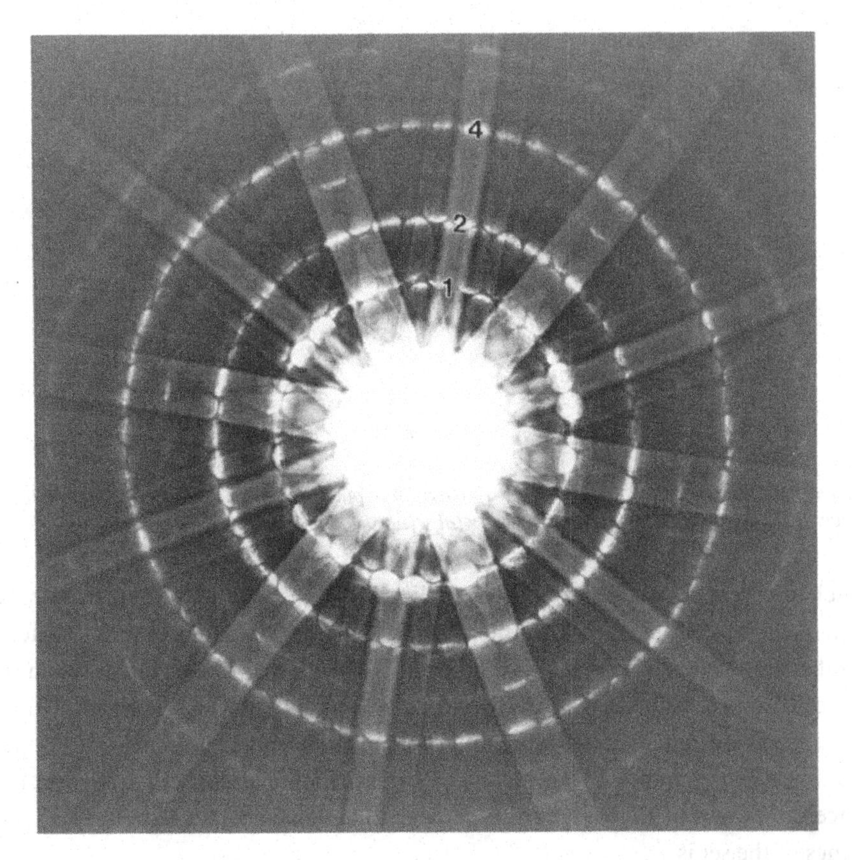

Figure 1.8 *HOLZ reflections of an α-Al_2O_3 crystal viewed along [0001]. The numbers indicate the corresponding orders of Laue circles. The diffraction pattern was recorded at 100 keV with converged beam in order to enhance the intensity of the HOLZ rings.*

direction. The ZOLZ reflections, in contrast, are the scattering of the projected crystal structure perpendicular to the beam. The Ewald sphere construction is a simple and direct method for illustrating the diffracting conditions of electron diffraction. The classification of Bragg beams as ZOLZ and HOLZ reflections leads to large differences in theoretical approaches.

Figure 1.8 shows an experimental diffraction pattern of [0001] α-Al_2O_3. The HOLZs are visible up to the sixth order. If the beam direction is parallel to c^*, then the radius of the mth Laue zone ring is $u_m = (2c^* m K_0)^{1/2}$. Thus $u_{m+1}^2 - u_m^2 = 2c^* K_0$, which can be applied to measure the reciprocal-lattice vector c^*. Besides HOLZs, Kikuchi bands are also observed, which are the result of electron inelastic scattering and will be discussed in Chapter 2.

We now return to the discussion of Bragg diffracting conditions. From Eq. (1.25), the projections of $\boldsymbol{u}$ on real space axes satisfy

$$\boldsymbol{u} \cdot \boldsymbol{a} = l, \qquad \boldsymbol{u} \cdot \boldsymbol{b} = m, \qquad \boldsymbol{u} \cdot \boldsymbol{c} = n, \tag{1.29}$$

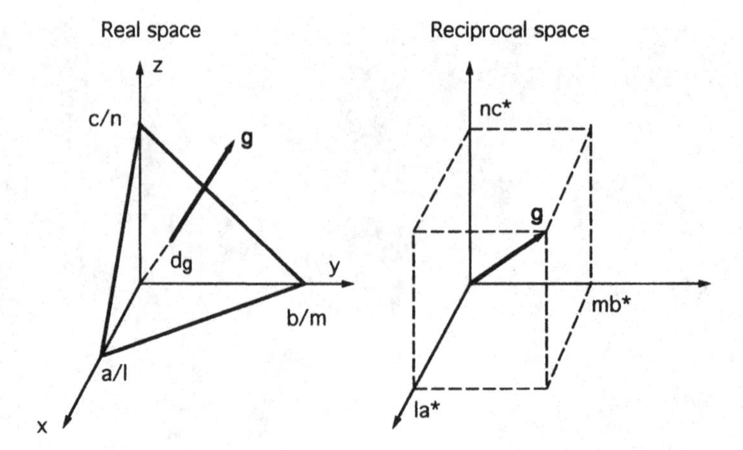

__Figure 1.9__ Relationships between real-space atomic planes and the corresponding reciprocal-space lattice vectors for an orthogonal crystal structure.

which represent the well-known Laue conditions for diffraction. In reciprocal space, each reciprocal-space lattice point $(l\ m\ n)$ represents a set of equally spaced, parallel atomic planes in real space. $(l\ m\ n)$ is known as a Miller index. When the unit cell origin is defined as the origin of the coordinate system and the $\boldsymbol{a}$, $\boldsymbol{b}$ and $\boldsymbol{c}$ axes are taken as the reference axes, the intercepts of the parallel plane of the set on the axes are a/l, b/m and c/n, respectively. For an orthogonal unit cell system in real space, as shown in Figure 1.9, the perpendicular distance between the adjacent planes of the set is

$$\frac{1}{d_g^2}=\frac{l^2}{a^2}+\frac{m^2}{b^2}+\frac{n^2}{c^2}. \tag{1.30}$$

On the other hand, the distance from the reciprocal-lattice origin to the $(l\ m\ n)$ reciprocal-lattice point is the inverse of the distance between the adjacent atomic planes,

$$g=1/d_g, \tag{1.31a}$$

and the vector perpendicular to the $(l\ m\ n)$ lattice planes is the vector from the origin to the $(l\ m\ n)$ reciprocal-lattice point in reciprocal space. The $\boldsymbol{g}$ vector is the normal direction of the $(l\ m\ n)$ lattice planes. These relationships also hold for non-orthogonal axes. Equation (1.31a) means that the distance between the origin and the reciprocal lattice point in the electron diffraction pattern is the inverse of the interplanar distance in real space. The larger the g value in reciprocal space, the smaller d_g in real space.

Using Eq. (1.31a), the modulus of Eq. (1.28) is $u=1/d_g$. Using the scattering angle defined in Figure 1.1, Bragg's law is explicitly written as

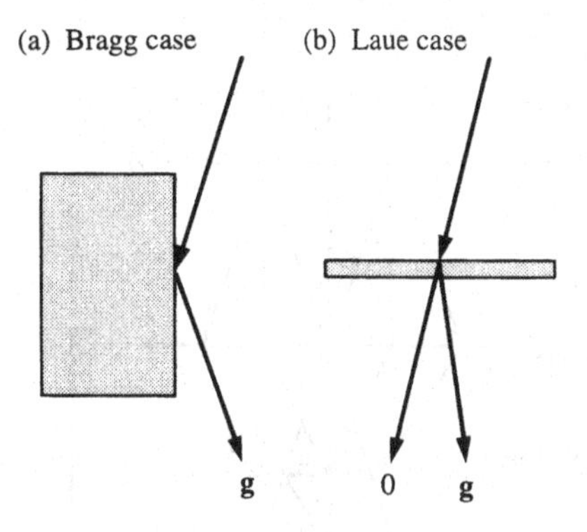

Figure 1.10 *Diagrams showing the diffraction of electrons in the (a) Bragg and (b) Laue cases.*

$$2d_g \sin\theta_g = \lambda. \tag{1.31b}$$

This condition is similar to the $2d_g \sin\theta_g = \lambda$ law for reflection of light from a semi-infinite medium, thus the well-defined diffracted beam given under condition (1.28b) is conveniently referred as a 'Bragg reflection'. Equation (1.31b) is usually called the Bragg condition, under which the corresponding reciprocal lattice point intersects the surface of the Ewald sphere with zero excitation error. Appendix D gives the general formulas for calculating the interplanar distances of crystals of different structures.

Diffraction of electrons is classified as taking place according to either the 'Bragg case' or the 'Laue case'. In the Bragg case an incident beam is diffracted from planes parallel or nearly parallel to a flat surface of a crystal that may be considered semi-infinite. Thus the diffracted beams observed are those that re-enter the vacuum on the same side of the crystal as the incident beam (Figure 1.10(a)). Reflection high-energy electron diffraction (RHEED) from a bulk crystal surface and low-energy electron diffraction (LEED) are typical examples of Bragg cases. Hence the Bragg case is basically 'backscattering'. This book is about electron diffraction and imaging in the Bragg case.

The Laue case is transmission diffraction (without backscattering), through a parallel-sided crystal plate of infinite extent in two dimensions (Figure 1.10(b)). This is the situation in transmission electron microscopy of thin specimens. A recent book by Wang (1995) specializes in this case.

1.9 Abbe's imaging theory

Figure 1.11 shows the simplest ray diagram for a TEM, in which only a single objective lens is considered for imaging and the intermediate lenses and projection

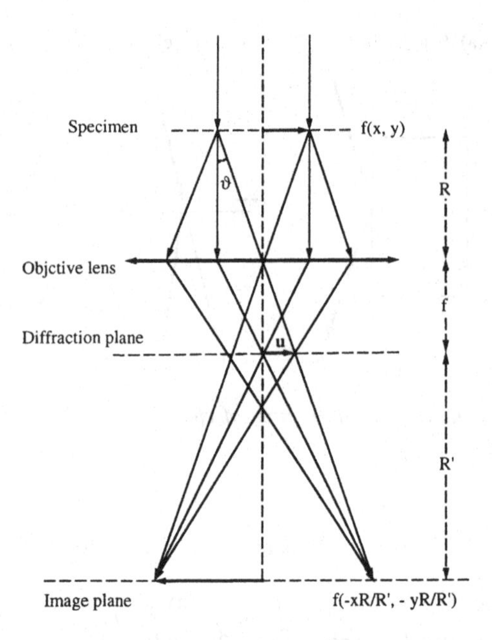

Figure 1.11 *Abbe's theory of image formation in a one-lens transmission electron microscope.*

lenses are omitted. An incident electron beam, which was emitted from an electron gun and made to converge by the condenser lens, illuminates a thin specimen. For simplification, we use the thin lens approximation without considering any aberration effect. The electron beam is diffracted by the lattices of the crystal, forming the diffracted beams propagating along different directions. The electron–specimen interaction results in phase and amplitude changes in the electron wave. For a thin specimen and high-energy electrons, the transmitted wave function $\Psi(x, y)$ at the exit face of the specimen can be assumed to be composed of a forward-scattered, axial, transmitted wave plus other scattered waves that proceed in directions slightly inclined to it. The diffracted beams will be focused in the back focal plane, where an objective aperture could be applied. In TEM micrographs, image contrast is the result of electron scattering by the atoms in the specimen and the transfer properties of the optic system, and characterizes the structure of the crystal. This image formation is usually described by Abbe's theory.

An ideal thin lens brings the parallel transmitted waves to a focus on the axis in the back focal plane. Waves leaving the specimen in the same direction (or angle θ with respect to the optic axis) are brought together at a point on the back focal plane (or diffraction plane) at a distance $X = f \tan\theta$, from the axis, where f is the focal length. Thus, on the back focal plane, the waves from all parts of the illuminated regions propagating in a given direction are added. This is equivalent to the case in which the observation point is at infinity. So, in the back focal plane, a Fraunhofer diffraction pattern is formed. The variable used for the distribution of the diffracted amplitude in the diffraction pattern for the one-dimensional case is $u = 2K\sin\theta \approx 2\theta/\lambda$. Mathematically, the formation of the Fraunhofer diffraction pattern is described by a Fourier transform (Cowley, 1981),

$$\Psi(\boldsymbol{u}) = \mathsf{FT}[\Psi(x, y, d_0)], \tag{1.32}$$

where d_0 is the thickness of the specimen.

The electron image is then described by the inverse Fourier transform of the diffracting amplitude $\Psi(\boldsymbol{u})$

$$\Psi_i(x, y, d_0) = \mathsf{FT}^{-1}[\Psi(\boldsymbol{u})] = \Psi(-x/M, -y/M, d_0), \tag{1.33}$$

where the negative sign denotes inversion of the object by the objective lens and M is the magnification of the lens. This equation is for the case of an ideal objective lens. In general, the chromatic and spherical aberrations of the lens as well as the size of the objective aperture introduce a transfer function $T_{obj}(\boldsymbol{u})$ for the optic system,

$$T_{obj}(\boldsymbol{u}) = A_{obj}(\boldsymbol{u}) \exp(i\chi), \tag{1.34a}$$

$$\chi = \tfrac{1}{2}\pi C_s \lambda^3 u^4 + \pi \Delta f \lambda u^2, \tag{1.34b}$$

where A_{obj} is the shape function of the objective aperture, C_s is the spherical aberration coefficient of the lens, and Δf is the defocus of the lens. Thus Eq. (1.33) is modified as

$$\begin{aligned}\Psi_i(x, y, d_0) &= \mathsf{FT}^{-1}[\Psi(\boldsymbol{u})\, T_{obj}(\boldsymbol{u})] = \Psi(\boldsymbol{r}) \otimes \mathsf{FT}^{-1}[T_{obj}(\boldsymbol{u})] \\ &= \int \mathrm{d}\boldsymbol{b}'\, \Psi(\boldsymbol{b}', d_0)\, T_{obj}(\boldsymbol{b} - \boldsymbol{b}'),\end{aligned} \tag{1.35}$$

where $\otimes$ indicates convolution, $\boldsymbol{b} = (x, y)$ and $\boldsymbol{b}' = (x', y')$. The diffraction pattern is given by

$$|\psi(\boldsymbol{u})|^2 = |\mathsf{FT}[\Psi_i(x, y)]|^2 = |\int \mathrm{d}\boldsymbol{b} \exp(-2\pi i \boldsymbol{u}\cdot\boldsymbol{b})\, \Psi_i(\boldsymbol{b}, d_0)|^2. \tag{1.36}$$

The defocus of the objective lens varies with the change in electron energy. For an electron that has lost energy ΔE, a relative focus shift Δf_c is introduced

$$\Delta f_c = C_c \frac{\Delta E}{E_0}, \tag{1.37}$$

where C_c is the chromatic aberration coefficient of the objective lens. The effect introduced by Δf_c is known as chromatic aberration, which is caused by the variation of focal length due to the change in electron energy.

In summary, Abbe's imaging theory is a two-step process – taking a Fourier transform of the electron wave function at the exit face of the crystal, then multiplying the lens transfer function and inversely Fourier-transforming the diffraction amplitude back to real space. The electron diffraction pattern is the modulus squared of the Fourier transformed wave function at the exit face of the crystal.

1.10 The phase object approximation

Electron–specimen interaction is a quantum scattering problem. The angular distribution of the diffracted electrons is determined by the solution of the Schrödinger equation and the boundary conditions. For very thin specimens, however, the diffraction problem can be treated in a simple way. If an electron of incident energy $E=eU_0$ is traveling in a potential field $V(\boldsymbol{r})$, the effective wave vector is related to the effective wavelength of the electron by

$$K_{\text{eff}}=\frac{1}{\lambda_{\text{eff}}}=\frac{1}{h}\left[2m_0e(U_0+V)\left(1+\frac{e(U_0+V)}{2m_0c^2}\right)\right]^{1/2}$$
$$=\frac{1}{\lambda}\left[\left(1+\frac{V}{U_0}\right)\left(1+\frac{eV}{2m_0c^2+eU_0}\right)\right]^{1/2}\approx\frac{1}{\lambda}\left(1+\frac{V(\boldsymbol{r})}{U_0}\right)^{1/2}. \quad (1.38)$$

Thus the relative phase shift of the wave traveling in the field relative to the wave traveling in the absence of a field for a distance Δz is

$$\int_{z_n}^{z_n+\Delta z} \mathrm{d}z\, 2\pi\left(\frac{1}{\lambda_{\text{eff}}}-\frac{1}{\lambda}\right)\approx \mathrm{i}\sigma V(\boldsymbol{b}), \quad (1.39)$$

where $\sigma=\pi/(\lambda U_0)$, and the projected potential of the nth slice is

$$V(\boldsymbol{b})=\int_{z_n}^{z_n+\Delta z} \mathrm{d}z\, V(\boldsymbol{r}).$$

A more rigorous derivation of (1.39) with consideration to the relativistic correction will be given in Section 3.2. Therefore, the effect of the potential field is represented by multiplying the wave function by a phase grating function

$$Q(\boldsymbol{b})=\exp(\mathrm{i}\sigma V(\boldsymbol{b})). \quad (1.40)$$

This is known as the *phase object approximation.*

If the specimen is a weak scatterer, so that $|\sigma V(\boldsymbol{b})|\ll 1$, then the phase grating function is approximated as

$$Q(\boldsymbol{b})\approx 1+\mathrm{i}\sigma V(\boldsymbol{b}). \quad (1.41)$$

This is the *weak phase object approximation*, which has been extensively applied to discuss the physics involved in electron phase contrast imaging.

1.11 Aberration and the contrast transfer function

In TEM, the aberration of the electron lenses limits the angular range of diffracted beams that can usefully contribute to the image. Under the weak phase object approximation (WPOA), the exit wave of the electron at the crystal exit face after excluding the term $\exp(2\pi i \boldsymbol{K}_0 \cdot \boldsymbol{r})$ is written as

$$\Psi(\boldsymbol{r}) = 1 + i\sigma V(\boldsymbol{b}), \tag{1.42}$$

where $\boldsymbol{b} = (x, y)$, and V is the projected crystal potential. The intensity distribution in a conventional phase contrast HRTEM image, based on Abbe's imaging theory, is (Cowley, 1988)

$$I_p = |(1 + i\sigma V(\boldsymbol{b})) \otimes T_{obj}(\boldsymbol{b})|^2 = |1 + i\sigma V(\boldsymbol{b}) \otimes T_{obj}(\boldsymbol{b})|^2. \tag{1.43}$$

If we separate the real and imaginary components of T_{obj} as $T_{obj}(\boldsymbol{b}) = T_c(\boldsymbol{b}) + iT_s(\boldsymbol{b})$, and ignore the V^2 term for a weakly scattering object, then (1.43) becomes

$$I_p \approx 1 - 2\sigma V(\boldsymbol{b}) \otimes T_s(\boldsymbol{b}). \tag{1.44}$$

Therefore, for thin crystals, $T_s(\boldsymbol{b})$ is known as the *contrast transfer function*, which determines the information that can be transferred by the objective lens. It is well known that the best phase contrast image can be obtained under *Scherzer focus* (Scherzer, 1949),

$$\Delta f_s = (\tfrac{4}{3} C_s \lambda)^{1/2}. \tag{1.45}$$

Figure 1.12 shows the calculated forms of $\cos\chi$ and $\sin\chi$ and the corresponding Fourier transforms T_c and T_s, respectively, for two different defocuses. Under the Scherzer defocus (Figure 1.12(a)), the width of the $\sin\chi$ band is maximized and all the reflections enclosed in the information band will be transmitted with the correct phases; thus, the image is normally called a *structural image*. The image recorded under this condition can be directly related to the crystal structure. Also, in this case, the T_c function is sharp. If the defocus is set at values different from the Scherzer value, then the shape of the transfer function $\sin\chi$ is changed (Figure 1.12(b)), and the corresponding electron image may not be easily linked with the atomic structure. Thus, image simulation becomes indispensable.

In this chapter, we have introduced the kinematical scattering theory of electron diffraction. Related physical quantities have been introduced in order to describe the scattering behavior of high-energy electrons. The preliminary knowledge provided in this chapter will be applied to illustrate RHEED.

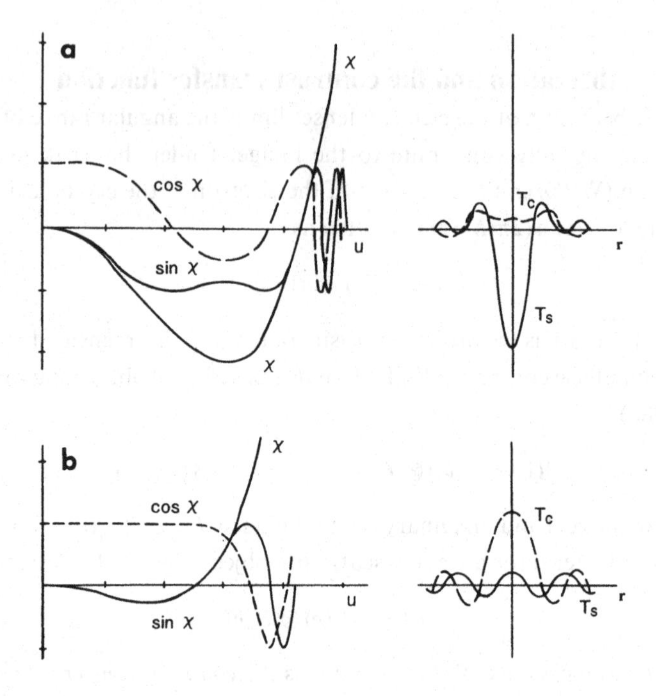

Figure 1.12 *The reciprocal-space forms of* $\mathrm{Re}(T_{\mathrm{obj}})=\cos\chi$ *and* $\mathrm{Im}(T_{\mathrm{obj}})=\sin\chi$ *of the objective lens transfer function and the corresponding Fourier transforms* $T_c = \mathrm{FT}^{-1}(\cos\chi)$ *and* $T_s = \mathrm{FT}^{-1}(\sin\chi)$ *into real space,* $C_s = 1$ *mm and* $E_0 = 200$ *keV. (a) For defocus* $\Delta f = -17$ *nm. (b) For defocus* $\Delta f = -58$ *nm (Scherzer defocus).*

PART A

Diffraction of reflected electrons

CHAPTER 2

Reflection high-energy electron diffraction

Reflection high-energy electron diffraction (RHEED) is a powerful technique for *in situ* observation of thin film growth in molecular beam epitaxy (MBE). The technique is particularly useful for providing *in situ* real-time structure evolution during thin film growth, and it is becoming an indispensable technique for characterizing phase transformation in synthesizing new materials of technological importance. In this chapter, we will apply the kinematical diffraction theory introduced in Chapter 1 to describe the fundamental characteristics of RHEED and its applications in determining surface structures. RHEED oscillation and its applications for monitoring film growth are systematically described. This remarkable phenomenon allows layer-by-layer monitoring of thin film growth.

2.1 The geometry of RHEED

RHEED was first used for oxide film studies by Miyake (1937), and it has been widely applied to monitor thin film growth during MBE deposition (Ino, 1987). A review of the RHEED geometry and the associated reciprocal space analyses has been given by Mahan *et al.* (1990) and Dobson (1987). A summary of surface structure determination using RHEED has been given by Ino (1987). Recent studies by Locquet and Machler (1994) have shown the sensitivity of the entire RHEED pattern to the layer-by-layer growth of $DyBa_2Cu_3O_{7-x}$ high-T_c superconductor thin films.

Figure 2.1 shows the scattering geometry of RHEED, in which the incident electron beam strikes the sample surface at a grazing angle of 1–3°. The electron energy can be as low as a few kilo-electron-volts and as high as 1000 keV (in TEM only). The electron gun and RHEED detector are located far from the film. The small incidence angle makes RHEED a surface-sensitive technique. The diffraction of the incident electrons by the surface lattice is used to define the surface crystallographic structures. The technique reveals almost instantaneous changes in either the coverage of the specimen surface by adsorbates or the surface structure of thin films. The electron gun and phosphorescent screen are positioned remotely from the substrate, so that they do not interfere with the growth process and are not exposed directly to an evaporation source or to the broad face of a hot substrate. In general, it is possible to rotate the growth stage, on which the film is attached; it is

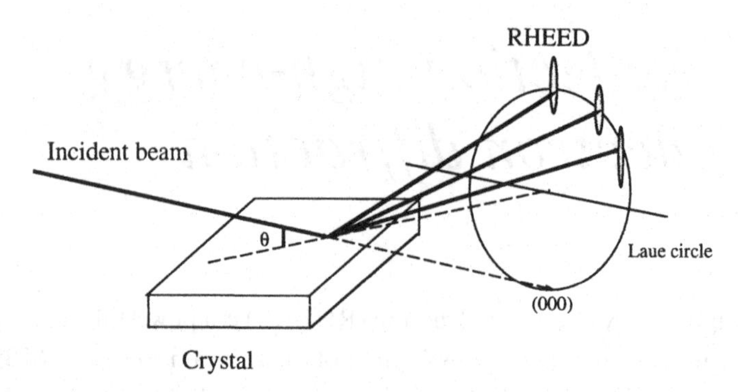

Figure 2.1 *A schematic diagram of reflection high-energy electron diffraction from a bulk crystal surface. The incidence angle θ is usually constrained within a few degrees in order to limit the penetration depth of the electrons into the bulk.*

thus possible to determine the surface structure from different crystallographic directions within the growth chamber.

If the incident beam is scanned across the surface by using a pair of scanning coils, then the intensity of a reflected beam detected for each incident beam position can be digitally displayed on a viewing screen (or TV screen) as a function of beam scanning position; thus a surface image is formed. The contrast of the image provides direct information about the real-space structure of the film's growth. This scanning imaging technique is an important application of RHEED and it will be introduced in detail in Section 12.1

Figure 2.2 shows a RHEED pattern of a clean Si(111) surface. Several features are seen. First, the pattern shows not only strong Bragg reflected beams O, A, B and C from bulk Si but also many superlattice reflections from the 7 × 7 surface structure. The 7 × 7 notation denotes a special structure of the surface, and a detailed introduction of this notation will be given in Section 2.2. Second, the reflected beams, particular those in the ZOLZ (the smallest circle in the pattern) are not spots but streaks in the direction perpendicular to the surface. The streaks are much more pronounced if there are some deposits on the surface (Ino, 1993). This feature will be described in Section 2.3. Third, the reflected beams tend to be located on some circles, which are usually described as Laue circles (or Laue rings), as described in Figure 1.7. Finally, sharp Kikuchi lines and bands are clearly seen. The formation of the Kikuchi pattern is related to elastic re-scattering of the inelastically scattered electrons. A detailed introduction of the inelastic scattering processes in RHEED will be given in Chapters 9–11. The pattern shown in Figure 2.2 exhibits the general characteristics of RHEED. The nature of each of these characteristics will be discussed in the following sections.

Reconstruction occurs not only at semiconductor surfaces but also at ceramic surfaces. Figure 2.3 shows a group of RHEED patterns recorded from an annealed

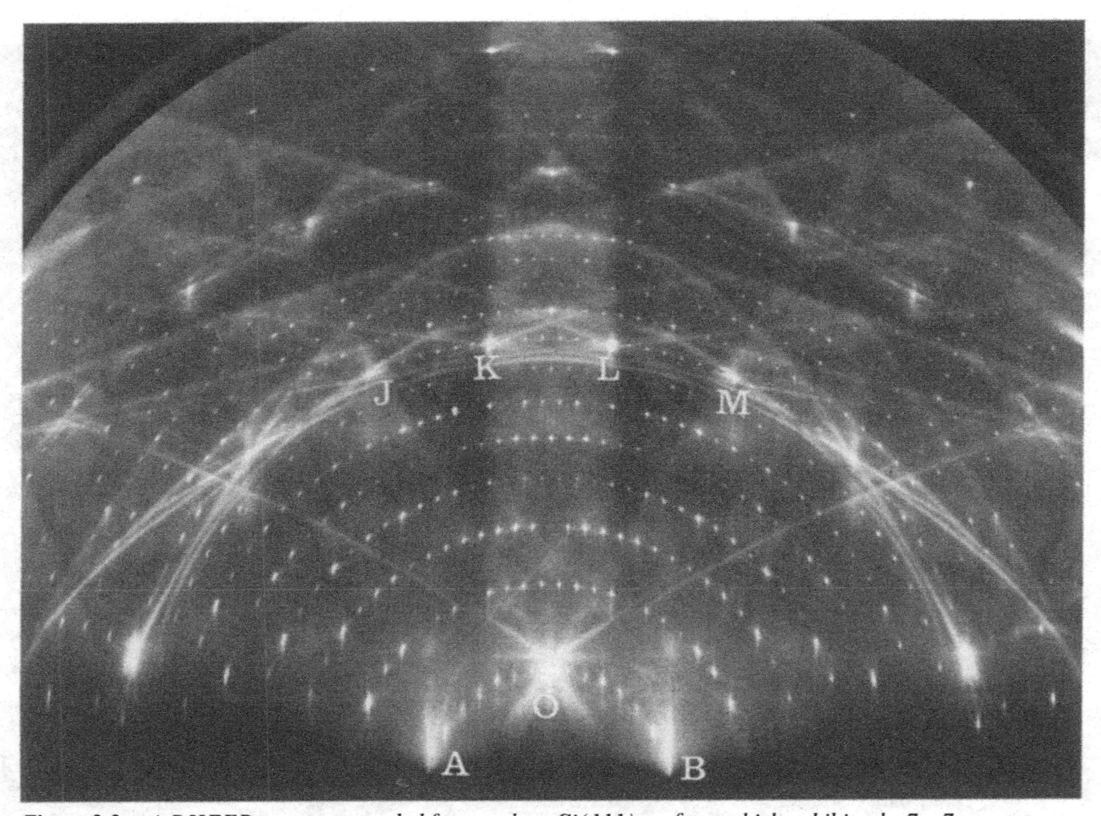

Figure 2.2 *A RHEED pattern recorded from a clean Si(111) surface, which exhibits the 7 × 7 reconstruction. Accelerating voltage and incident beam azimuth are 20 kV and [11$\bar{2}$]. O, A, B, J, K, L and M indicate the bulk reflections (courtesy of Dr S. Ino, 1993).*

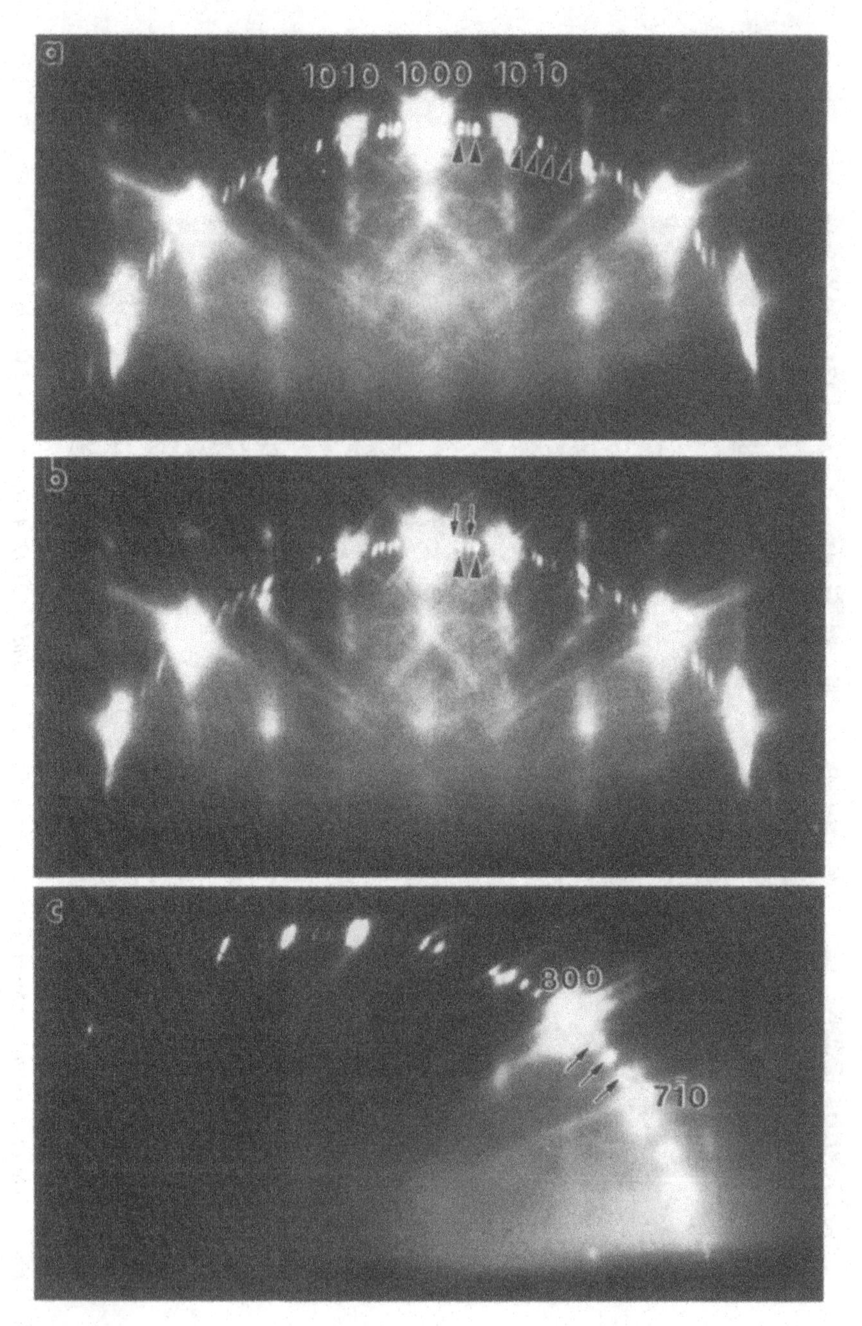

Figure 2.3 *RHEED patterns recorded at 300 kV from a thermal annealed* $LaAlO_3$ *(100) surface, exhibiting {100} twin and 5 × 5 surface reconstruction. The sample was prepared by annealing a polished (100) surface at 1500°C for 20 h. Besides the Bragg reflections from the periodic structures of the bulk and surface, a diffuse background decorated with Kikuchi lines and continuous resonance parabolas is seen. The beam energy is 120 keV (Wang and Shapiro, 1995a, b).*

$LaAlO_3(100)$ surface when the incident beam strikes the surface near the [001] azimuth. In Figure 2.3(a), two strong superlattice reflections are seen between bulk reflections (10 0 0) and (10 $\bar{1}$ 0). The two superlattice reflections are separated by $\frac{1}{5}(020)$ and $\frac{1}{5}(020)$, respectively, from the (10 0 0) and (10 $\bar{1}$ 0) bulk reflections. On the right-hand side of the (10 $\bar{1}$ 0) reflection, four equally-spaced superlattice reflections are seen. In Figure 2.3(a), a weak reflection located half-way between the two superlattice reflections is seen. This is because of the {100} twin structure of $LaAlO_3$, which causes a double splitting of the Bragg reflected spots due to crystal rotation by the twin angle of 0.12°. We now slightly change the illumination area and increase the convergence of the incident beam, so that the diffractions from the twin grains are equally seen in the pattern, as shown in Figure 2.3(b), in which the superlattice reflections are doubled in comparison with those shown between the (10 0 0) and (10 $\bar{1}$ 0) bulk reflections. This pattern proves that the weak spot seen between the two superlattice reflections in Figure 2.3(a) came from one of the twinned grains that was not strongly excited. To see the equivalent excitation of the superlattice reflection along [010], a direction parallel to the (100) surface, the crystal was slightly rotated and the corresponding pattern is shown in Figure 2.3(c). Three superlattice reflections are clearly seen. The superlattice reflection next to the (800) reflection is shadowed by the strong intensity of the (800) beam. The superlattice reflection next to the $(7\bar{1}0)$ one is seen. Therefore, $LaAlO_3(100)$ exhibits 5×5 surface reconstruction.

The last two examples have clearly demonstrated the surface sensitivity of RHEED, which can be applied to reveal the structure of the top surface layer. The following sections will give a comprehensive introduction regarding the theory, physics and applications of the RHEED technique.

2.2 Surface crystallography

The description of symmetry and structures of crystalline materials requires a reasonable understanding of crystallography; notably of the restricted number of types of translational and rotational symmetries that the crystals can possess and the finite numbers of point and space groups that can define the additional symmetry properties of all possible crystals. A surface is an interface between a vacuum and a crystal, and it is the termination of the bulk crystal. The distribution of surface atoms can be conveniently described by two-dimensional crystallography (or surface crystallography). A general understanding of two-dimensional crystallography is essential for quantifying X-ray and electron diffraction data. However, it must be pointed out that the surface two-dimensional structure refers to the structure in the immediate vicinity of the two-dimensional plane.

In addition to the translation symmetry parallel to the surface, which character-

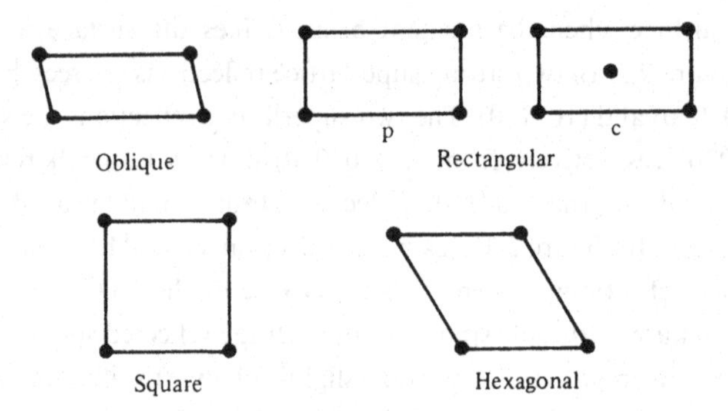

Figure 2.4 *The five two-dimensional Bravais lattices described in Table 2.1.*

izes the crystallinity of the solid surface, there are a few point and line symmetry operations that involve rotation and reflection within planes parallel to the surface. This whole subject is fully described in the International Tables for Crystallography. These symmetry operations are: one-, two-, three-, four- and six-fold rotation axes (note that a five-fold rotation axis or more than six-fold rotation axes are not compatible with the two-dimensional translational symmetry); mirror reflection in a plane perpendicular to the surface; and glide reflection (involving reflection in a line combined with translation along the direction of the line by half of the translational periodicity in this direction). Consideration of the symmetry properties of two-dimensional lattices leads to just five symmetry different Bravais nets, as shown in Figure 2.4. These are hexagonal, characterized by a six-fold rotation axis; square, characterized by a four-fold rotation axis; primitive (p) or centered (c) rectangular, which are the two symmetrically inequivalent lattices characterized by mirror symmetry; and oblique, which lacks all these symmetries. Note that the centered rectangular net is the only non-primitive net. The characteristics and notations of these Bravais nets are summarized in Table 2.1.

Combining the five Bravais lattices with the ten different possible point groups (Figure 2.5) leads to a possible 17 two-dimensional space groups (Figure 2.6). These symmetry-required equivalent points are useful in quantifying surface structures based on diffraction data.

2.2.1 Surface reconstruction

As shown in Figures 2.2 and 2.3, superlattice surface reflections are observed in RHEED. It is thus useful to introduce a notation that can be applied to identify each of these structures. Since the structure of the bulk crystal is the initially known structure, its net parallel to the surface is the reference net. For example, if the surface of a cubic bulk crystal is the (111) surface, then the base net on the (111) surface is hexagonal rather than cubic, and the surface net is thus referenced to the

Table 2.1. *Description of the five two-dimensional Bravais lattices.*

Shape of net	Net symbol	Choice of axes	Axis and angles	Name
General parallelogram	p	None	$A \neq B, \gamma \neq 90°$	Oblique
Rectangle	p	Two shortest, mutually perpendicular vectors	$A \neq B, \gamma = 90°$	Rectangular
Rectangle	c	Two shortest, mutually perpendicular vectors	$A \neq B, \gamma = 90°$	Rectangular
Square	p	Two shortest, mutually perpendicular vectors	$A = B, \gamma = 90°$	Square
60° angle rhombus	p	Two shortest vectors at 120° to each other	$A = B, \gamma \neq 120°$	Hexagonal

hexagonal net. If the two basis vectors for the reference net are $\boldsymbol{A}$ and $\boldsymbol{B}$, and the surface mesh basis vectors are $\boldsymbol{A}_s = \xi \boldsymbol{A}$ and $\boldsymbol{B}_s = \zeta \boldsymbol{B}$, then a shorthand notation of the surface structure is denoted by $\xi \times \zeta$ reconstruction. In reciprocal space, the basis vectors of the surface and the reference are related by $\boldsymbol{A}_s^* = (1/\xi)\boldsymbol{A}^*$ and $\boldsymbol{B}_s^* = (1/\zeta)\boldsymbol{B}^*$. Figure 2.7 shows an example of a 3×1 surface structure in real space and the corresponding diffraction pattern in reciprocal space. A review of these notations has been given by Wood (1964).

In some cases the mesh of the surface structure may be rotated with respect to the reference net. A shorthand notation for this case is $\xi \times \zeta$–Rϕ°, where ϕ is the rotation angle. Figure 2.8 shows a schematic diagram of $\sqrt{2} \times 2\sqrt{2}$–R45° surface structure. This notation only describes the geometry of the two-dimensional surface unit cell. For a more general case in which the superlattice mesh is expressed as $\boldsymbol{A}_s = \xi_1 \boldsymbol{A} + \zeta_1 \boldsymbol{B}$ and $\boldsymbol{B}_s = \xi_2 \boldsymbol{A} + \zeta_2 \boldsymbol{B}$, an alternative notation of $(\xi_1, \zeta_1) \times (\xi_2, \zeta_2)$ is also useful (Yamamoto, 1993).

The surface structure induced by adsorption can be denoted as follows. If adsorbate P on the $\{hkl\}$ surface of material X causes the formation of a structure having a primitive translation of $|\boldsymbol{A}_s| = \xi|\boldsymbol{A}|$ and $|\boldsymbol{B}_s| = \zeta|\boldsymbol{B}|$ with a unit mesh rotation of ϕ, then the structure is referred to as X$\{hkl\}$ $\xi \times \zeta$–R ϕ°–P.

To summarize, atoms on the surfaces of materials can arrange themselves into a net that is different from the corresponding face of the terminated bulk lattice. The arrangements are classified as either surface reconstruction or surface relaxation. Both involve displacements of some surface atoms in the direction normal to the surface. Reconstruction leads to a unit mesh that is normally larger than the unit mesh cell of the crystal on the surface. The unit mesh of the surface's reciprocal net is consequently reduced in size, and new features appear in the RHEED pattern,

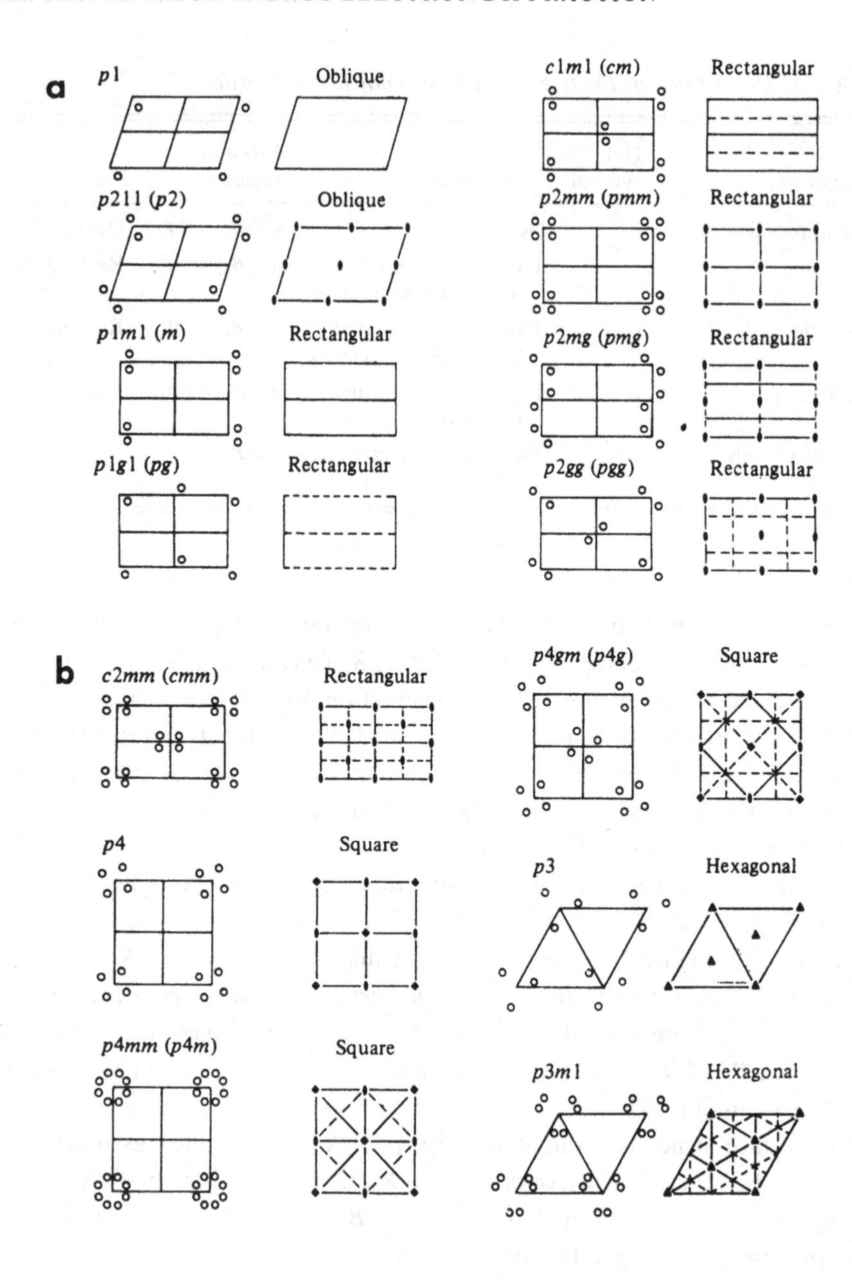

namely higher order streaks and higher order Laue zones. Sometimes there is a displacement of the surface atoms without an enlargement of the projected bulk unit mesh; this is termed surface relaxation. The surface atoms can be relaxed inward or outward depending on the electronic structure of the material.

The two-dimensional description of surface structure follows the mathematical assumption that a surface is a plane with zero thickness. This mathematical

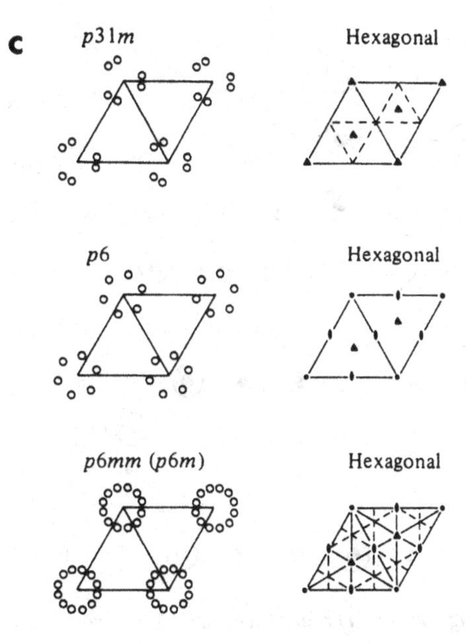

Figure 2.5 *Stereograms of two-dimensional point groups. On the left-hand side are shown the equivalent positions, on the right the symmetry operations. The names follow the full and abbreviated international standards (Woodruff and Delchar, 1994).*

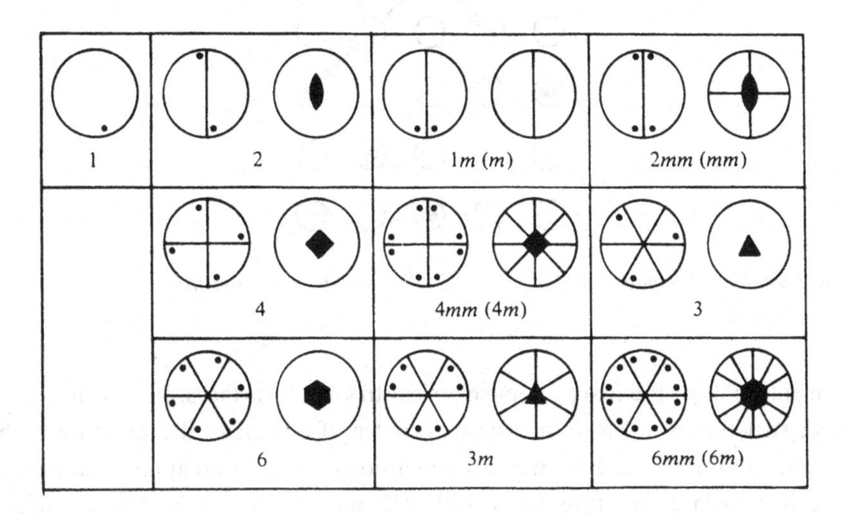

Figure 2.6 *Equivalent positions, symmetry operations and long and short international standard notations for the 17 two-dimensional space groups.*

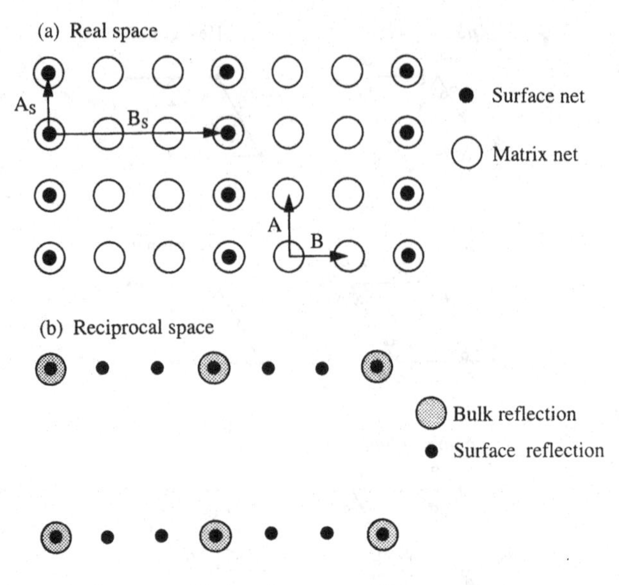

Figure 2.7 *Schematic diagrams of (a) real-space and (b) reciprocal-space nets of 3 × 1 surface reconstruction.*

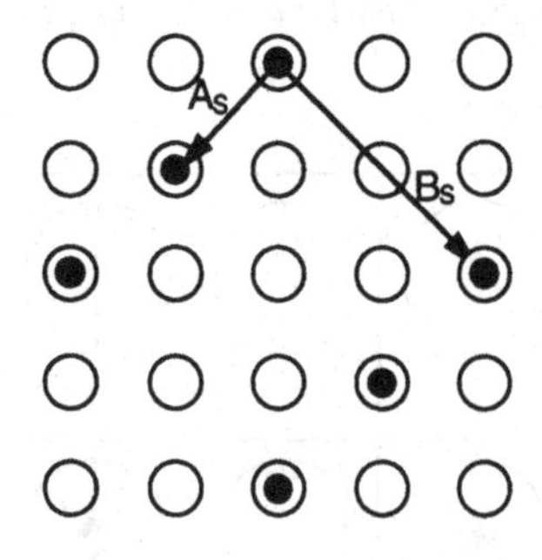

Figure 2.8 *The real-space net of a $\sqrt{2} \times 2\sqrt{2}$–R45° surface reconstruction.*

treatment holds in the macroscopic classical theory of surfaces. In fact, however, surface structure is a three-dimensional problem if one views the structure in the atomic dimension. Therefore, the two-dimensional model is an approximated case of the real surface structure. Since RHEED may not be sensitive to the minor displacement of the atoms along the direction perpendicular to the surface, the two-dimensional model is adequate for treating surface diffraction problems.

2.2.2 Two-dimensional reciprocal space

Since surface structure can be described by two basis vectors, it is convenient to use two basis reciprocal-space vectors to describe the diffraction of surfaces. This is equivalent to treating the surface diffraction as two-dimensional diffraction, so that each reflection is not a spot but rather a rod. The surface unit cell is defined by two real-space lattice basis vectors $\boldsymbol{A}_s$ and $\boldsymbol{B}_s$. The corresponding reciprocal-space lattice basis vectors are $\boldsymbol{A}_s^*$ and $\boldsymbol{B}_s^*$

$$\boldsymbol{A}_s^* = \frac{\boldsymbol{B}_s \times \boldsymbol{n}}{|\boldsymbol{A}_s \times \boldsymbol{B}_s|}, \qquad \boldsymbol{B}_s^* = \frac{\boldsymbol{n} \times \boldsymbol{A}_s}{|\boldsymbol{A}_s \times \boldsymbol{B}_s|}, \tag{2.1}$$

where $\boldsymbol{n}$ is an outward unit vector of the surface normal direction. The reciprocal-space lattice basis vectors of the surface reference net are

$$\boldsymbol{A}^* = \frac{\boldsymbol{B} \times \boldsymbol{n}}{|\boldsymbol{A} \times \boldsymbol{B}|}, \qquad \boldsymbol{B}^* = \frac{\boldsymbol{n} \times \boldsymbol{A}}{|\boldsymbol{A} \times \boldsymbol{B}|}. \tag{2.2}$$

Thus the surface lattice is denoted by $\boldsymbol{R}_s = h\boldsymbol{A}_s + k\boldsymbol{B}_s$, and is in the Miller index (hk). In this case, each reflection is a rod in the direction perpendicular to the surface.

For a general case of surface reconstruction denoted by $\boldsymbol{A}_s = \xi_1 \boldsymbol{A} + \zeta_1 \boldsymbol{B}$ and $\boldsymbol{B}_s = \xi_2 \boldsymbol{A} + \zeta_2 \boldsymbol{B}$, the corresponding reciprocal-space lattice basis vectors are

$$\boldsymbol{A}_s^* = \frac{1}{|\xi_1\zeta_2 - \xi_2\zeta_1|}(\zeta_2 \boldsymbol{A}^* - \xi_2 \boldsymbol{B}^*), \qquad \boldsymbol{B}_s^* = \frac{1}{|\xi_1\zeta_2 - \xi_2\zeta_1|}(\zeta_1 \boldsymbol{A}^* - \xi_1 \boldsymbol{B}^*). \tag{2.3}$$

In practice, the RHEED pattern is usually indexed first in a notation of $\boldsymbol{A}_s^* = \chi_1 \boldsymbol{A}^* + \eta_1 \boldsymbol{B}^*$, and $\boldsymbol{B}_s^* = \chi_2 \boldsymbol{A}^* + \eta_2 \boldsymbol{B}^*$, the real-space parameters (ξ_1, ζ_1) and (ξ_2, ζ_2) are thus calculated from

$$\xi_1 = -\,\mathfrak{p}\eta_2, \qquad \zeta_1 = \mathfrak{p}\chi_2, \qquad \xi_2 = -\,\mathfrak{p}\eta_1, \qquad \zeta_2 = \mathfrak{p}\chi_1, \tag{2.4}$$

where $\mathfrak{p} = 1/|\chi_1\eta_2 - \chi_2\eta_1|$. These relations are useful in Section 2.4 for determining surface structures using RHEED.

2.3 Streaks and Laue rings in RHEED

For three-dimenstional perfect crystals, the diffraction maxima are sharp spots if the crystal is large. In RHEED, particularly for ZOLZ reflections, the diffraction maxima are streaks in the direction perpendicular to the surface. There are two possible sources for producing streaks perpendicular to the surface. The small penetration of the beam into the surface is equivalent to sampling a thin-sheet crystal, which is thin in the direction nearly perpendicular to the incident beam. The shape factor introduced by the thin sheet produces streaking in the direction perpendicular to the sheet (or surface), as shown earlier in Figure 1.4. The possible streaking in RHEED is schematically shown in Figure 2.9. The other source is wavy surface morphology and surface disorder. The disordered structure produces

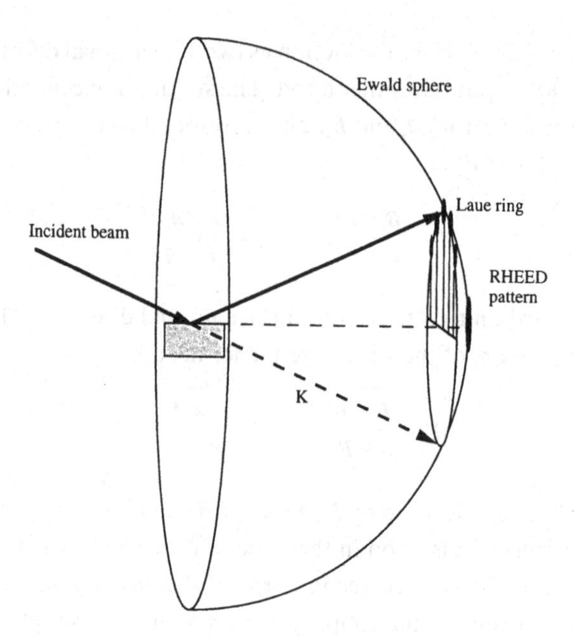

Figure 2.9 *Ewald sphere construction of a Laue circle in RHEED.*

reflections that do not fall on the position of Bragg beams. The diffuse scattering produced by short-range ordering in TEM is a typical example (Cowley, 1981).

The streaking of RHEED spots may complicate the analysis of the pattern, because the streaks belonging to different reflection indexes tend to be mixed and lined up, making it difficult to index the streaked rods. This effect will be seen in Figure 2.13 later.

Regularly spaced surface steps can also introduce streaks in RHEED. The full width at half maximum of the reflected beam is inversely proportional to the lateral extent of the terraces (see Section 2.6). The beam convergence is also a source for producing spreading of intensity in RHEED.

In RHEED, the curvature of the Ewald sphere for electron energies of a few tens of kilo-electron-volts is rather significant, giving rise to curved streaks in the RHEED pattern, as shown in Figure 2.10. This effect is severe if the incident beam energy is lower, particular in a RHEED facility on an MBE system. For TEM of energy larger than 100 keV, this problem does not occur. Therefore, in general, caution must be exercised if one intends to index RHEED patterns based on the knowledge of transmission electron diffraction.

2.4 Determination of surface structures

RHEED is a powerful technique that can be applied to determining the structures of surfaces. The first step is to index the RHEED pattern. Here we take Si(111) as an

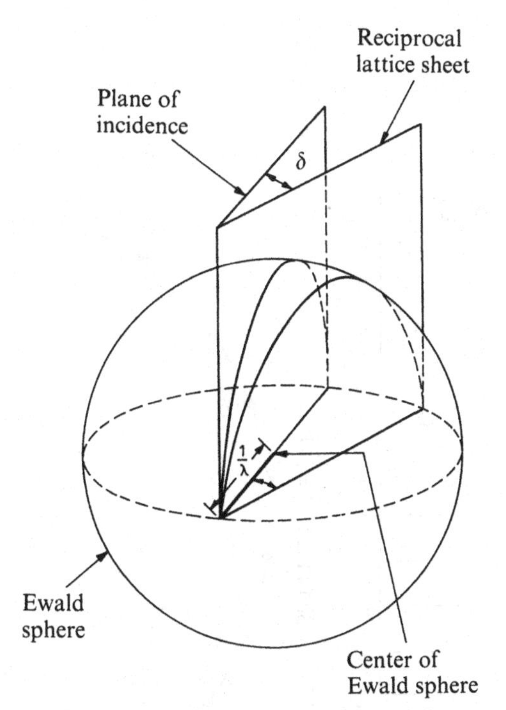

Figure 2.10 Reciprocal-lattice–Ewald-sphere construction showing the intersection of a reciprocal lattice sheet at an angle δ with respect to the incident plane. This gives rise to a curved streak in the RHEED pattern. This case occurs only for electrons with energies about 10–30 keV.

example. The crystalline net of the Si(111) surface is a close-packed hexagonal array. This surface's net and its reciprocal-space net are shown in Figures 2.11(a) and (b), respectively. We choose the basis vectors of the real-space net as follows: $\boldsymbol{A}=(a/2)[10\bar{1}]$ and $\boldsymbol{B}=(a/2)[\bar{1}10]$, where $a=0.543$ nm is the lattice constant of Si. The net point spacing in the (111) plane is 0.384 nm. The unit vector normal to this surface is $\boldsymbol{n}=(1/\sqrt{3})$ [111]. The corresponding basis vectors of the reciprocal-lattice net are calculated according to Eq. (2.2) as $\boldsymbol{A}^*=(3a/2)$ $[11\bar{2}]$ and $\boldsymbol{B}^*=(3a/2)$ $[\bar{1}2\bar{1}]$. Note that the reciprocal net is a hexagonal close-packed array of points as well, but it is rotated by 30° with respect to the real-space net. The two-dimensional reciprocal rods are constructed using $\boldsymbol{A}^*$ and $\boldsymbol{B}^*$ basis vectors, as shown in Figure 2.11(b). The incident beam direction is $\boldsymbol{K}$; a RHEED pattern consists of the intersections of the Ewald sphere with the reciprocal lattice rods, as viewed along the beam direction. For simplicity, the incident angle of the beam is assumed to be zero.

Figure 2.11(c) is a section of the circle made by the intersection of the Ewald sphere surface and the two-dimensional reciprocal rods of Si(111). The beam direction is $[1\bar{1}0]$. The rods in the ZOLZ are streaks and the beams in the FOLZ are spots. For an incident beam along $[\bar{1}\bar{1}2]$, the streak spacing is larger by a factor of $\sqrt{3}$

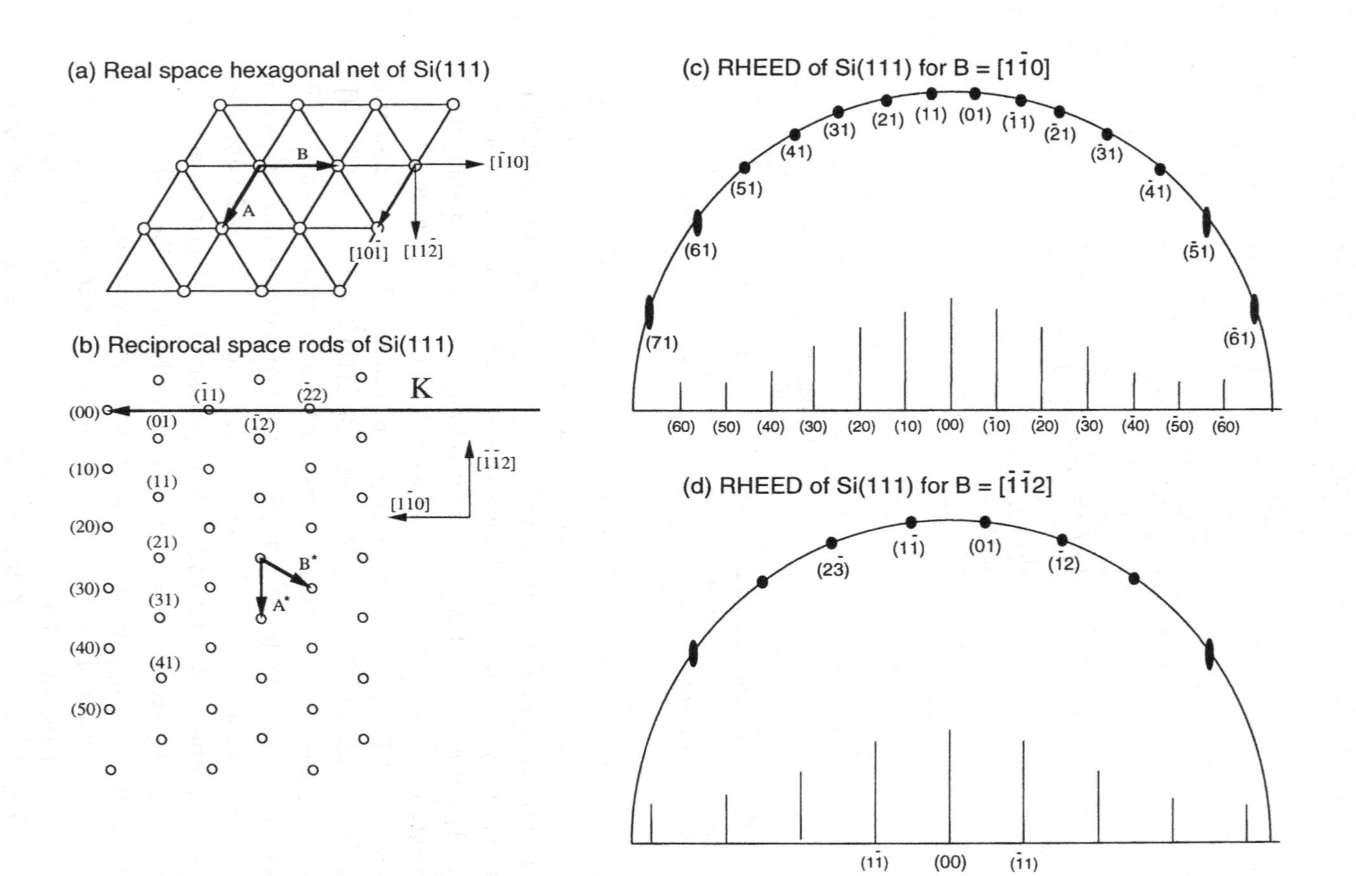

Figure 2.11 *(a) Real- and (b) reciprocal-lattice nets of the Si(111) face. No surface reconstruction is considered here. The predicted RHEED patterns from Si(111), with the incident beam along (c) $[1\bar{1}0]$ and (d) $[\bar{1}\bar{1}2]$. A zero angle of incidence was assumed. The RHEED patterns are shown for 10 kV incident electrons (Mahan et al., 1990).*

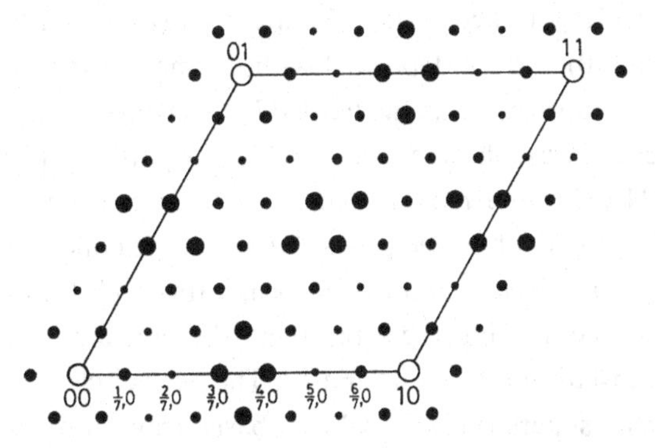

Figure 2.12 Reciprocal-lattice rods of the 7 × 7 Si(111) surface. The spot size schematically represents the integrated intensity of each rod (Ino, 1987). The open circles are the reflections from the bulk.

when compared with the $[1\bar{1}0]$ pattern. The predicted RHEED pattern is shown schematically in Figure 2.11(d), and the corresponding experimental observation was shown in Figure 2.2 (note that the superlattice reflections are the result of surface reconstruction).

The Si(111) surface can exhibit the 7 × 7 reconstruction. This means that the Si(111) 7 × 7 reconstructed surface, as it is customarily specified, possesses a crystalline unit mesh whose edges are parallel to those of the projected bulk unit cell, and that they are seven times as long. Whether the surface of a sample reconstructs, and the specific form that a reconstruction might take, depend upon the chemical and thermal history of the sample. The Si(111) 7 × 7 reconstruction introduces superlattice reflection rods into the RHEED pattern (see Figure 2.2). The diffraction rods are modified to include the extra reflections, and the result is shown in Figure 2.12.

Indexes of bulk reflections can usually be determined from the kinematically simulated diffraction pattern for the bulk crystal. The indexes of surface superlattice reflections must be determined with consideration given to the streaking effect and the geometry of the Ewald sphere. In this case, the spherical method introduced by Ino (1977, 1993) can be very useful, and the recorded RHEED patterns can then be directly compared with LEED patterns.

In RHEED, a large surface area is illuminated by the beam and the diffraction from the entire surface is represented in the pattern. Thus, the domain structure and inhomogeneous growth are accumulated in a single pattern. On the other hand, the reflections from HOLZs are pronounced. Therefore, for accuracy, it is recommended that the surface structure be determined from a group of RHEED patterns acquired from different zone axes and indexed consistently.

We now show two examples to illustrate the application of RHEED to thin film growth. Ge/Si is a widely studied system. Figure 2.13(a) shows a RHEED pattern recorded from a Si(110) surface deposited with 0.5 monolayer (ML) Ge (Yamamoto, 1993). The Si substrate was heated to 700 °C when Ge was deposited. A $(10,1)\times(\bar{3},4)$ structure was observed. The RHEED pattern is interpreted as a superposition of RHEED patterns from two domains in mirror relation with respect to Si[001]. The arrangement of the reciprocal-lattice rods for one domain is shown in Figure 2.13(b). The solid circles are the reciprocal-lattice rods for the Si(110) plane and the broken lines connecting them represent the unit mesh of the reciprocal lattice with basis translation vectors $\boldsymbol{A}^*$ and $\boldsymbol{B}^*$. The open circles are the reciprocal-lattice rods for the superstructure, whose basis translation vectors are $\boldsymbol{A}_s^* = (4/43)\boldsymbol{A}^* + (3/43)\boldsymbol{B}^*$ and $\boldsymbol{B}_s^* = -(1/43)\boldsymbol{A}^* + (10/43)\boldsymbol{B}^*$. Thus, the basis translation vectors of the real-space lattice basis vectors of the surface are $\boldsymbol{A}_s = 10\boldsymbol{A} + \boldsymbol{B}$ and $\boldsymbol{B}_s = -3\boldsymbol{A} + 4\boldsymbol{B}$ according to Eq. (2.4).

After 1 ML Ge had been deposited on the Si(110) surface, the surface was heated to 600 °C. The RHEED pattern shown in Figure 2.14(a) was observed (Yamamoto, 1993). This pattern is interpreted as a superposition of the two domains in mirror relation with respect to Si[001]. The arrangement of the reciprocal-lattice rods for one domain is shown in Figure 2.14(b). This has basis translation vectors $\boldsymbol{A}_s^* = (1/8)\boldsymbol{A}^* + (1/4)\boldsymbol{B}^*$ and $\boldsymbol{B}_s^* = -(3/8)\boldsymbol{A}^* + (1/4)\boldsymbol{B}^*$ as shown in Figure 2.14(b), and leads to a unit mesh of $(2,3)\times(\bar{2},1)$.

In general, RHEED can sensitively detect the periodic reconstruction of the surface, but it is usually difficult to apply to the determination of the elements and atom positions involved in the reconstruction. In this case, other spectroscopic techniques can be very useful. The real-space models of the observed $(10,1)\times(\bar{3},4)$ and $(2,3)\times(\bar{2},1)$ reconstruction of Ge/Si(110) will be shown in Section 12.7 after the surfaces have been studied by some spectroscopic techniques.

2.5 RHEED oscillation and its application in MBE crystal growth

One of the most exciting and useful phenomena in RHEED is the intensity oscillation during epitaxial growth of thin films. RHEED oscillation was first observed by Harris *et al.* (1981) during growth of GaAs. This phenomenon has been studied in detail by Neave *et al.* (1983) and Van Hove *et al.* (1983). Figure 2.15 shows a RHEED intensity oscillation during successive growth of GaAs/GaAlAs/AlAs etc. (Sakamoto *et al.*, 1984). The oscillation period of each growth layer is significantly different, indicating the difference in growth rate. RHEED oscillation has attracted considerable interest because it can be applied to studying many fundamental surface problems, such as electron diffraction, epitaxial growth and atomic dynamics. The mechanism of RHEED oscillation is described below.

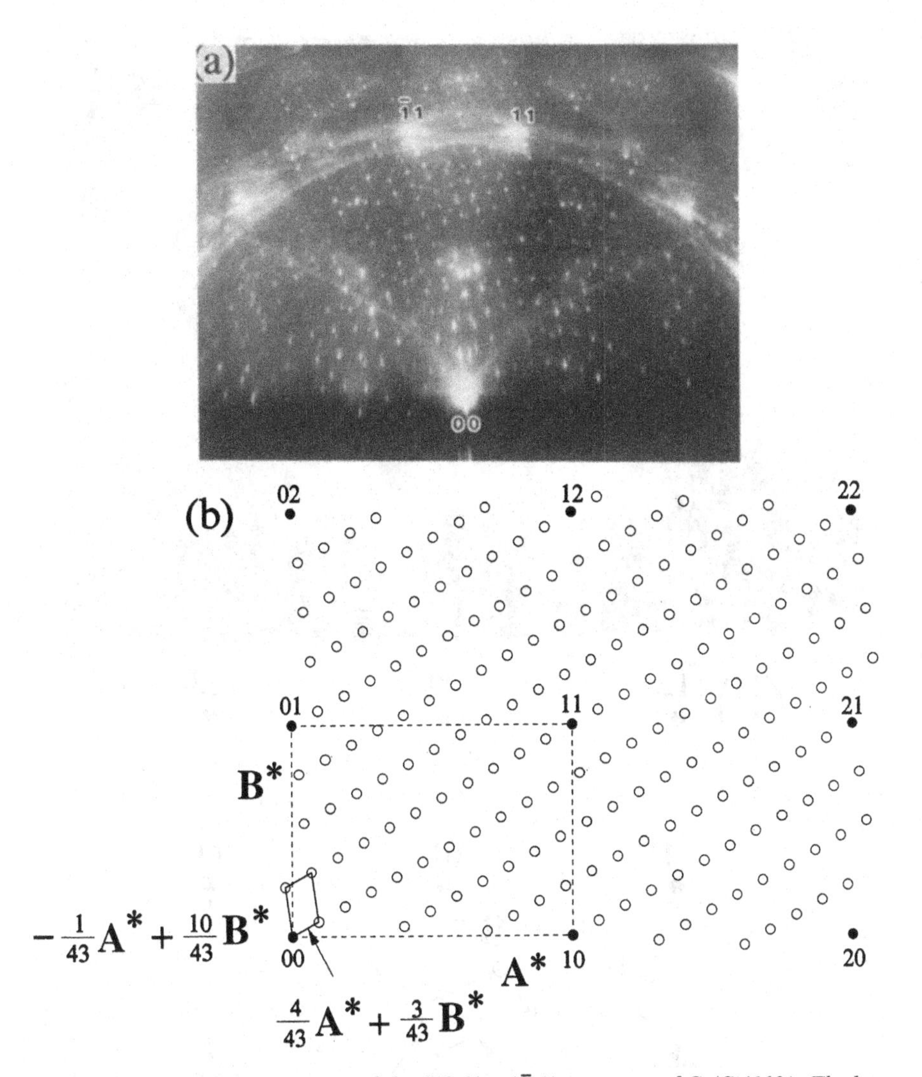

Figure 2.13 *(a) RHEED pattern of the (10,1) × ($\bar{3}$,4) structure of Ge/Si(110). The beam energy is 7 keV and the beam azimuth is [001]. (b) Reciprocal-lattice rods for the (10,1) × ($\bar{3}$,4) structure. The solid circles are the reciprocal-lattice rod positions for the Si(110) plane (surface normal) and the broken lines connecting them represent the unit mesh with basic translation vectors* $\boldsymbol{A}^*$ *and* $\boldsymbol{B}^*$. *The open circles represent the reciprocal-lattice rod positions and the superstructure, and the solid line corresponding to them represents the unit mesh with basic translation vectors of* $(4/43)\boldsymbol{A}^* + (3/43)\boldsymbol{B}^*$ *and* $-(1/43)\boldsymbol{A}^* + (10/43)\boldsymbol{B}^*$ *(courtesy of Dr Y. Yamamoto, 1993).*

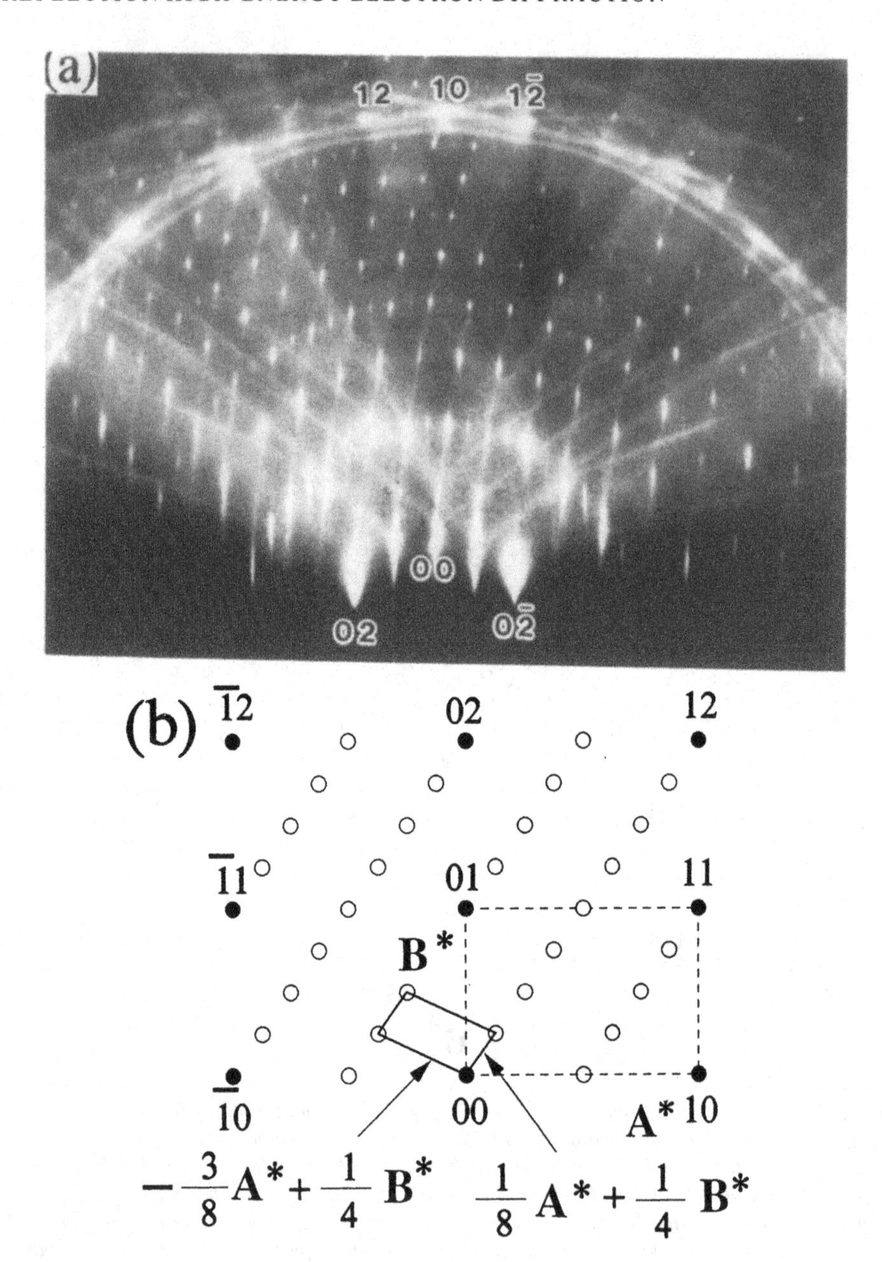

Figure 2.14 *(a) The RHEED pattern of the (2,3) × ($\bar{2}$,1) structure of Ge/Si(110). The beam energy is 7 keV and the beam azimuth is [001]. (b) Reciprocal-lattice rods for the (2,3) × ($\bar{2}$,1) structure. The solid circles are the reciprocal-lattice rod positions for the Si(110) plane (surface normal) and the broken lines connecting them represent the unit mesh with basic translation vectors* A^* *and* B^**. The open circles represent the reciprocal-lattice rod positions and the superstructure, and the solid line corresponding to them represents the unit mesh with basic translation vectors of* $\frac{1}{8}A^* + \frac{1}{4}B^*$ *and* $-\frac{3}{8}A^* + \frac{1}{4}B^*$ *(Courtesy of Dr Y. Yamamoto, 1993).*

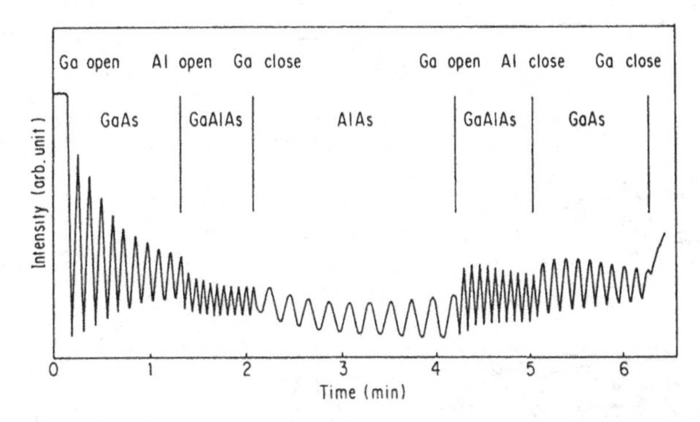

Figure 2.15 The RHEED intensity oscillation observed during successive depositions of GaAs, GaAlAs, AlAs ... multilayer material at 560°C (Sakamoto et al., 1984).

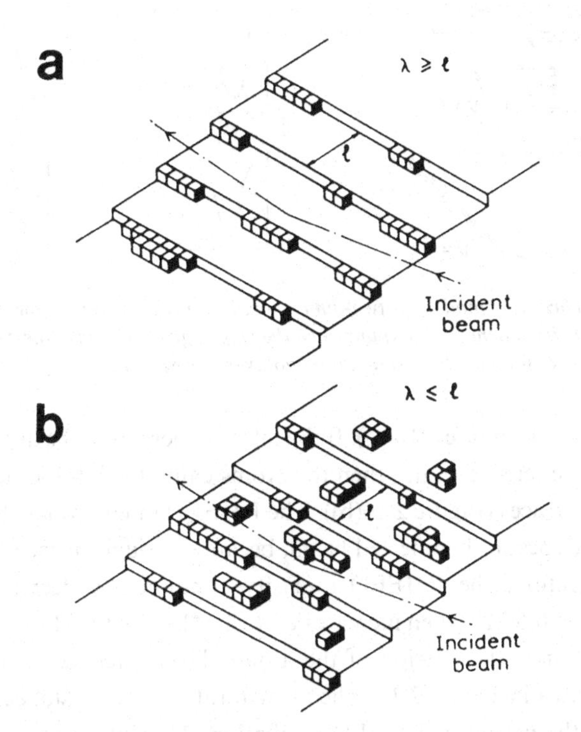

Figure 2.16 Schematic illustrations showing the change in RHEED information as the growth mode changes from (a) step propagation to (b) two-dimensional growth (Joyce et al., 1987).

For simplification, we consider the growth process on a clean and flat crystal surface that has some steps. If the diffusion length of the deposited atoms is sufficiently larger than the terrace width (i.e., the spacing between steps), then all the deposited atoms may reach the steps by diffusion and may be incorporated into the steps. In this case, the steps move (or grow) during deposition, preserving a flat surface terrace with approximately the same width, and no change in RHEED

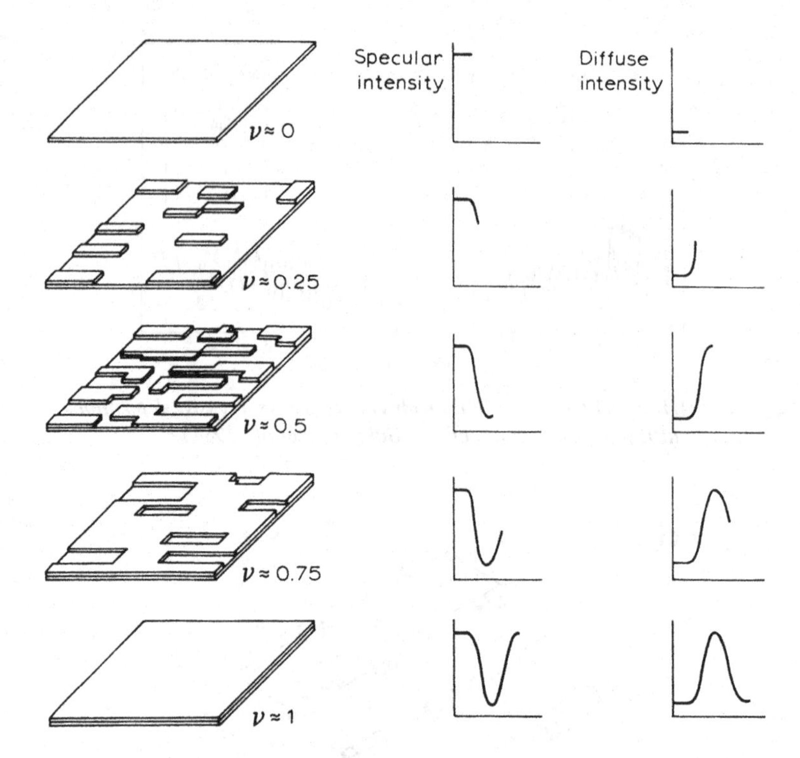

Figure 2.17 *A schematic model of two-dimensional growth showing how the specularly reflected beam intensity reaches a minimum when the step edge density is highest and the diffuse intensity reaches a maximum. The fractional monolayer coverage is* υ.

intensity is produced (Figure 2.16(a)). If the diffusion length of the deposited atoms is smaller than the terrace width, then the atoms cannot reach the steps and may nucleate on the terraces (Figure 2.16(b)). The RHEED intensity may be decreased, especially for the specularly reflected beam, because a rough surface has a smaller reflectance. Therefore, the RHEED intensity continuously decreases until the deposition reaches 0.5 ML, then increases up to 1 ML growth. This is the RHEED intensity oscillation with a period of the atomic layer thickness. This process is schematicaly shown in Figure 2.17. The growth rate of the crystal can be directly measured from the period of RHEED oscillation. This information is useful for molecular dynamic calculations. The number of layers grown is the number of oscillations. The features observed in RHEED oscillation curves directly reflect the surface structures. A detailed review has been given by Dobson *et al.* (1987) and Joyce *et al.* (1987).

The damped nature of RHEED oscillation implies that the step density is periodically changing and reducing with time. When the oscillations are completely damped this corresponds to a constant step density, which would occur when the arriving adatoms migrate to the step edges rather than nucleate new two-dimensional growth centers on the terraces. Under these conditions, growth occurs by regular step movement. In addition, the locally different growth rates on the surface can produce a damping and beating effect.

An absence of oscillations may indicate that the step density is constant and growth occurs by regular step movement. On the other hand, if three-dimensional nucleation is occurring, there will be an absence of oscillations, and this situation can usually be recognized by the overall appearance of a spotty pseudo-transmission RHEED pattern.

2.6 The kinematical diffraction theory of RHEED

The positions of reflected beams in RHEED can be predicted using kinematical scattering (or single-scattering) theory. Using the theory illustrated in Chapter 1, we now consider the scattering amplitude from a crystal surface. For simplification, we will only consider the contribution made by the top atomic layer of the surface. Under the rigid-body approximation, the potential distribution in the surface can be generally written as a superposition of the potential distribution from each atom site $\boldsymbol{r}_i$,

$$V_s(\boldsymbol{r}) = \sum_i V_i(\boldsymbol{r} - \boldsymbol{r}_i), \tag{2.5}$$

where i refers to the ith atom site on the surface. The scattering amplitude is a Fourier transform of V_s according to the first Born approximation

$$V_s(\boldsymbol{u}) = \mathrm{FT}[V_s(\boldsymbol{r})] = \sum_i f_i^e(\boldsymbol{u}) \exp(-2\pi i \boldsymbol{u} \cdot \boldsymbol{r}_i). \tag{2.6}$$

The kinematically diffracted intensity is thus

$$I_s(\boldsymbol{u}) = |V_s(\boldsymbol{u})|^2 = \sum_i \sum_j f_i^e(\boldsymbol{u}) f_j^{e*}(\boldsymbol{u}) \exp[-2\pi i \boldsymbol{u} \cdot (\boldsymbol{r}_i - \boldsymbol{r}_j)]. \tag{2.7}$$

This is a general equation for RHEED. We now look into a few special cases.

2.6.1 Perfectly ordered surfaces

For a perfect surface without steps and defects, the atom position $\boldsymbol{r}_i$ can be expressed as a sum of the unit cell position $\boldsymbol{R}_n$ and the position of the atom in the unit cell $\boldsymbol{r}_\alpha$, $\boldsymbol{r}_i = \boldsymbol{R}_n + \boldsymbol{r}_\alpha$,

$$V_s(\boldsymbol{r}) = \sum_n \sum_\alpha V_i(\boldsymbol{r} - \boldsymbol{R}_n - \boldsymbol{r}_\alpha)\delta(x), \tag{2.8}$$

where $\delta(x)$ specifies the 'ideal' two-dimensional surface plane. Thus

$$\begin{aligned} I_s(\boldsymbol{u}) &= \sum_n \sum_{n'} \exp[-2\pi i \boldsymbol{u}_b \cdot (\boldsymbol{R}_n - \boldsymbol{R}_{n'})] \sum_\alpha \sum_{\alpha'} f_\alpha^e(\boldsymbol{u}_b) f_{\alpha'}^{e*}(\boldsymbol{u}_b) \exp[-2\pi i \boldsymbol{u}_b \cdot (\boldsymbol{r}_\alpha - \boldsymbol{r}_{\alpha'})] \\ &= N_s \sum_{\boldsymbol{g}_s} \delta(\boldsymbol{u}_b - \boldsymbol{g}_s) \left| \sum_\alpha f_\alpha^e(\boldsymbol{u}) \exp(-2\pi i \boldsymbol{u} \cdot \boldsymbol{r}_\alpha) \right|^2 \\ &= N_s \sum_{\boldsymbol{g}_s} \delta(\boldsymbol{u}_b - \boldsymbol{g}_s) |V_{\boldsymbol{g}_s}|^2, \end{aligned} \quad (2.9)$$

where $\boldsymbol{u}_b = (u_y, u_z)$ specifies the reciprocal vector parallel to the surface, $\boldsymbol{g}_s$ is a reciprocal lattice vector of the surface net, $V_{\boldsymbol{g}_s}$ is the surface structure factor, N_s is the number of unit cells on the surface, and the equality

$$\sum_n \exp(-2\pi i \boldsymbol{u}_b \cdot \boldsymbol{R}_n) = \sum_{\boldsymbol{g}_s} \delta(\boldsymbol{u}_b - \boldsymbol{g}_s)$$

was used. The delta function determines the position of the reflections and $|V_{\boldsymbol{g}_s}|^2$ determines the scattering intensity. Thus, sharp reflection rods located at reciprocal-lattice position $\boldsymbol{g}_s$ are produced.

2.6.2 Completely disordered surfaces

For a completely disordered surface, it is necessary to average the result over the ensemble describing the disorder. For a completely random distribution of surface atoms, we have

$$I_s(\boldsymbol{u}) = \left\langle \sum_i \sum_j f_i^e(\boldsymbol{u}) f_j^{e*}(\boldsymbol{u}) \exp[-2\pi i \boldsymbol{u} \cdot (\boldsymbol{r}_i - \boldsymbol{r}_j)] \right\rangle_c = N_s |f^e(\boldsymbol{u})|^2, \quad (2.10)$$

where f^e is the average scattering factor of the surface atoms, and $\langle\,\rangle_c$ refers to an average over configurations of atom arrangements on the surface. Therefore, a uniform distribution of intensity is expected in the RHEED pattern. No Bragg reflection is expected to be seen except for those from the bulk crystal.

2.6.3 Surfaces with islands

If the top layer structure is between the limits of order and disorder, then the calculation is usually somewhat more difficult. It is, however, possible to do some analysis if the atoms are clustered into crystallites that are randomly distributed on the surface. For simplicity it is assumed that all the islands are similar. A typical atom position can be written as $\boldsymbol{r}_i = \boldsymbol{R}_I + \boldsymbol{r}_{Ik}$, where $\boldsymbol{R}_I$ is the position of the center of the Ith cluster and $\boldsymbol{r}_{Ik}$ is the kth atom displacement relative to the center of the Ith island. The scattering intensity is

$$\begin{aligned} I_s(\boldsymbol{u}) &= \left\langle \sum_I \sum_{I'} \exp[-2\pi i \boldsymbol{u} \cdot (\boldsymbol{R}_I - \boldsymbol{R}_{I'})] \right\rangle_c \sum_k \sum_{k'} f_k^e(\boldsymbol{u}) f_{k'}^{e*}(\boldsymbol{u}) \exp[-2\pi i \boldsymbol{u} \cdot (\boldsymbol{r}_{Ik} - \boldsymbol{r}_{I'k'})]. \\ &\approx N_I \left| \sum_k f_k^e(\boldsymbol{u}) \exp(-2\pi i \boldsymbol{u} \cdot \boldsymbol{r}_k) \right|^2, \end{aligned} \quad (2.11)$$

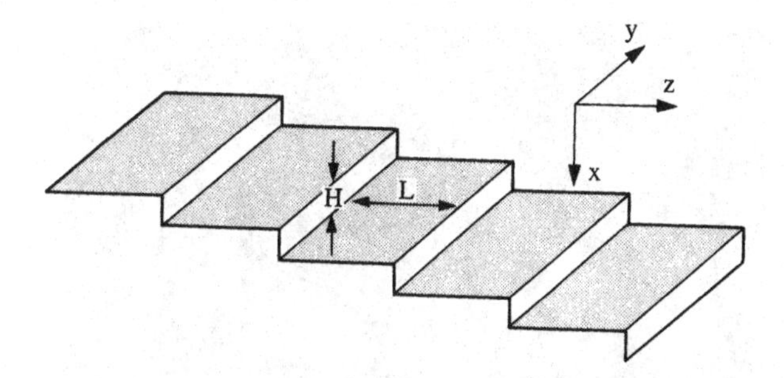

Figure 2.18 *A schematic model of equally spaced one-dimensional straight surface steps.*

where N_I is the number of islands and $\boldsymbol{R}_I$ is the position of the island. The ordered distribution of islands on the surface would give some additional structure in (2.11) arising from the sum over k, but this is qualitatively unimportant. The surface areas not covered by the islands exhibit periodic atomic structure and will give diffracted beams. However, the Bragg peaks will be broadened by a factor inversely proportional to the average size of the islands. This effect may be applied to estimate the size distribution of the islands.

2.6.4 Stepped surfaces

We now consider a case in which there are equally spaced, equal-height, one-dimensional surface steps on the surface, as shown in Figure 2.18. The position of an atom located in the mth terrace is expressed as $\boldsymbol{r}_i = mL\hat{z} + mH\hat{x} + \boldsymbol{r}_j$, where $\hat{z}$ and $\hat{x}$ are the unit vectors of the z and x axes, respectively, H is the step height and L the terrace width, and $\boldsymbol{r}_j$ is the position of the atom in the terrace. Thus

$$I_s(\boldsymbol{u}) = \left| \sum_m \exp[-2\pi i(u_z L + u_x H)m] \sum_j f_j^e(\boldsymbol{u}_b) \exp(-2\pi i \boldsymbol{u}_b \cdot \boldsymbol{r}_j) \right|^2$$
$$= N_j \sum_{n_1} \sum_{n_2} \delta(u_z - n_1/L)\, \delta(u_x - n_2/H) \left| \sum_j f_j^e(\boldsymbol{u}_b) \exp(-2\pi i \boldsymbol{u}_b \cdot \boldsymbol{r}_j) \right|^2, \quad (2.12)$$

where the sum over j gives the Bragg scattering from the surface terrace. The reflections, however, are split at intervals of $1/L$ and $1/H$ along the u_z and u_x directions, respectively.

Figure 2.19 shows a RHEED pattern recorded from a stepped Au(111) surface. Multiple splitting of the reflected beam is seen. The width of the step terrace can be directly calculated from the separation of the splitting and the results show $L \approx 2.7$ nm. Splitting in the direction parallel to the surface does not appear because the step is one atom high. The mis-cut angle α, an angle between the surface normal and the normal of the atomic planes, can also be estimated from the RHEED pattern.

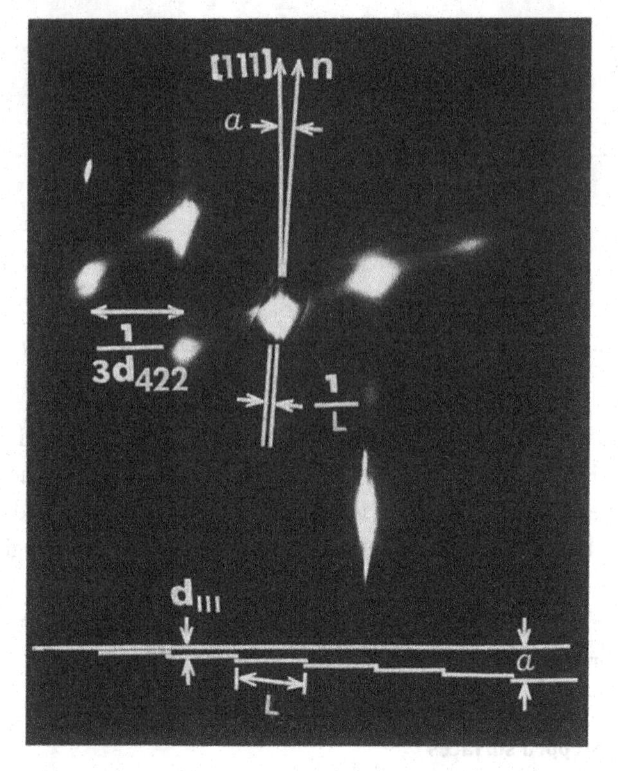

Figure 2.19 *The RHEED pattern of a stepped Au(111) surface, showing multiple splitting of the reflected spots (courtesy of Dr T. Hsu).*

2.6.5 Surfaces with randomly distributed coverage

Studying surface structural evolution during crystal growth is important for understanding the growth process of crystals. As illustrated in Section 2.5, RHEED oscillations have been observed as a function of surface coverage (or growth time at constant growth rate). We now consideer a case in which the surface is partially covered by A atoms on a B atom substrate. The coverage is assumed to be smaller than one monolayer with the A atoms distributed randomly on the surface. The diffraction intensity can be generally written as (Lagally *et al.*, 1987),

$$\begin{aligned}\langle I_s(\boldsymbol{u})\rangle &= N_s\langle f^2\rangle_c + \langle f\rangle_c^2 \sum_{i\neq j}\sum \exp[-2\pi i\boldsymbol{u}\cdot(\boldsymbol{r}_i-\boldsymbol{r}_j)]\\ &= N_s[\langle f^2\rangle_c - \langle f\rangle_c^2] + \langle f\rangle_c^2 \sum_i\sum_j \exp[-2\pi i\boldsymbol{u}\cdot(\boldsymbol{r}_i-\boldsymbol{r}_j)],\end{aligned} \tag{2.13}$$

where

$$\langle f^2\rangle_c = \upsilon(f_B^e)^2 + (1-\upsilon)(f_A^e)^2, \qquad \langle f^2\rangle_c^2 = [\upsilon f_B^e + (1-\upsilon) f_A^e]^2, \tag{2.14}$$

$\upsilon(<1)$ is the surface coverage. The first term in Eq. (2.13) is a uniform background and the second contains the interference between all the scatters. The scattering factors of atoms in the two layers, f^e_B and f^e_A, differ in general in their absolute magnitude, $|f^e_B| \neq |f^e_A|$, in the phase shift upon scattering ϕ, and the phase shift $u_x H$, caused by the height difference. The relationships between them are

$$f^e_A = |f^e_A| \exp(i\phi_A), \tag{2.15a}$$

$$f^e_B = |f^e_B| \exp(i\phi_B) \exp(-2\pi i u_x H). \tag{2.15b}$$

Substituting Eq. (2.15a) and Eq. (2.15b) into Eq. (2.13) yields

$$\begin{aligned}\langle I_s(\boldsymbol{u})\rangle_c = {} & N_s\upsilon(1-\upsilon)[(f^e_B)^2 + (f^e_A)^2 - 2|f^e_A f^e_B| \cos(\phi_B - \phi_A - 2\pi u_x H)] \\ & + \delta(\boldsymbol{u}_b)[\upsilon^2 (f^e_B)^2 + (1-\upsilon)^2 (f^e_A)^2 + 2\upsilon(1-\upsilon)|f^e_A f^e_B| \cos(\phi_B - \phi_A - 2\pi u_x H)].\end{aligned} \tag{2.16}$$

If the two atoms in the two layers are identical, so that $|f^e_B| = |f^e_A|$ and $\phi_B = \phi_A$, Eq. (2.16) reduces to

$$\begin{aligned}\langle I_s(\boldsymbol{u})\rangle_c = {} & 2N_s(f^e_A)^2\upsilon(1-\upsilon)[1 - \cos(2\pi u_x H)] \\ & + \delta(\boldsymbol{u}_b)(f^e_A)^2\{\upsilon^2 + (1-\upsilon)^2 - 2\upsilon(1-\upsilon)[1 - \cos(2\pi u_x H)]\}.\end{aligned} \tag{2.17}$$

The measurement of the step height can only be made from the interference term.

In RHEED, it is rather difficult using reciprocal-space scattering data, especially for partially disordered surfaces, to extract the real-space structure information. It is thus necessary to explore some imaging techniques which can be applied to directly image the surface structure. This will be the purpose of Part B of this book. For simple cases, it is still possible to qualitatively predict surface morphology based on RHEED patterns. Figure 2.20 summarizes the RHEED patterns for the corresponding surfaces with different structures.

2.7 Kikuchi patterns in RHEED

For perfect crystals, Kikuchi lines and bands are normally seen in electron diffraction patterns. The general characteristics of Kikuchi patterns can be illustrated using the transmission electron diffraction pattern shown in Figure 2.21. The Kikuchi lines are sharp, and the intensities of the Bragg beams are greatly enhanced when the beams intersect the Kikuchi lines, indicating the Bragg condition. The bright–dark pairs of Kikuchi lines are identified.

Kikuchi patterns are formed by electrons that have been inelastically scattered by the crystal. When an electron beam strikes a crystal, various diffracted beams are generated due to elastic scattering. Each of the beams has equal probability to be inelastically scattered afterward. The center (000) beam, for example, can be inelastically scattered at some depth within the crystal, so that the electron angular

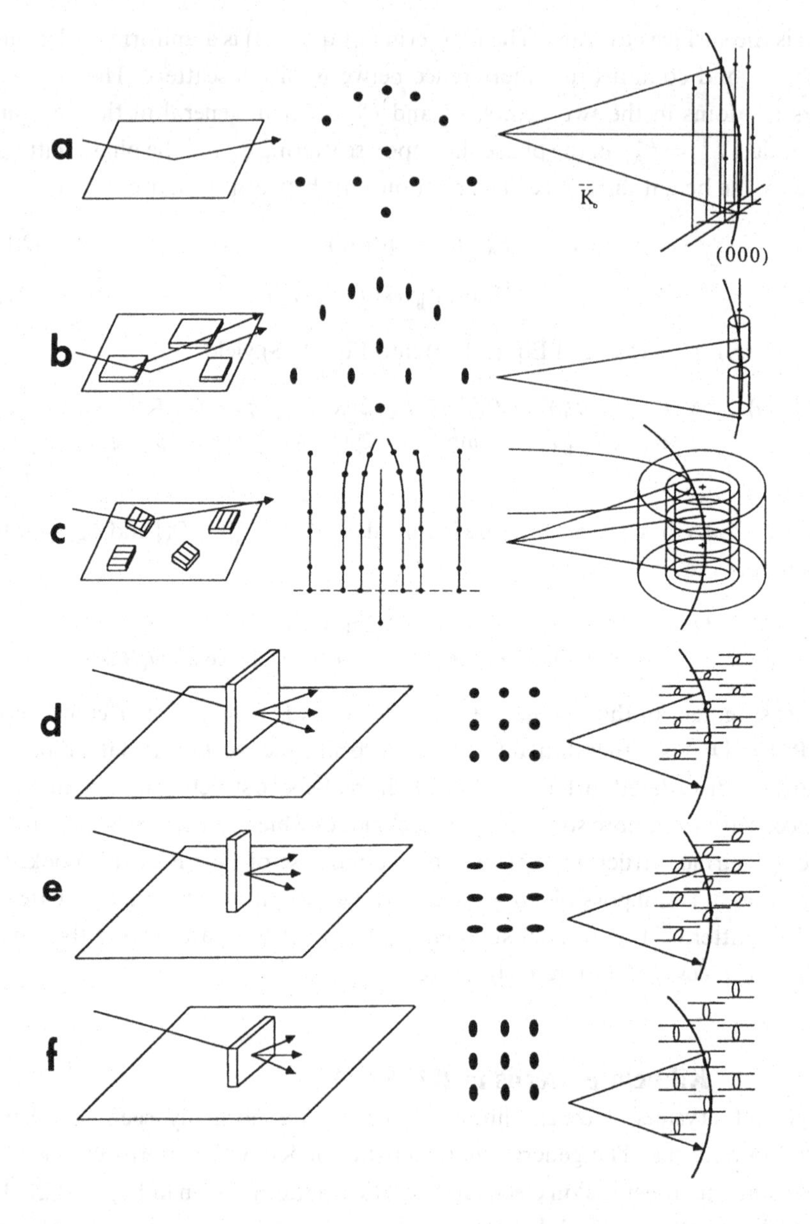

Figure 2.20 *A schematic diagram of RHEED patterns for different surface microstructures. In each case, the real-space picture, the diffraction pattern, and the reciprocal-lattice and Ewald-sphere construction are shown. (a) A perfect flat surface. (b) A surface with two-dimensional extended defects, such as terraces and adsorbed-monolayer islands. Islands are assumed to have random azimuth orientation relative to the substrate. (c) Three-dimensional islands with a particular contact plane with the substrate but with random azimuth orientation. Sharp transmission diffraction spots may be formed. (d) Tall, wide, and thin asperities. (e) Tall, narrow, and thin asperities. (f) Short, wide, and thin asperities. The situations shown in (d–f) are primarily dominated by transmission electron diffraction, and the spot elongation is produced by the shape factor of the surface object.*

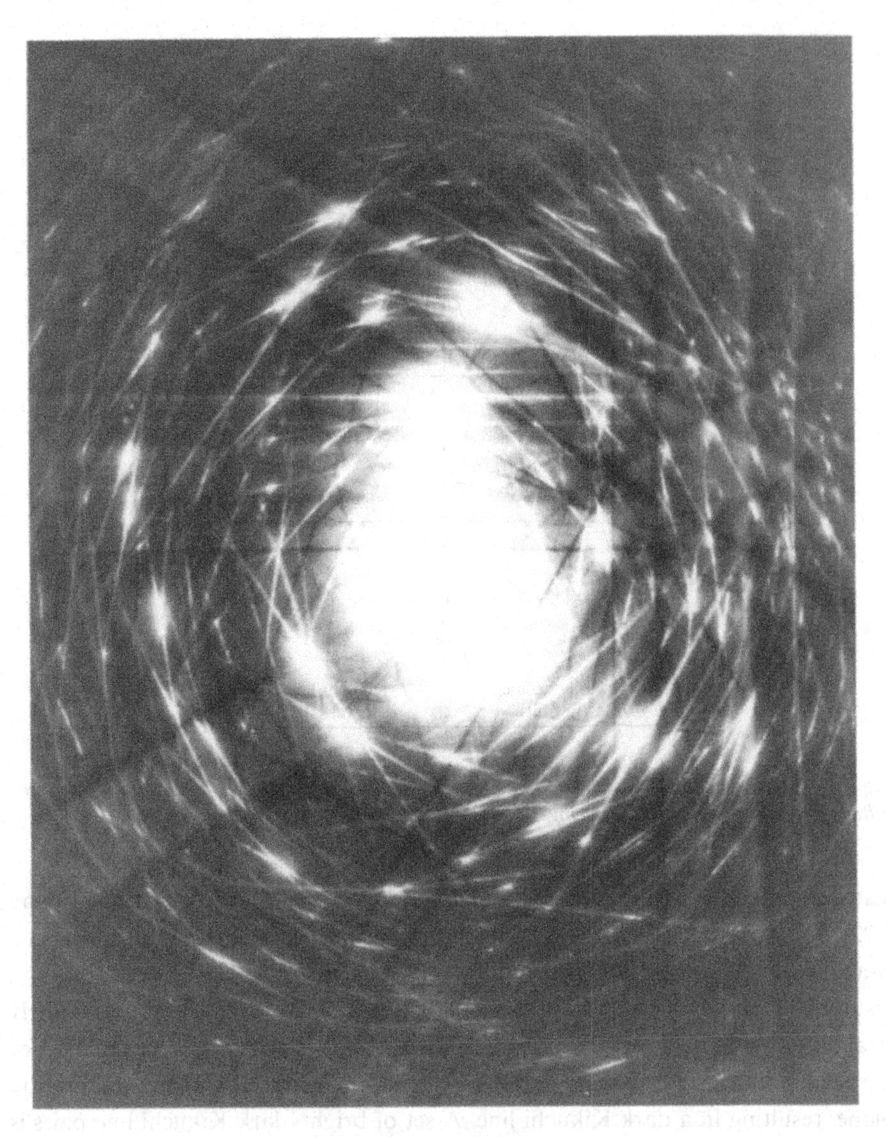

Figure 2.21 *The transmission electron diffraction pattern recorded at 100 kV from an α-Al_2O_3 crystal near the [0001] zone, showing the formation of Kikuchi lines and bands.*

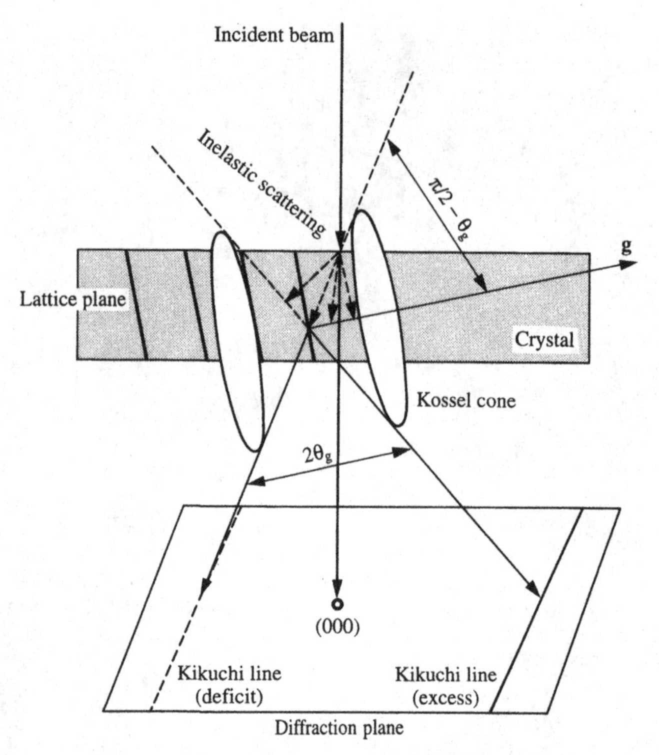

Figure 2.22 *A schematic diagram showing the formation of a Kikuchi pattern in electron diffraction due to inelastic scattering.*

distribution can be approximately described by a cone-shaped function (Figure 2.22). Some of the inelastically scattered electrons that strike the crystal planes at a Bragg angle θ_g will be elastically diffracted (i.e., undergo elastic re-scattering), resulting in a bright excess line in the observation plane. This line is approximately straight owing to the low value of θ_g. The Bragg reflection decreases the intensity of the beam propagating along $\boldsymbol{K}_0$, and the incident core intersects with the observation plane, resulting in a dark Kikuchi line. A set of bright–dark Kikuchi line pairs is generated if the mechanism is applied to every set of crystal planes. The cone of semi-conical angle $\pi/2 - \theta_g$, called the Kossel cone, is fixed to the crystal, which means that the Kikuchi lines move only if the crystal is tilted. This characteristic is employed to orient crystals in TEM and REM. The positions of Bragg-reflected beams, however, move with the variation in incident beam direction. The Kikuchi bands are produced by the channeling diffraction of the inelastically scattered electrons.

In this chapter, we have shown the fundamental characteristics of RHEED and its applications in surface structure determination. The most important step in this process is indexing the RHEED pattern. It is generally useful if one can use the transmission electron diffraction patterns shown in Appendix D as a guide, with due consideration being given to the streaking of RHEED spots in the surface normal direction. It is highly recommended that the nature of superlattice reflections observed in RHEED be determined first using some imaging techniques before a conclusion is made about the surface reconstruction, because the formation of second phase(s) and surface domains can complicate the spot distribution RHEED patterns.

CHAPTER 3

Dynamical theories of RHEED

Reflection high-energy electron diffraction is a Bragg case, in which the beams to be detected are those reflected from the bulk crystal surface. Owing to the strong interaction between the incident electron and the crystal atoms, multiple (or dynamical) scattering is very strong. Although the positions of RHEED beams can generally be predicted by kinematical scattering theory, quantitative analysis of RHEED patterns relies on dynamical calculations. In this chapter, we will first illustrate the quantum mechanical approach for electron diffraction. Then we will outline a few commonly used dynamical theories accompanied by some calculated results.

Based on the first-principles approach, we will consider the fundamental equation that governs the scattering of high-energy electrons in crystals. Before we show the mathematical description, it is important to consider the nature of the events that we are studying. The average distance between successive electrons that strike the crystal in a transmission electron microscope is about 0.2 mm (for 100 keV electrons) if the electron flux is on the order of $10^{12}\ s^{-1}$. This distance is much larger than the thickness (typically less than 0.5 μm) of the specimen, thus the interaction between any successive incident electrons is extremely weak. Therefore, the interaction between the incident beam and the crystal can be treated one electron at a time. In other words, electron diffraction theory is basically a single-electron scattering theory.

Strictly speaking, scattering of high-energy electrons obeys the Dirac equation. The Dirac equation contains not only the relativistic effects but also electron spin. It has been shown by Fujiwara (1961) and Howie (1962) that the effect of electron spin is negligible in transmission electron diffraction, but the relativistic corrections in the electron mass and wavelength must be considered in both kinematical and dynamical scattering theories (see Gevers and David (1982) for a review). In general, the solution of the Dirac equation for high-energy electron diffraction is quite complex (Fujiwara, 1961). For high-energy electron diffraction, a Schrödinger-like equation (Humphreys, 1979; Spence, 1988a) is introduced:

$$-\frac{\hbar^2}{2m_0}\nabla^2\Psi - \gamma e V\Psi = E\Psi, \qquad (3.1a)$$

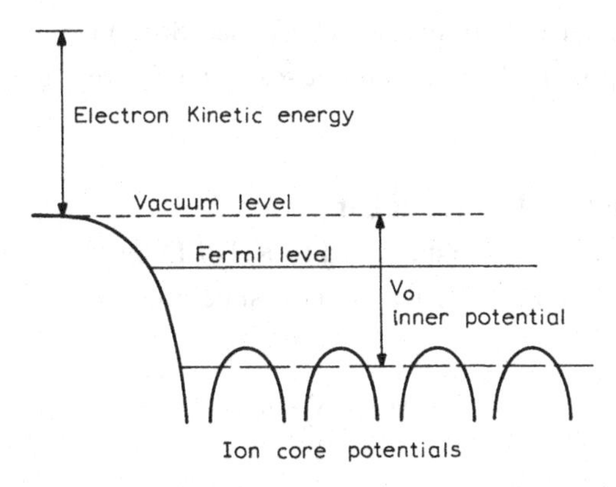

Figure 3.1 *A schematic diagram showing the change in energy that an electron experiences when it enters a solid.*

where

$$E = eU_0 \left(1 + \frac{eU_0}{2m_0c^2}\right) \tag{3.1b}$$

and the relativistic factor $\gamma = (1 - v^2/c^2)^{-1/2}$. By defining a modified crystal potential

$$U(\mathbf{r}) = \frac{2\gamma m_0 e}{h^2} V(\mathbf{r}) \tag{3.2}$$

and the electron wave number as

$$K = \frac{(2m_0E)^{1/2}}{h}, \tag{3.3}$$

Eq. (3.1a) is rewritten as

$$[\nabla^2 + 4\pi^2(U(\mathbf{r}) + K^2)]\Psi(\mathbf{r}) = 0. \tag{3.4}$$

This is the *fundamental equation*, which governs the scattering behavior of electrons in crystals. The various RHEED theories are actually different methods for solving Eq. (3.4) under some approximations.

In RHEED calculations, the potential U is no longer periodic, at least in the direction perpendicular to the surface. Figure 3.1 shows a one-dimensional profile of the surface potential. Since RHEED is a surface-sensitive technique, it is important that the variation of potential at the surface is properly considered in the calculation. The surface potential is greatly affected by surface relaxation and reconstruction as well as atom ionization. An important procedure in RHEED theory is to accurately include the lateral variation of surface potential in the numerical calculation. The average potential of the crystal, which is normally the inner potential, is a quantity for describing the surface refraction effect (see Section 5.5). The relationship between the inner potential and the atomic scattering factor is given in Appendix B.

3.1 The Bloch wave theory

The Bloch wave theory was first applied to RHEED by Colella (1972) and Moon (1972). The full solution is written as a linear superposition of Bloch waves (Bethe, 1928),

$$\Psi(\mathbf{r}) = \sum_i \alpha_i B_i(\mathbf{r}), \tag{3.5}$$

each Bloch wave $B_i(\mathbf{r})$ is an eigensolution of Eq. (3.4), and the coefficients α_i are determined by the boundary conditions. The meaning of Eq. (3.5) is that each Bloch wave is an eigenstate of the electron–crystal system, the electron wave function is a linear superposition of the Bloch waves. The probability for the ith Bloch wave to be excited is determined by the superposition coefficient α_i. Although there are many Bloch wave states in the crystal, the boundary conditions determine which of them are to be excited. This is similar to the selection of Bragg beams by the diffracting condition in electron diffraction. Even though there are many possible Bragg reflections, those that will be seen in the diffraction pattern are selected by the initial diffracting condition as illustrated by the Ewald sphere construction in Figure 1.7.

The Bloch wave theory is usually convenient for examining the diffraction of a periodically structured crystal. In this case, the modified crystal potential U can be expanded as a Fourier series based on the reciprocal lattice vectors,

$$U(\mathbf{r}) = \sum_g U_g \exp(2\pi i \mathbf{g} \cdot \mathbf{r}), \tag{3.6a}$$

with

$$U_g = \frac{2\gamma m_0 e}{h^2} \sum_\alpha V_g \exp(-2\pi i \mathbf{g} \cdot \mathbf{r}_\alpha), \tag{3.6b}$$

where $\mathbf{r}_\alpha$ represents the atom's position in the unit cell. Similarly, the Bloch wave within the crystal may also be written as a Fourier series

$$B(\mathbf{r}) = \sum_g C_g \exp[2\pi i (\mathbf{k} + \mathbf{g}) \cdot \mathrm{r}], \tag{3.7}$$

where C_g is the Bloch wave coefficient for Bragg reflection $\mathbf{g}$. For high-energy electron diffraction, a Bloch wave is neither a spherical wave nor a single plane wave but a linear superposition of plane waves with wavevectors $(\mathbf{k} + \mathbf{g})$. A Bloch wave contains many plane wave components.

Figure 3.2 shows the coordinate system and the constructed surface model in the

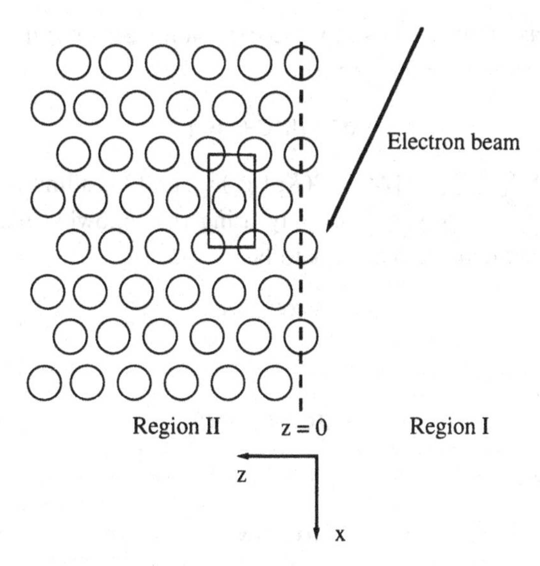

Figure 3.2 Construction of a crystal surface for Bloch wave calculation in RHEED. The coordinate system is redefined with the z axis perpendicular to the surface and the x axis parallel to the surface. The incident beam strikes the surface along a direction nearly parallel to the x axis.

Bloch wave calculation. The left-hand half-space is filled with atoms of periodic structure (region II), and the right-hand half is vacuum (region I), so that the surface is a sharp termination of the bulk crystal at $z=0$. No surface reconstruction is allowed. The z axis direction is defined to be anti-parallel to the normal direction of the surface. Thus the potential distribution is $V=V(\mathbf{r})$ for $z>0$ and $V=0$ for $z<0$. The electron wave function inside the bulk crystal can be written as a superposition of Bloch waves for $\mathbf{K}=\mathbf{K}+v_z$

$$\Psi_{\text{II}}(\mathbf{r})=\sum_i \alpha_i B_i(\mathbf{r})=\sum_i \sum_g \alpha_i C_g^{(i)} \exp[2\pi i(\mathbf{K}+\mathbf{g})\cdot\mathbf{r}+2\pi i v_i z], \tag{3.8}$$

where α_i are the summation coefficients of different Bloch waves and are determined by the boundary conditions and v_i is an eigenvalue to be determined. Substituting Eq. (3.6a) and Eq. (3.7) into Eq. (3.4), the C_g coefficients are determined by

$$[2KS_g-2(K_z+g_z)v-v^2]C_g+\sum_h U_{g-h}C_h=0, \tag{3.9}$$

where K_z and g_z are the components of the electron wavevector $\mathbf{K}$ and the reciprocal-lattice vector $\mathbf{g}$ in the direction parallel to the surface normal, respectively.

In RHEED, since $K_z \ll K$, v may not be small in comparison with K_z, so that the v^2 term cannot be neglected. Thus, for the N-beam case, Eq. (3.9) has $2N$ Bloch wave solutions, twice as many as those for transmission electron diffraction. Equation (3.9) may be written into a matrix form

$$\mathbf{AC} = \nu[\mathbf{ZC} + \nu\mathbf{IC}], \tag{3.10}$$

where $[\mathbf{A}]_{gh} = 2KS_g\delta_{gh} + U_{g-h}$, $[\mathbf{Z}]_{gh} = 2(K_z + g_z)\delta_{gh}$, and $\mathbf{I}$ is a unit matrix.

Equation (3.10) can be solved exactly using the following method (Kim and Shenin, 1982). If we define a matrix $\mathbf{X}$ to be

$$\mathbf{X} = \mathbf{ZC} + \nu\mathbf{IC}, \tag{3.11a}$$

or rewrite it as

$$-\mathbf{ZC} + \mathbf{IX} = \nu\mathbf{C}, \tag{3.11b}$$

then Eq. (3.10) becomes

$$\mathbf{AC} = \nu\mathbf{X}. \tag{3.12}$$

Equations (3.11b) and (3.12) are combined into one matrix form,

$$\mathbf{A}_x\mathbf{C}_x = \nu\mathbf{C}_x, \tag{3.13}$$

where

$$\mathbf{A}_x = \begin{pmatrix} -\mathbf{Z} & \mathbf{I} \\ \mathbf{A} & \mathbf{0} \end{pmatrix}, \qquad \mathbf{C}_x = \begin{pmatrix} \mathbf{C} \\ \mathbf{X} \end{pmatrix}.$$

Equation (3.13) can be solved using the conventional method of matrix diagonalization.

For convenience in the following discussion, we defined the tangential components of $\boldsymbol{K}$ and $\boldsymbol{g}$ parallel to the surface to be $\boldsymbol{K}_t$ and $\boldsymbol{g}_t$, respectively. In the vacuum the electron wave is a summation of the incident wave and the reflected wave,

$$\Psi_{\mathrm{I}}(\boldsymbol{r}) = \exp(2\pi i\boldsymbol{K}\cdot\boldsymbol{r}) + \sum_g R_g \exp[2\pi i(\boldsymbol{\chi}_{gt} - \boldsymbol{\chi}_{gz})\cdot\boldsymbol{r}], \tag{3.14}$$

where χ_{gt} and χ_{gz} are the wavevector components of the reflected beams in directions parallel and perpendicular to the surface plane, respectively, and they are to be determined by the boundary conditions; R_g are the amplitudes of the reflected beams. The boundary conditions require

$$\Psi_{\mathrm{I}}(\boldsymbol{b},0) = \Psi_{\mathrm{II}}(\boldsymbol{b},0), \tag{3.15a}$$

$$\left.\frac{\partial\Psi_{\mathrm{I}}(\boldsymbol{r})}{\partial z}\right|_{z=0} = \left.\frac{\partial\Psi_{\mathrm{II}}(\boldsymbol{r})}{\partial z}\right|_{z=0}. \tag{3.15b}$$

Equations (3.15a) and (3.15b) provide a total of $2N$ equations, but the unknown variables to be determined are R_g, χ_{gz} and α_i, for a total of $4N$ ($g = 1, 2, \ldots, N$, and

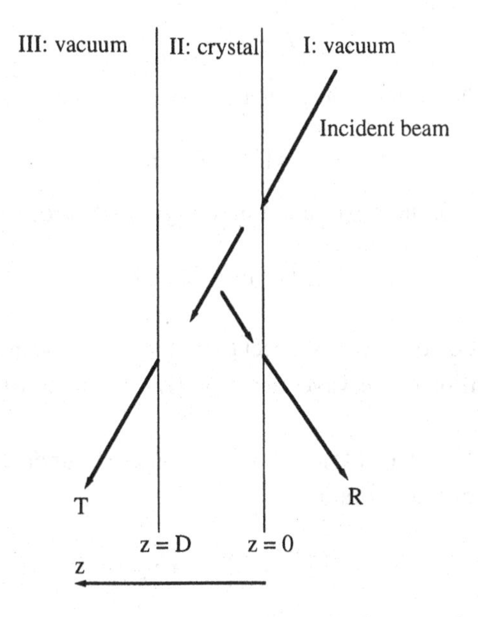

Figure 3.3 *Truncation of a virtual surface located at $z = D$ for uniquely determining the Bloch wave solution. Regions I and III are vacuum, and region II is the bulk crystal of thickness D.*

$i = 1, 2, \ldots, 2N$). Thus the equations that we obtained from Eq. (3.15a) and Eq. (3.15b) are not sufficient to uniquely determine the required physical quantities.

In order to solve this problem, a virtual surface is defined, called the back surface. It is parallel to the front surface, and is created at a distance D from the front surface (Colella, 1972; Moon, 1972). Thus the crystal is an erected slab defined by two parallel faces at $z = 0$ and $z = D$. This arrangement is schematically shown in Figure 3.3. The artificially created vacuum part is called region III. The beams propagating in region III can only be transmitted waves if the crystal slab is sufficiently thick that there are no reflected waves from the front surface, thus

$$\Psi_{III}(\boldsymbol{r}) = \sum_g T_g \exp[2\pi i(\boldsymbol{\kappa}_{gt} + \boldsymbol{\kappa}_{gz}) \cdot \boldsymbol{r}], \tag{3.16}$$

where $\boldsymbol{\kappa}_{gt}$ and κ_{gz} are the components of the reflected beam wavevector in directions parallel and perpendicular to the surface plane, respectively; T_g are the amplitudes of the transmitted beams. The determination of the various quantities can be separated into three steps.

1. In order to match the waves at the boundary, the tangential components of the wavevectors satisfy

$$\chi_{gt} = \boldsymbol{K}_t + \boldsymbol{g}_t = \boldsymbol{\kappa}_{gt}. \tag{3.17a}$$

Conservation of energy for elastic scattering also requires

$$K = \chi_g = \kappa_g. \tag{3.17b}$$

Using Eq. (3.17a), the normal components of the wavevectors are

$$\chi_{gz} = \kappa_{gz} = \pm[K^2 - (\boldsymbol{K}_t + \boldsymbol{g}_t)^2]^{1/2}. \tag{3.17c}$$

2. Using Eq. (3.17a–c), the wave function in region III is rewritten as

$$\Psi_{III}(\boldsymbol{r}) = \sum_g T_g \exp[2\pi i(\chi_{gt} + \chi_{gz}) \cdot \boldsymbol{r}], \tag{3.18}$$

where the sign reversion of χ_{gz} with respect to that in Eq. (3.14) indicates the opposite direction of propagation of the waves in region III (along $+z$) with respect to that in region I (along $-z$).

3. Matching Eq. (3.8), Eq. (3.14) and Eq. (3.18) at the boundaries according to Eq. (3.15a) and Eq. (3.15b), we obtain

$$\delta_{g0} + R_g = \sum_i \alpha_i C_g^{(i)}, \tag{3.19a}$$

$$K_z \delta_{g0} - \chi_{gz} R_g = \sum_i \alpha_i C_g^{(i)} [K_z + g_z + v_i], \tag{3.19b}$$

$$T_g \exp(2\pi i \chi_{gz} D) = \sum_i \alpha_i C_g^{(i)} \exp[2\pi i(K_z + g_z + v_i)D], \tag{3.19c}$$

$$\chi_{gz} T_g \exp(2\pi i \chi_{gz} D) = \sum_i \alpha_i C_g^{(i)} [K_z + g_z + v_i] \exp[2\pi i(K_z + g_z + v_i)D], \tag{3.19d}$$

These $4N$ equations are sufficient to give a set of unique solutions of all the variables.

However, it has been pointed out by Marks and Ma (1988) that problems are encountered when the absorption effect is considered. In this case, the electron wavevector has an imaginary component so that the $\exp[2\pi i(K_z + g_z + v_i)D]$ factor gets damped very quickly, especially for large slab thickness D. Thus Eq. (3.19c) and Eq. (3.19d) are not useful for determining the required parameters. The so-called evanescent wave is thus generated, and it propagates along a direction parallel or nearly parallel to the surface. An alternative method to find the unique solution is to use the energy flow concept (Ma and Marks, 1989; Marks and Ma, 1988), which is described below.

In wave mechanics, the direction of energy flow is characterized by a vector $\boldsymbol{j}$, which is proportional to the expectation value of the Bloch-wave momentum and the group velocity of the Bloch wave. For a single Bloch wave, the energy flow vector can be directly derived by substituting the Bloch wave solution (Eq. (2.8)) into the equation of current-flow density $\boldsymbol{j}$ in quantum mechanics

$$\boldsymbol{j} = \frac{\hbar}{2im_0}[\Psi^* \nabla \Psi - \Psi \nabla \Psi^*] = \frac{h}{m_0} \exp(-4\pi i \boldsymbol{k}_i^{I} \cdot \boldsymbol{r}) \sum_g |C_g^{(i)}|^2 (\boldsymbol{k}_i^{R} + \boldsymbol{g}), \tag{3.20}$$

where $\boldsymbol{k}_i^{\mathrm{R}}$ and $\boldsymbol{k}_i^{\mathrm{I}}$ are the real and imaginary components, respectively, of the wavevector $\boldsymbol{j}_i$. The physical meaning of this equation can be understood as that the average momentum of a Bloch wave is the sum of the momenta of its plane wave components, weighted by a probability function $|C_g|^2$. The vector $\boldsymbol{j}_i$ defines the true path of the electron wave. The z component of the energy flow is

$$j_z = \frac{h}{m_0} \exp(-4\pi \mathrm{i} \boldsymbol{k}_i^{\mathrm{I}} \cdot \boldsymbol{r}) \sum_g |C_g^{(i)}|^2 (k_{iz}^{\mathrm{R}} + g_z). \tag{3.21}$$

We can now describe two cases. First, if the energy flow of a Bloch wave is not toward the crystal, then it is non-physical to consider this Bloch wave in the solution. Second, the waves whose momenta increase with increasing crystal thickness cannot be excited because they vanish with deep penetration due to the absorption effect. Combining the above two points, the Bloch waves that can be excited are those satisfying $j_z \geq 0$ and $k_{iz}^{\mathrm{I}} \geq 0$ (note that the z axis is anti-parallel to the surface normal). Thus we only have N possible Bloch waves that can be excited inside the crystal after applying these selection rules, making it possible to produce a set of unique solutions using only the boundary conditions for the surface located at $z = 0$. In this case, the virtual surface truncated at $z = D$ is unnecessary.

The Bloch wave theory has been applied by Zuo and Liu (1992) to investigate the surface resonance effect. Their calculated results together with experimental observations have shown that the anomalous intensity enhancement of the specularly reflected beam at the resonance condition is due to multiple scattering in the top surface atomic layer.

There are two problems with the Bloch wave approach for RHEED. First, the crystal surface is assumed to be a sharp cut-off of the perfect lattice, so that the scattering of the extended surface potential in the region $z < 0$ is ignored. The surface potential is particularly important in RHEED because of its high sensitivity to the structure of the topmost surface layer. For densely packed surface structures, such as Au(110), the surface formed by sharply cutting the crystal has unavoidably split the atoms located at or very near the cutting plane into two parts, resulting in a non-physical model with 'broken' atoms. In this case, the Bloch wave method fails (Ma and Marks, 1992). It is therefore recommended that the applicability of the Bloch wave theory in RHEED be examined case by case. Second, the Bloch wave method, in principle, is only suitable for a perfect and periodic crystal. Thus, it is difficult to incorporate surface imperfections such as surface steps, dislocations, reconstruction, or relaxation into calculations. In practice, the contrast from the imperfect structures is the most interesting feature. Therefore, application of the Bloch wave approach in RHEED calculations may be limited. However, an alternative method is to combine the Bloch wave theory with the parallel-to-surface multislice method.

3.2 Parallel-to-surface multislice theories I

To incorporate the surface potential effects and surface relaxation in the RHEED calculations, other approaches have been proposed in which a crystal is considered to be periodic in the plane parallel to the surface, and non-periodic modulations of the potential only occur in the direction normal to the surface (Maksym and Beeby, 1981; Ichimiya, 1983; Lynch and Moodie, 1972; Zhao *et al.*, 1988). These are the parallel-to-surface multislice (PaTSM) theories. In this model, the crystal potential is written as

$$U(\boldsymbol{b},z)=\sum_{g} U_g(z)\exp(2\pi i\boldsymbol{g}\cdot\boldsymbol{b}). \tag{3.22}$$

The solution of the Schrödinger equation can be written as

$$\Psi(\boldsymbol{b},z)=\sum_{g} \Psi_g(z)\exp[2\pi i(\boldsymbol{K}_t+\boldsymbol{g}_t)\cdot\boldsymbol{b}]. \tag{3.23}$$

Substituting Eq. (3.22) and Eq. (3.23) into Eq. (3.4), we obtain

$$\frac{d^2\Psi_g(z)}{dz^2}+\Gamma_g^2\Psi_g(z)+4\pi^2\sum_{h} U_{g-h}(z)\Psi_h(z)=0, \tag{3.24}$$

where

$$\Gamma_g^2=4\pi^2[K^2-(\boldsymbol{K}_t+\boldsymbol{g}_t)^2]. \tag{3.25}$$

Equation (3.24) is a coupled second-order differential equation, the solutions of which are the amplitudes of the reflected beams. Equation (3.24) has been solved analytically under the two-beam approximation (Dudarev and Whelan, 1994). As is known in RHEED, many-beam dynamical diffraction has to be considered. For general purposes, two methods have been proposed to solve Eq. (3.24) for RHEED. They are based on the same physical approach of cutting the crystal into slices parallel to the surface, but they are slightly different in mathematical treatments. Figure 3.4 shows a schematic diagram of the slice cutting for RHEED calculations.

The first method was developed independently by Ichimiya (1983) and Maksym and Beeby (1981). It has generally been applied to the calculation of rocking curves, relations between the Bragg-reflected intensities, and the rocking of the incident beam direction. In solving Eq. (3.24) the first step is to transform the second-order differential equation into a first-order differential equation. For this purpose, the following transformations using G_g and H_g are introduced

$$\left(\frac{d}{dz}+i\Gamma_g\right)\Psi_g(z)=iG_g(z), \tag{3.26a}$$

$$\left(\frac{d}{dz}-i\Gamma_g\right)\Psi_g(z)=-iH_g(z). \tag{3.26b}$$

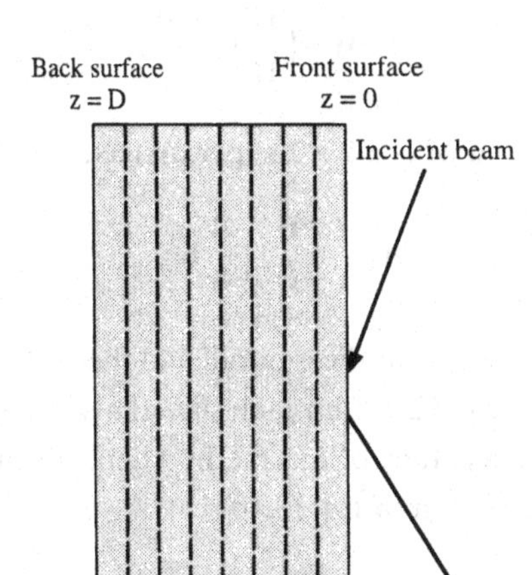

Figure 3.4 *The multislice method with the slice being cut parallel to the surface. Again a back surface is truncated for uniquely determining the two-dimensional Bloch wave solution.*

Using Eq. (3.26), Eq. (3.24) becomes

$$\frac{\mathrm{d}G_g(z)}{\mathrm{d}z} - \mathrm{i}\Gamma_g G_g(z) - 4\pi^2\mathrm{i}\sum_h \frac{U_{g-h}(z)}{2\Gamma_h}[G_h(z) + H_h(z)] = 0, \tag{3.27a}$$

$$\frac{\mathrm{d}H_g(z)}{\mathrm{d}z} + \mathrm{i}\Gamma_g H_g(z) + 4\pi^2\mathrm{i}\sum_h \frac{U_{g-h}(z)}{2\Gamma_h}[G_h(z) + H_h(z)] = 0. \tag{3.27b}$$

Equations (3.27a) and (3.27b) are expressed in matrix form as

$$\frac{\mathrm{d}\mathbf{G}(z)}{\mathrm{d}z} - \mathbf{B}[\mathbf{G}(z) + \mathbf{H}(z)] = 0, \tag{3.28a}$$

$$\frac{\mathrm{d}\mathbf{H}(z)}{\mathrm{d}z} + \mathbf{B}[\mathbf{G}(z) + \mathbf{H}(z)] = 0. \tag{3.28b}$$

where $\mathbf{B}$ is an $N \times N$ matrix with elements

$$[\mathbf{B}]_{gh} = \mathrm{i}\Gamma_g\delta_{gh} + 4\pi^2\mathrm{i}\,\frac{U_{g-h}(z)}{2\Gamma_h}. \tag{3.29}$$

Equation (3.28) can be written into a more convenient form,

$$\frac{\mathrm{d}\mathbf{W}(z)}{\mathrm{d}z} = \mathbf{M}(z)\mathbf{W}(z), \tag{3.30}$$

where

$$\mathbf{W}=\begin{pmatrix}\mathbf{G}\\ \mathbf{H}\end{pmatrix} \tag{3.31}$$

is a super column vector, and **M** is a 2 × 2 super matrix,

$$\mathbf{M}=\begin{pmatrix}\mathbf{B} & \mathbf{B}\\ -\mathbf{B} & -\mathbf{B}\end{pmatrix}. \tag{3.32}$$

If the crystal is cut into many thin slices parallel to the surface, then the matrix **M** is approximately independent of z within each slice. This is the model to be used for solving Eq. (3.30), and the accuracy of this method depends on the thickness of the slice. For a thin slice the solution of Eq. (3.30) is

$$\mathbf{W}(z+\Delta z)=\exp\left(\int_{z}^{z+\Delta z} \mathrm{d}z'\mathbf{M}(z')\right)\mathbf{W}(z). \tag{3.33}$$

The successive application of Eq. (3.33) to each slice leads to the solution

$$\mathbf{W}(z)=\mathbf{PW}(0), \tag{3.34a}$$

where

$$\mathbf{P}=\mathbf{P}_N\ldots\mathbf{P}_n\ldots\mathbf{P}_1, \tag{3.34b}$$

$$\mathbf{P}_i=\exp\left(\int_{z_i}^{z_i+\Delta z} \mathrm{d}z'\mathbf{M}(z')\right). \tag{3.34c}$$

P is the transfer matrix that correlates the scattering amplitude at the surface with that in the bulk.

The next major task is to match the solution (3.34a) with the boundary condition. For easy treatment, we assume that the bulk crystal has thickness D, so that the boundary conditions must be matched at $z=0$ and $z=D$ (Figure 3.4). If D is sufficiently large, the backscattering from the surface located at $z=D$ does not affect the wave distribution at the front surface. From Eq. (3.14), vacuum region I contains only the incident wave and the reflected wave, so that

$$[\Psi_g(z)]_{z<0}=\exp(\mathrm{i}\Gamma_0 z)\delta_{g0}+R_g\exp(-\mathrm{i}\Gamma_g z), \tag{3.35}$$

where the amplitude of the incident beam is assumed as unity. In region III outside the crystal, there are only transmitted waves, so that

$$[\Psi_g(z)]_{z>D}=T_g\exp(\mathrm{i}\Gamma_g z). \tag{3.36}$$

Substituting Eq. (3.35) and Eq. (3.36) into Eq. (3.15a) and Eq. (3.15b) leads to the following relations:

$$G_g(0)=2\Gamma_0\delta_{g0}, \tag{3.37a}$$

$$H_g(0)=2R_g\Gamma_g, \tag{3.37b}$$

for the front surface at $z=0$; and

$$G_g(D)=2T_g\Gamma_g\exp(\mathrm{i}\Gamma_g D), \tag{3.38a}$$

$$H_g(D)=0, \tag{3.38b}$$

for the back surface located at $z=D$. Therefore, the solution of Eq. (3.28a) and Eq. (3.28b) must satisfy the boundary conditions listed in Eq. (3.27) and Eq. (3.38). We begin now by applying the condition (3.38) at the last slice N at $z=D$. From Eq. (3.34a), the vector matrix is related by

$$\mathbf{W}(z_{N-1})=(\mathbf{P}_N)^{-1}\mathbf{W}(D), \tag{3.39}$$

where $(\mathbf{P}_N)^{-1}$ is the inverse of $\mathbf{P}_N$ and is written in a 2×2 super matrix form

$$(\mathbf{P}_N)^{-1}=\begin{pmatrix}\boldsymbol{\beta}_{11}(z_N) & \boldsymbol{\beta}_{12}(z_N)\\ \boldsymbol{\beta}_{21}(z_N) & \boldsymbol{\beta}_{22}(z_N)\end{pmatrix}. \tag{3.40}$$

If we rewrite Eq. (3.38) into a matrix form

$$\mathbf{W}(D)=\begin{pmatrix}\mathbf{0}\\ \mathbf{T}\end{pmatrix}, \tag{3.41}$$

where $\mathbf{0}$ is a zero vector, then substitution of Eq. (3.40) and Eq. (3.41) into Eq. (3.39) gives Eq. (3.42) after eliminating the $\mathbf{T}$ matrix,

$$\mathbf{W}(z_{N-1})=\begin{pmatrix}\mathbf{G}(z_{N-1})\\ \mathbf{H}(z_{N-1})\end{pmatrix}=\begin{pmatrix}\mathbf{G}(z_{N-1})\\ \boldsymbol{\beta}_{21}(z_N)[\boldsymbol{\beta}_{12}(z_N)]^{-1}\mathbf{G}(z_{N-1})\end{pmatrix}. \tag{3.42}$$

For the $(N-1)$th slice, from Eq. (3.34a) we have

$$\mathbf{W}(z_{N-2})=(\mathbf{P}_{N-1})^{-1}\mathbf{W}(z_{N-1}), \tag{3.43}$$

where $(\mathbf{P}_{N-1})^{-1}$ is the inverse of $\mathbf{P}_{N-1}$ and is written in a general form

$$\begin{pmatrix}\mathbf{G}(z_{N-2})\\ \mathbf{H}(z_{N-2})\end{pmatrix}=\begin{pmatrix}\boldsymbol{\beta}_{11}(z_{N-1}) & \boldsymbol{\beta}_{12}(z_{N-1})\\ \boldsymbol{\beta}_{21}(z_{N-1}) & \boldsymbol{\beta}_{22}(z_{N-1})\end{pmatrix}\begin{pmatrix}\mathbf{G}(z_{N-1})\\ \mathbf{H}(z_{N-1})\end{pmatrix}. \tag{3.44}$$

Solving Eq. (3.44) for $\mathbf{G}$ and $\mathbf{H}$, and using Eq. (3.42), we have

$$\mathbf{G}(z_{N-2})=\{\boldsymbol{\beta}_{11}(z_{N-1})+\boldsymbol{\beta}_{12}(z_{N-1})\boldsymbol{\beta}_{21}(z_N)[\boldsymbol{\beta}_{12}(z_N)]^{-1}\}\,\mathbf{G}(z_{N-1})\equiv\mathbf{B}_1(N-2)\,\mathbf{G}(z_{N-1}), \tag{3.45a}$$

$$\mathbf{H}(z_{N-2})=\{\boldsymbol{\beta}_{21}(z_{N-1})+\boldsymbol{\beta}_{22}(z_{N-1})\boldsymbol{\beta}_{21}(z_N)[\boldsymbol{\beta}_{12}(z_N)]^{-1}\}\,\mathbf{G}(z_{N-1})\equiv\mathbf{B}_2(N-2)\,\mathbf{G}(z_{N-1}). \tag{3.45b}$$

Solving for $\mathbf{G}(z_{N-1})$ from Eq. (3.45a) and then substituting it into Eq. (3.45b) yields

$$\mathbf{H}(z_{N-2}) = \mathbf{B}_2(N-2)[\mathbf{B}_1(N-2)]^{-1}\,\mathbf{G}(z_{N-2}). \tag{3.46}$$

For the nth slice, we obtain successively

$$\mathbf{G}(z_{n-1}) = \{\boldsymbol{\beta}_{11}(z_n) + \boldsymbol{\beta}_{12}(z_n)\mathbf{B}_2(n)[\mathbf{B}_1(n)]^{-1}\}\,\mathbf{G}(z_n) \equiv \mathbf{B}_1(n-1)\mathbf{G}(z_n), \tag{3.47a}$$

$$\mathbf{H}(z_{n-1}) = \{\boldsymbol{\beta}_{21}(z_n) + \boldsymbol{\beta}_{22}(z_n)\mathbf{B}_2(n)[\mathbf{B}_1(n)]^{-1}\}\,\mathbf{G}(z_n) \equiv \mathbf{B}_2(n-1)\mathbf{G}(z_n). \tag{3.47b}$$

When we reach the first slice (i.e., the front surface at $z=0$), the boundary condition Eq. (3.41) requires

$$\mathbf{G}(0) = \{\boldsymbol{\beta}_{11}(z_1) + \boldsymbol{\beta}_{12}(z_1)\mathbf{B}_2(1)[\mathbf{B}_1(1)]^{-1}\}\,\mathbf{G}(z_1) \equiv \mathbf{B}_1(0)\mathbf{G}(z_n), \tag{3.48a}$$

$$\mathbf{H}(0) = \{\boldsymbol{\beta}_{21}(z_1) + \boldsymbol{\beta}_{22}(z_1)\mathbf{B}_2(1)[\mathbf{B}_1(1)]^{-1}\}\,\mathbf{G}(z_1) \equiv \mathbf{B}_2(0)\mathbf{G}(z_n). \tag{3.48b}$$

where

$$[\mathbf{G}(0)]_g = 2\Gamma_0\delta_{g0}, \qquad [\mathbf{H}(0)]_g = 2R_g\Gamma_g. \tag{3.49}$$

Solving Eq. (3.48) yields

$$\mathbf{G}(0) = \mathbf{B}_1(0)[\mathbf{B}_2(0)]^{-1}\mathbf{H}(0). \tag{3.50}$$

Using Eq. (3.49), the gth column of Eq. (3.50) is

$$R_g = \frac{\Gamma_0}{\Gamma_g}\sum_h [\mathbf{B}_1(0)]_{gh}\{[\mathbf{B}_2(0)]^{-1}\}_{h0}. \tag{3.51}$$

Therefore, the reflected intensity of $\boldsymbol{g}$ is $|R_g|^2$. The numerical calculation starts at the Nth slice in order to get rid of the transmission matrix $\mathbf{T}$. For sufficiently large D, the calculated result of Eq. (3.51) is independent of the crystal's thickness.

In the above discussion, the transfer matrix of each slice may be different. It is possible that the theory can be modified for calculating the reflected intensity of a layered material, but caution must be exercised in order to satisfy the expansion form of crystal potential (Eq. (3.22)) and the relevant boundary conditions at the interfaces. In general, for an incident angle of about 3°, the thickness D of the crystal is no more than about 2 nm.

Figure 3.5 shows a comparison of the calculated RHEED rocking curve and the observed curves of the Si(111) 1 × 1 surface. This surface is considered to include disordered structures such as a randomly relaxed bulk-like structure (the bulk-like model), a relaxed bulk-like structure with random vacancies (the vacancy model), a relaxed bulk-like structure with random adatoms (the adatom model), or a random dimer–adatom–stacking-fault structure (the random DAS model). In Figure 3.5, it can be seen that the curves for the adatom model and the random DAS model are in good agreement with the experimental curve. For the vacancy model, the peak position at the (222) reflection of the curve shifts about 0.2° to a lower angle

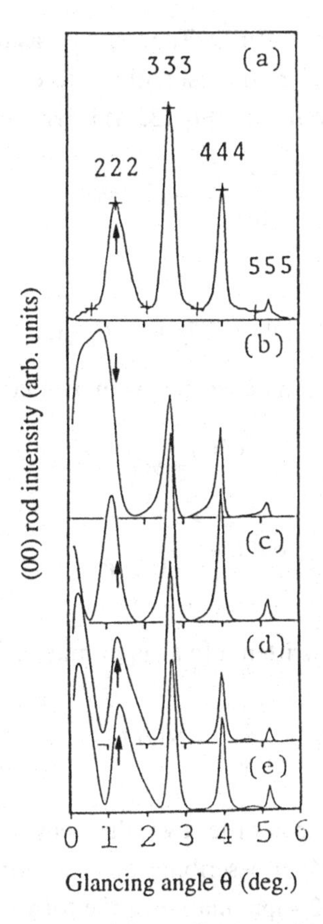

Figure 3.5 *Experimental and dynamical parallel-to-surface multislice calculated rocking curves of the Si(111) surface for beam energy 10 keV. (a) The experimental curve. In (b), (c), (d) and (e) are shown the calculated curves based on the bulk-like model, vacancy model, adatom model, and random DAS model, respectively. The arrows indicate the peak position of the (222) reflection in the experimental curve (a) (courtesy of Dr A. Ichimiya, 1993).*

compared with the peak of the experimental one indicated by the arrowhead in Figure 3.5(a). In order to finalize the structural model, the crystal was rotated to a different zone axis direction, since one only sees the projected structure along the beam direction for the ZOLZ reflections. The result showed that the adatom model fits the experimental observation best (Ichimiya, 1993). This example demonstrates clearly the applications of dynamical simulations for determining the surface structures.

3.3 Parallel-to-surface multislice theories II

The second method was originally proposed for calculating the diffracted intensity of low-energy electrons (Lynch and Moodie, 1972; Lynch and Smith, 1983) and has been successfully applied for RHEED. The theory was derived based on the same scheme as method I except for some slightly different mathematical treatments (Zhao *et al.*, 1988). For convenience, Eq. (3.24) is written into matrix form

$$\frac{d^2\boldsymbol{\Psi}(z)}{dz^2} = -\mathbf{L}(z)\boldsymbol{\Psi}(z), \tag{3.52a}$$

with

$$[\mathbf{L}(z)]_{gh} = \Gamma_g^2\delta_{gh} + 4\pi^2 U_{g-h}(z). \tag{3.52b}$$

Mathematically, Eq. (3.52a) can be written as two equations,

$$\frac{d\boldsymbol{\Psi}(z)}{dz} = \boldsymbol{\Psi}'(z), \tag{3.53a}$$

$$\frac{d\boldsymbol{\Psi}'(z)}{dz} = -\mathbf{L}(z)\boldsymbol{\Psi}(z). \tag{3.53b}$$

These two equations can be written into a super matrix

$$\frac{d\boldsymbol{\Psi}(z)}{dz} = \mathbf{N}(z)\boldsymbol{\Psi}(z), \tag{3.54}$$

where $\boldsymbol{\Psi}(z)$ is a column vector of order two, the elements of which are two infinite column vectors $\boldsymbol{\Psi}(z)$ and $\boldsymbol{\Psi}'(z)$ describing the scattering amplitudes and their z direction 'slopes'; $\mathbf{N}$ is a 2×2 super matrix of the form

$$\mathbf{N}(z) = \begin{pmatrix} \mathbf{0} & \mathbf{I} \\ -\mathbf{L}(z) & \mathbf{0} \end{pmatrix}, \tag{3.55}$$

$\mathbf{I}$ is a unit matrix. If the $\mathbf{N}(z)$ matrix is approximated to be independent of z within each slice, then Eq. (3.54) can be solved in the multislice method as

$$\begin{pmatrix} \boldsymbol{\Psi}(z) \\ \boldsymbol{\Psi}'(z) \end{pmatrix} \approx \boldsymbol{\Omega}(z) \begin{pmatrix} \boldsymbol{\Psi}(0) \\ \boldsymbol{\Psi}'(0) \end{pmatrix}, \tag{3.56}$$

where the propagation operator $\boldsymbol{\Omega}(z)$ is a product series

$$\boldsymbol{\Omega}(z) = \boldsymbol{\Omega}_N(\Delta z_N) \ldots \boldsymbol{\Omega}_{n_c}(\Delta z_{n_c}) \ldots \boldsymbol{\Omega}_1(\Delta z_1), \tag{3.57a}$$

$$\boldsymbol{\Omega}_{n_c}(\Delta z_{n_c}) = \exp\left[\Delta z_{n_c} \begin{pmatrix} \mathbf{0} & \mathbf{I} \\ -\mathbf{L}(z_{n_c}) & \mathbf{0} \end{pmatrix}\right] \tag{3.57b}$$

is the matrix operator for a thin slice. The calculations of Eq. (3.57) can be simplified using the diagonalization transform (Lynch and Moodie, 1972; Lynch and Smith,

1983). We now match Eq. (3.56) to the boundary conditions. For convenience, the general form of the $\boldsymbol{\Omega}$ matrix can be written as

$$\boldsymbol{\Omega}=\begin{pmatrix}\boldsymbol{\Omega}_{11} & \boldsymbol{\Omega}_{12}\\ \boldsymbol{\Omega}_{21} & \boldsymbol{\Omega}_{22}\end{pmatrix}. \tag{3.58}$$

The amplitudes of the reflected waves are determined by the boundary conditions. In region I,

$$[\Psi_g(z)]_{z<0}=\exp(\mathrm{i}\Gamma_0 z)\,\delta_{g0}+R_g\exp(-\mathrm{i}\Gamma_g z), \tag{3.59}$$

so that the boundary conditions are

$$\boldsymbol{\Psi}(0)=\boldsymbol{a}_i+\boldsymbol{R}, \tag{3.60a}$$

$$\boldsymbol{\Psi}'(0)=\mathrm{i}\boldsymbol{\Gamma}\boldsymbol{a}_i-\mathrm{i}\boldsymbol{\Gamma}\boldsymbol{R}, \tag{3.60b}$$

where $(\boldsymbol{a}_i)_g=\delta_{g0}$ is the incident wave amplitude vector, $\boldsymbol{R}$ is the vector of the reflected wave, and $\boldsymbol{\Gamma}$ is the diagonal matrix representation of the z components of the momenta of the various waves. In region III, there are only transmitted waves,

$$[\Psi_g(z)]_{z>D}=T_g\exp(\mathrm{i}\Gamma_g z). \tag{3.61}$$

The boundary conditions are

$$\boldsymbol{\Psi}(D)=\exp(\mathrm{i}\boldsymbol{\Gamma}D)\,\mathbf{T} \tag{3.62a}$$

and

$$\boldsymbol{\Psi}'(D)=\mathrm{i}\boldsymbol{\Gamma}\exp(\mathrm{i}\boldsymbol{\Gamma}D)\,\mathbf{T}. \tag{3.62b}$$

We now use Eq. (3.58); matching Eq. (3.56) to Eq. (3.59) and Eq. (3.61) at $z=0$ and $z=D$, respectively, the following equations are obtained:

$$\exp(\mathrm{i}\boldsymbol{\Gamma}D)\,\mathbf{T}=\boldsymbol{\Omega}_{11}(\boldsymbol{a}_i+\boldsymbol{R})+\boldsymbol{\Omega}_{12}(\mathrm{i}\boldsymbol{\Gamma}\boldsymbol{a}_i-\mathrm{i}\boldsymbol{\Gamma}\boldsymbol{R}), \tag{3.63a}$$

$$\mathrm{i}\boldsymbol{\Gamma}\exp(\mathrm{i}\boldsymbol{\Gamma}D)\,\mathbf{T}=\boldsymbol{\Omega}_{21}(\boldsymbol{a}_i+\boldsymbol{R})+\boldsymbol{\Omega}_{22}(\mathrm{i}\boldsymbol{\Gamma}\boldsymbol{a}_i-\mathrm{i}\boldsymbol{\Gamma}\boldsymbol{R}). \tag{3.63b}$$

Finally, the reflection vector is found to be

$$\boldsymbol{R}=(\chi+\upsilon)^{-1}(\chi-\upsilon)\boldsymbol{a}_i, \tag{3.64a}$$

with

$$\chi=\mathrm{i}\boldsymbol{\Omega}_{22}\boldsymbol{\Gamma}+\boldsymbol{\Gamma}\boldsymbol{\Omega}_{12}\boldsymbol{\Gamma}, \tag{3.64b}$$

$$\upsilon=\mathrm{i}\boldsymbol{\Gamma}\boldsymbol{\Omega}_{11}-\boldsymbol{\Omega}_{21}. \tag{3.64c}$$

The results have been obtained through various matrix multiplications and inversions. The calculation of Eq. (3.57a) is perhaps the most difficult one, because of the

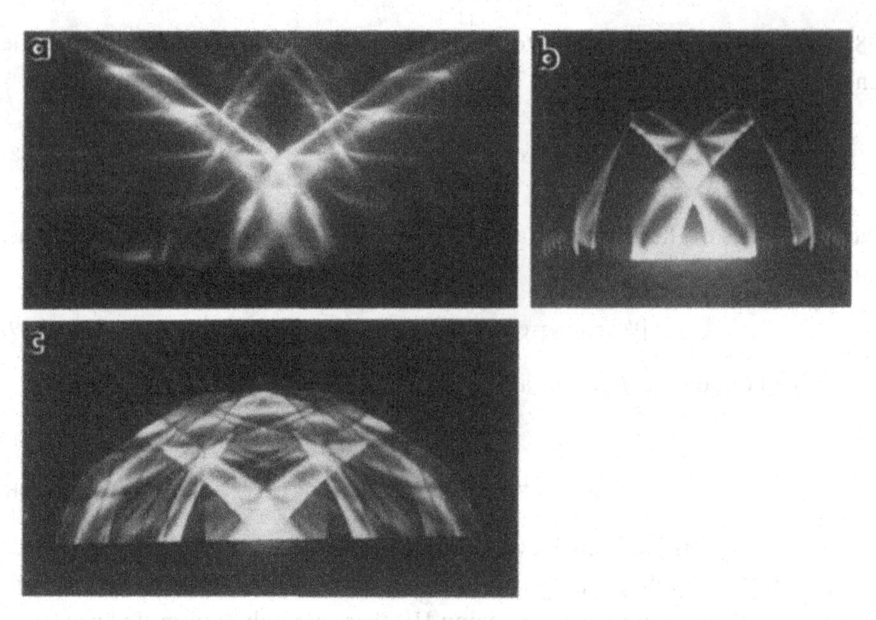

Figure 3.6 *(a) The experimental large-angle convergent beam RHEED pattern from Pt(111) near [11$\bar{2}$]. (b) The simulated RHEED pattern using the parallel-to-surface multislice theory (method II). (c) A simulation for larger tilt angle. The beam energy is 100 keV (courtesy of Dr A. E. Smith et al., 1992).*

requirement of high-order matrix multiplication. The following identity may be useful to reduce the amount of calculations (Lynch and Smith, 1983):

$$\exp\left[\Delta z_n\begin{pmatrix} \mathbf{0} & \mathbf{I} \\ -\mathbf{L}(z) & \mathbf{0}\end{pmatrix}\right]=\frac{1}{2}\begin{pmatrix} (\mathbf{E}_L+\mathbf{E}_L^{-1}) & -\mathrm{i}\mathbf{M}_L^{-1}(\mathbf{E}_L-\mathbf{E}_L^{-1}) \\ \mathrm{i}\mathbf{M}_L^{-1}(\mathbf{E}_L-\mathbf{E}_L^{-1}) & (\mathbf{E}_L+\mathbf{E}_L^{-1})\end{pmatrix}, \tag{3.65}$$

where $\mathbf{M}_L$ denotes a matrix that satisfies

$$\mathbf{M}_L\mathbf{M}_L=-\mathbf{L}, \tag{3.66a}$$

$$\mathbf{E}_L=\exp(\mathrm{i}\mathbf{M}_L\Delta z). \tag{3.66b}$$

The theory presented here has been applied to simulating RHEED patterns (Lynch and Smith, 1983; Smith *et al.*, 1992). Figure 3.6(a) is an experimentally observed convergent beam RHEED pattern for Pt(111) near the [11$\bar{2}$] zone axis. The pattern is called a Kossel pattern and is similar to the Kikuchi pattern. This pattern shows the reflected intensity distribution for a large range of beam incidence and also exhibits double-resonance parabolas. The arrow head indicates the (333) reflection beam. Intensity enhancement can be seen where the parabolas intersect the Bragg beams. Figure 3.6(b) is a simulated RHEED pattern for the incident beam geometry of Figure 3.7(a). It is apparent that good agreement has been achieved.

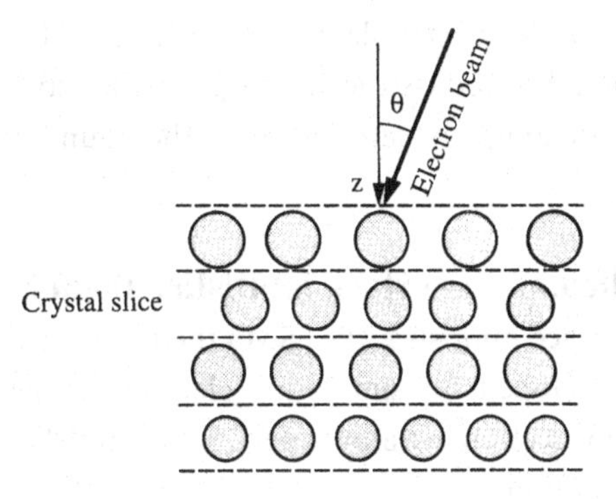

Figure 3.7 *The slice cutting method and incident beam geometry of the multislice theory for the calculation of transmission electron diffraction.*

Figure 3.6(c) shows simulations for larger angles of incidence starting with (333) up to approximately (666).

An alternative method for solving Eq. (3.24) has been proposed by Meyer-Ehmsen (1989), who decompose the electron waves into forward- and backward-traveling components. The forward- and backward-scattered waves are related by a reflection matrix. Then, a subsidiary condition is introduced in order to convert the double differential equation into a single differential equation. The numerical calculated results of this theory have shown excellent agreement with the result of the theory presented in Section 3.2 (Korte and Meyer-Ehmsen, 1992a and b).

As pointed out at the beginning of Section 3.2, the parallel-to-surface multislice theories were proposed based on the assumption that the atomic potential can be non-periodic in the direction perpendicular to the surface but not within the surface plane, making it possible to include the surface potential effect and atomic relaxation in the surface normal direction. Surface steps, dislocations, and reconstruction within the plane parallel to the surface may not be easily included in this approach. An alternative method is to introduce an artificial super cell, which is so large that the surface structures of interest can be incorporated. This has been shown to be possible for calculations of surface step images in REM (McCoy and Maksym, 1993 and 1994).

The Bloch wave and the parallel-to-surface multislice calculations are usually performed in reciprocal space for convenience in calculating RHEED rocking curves via the variation of beam incident angle. However, the major disadvantage of these methods is the difficulty of including surface steps and defects, so they have only rarely been applied to calculate REM images.

The combination of the parallel-to-surface multislice method and the Bloch wave theory has been applied to treat cases involving surface potential tail effects or surface reconstruction in RHEED (see Peng (1995) for a review). The main scheme is to cut the crystal into thin slices parallel to the surface, so that each slice can be treated as a two-dimensional periodic structure based on the Bloch wave theory. The waves in adjacent slices are uniquely determined by the boundary conditions.

3.4 Perpendicular-to-surface multislice theory

The perpendicular-to-surface multislice (PeTSM) theory is based on the Cowley–Moodie formulation, developed for simulating electron images in transmission electron diffraction. In this section, we aim to derive the multislice equation directly from the Schrödinger equation.

In order to illustrate the quantum mechanical basis of the multislice theory, we consider a general case in which the electron strikes the crystal entrance surface at a small angle θ, as shown in Figure 3.7. The crystal slice is usually cut parallel to a crystallographic plane. This case occurs in either dark-field or bright-field HRTEM of small-angle wedge-shaped crystals. We now follow Ishizuka's method to derive the multislice equation based on the first-principles approach (Ishizuka and Uyeda, 1977; Ishizuka, 1982).

The first approximation that we make is that of forward scattering of high-energy electrons. It is always assumed that there is no backscattering in electron diffraction. The validity of this approximation has been examined by Van Dyck (1985), who found that the amplitude ratio of the backscattered wave to the transmitted wave is on the order of $(V/U_0)(1-V/U_0)^{-1}$, which tends to be very small for high-energy electrons. Based on this approximation, Eq. (3.4) is first written into an integral equation

$$\Psi(\mathbf{r})=\exp(2\pi \mathrm{i}\mathbf{K}\cdot\mathbf{r})+\pi\int \mathrm{d}\mathbf{r}'\,\frac{\exp(2\pi \mathrm{i}K|\mathbf{r}-\mathbf{r}'|)}{|\mathbf{r}-\mathbf{r}'|}\,U(\mathbf{r}')\Psi(\mathbf{r}'), \tag{3.67}$$

where the incident wave is assumed to be a plane wave. This assumption does not affect the final result, however, even if the incident wave is a converged wave packet. The plane-wave component can be removed in the following discussion by defining $\Psi(\mathbf{r})=\exp(2\pi \mathrm{i}\mathbf{K}\cdot\mathbf{r})\,\Phi(\mathbf{r})$, such that

$$\begin{aligned}\Phi(\mathbf{r})&=1+\mathrm{i}\pi\lambda\int \mathrm{d}\mathbf{r}'\,P'(\mathbf{r}-\mathbf{r}')U(\mathbf{r}')\Phi(\mathbf{r}')\\&=1+\mathrm{i}\sigma\int \mathrm{d}\mathbf{r}'\,P'(\mathbf{r}-\mathbf{r}')V(\mathbf{r}')\Phi(\mathbf{r}'),\end{aligned} \tag{3.68}$$

where P' is a propagation function for a general incident beam case,

$$P'(\mathbf{r})=\frac{1}{\mathrm{i}\lambda}\,\frac{\exp(2\pi \mathrm{i}Kr-2\pi \mathrm{i}\mathbf{K}\cdot\mathbf{r})}{r}, \tag{3.69}$$

and $\sigma = \pi e\gamma/(\lambda E) = 1/(\hbar v)$, with $E = eU_0[1 + eU_0/(2m_0c_0^2)]$. The constant σ given here has taken into consideration the relativistic correction.

Before starting to find the multislice solution of Eq. (3.68), we consider two properties of the P' function. The first of these is that

$$\int_{\Sigma} \mathrm{d}\boldsymbol{b}\, P'(\boldsymbol{b}-\boldsymbol{b}_1, z-z_1)P'(\boldsymbol{b}_2-\boldsymbol{b}, z_2-z) \approx \frac{K}{K_z} P'(\boldsymbol{b}_2-\boldsymbol{b}_1, z_2-z_1), \quad (3.70a)$$

or in reciprocal space

$$P'(\boldsymbol{u}, z-z_1)P'(\boldsymbol{u}, z_2-z) \approx \frac{K}{K_z} P'(\boldsymbol{u}, z_2-z_1), \quad (3.70b)$$

where the plane of integration Σ is arbitrarily located between z_1 and z_2. To prove this relation, we take a two-dimensional Fourier transform of P',

$$P'(\boldsymbol{u}, z) = \int \mathrm{d}\boldsymbol{b}\, \frac{1}{\mathrm{i}\lambda} \frac{\exp(2\pi\mathrm{i}Kr - 2\pi\mathrm{i}\boldsymbol{K}\cdot\boldsymbol{r})}{r} \exp(-2\pi\mathrm{i}\boldsymbol{u}\cdot\boldsymbol{b}).$$

By carrying the integration successively over x and y, the result is (Ishizuka, 1982)

$$P'(\boldsymbol{u}, z) = \frac{K}{[K^2 - (\boldsymbol{K}_\mathrm{b} + \boldsymbol{u})^2]^{1/2}} \exp\{-2\pi\mathrm{i}z[K_z - |\boldsymbol{K}_\mathrm{b} + \boldsymbol{u}|]\}. \quad (3.71)$$

Using Eq. (3.71) it can be readily shown that

$$P'(\boldsymbol{u}, z-z_1)P'(\boldsymbol{u}, z_2-z) = \frac{K}{(K^2 - |\boldsymbol{K}_\mathrm{b} + \boldsymbol{u}|^2)^{1/2}} P'(\boldsymbol{u}, z_2-z_1) \approx \frac{K}{K_z} P'(\boldsymbol{u}, z_2-z_1), \quad (3.72)$$

where an approximation of $K^2 \gg |\boldsymbol{K}_\mathrm{b} + \boldsymbol{u}|^2$ has been made. This requires that the angle θ between wavevector $\boldsymbol{K}$ and the normal direction z of the slices is small. This condition is usually satisfied in dark-field HRTEM imaging.

The second property is that

$$\int \mathrm{d}\boldsymbol{b}\, P'(\boldsymbol{b}, \Delta z) = \frac{K}{K_z}. \quad (3.73)$$

For small-angle scattering so that $|\boldsymbol{r} - \boldsymbol{r}'| \approx z - z'$, P' becomes

$$P'(\boldsymbol{b}, \Delta z) = \frac{1}{\mathrm{i}\lambda\Delta z} \exp[2\pi\mathrm{i}(K - K_z)\Delta z] \exp\left(\mathrm{i}\frac{\pi b^2}{\lambda\Delta z} - 2\pi\mathrm{i}\boldsymbol{K}_h\cdot\boldsymbol{b}\right)$$
$$= \exp[2\pi\mathrm{i}(K - K_z)\Delta z)] \exp(-2\pi\mathrm{i}\boldsymbol{K}_\mathrm{b}\cdot\boldsymbol{b})\, P(\boldsymbol{b}, \Delta z). \quad (3.74)$$

In general, the constant factor $\exp[2\pi\mathrm{i}(K - K_z)\Delta z]$ can be omitted for slices of equal thickness. The $\exp(-2\pi i \boldsymbol{K}_b \cdot \boldsymbol{b})$ factor corresponds to the shift of the electron wave function in the $\boldsymbol{b}$ plane due to the inclined propagation of the electrons.

3.4.1 Multislice solution of the Schrödinger equation for transmission electron diffraction

We now return to Eq. (3.10). In order to find its multislice solution, we need to derive a relationship between the wave $\Phi(\boldsymbol{b}, z_0)$ going into a crystal slice at $z = z_0$ and that $\Phi(\boldsymbol{b}, z)$ after its penetration through the slice at $z = z$. The first step is to separate the entrance wave and the newly scattered wave of the slice:

$$\begin{aligned}\Phi(\boldsymbol{b}, z) &= 1 + i\sigma \int d\boldsymbol{b}' \int_{z'=-\infty}^{z'=z} dz'\, P'(\boldsymbol{b}-\boldsymbol{b}', z-z') V(\boldsymbol{b}', z') \Phi(\boldsymbol{b}, z') \\ &= 1 + i\sigma \int d\boldsymbol{b}' \int_{z'=-\infty}^{z'=z_0} dz'\, P'(\boldsymbol{b}-\boldsymbol{b}', z-z') V(\boldsymbol{b}', z') \Phi(\boldsymbol{b}, z') \\ &\quad + i\sigma \int d\boldsymbol{b}' \int_{z'=z_0}^{z'=z} dz'\, P'(\boldsymbol{b}-\boldsymbol{b}', z-z') V(\boldsymbol{b}', z') \Phi(\boldsymbol{b}, z'). \end{aligned} \tag{3.75}$$

Using the properties of function P' introduced in Eq. (3.70a) and Eq. (3.73), which are

$$1 = \frac{K_z}{K} \int d\boldsymbol{b}_0\, P'(\boldsymbol{b}-\boldsymbol{b}_0, z-z_0), \tag{3.76a}$$

$$P'(\boldsymbol{b}-\boldsymbol{b}', z-z') = \frac{K_z}{K} \int d\boldsymbol{b}_0\, P'(\boldsymbol{b}-\boldsymbol{b}_0, z-z_0) P'(\boldsymbol{b}_0-\boldsymbol{b}', z_0-z'), \tag{3.76b}$$

replacing the first and the second terms on the right-hand side of Eq. (3.75) by Eq. (3.76a) and Eq. (3.76b), respectively, yields

$$\begin{aligned}\Phi(\boldsymbol{b}, z) &= \frac{K_z}{K} \int d\boldsymbol{b}_0 \left(1 + i\sigma \int d\boldsymbol{b}' \int_{z'=-\infty}^{z'=z_0} dz'\, P'(\boldsymbol{b}_0-\boldsymbol{b}', z_0-z') V(\boldsymbol{b}', z') \Phi(\boldsymbol{b}', z') \right) P'(\boldsymbol{b}-\boldsymbol{b}_0, z-z_0) \\ &\quad + i\sigma \int d\boldsymbol{b}' \int_{z'=z_0}^{z'=z} dz'\, P'(\boldsymbol{b}-\boldsymbol{b}', z-z') V(\boldsymbol{b}', z') \Phi(\boldsymbol{b}', z'). \end{aligned} \tag{3.77}$$

By comparing the terms inside the large brackets in (3.77) with Eq. (3.75) for $z = z_0$, (3.77) can be written as

$$\Phi(\boldsymbol{b},z)=\frac{K_z}{K}\int \mathrm{d}\boldsymbol{b}_0\Phi(\boldsymbol{b}_0,z_0)P'(\boldsymbol{b}-\boldsymbol{b}_0,z-z_0)+\mathrm{i}\sigma\int \mathrm{d}\boldsymbol{b}'\int_{z'=z_0}^{z'=z}\mathrm{d}z'\,P'(\boldsymbol{b}-\boldsymbol{b}',z-z')V(\boldsymbol{b}',z')\Phi(\boldsymbol{b}',z'). \tag{3.78}$$

In order to solve Eq. (3.78), one expands Φ in the order of $(\mathrm{i}\sigma)$,

$$\Phi(\boldsymbol{b},z)=\sum_{L=0}^{\infty}(\mathrm{i}\sigma)^L f_L(\boldsymbol{b},z). \tag{3.79}$$

On substituting (3.79) into (3.78), and assembling the terms with the same orders of $(\mathrm{i}\sigma)^L$,

$$f_0(\boldsymbol{b},z)=\frac{K_z}{K}\int \mathrm{d}\boldsymbol{b}_0\,P'(\boldsymbol{b}-\boldsymbol{b}_0,z-z_0)\Phi(\boldsymbol{b}_0,z_0), \tag{3.80a}$$

$$f_L(\boldsymbol{b},z)=\int \mathrm{d}\boldsymbol{b}'\int_{z'=z_0}^{z'=z}\mathrm{d}z'\,P'(\boldsymbol{b}-\boldsymbol{b}',z-z')V(\boldsymbol{b}',z')f_{L-1}(\boldsymbol{b}',z'). \tag{3.80b}$$

If V varies slowly in the region $\Delta z=(z-z_0)$, then using (3.70a)

$$f_1(\boldsymbol{b},z)=\frac{K_z}{K}\int \mathrm{d}\boldsymbol{b}_0\,\Phi(\boldsymbol{b}_0,z_0)\int_{z'=z_0}^{z'=z}\mathrm{d}z'\int \mathrm{d}\boldsymbol{b}'\,P'(\boldsymbol{b}-\boldsymbol{b}',z-z')V(\boldsymbol{b}',z)P'(\boldsymbol{b}'-\boldsymbol{b}_0,z'-z_0)$$

$$\approx\frac{K_z}{K}\int \mathrm{d}\boldsymbol{b}_0\,\Phi(\boldsymbol{b}_0,z_0)\left(\frac{K}{K_z}\int_{z'=z_0}^{z'=z}\mathrm{d}z'\,V(\boldsymbol{b}_0,z')\right)P'(\boldsymbol{b}-\boldsymbol{b}_0,z-z_0). \tag{3.80c}$$

By applying the same method, it can be easily proved that

$$f_L(\boldsymbol{b},z)\approx\frac{K_z}{K}\int \mathrm{d}\boldsymbol{b}_0\,\Phi(\boldsymbol{b}_0,z_0)\frac{1}{L!}\left(\frac{K}{K_z}\int_{z'=z_0}^{z'=z}\mathrm{d}z'\,V(\boldsymbol{b}_0,z')\right)^L P'(\boldsymbol{b}-\boldsymbol{b}_0,z-z_0). \tag{3.80d}$$

Defining the projected potential as

$$V(\boldsymbol{b})=\int_{z'=z_0}^{z'=z}\mathrm{d}z'\,V(\boldsymbol{b},z'), \tag{3.81a}$$

the relation that governs the wave before and after being scattering by a crystal slice is

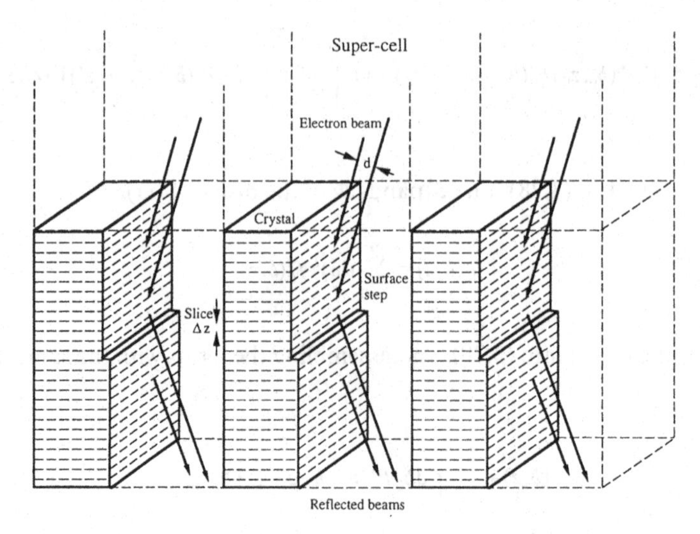

Figure 3.8 *Construction of a super cell for the perpendicular-to-surface multislice calculation. The slice thickness Δz is usually chosen to be no more than 0.3 nm. A surface step can be easily introduced into the calculation by varying the phase grating function of the slices at different depths. The entire space is periodically filled with the super cell. The super cell is assumed to be very large so that the scattering within each cell is considered to be independent of that in other cells. The calculated result of each cell corresponds to the experimentally observed result.*

$$\begin{aligned}\Phi(\boldsymbol{b},z)&=\int \mathrm{d}\boldsymbol{b}_0\,\Phi(\boldsymbol{b}_0,z_0)\exp\left(\mathrm{i}\sigma\frac{K}{K_z}V(\boldsymbol{b}_0)\right)\frac{K_z}{K}P'(\boldsymbol{b}-\boldsymbol{b}_0,z-z_0)\\&=\left[\Phi(\boldsymbol{b},z_0)\exp\left(\mathrm{i}\sigma\frac{K}{K_z}V(\boldsymbol{b})\right)\right]\otimes\left(\frac{K_z}{K}P'(\boldsymbol{b},z-z_0)\right).\end{aligned}\qquad(3.81\text{b})$$

Equation (3.22) is the multislice solution of the Schrödinger equation for an inclined incident electron beam. It is important to point out that the potential used in Eq. (3.81) for the inclined incident beam case is projected along the z axis direction (i.e., the normal direction of the slices) instead of the beam direction. This is particularly convenient for RHEED calculation.

In the multislice calculation, the major error source is the thickness of the slice. However, with the availability of super-fast computers, the slice thickness can be taken to be the thickness of one atomic layer.

3.4.2 Applications in RHEED calculations

The multislice theory presented in Section 3.4.1 was originally developed for calculating the images and diffraction patterns of transmission electrons. The theory can also be adopted for RHEED calculation if the slice is cut normal to the surface and the electron beam azimuth (Peng and Cowley, 1986), as shown in Figure 3.8. The crystal and the vacuum part, in which the incident beam approaches the surface, are defined as a super cell. The entire space is filled with the repetition of the super cell. In RHEED geometry, if we look along the incident beam direction, then the scattering angle of the electrons is no more than 5°, thus the backscattering can be ignored. If the crystal slices are cut in the plane perpendicular or nearly perpendicular to the crystal surface plane and the direction of the incident beam, then the transmission of the electrons through each slice is analogous to that in TEM. The size of the super cell has to be chosen so as to avoid the interference effects from waves scattered by neighboring cells. The slice is partly filled with atoms and the other part is vacuum. The transmitted waves inside the crystal and the reflected waves in the vacuum are calculated simultaneously slice by slice. In the vacuum, the output waves are the incident waves and the reflected waves. The former can be filtered out using the Fourier transform technique because they are propagating in different directions.

The incident wave is defined by a one-dimensional window function, the width of which can be artificially controlled. The beam is assumed to be infinitely large in the direction parallel to the surface, but it is a small window function in the direction perpendicular to the surface. This configuration is designed to reduce the number of sampling points, so that the width of the super cell in the direction perpendicular to the surface is the size of the crystal unit cell. It is possible, in principle, to introduce a small two-dimensional electron probe defined by a circular condenser aperture, provided that the super cell is constructed large enough in the direction parallel to the surface (i.e., into the plane of the paper in Figure 3.8).

The theory has three important advantages. First, there is no restriction on the method for constructing the phase grating function of each slice. It is thus possible, in principle, to include any desired modification of the surface structure such as steps, defects, and surface potential in the calculation. Second, this real-space method simply provides an easy way of tracking the propagation process of electrons slice by slice, giving the possibility of viewing the build-up process of electrons waves near the surface. This feature is unique to this method. Finally, there is no restriction on the form of the incident electrons, so that it is feasible to calculate the scattering of a nano-probe electron beam at the surface. This type of calculation, however, may not be so easily performed with the theories to be outlined in Sections 3.2 and 3.3, in which a plane wave incident beam is assumed in the mathematical derivation.

On the other hand, there are three difficulties with this theory. First, a large super cell needs to be truncated to isolate all the scattering occurring in one cell from that in neighboring cells in order to avoid 'interference' between cells. The steady state solution of the wave function starts to emerge after sampling thicknesses larger than 70 nm (for GaAs at 100 kV) (Wang *et al.*, 1989a). A surface step contrast can only be calculated after the reflected wave has arrived at a steady state (i.e., an intensity

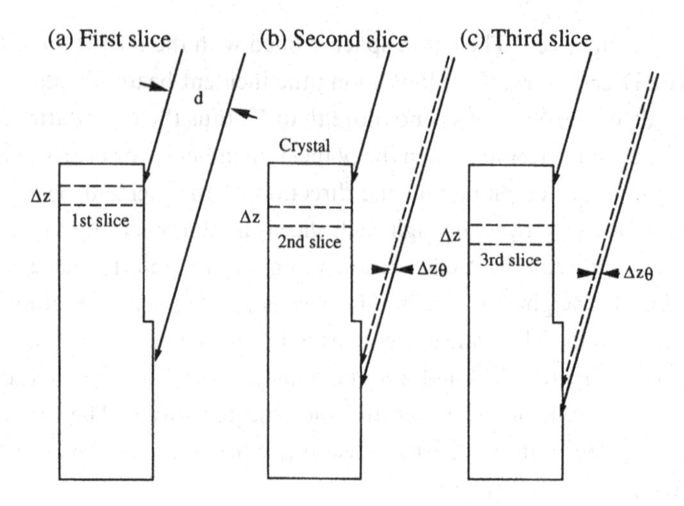

Figure 3.9 *The edge-patching method for calculating the RHEED pattern and REM images of a broad incident beam using a finite super cell. For the first slice, the incident beam width is assumed as a window function which is narrower than the half-width of the super cell. For the second slice, the right-hand-side edge of the window function is modified and its width is increased by $\Delta z\theta$. This procedure is carried out for each successive slice, so that the right-hand side of the window function is constantly kept at distance d from the crystal surface.*

distribution that is no longer affected by the 'edge' effect of the incident beam window and is purely determined by the properties of the crystal surface). This is a rather time-consuming process. Second, the incident electron wave is usually assumed to be a window function, so that the sharp edges of the effective incident wave will introduce severe edge effects due to Fresnel diffraction as the wave propagates along the surface (Peng and Cowley, 1988b). Finally, this theory is not effective for calculating RHEED rocking curves, because the entire calculation has to be repeated for each incident beam direction.

To overcome the first difficulty, Ma and Marks (1990) examined the numerical consistency between Bloch wave calculations and the PeTSM calculations under identical scattering potentials, and have combined the two approaches in the following way. The steady state wave is calculated first by the Bloch wave method for a perfect periodic surface, and then the interaction of the steady state wave with surface steps is calculated according to the PeTSM theory. This combined method attempts to eliminate the disadvantages of the two individual methods, but caution has to be exercised in Bloch wave calculations for close-packed surfaces. To overcome the second difficulty, an edge-patching method has been introduced (Ma, 1991) to modify the incident wave (in the vacuum part) slice-by-slice by considering a constant phase shift, so that the vacuum part in the super cell is covered by a plane wave (see Figure 3.9). This method makes it possible to calculate the reflection of a large plane wave from the surface so that the calculated results can be directly compared with REM observations. This method has recently been applied to quantifying RHEED patterns recorded from GaAs(001) 2×4 surfaces grown by molecular beam epitaxy (Ma *et al.*, 1993) and the order–disorder transition of the $Cu_3Au(111)$ surface (Ma *et al.*, 1994).

The PeTSM theory is most useful for REM image and fixed-beam RHEED pattern calculations. This approach provides a real-space picture of how the resonance wave is built-up along the surface, how far the electrons penetrate into the surface, and how far they travel along the surface before being re-diffracted into the vacuum.

The calculation of REM images is usually difficult, especially if there are steps on the surface. The calculations by Anstis (1989) using the Bloch wave and PeTSM theory have given some interesting results. Although the PeTSM theory appears different from the PaTSM theory, the two theories are equivalent. When applied to Si(111), the calculated results of the two theories are in agreement (Anstis and Gan, 1992 and 1994). It was concluded that the PeTSM theory is best suited for calculating RHEED patterns from surfaces with long periods or disordered surface structures.

3.5 Diffraction of disordered and stepped surfaces

An important application of RHEED is to the study of surfaces containing imperfect structures, such as steps and adsorption. The background and Kikuchi patterns observed in RHEED are produced by surface disorder and thermal diffuse scattering. While the perpendicular-to-surface multislice theory can be applied, in principle, to surfaces that contain any desired defect or disorder structures, the Bloch wave theory presents technical difficulties when performing this type of calculation. In this section we introduce some modified theories of the parallel-to-surface multislice theory to deal with surfaces containing imperfect or non-periodic structures.

3.5.1 A perturbation theory

The theory is based on the first-order perturbation method (Korte and Meyer-Ehmsen, 1992a and b, 1993a). The crystal potential is written as a sum of a periodic function (V_0) and a non-periodic function (ΔV). The Bragg reflections are considered to come only from the periodic potential V_0. The dynamical scattering is assumed to arise from V_0, and the scattering from ΔV is treated kinematically. We now consider the TDS intensity at a non-Bragg position $\boldsymbol{u} = \boldsymbol{g} + \boldsymbol{q}$. For each $\boldsymbol{q}$, the reflection beams to be included in the calculations are: (1) the periodic crystal reciprocal lattice array $\{\boldsymbol{g}\}$ and (2) the 'diffuse' set $\{\boldsymbol{u} = \boldsymbol{g} + \boldsymbol{q}\}$. For the $N\boldsymbol{g}$ beam case, the total number of beams to be included in the calculations is $2N$, as shown in

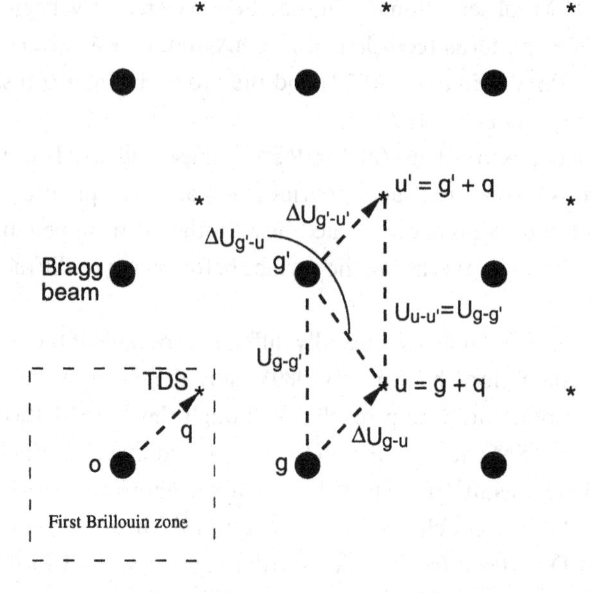

Figure 3.10 *Beam interactions in dynamical calculations of thermal diffusely scattered electrons using the parallel-to-surface multislice theory.* $U_g = (2\gamma m_0 e/h^2)\, V_g$ *and* $\Delta V_u = \mathrm{FT}(\Delta V)$. *Here, nine Bragg beams corresponding to the reflections from the periodic structure are accompanied by nine TDS 'beams'. The calculation is performed for a total of 9+9=18 beams for each* $\boldsymbol{q}$ *point, where* $\boldsymbol{q}$ *is restricted to the first Brillouin zone.*

Figure 3.10. The waves corresponding to the sharp reflections ($\boldsymbol{g}$ and $\boldsymbol{g}'$) are coupled by $U_{g-g'}$. The diffuse beams ($\boldsymbol{u}$ and $\boldsymbol{u}'$) interact via $U_{u-u'}$. The interaction between the $\boldsymbol{g}$ beam and the diffuse beam $\boldsymbol{u}$ is ΔU_{g-u}, where $\Delta V_u = \mathrm{FT}[\Delta V_u(\boldsymbol{r})]$. Therefore, the **B** matrix defined in Eq. (3.29) is redefined as

$$[\mathbf{B}]_{gh} = \mathrm{i}\Gamma_g \delta_{gh} + 4\pi^2 \mathrm{i}\, \frac{U_{g-h}(z)}{2\Gamma_h}, \tag{3.82a}$$

$$[\mathbf{B}]_{uu'} = \mathrm{i}\Gamma_u \delta_{uu'} + 4\pi^2 \mathrm{i}\, \frac{U_{u-u'}(z)}{2\Gamma_u}, \tag{3.82b}$$

$$[\mathbf{B}]_{gu} = 4\pi^2 \mathrm{i}\, \frac{\Delta U_{g-u}(z)}{2\Gamma_u}. \tag{3.82c}$$

The numerical calculation follows the same equation as that given in (3.33) or (3.51). The entire calculation has to be repeated for each $\boldsymbol{q}$, in analogy to the calculation of the rocking surface in RHEED. The perturbation theory introduced here can also be applied to calculate the diffraction of disordered surfaces (Korte and Meyer-Ehmsen, 1992a and b, 1993b).

3.5.2 Stepped surfaces

A theoretical scheme has been recently proposed by Beeby (1993) to calculate electron diffraction from stepped surfaces. The surface is assumed to be of a layered structure in which all layers except the outermost are perfectly periodic in the direction parallel to the surface. The outermost layer will be taken to have atoms distributed in some way over lattice sites and all lying in the same plane. The derivation is divided into a discussion of the scattering from an averaged form of the surface, which follows a standard dynamical calculation and can be carried out to arbitrary accuracy, and the additional effects of the disordered surface layer. The disorder can only be treated by the perturbation method.

Because of its periodicity, the substrate can be treated by the standard dynamical approaches so that the diffraction of any incident wave can be assumed known. If the surface atoms are always at lattice points (i.e. at positions that would be occupied were the substrate to be extended by another layer), then the average surface potential is also periodic, with the same symmetries as the substrate. Thus the problem could be treated as a single diffraction calculation within the normal fully dynamical theory. However, to clarify the analysis it is useful to split the scattering into the part from the surface layer and the part from the substrate, matching them along a plane dividing the two. In this way the effects of the average surface layer arise through the matrix that describes the transmission and reflection from that layer.

The full surface layer potential, which, before averaging, is not periodic, will scatter electrons in directions other than those of the diffracted beams. Electrons that pass into the substrate will be diffracted by it, and the emergent beams can be dynamically calculated in the usual way. The perturbing potential that leads to diffuse scattering is given by

$$\Delta V = V - \langle V \rangle. \tag{3.83}$$

The diffuse scattering wave is given by

$$\Delta \Psi(r) = \int \mathrm{d}\boldsymbol{r}'\, G(r, \boldsymbol{r}')\, \Delta V(\boldsymbol{r}')\, \Psi_0(\boldsymbol{r}'), \tag{3.84}$$

where $\Psi_0(\boldsymbol{r}')$ is the wave produced by the average structure, G is the Green function.

RHEED is an important technique for examining the *in situ* surface structure evolution in epitaxial growth of thin films. We have introduced the several commonly used dynamical theories that have been applied to calculate the diffraction and image intensities of the reflected high-energy electrons. The parallel-to-surface multislice theories are best suited for calculating RHEED rocking curves. The perpendicular-to-surface multislice theory is powerful for simulating the images formed by the reflected electrons, particularly in the cases in which there are steps and defects. The merits and disadvantages of each theory are outlined.

It must be pointed out that the Bloch wave (or Bethe) theory can also be applied to low-energy electron diffraction, provided that the crystal potential is modified to include the exchange effect. The perpendicular-to-surface multislice theory, however, can only be applied for RHEED calculation because the backscattering effect is ignored.

CHAPTER 4

Resonance reflections in RHEED

Electrons reflected from a crystal surface are generated by two scattering mechanisms. Bragg reflection, which is purely an elastic scattering process, is responsible for producing sharp peaks governed by Bragg's law. Inelastic scattering, which is dominated by valence-loss and thermal diffuse scattering (TDS), contributes a Kikuchi pattern background in the electron angular distribution and results in electron energy losses and momentum transfers. The excitation of Bragg reflections critically depends on the diffracting conditions, and the intensity distribution in a RHEED pattern is the result of dynamical scattering of electrons by the crystal surface. It is thus important to understand how electrons are reflected from the surface in order to illustrate the surface sensitivity of REM.

In this chapter, the physics of surface resonance in RHEED are systematically illustrated. Experimental results will be shown to exhibit the nature of surface resonance. It will be demonstrated that the surface sensitivity of RHEED will be dramatically enhanced by resonance. Finally, the effect of inelastic scattering on resonance reflection will be discussed.

4.1 The phenomenon

Surface resonance is one of the most important scattering processes in RHEED and REM. The surface resonance effect was first observed by Kikuchi and Nakagawa (1933); the reflected-electron intensity suddenly increases when the diffraction spot crosses an oblique Kikuchi line. This is illustrated by the RHEED pattern shown in Figure 4.1. When the resonance condition is not satisfied, there is no strongly reflected beam (Figure 4.1(a)). The intensity is increased by almost an order of magnitude if the surface resonance is excited (Figure 4.1(b)). The phenomenon was explained by Miyake and Hayakawa (1970) as a surface resonance effect in which a Bragg-diffracted beam is excited in the direction parallel or nearly parallel to the surface of the specimen. The increased intensity results from the re-scattering of the surface resonance beams to the specularly reflected beam by the atoms located near the surface. The intensity in the entire RHEED pattern is enhanced when the resonance condition is met, particularly for the Bragg-reflected beams.

Figure 4.2 shows a group of RHEED patterns recorded from a cleaved GaAs(110) surface under the (880) specular reflection condition. The strongest

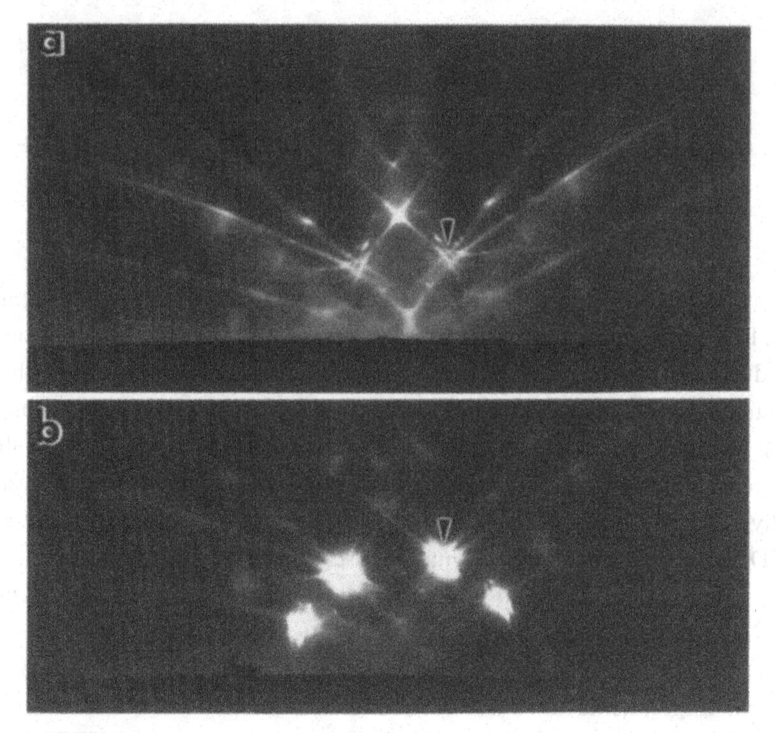

Figure 4.1 RHEED patterns recorded from an InP(110) surface showing the dramatic increase in intensity of the Bragg reflection beams when the resonance conditions are not satisfied (a) and satisfied (b). The beam energy is 100 keV.

intensity is observed when the (880) Bragg beam falls on the intersection of the inclined parabola curve with the Kikuchi line parallel to the surface (Figure 4.2(b)). Under the resonance condition, the intensity of the entire pattern is enhanced. A dramatic change in the reflected beam intensity is seen between Figures 4.2(b) and 4.2(c) in spite of there being only a minor change in diffracting condition. This figure clearly demonstrates the important effect of the diffraction condition on the excitation of surface resonance.

In REM, Bragg reflections are generated by two processes. One is the scattering of the incident beam by the crystal surface directly towards the direction of the specularly reflected beam, similar to the reflection of light from a mirror surface. This is so-called 'kinematical' Bragg reflection. In the other process, electrons are scattered to an intermediate state, followed by propagation parallel or nearly parallel to the surface, finally ending with another Bragg scattering to join the specularly reflected wave. This is simply a double-Bragg-scattering process, which is usually called resonance reflection. The double-Bragg-diffraction process is dominant if the resonance condition is met.

The resonance condition has a remarkable impact on the resolution, intensity and

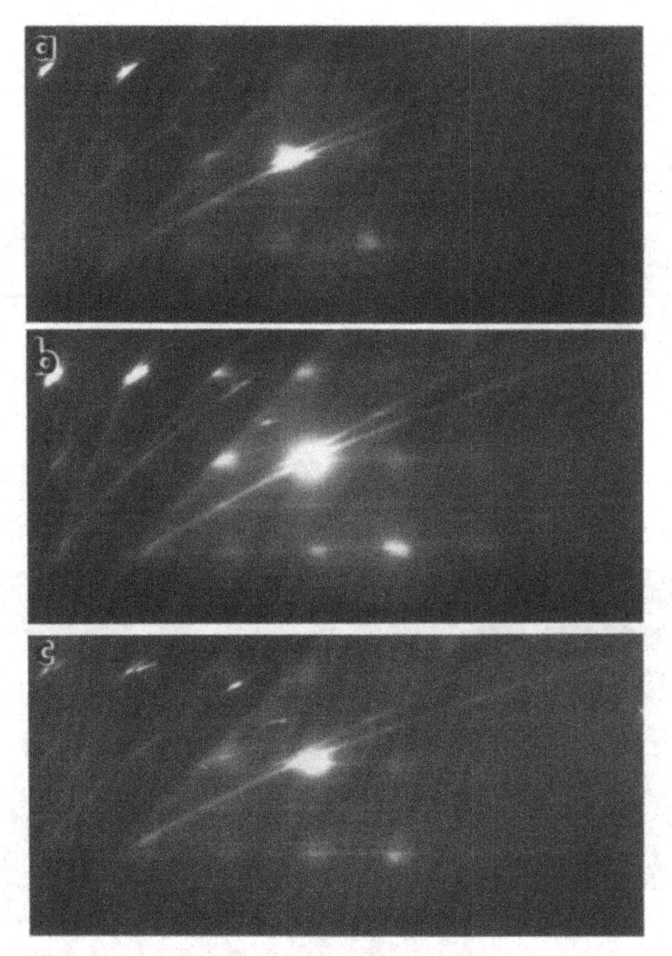

Figure 4.2 *RHEED patterns recorded from a GaAs(110) surface showing the change in the specularly reflected beam's intensity under slightly different diffraction conditions. The beam energy is 120 keV.*

contrast of REM images. Details regarding REM images will be given in Chapter 5. A REM image recorded under Bragg and resonance (BR) conditions has a stronger signal and better contrast than an image taken under Bragg but not resonance (B$\bar{\text{R}}$) conditions (Uchida *et al.*, 1984a). Figures 4.3(a) and 4.3(b) show REM images recorded from the same surface area of Pt(111) under BR and B$\bar{\text{R}}$ conditions, respectively. The reflection beams, as arrowed, were used for taking the corresponding images. It is obvious that the REM image contrast is critically affected by the diffracting conditions (Hsu and Peng, 1987b). Under the BR condition shown in Figure 4.3(a), the surface steps are clearly resolved. In Figure 4.3(b), taken under the B$\bar{\text{R}}$ condition, the image contrast is greatly degraded, but the image resolution is not necessarily lower than that of the image shown in Figure 4.3(a) taken under the

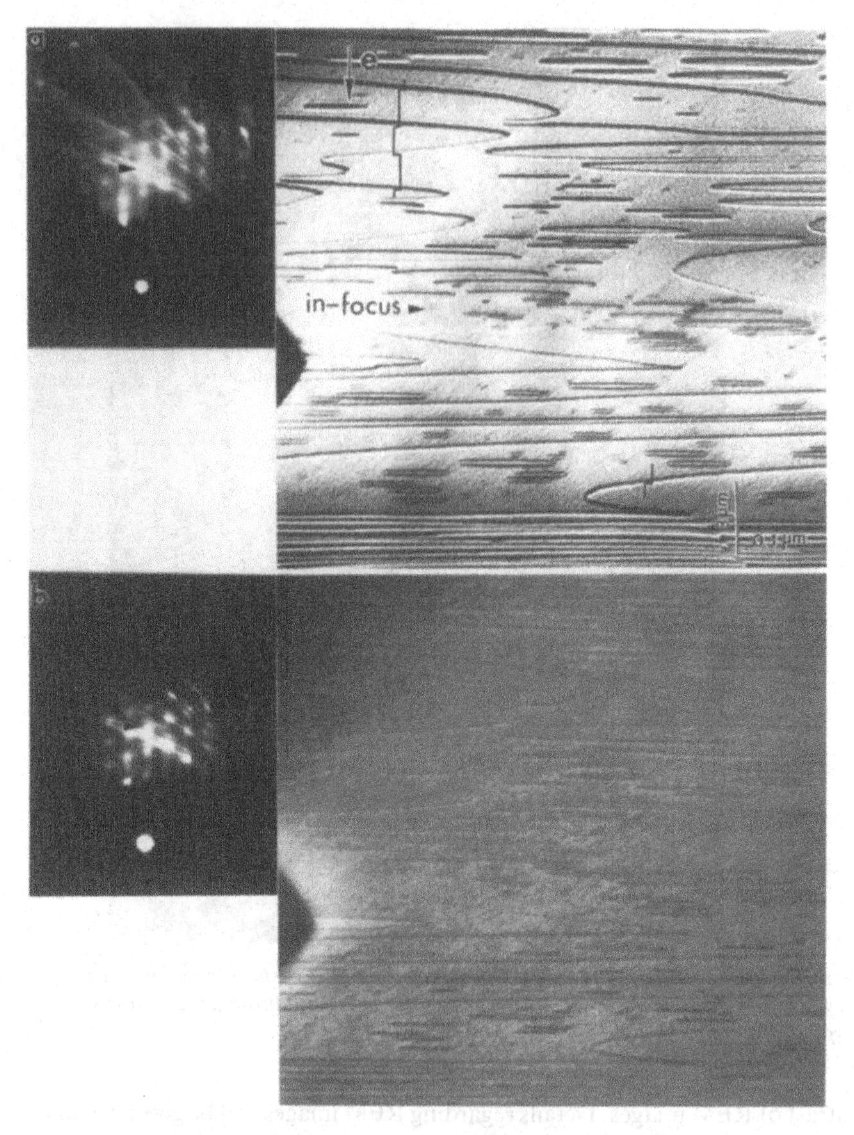

Figure 4.3 *A comparison of REM images recorded under two different diffraction conditions from the same area of the Pt(111) surface showing the dependence of image contrast on resonance conditions: (a) is taken under the Bragg resonance (BR) condition, (b) is taken in the absence of resonance (B$\bar{R}$). The incident beam azimuth is [1$\bar{1}$0]. The in-focus area is indicated, above and below which are over-focus and under-focus surface areas, respectively. The variation in step contrast from under-focus to over-focus can be used to identify the nature of the steps. In this image and the REM images shown in all the following figures, the incident beam direction is from the top to the bottom of the figure (i.e., the vertical direction in the displayed image).*

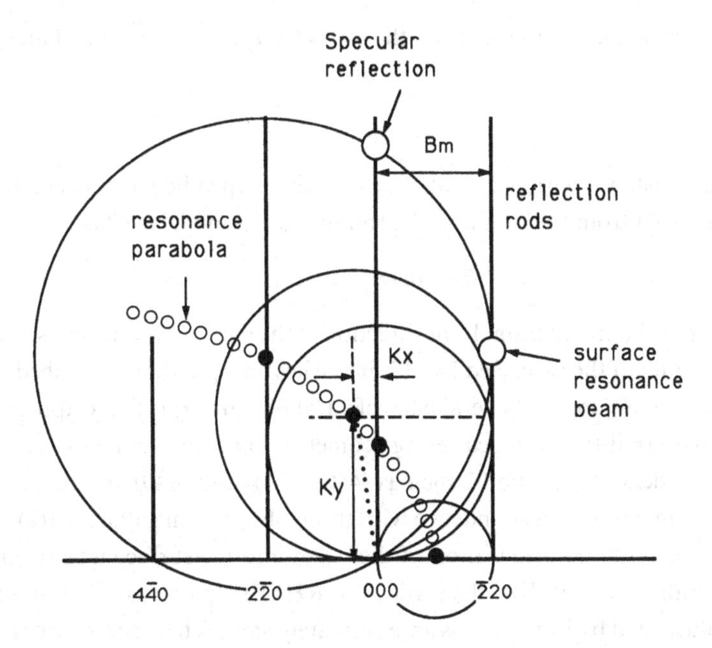

Figure 4.4 *Ewald sphere construction showing the diffraction condition under which surface resonance is excited. An enhancement of the surface resonance beam will increase the reflected intensity of other Bragg beams due to the dynamical scattering effect. The trace of the Ewald sphere is the so-called resonance parabola.*

resonance condition. This observation shows the remarkable effect of surface resonance on REM imaging. Therefore, studies of the resonance phenomenon are critical for understanding the contrast of REM images.

4.2 The resonance parabola and the resonance condition

As shown in Figures 4.1 and 4.2, surface resonance occurs when the reflected beam falls on a parabola-shaped intensity curve, which is normally called the resonance parabola. The formation of this parabola is illustrated below. The diffraction condition under which surface resonance is excited can be obtained from a simple geometrical consideration (Ichimiya *et al.*, 1980). For simplification, we neglect the influence of the mean crystal inner potential. We assume that, because of the small penetration of the beam into the surface, the RHEED pattern can be described by the kinematical scattering of a two-dimensional reciprocal lattice. Surface resonance is expected to occur when the Ewald sphere touches a reflection rod at a point intersecting the surface shadow (Figure 4.4). Under these conditions, a wave propagating parallel or nearly parallel to the surface is strongly excited. The excitation of the wave parallel to the surface may enhance the intensity of the specularly reflected beam due to dynamical multiple scattering effects. In this case, the coordinates (K_x, K_y) of the center of the projected Ewald sphere (i.e., Laue circle) satisfy

$$K_x^2 + K_y^2 = (K_x + B_m)^2, \tag{4.1}$$

where B_m is a distance within the diffraction plane perpendicular to the reflection rods and pointing from the origin to the rod m. Equation (4.1) yields

$$K_y^2 = 2B_m(K_x + B_m/2). \tag{4.2}$$

This is just a parabolic equation. If the direction of the incident beam varies, then the trace of the center of the projected Ewald sphere lies on a parabola described by Eq. (4.2), as shown in Figure 4.4. Several pairs of parabolas are expected to appear in the diffraction pattern if the same scattering geometry shown in Figure 4.4 is repeated for different orders of reflections, such as $(\bar{4}40)$, $(2\bar{2}0)$ and $(4\bar{4}0)$. These resonance parabolas are in good correspondence with those observed in Figure 4.1(a).

In RHEED, the finite penetration of the beam into the surface would generate sharp reflection spots rather than rods in RHEED patterns. The resonance condition illustrated by Figure 4.4 is approximately valid. The resonance parabolas observed experimentally always appear in continuous curves, even for a fixed-parallel incident beam case (i.e., there is no incident beam rocking effect). This is because the thermal diffuse scattering in RHEED is equivalent to producing a large-angle beam convergence, and it is thus similar to rocking the incident beam direction. This is simply illustrated by a calculated large-angle convergent beam electron diffraction pattern (or Kossel pattern) for a small electron probe striking a Si(110) surface (Figure 4.5). The diffraction pattern is formed by all the transmitted (inside the crystal) and reflected (in vacuum) electrons. Electron diffraction within the crystal produces the Kossel pattern (the lower part in Figure 4.5), which is associated with the symmetry of the crystal. In vacuum, the reflected waves exhibit resonance parabolas. These parabolas are the traces of the centers of the Laue circles for the electrons incident along different directions. It should also be noticed that similar parabolic traces are formed in the transmitted electron diffraction pattern, as indicated in the lower part of Figure 4.5. This shows the similarity of RHEED and transmission high-energy electron diffraction (THEED), in agreement with the experimental observations (Lehmpfuhl and Dowell, 1986). Parabolas constructed according to Eq. (4.2) have successfully explained the reflected intensity resonance observed in LEED patterns from MgO(100) surfaces (Hayakawa and Miyake, 1974).

In addition, a strong channeling effect is observed at the crystal surface (Figure 4.5), which corresponds to the intensity of the surface resonance beams parallel or nearly parallel to the surface. This effect is observed in the RHEED pattern recorded from a down-stepped Pt(111) surface (Figure 4.6), where the surface resonance

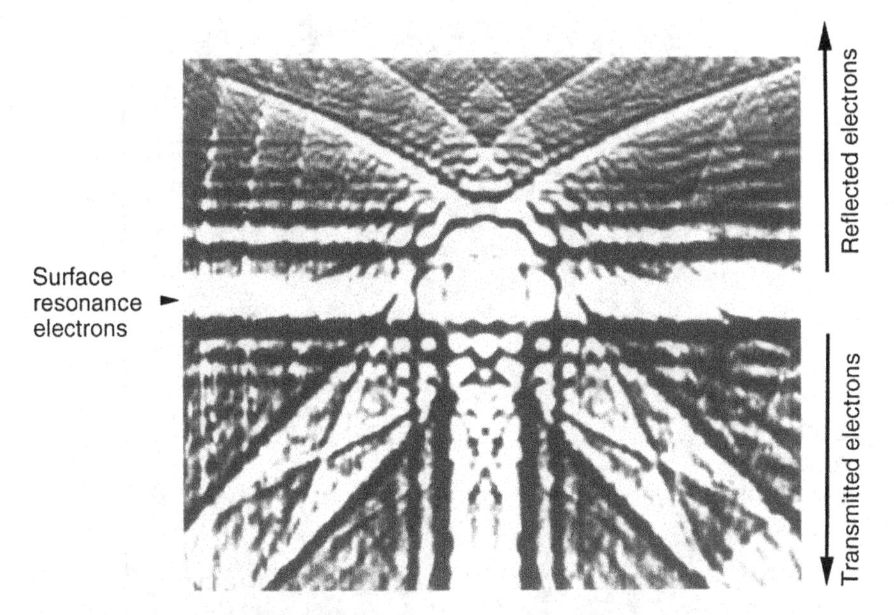

Figure 4.5 *Dynamical calculated 100 kV Kossel pattern of Si(110) in RHEED geometry showing resonance parabolas of the reflected electrons in vacuum (upper part, above the arrow) and K lines of the transmitted electrons (lower part, below the arrow). The beam convergence is 37 mrad, the incidence angle is 23.5 mrad, and the beam azimuth is 8.5 mrad with [001].*

waves are interrupted by the down steps so as to propagate outside the crystal, resulting in strong transmitted intensity below the surface shadow.

Resonance reflection can also be excited if the specularly reflected beam falls on the Kikuchi line parallel to the surface (Kikuchi and Nakagawa, 1933; Miyake and Hayakawa, 1970). This is shown by the experimental RHEED pattern recorded from GaAs(110) (Figure 4.7). This resonance reflection is commonly used to form REM images when the incident beam azimuth is far from any low-index crystal zone axis. A simple interpretation of this resonance condition is given in Figure 4.8. When the incident beam falls on the Kikuchi line, the exact Bragg reflection condition is met. Thus, between the two adjacent atomic planes, Bragg reflection is dominant and the beam tends to propagate along the surface in a way similar to light reflection inside an optic fiber tube, corresponding to the 'total' internal reflection phenomenon. Multiple reflection leads to an increase in the reflected intensity.

4.3 The width of the resonance parabola

Surface resonance occurs within a certain angular width. To find the quantity that affects the parabola width, we now consider the effects of the crystal mean potential on the construction of the resonance parabolas. If the diffracted wave of wavevector

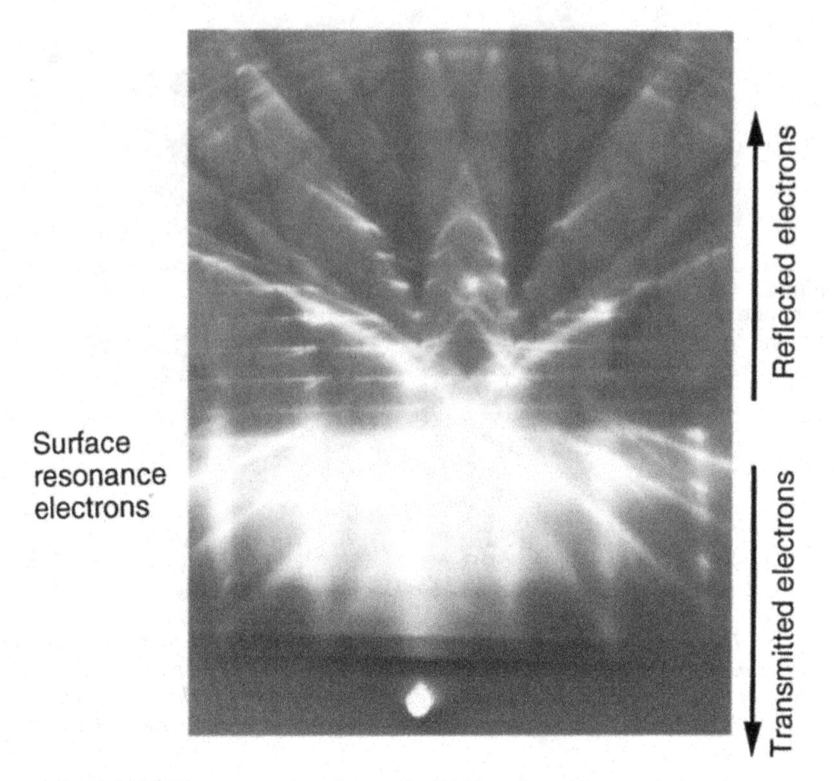

Figure 4.6 A RHEED pattern taken from a Pt(111) surface comprised of many down steps, showing strong intensity distribution beneath the surface shadow in correspondence to the surface resonance beams.

Figure 4.7 A RHEED pattern recorded from GaAs(110) showing the enhancement of specularly reflected beam intensity when it intersects a Kikuchi line parallel to the surface. The beam energy is 120 keV.

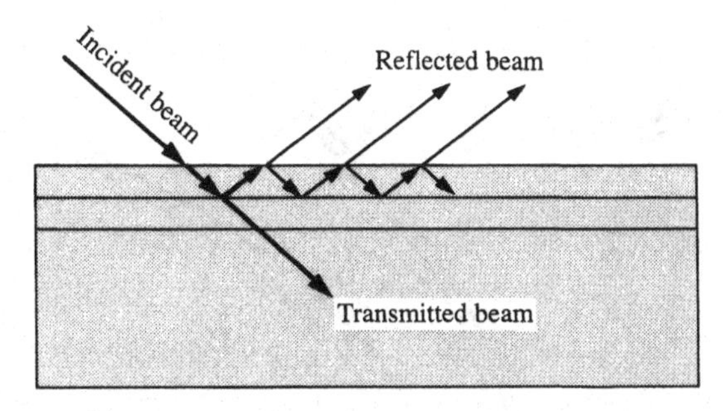

Figure 4.8 *A model for interpretation of the resonance scattering observed in Figure 4.7.*

$\boldsymbol{K}_m$ in vacuum is regarded as a continuation of the diffracted wave inside the crystal with wavevector $\boldsymbol{k}_m$, based on the conservation of energy for elastic scattering, then one has

$$|\boldsymbol{K}_m|^2 = |\boldsymbol{k}_m|^2 - \frac{2m_0 e}{h^2} \bar{V}_0, \tag{4.3}$$

where $\bar{V}_0$ is the crystal mean inner potential, which can be directly calculated from the atom scattering factor (see Appendix B). Also, the conservation of momentum requires the tangential components parallel to the surface of the wavevectors to be continuous,

$$k_{mt} = K_{mt}. \tag{4.4}$$

A surface resonance wave is excited when the mth diffracted wave is propagating inside the crystal, but it is evanescent in vacuum. This implies the conditions

$$|k_{mt}|^2 < |k_m|^2, \tag{4.5}$$

$$K_{mt}^2 > K_m^2 = |K_0|^2, \tag{4.6}$$

which means that the normal component of the crystal wave k_{mn} is real and that of the vacuum wave K_{mn} is imaginary because $K_{mn} = (K_m^2 - K_{mt}^2)^{1/2}$. Combining relations (4.3)–(4.6) and using the coordinate system (K_x, K_y) introduced earlier in Figure 4.4, one obtains (Ichimiya *et al.*, 1980)

$$2B_m(K_x + B_m/2) > K_y^2 > 2B_m(K_x + B_m/2) - \frac{2m_0 e}{h^2} \bar{V}_0. \tag{4.7}$$

This relation sets a range (as shown in Figure 4.9(a)) for the wavevector of the incident electron beam within which the surface resonance occurs. The resonance regions are those in the reciprocal space bounded by the two sets of parbolas (Figure 4.9(a)) (Peng and Cowley, 1987). In practice, the double parabolas observed in

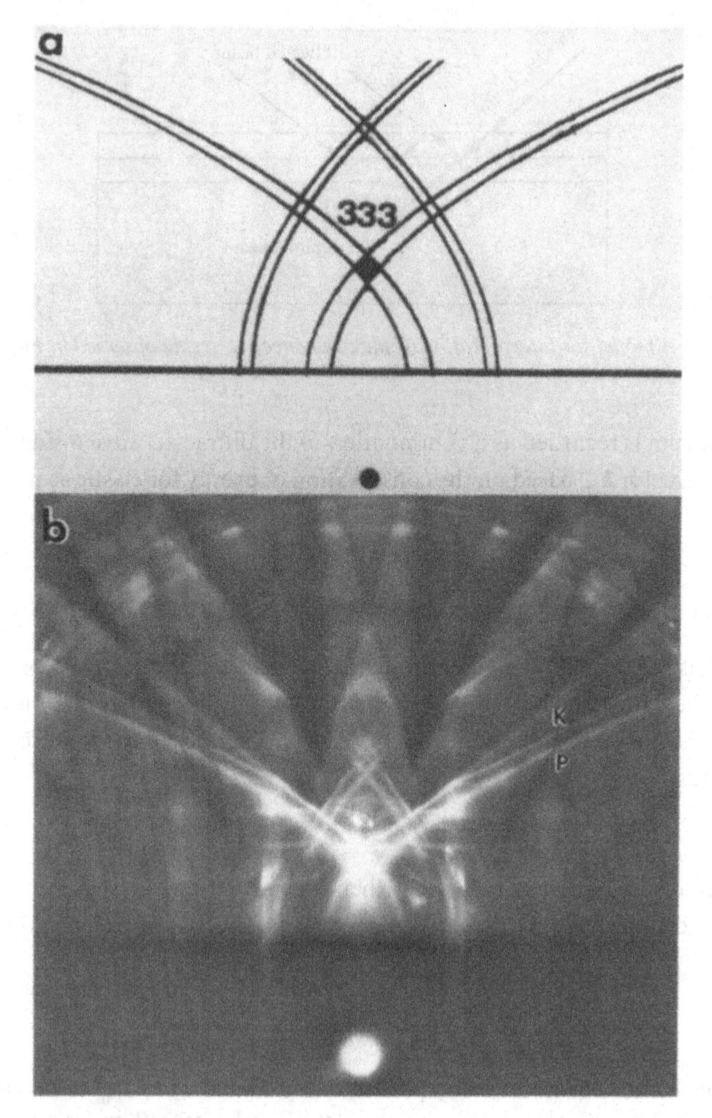

Figure 4.9 *(a) Simulated angular ranges for exciting surface resonance. (b) The observed splitting of the resonance parabola from the Kikuchi envelope of Pt(111), K and P represent the Kikuchi envelope and the resonance parabola, respectively. The beam energy is 100 kV. The splitting effect is mostly observed in the diffraction patterns of large-Z materials.*

Figure 4.9(b) are the Kikuchi line envelope (the top one) and the resonance parabola (the bottom one); the gap between the two is produced by the difference in average potentials experienced by the surface resonance wave channeling along the rows of atoms and the waves scattered by the bulk crystal. For low-Z materials, such as GaAs, the resonance range is so narrow that Eq. (4.7) is simplified to Eq. (4.2). The predicted resonance parabolas are in good agreement with the dynamical calculated convergent beam RHEED patterns for Pt(111) (Smith *et al.*, 1992). For GaAs(110) at 100 keV, the effective resonance region for the (880) reflection is about 2 mrad about the azimuth angle and about 1 mrad about the glancing angle (Wang *et al.*, 1989b; Lu *et al.*, 1991).

4.4 The Kikuchi envelope

On carefully examining the shapes of the parabolas observed in Figure 4.9(b), one sees that the low curve is really a continuous, smooth parabola. The upper curve, however, is composed of straight line segments, and it is thus referred to as the Kikuchi envelope. This 'parabola'-like Kikuchi envelope is formed by the segments of Kikuchi lines, which are tangential to the resonance parabola if the crystal inner potential is ignored. The Kikuchi lines originate from inelastic scattering, which is responsible for scattering the electrons into a wide angular range. This is equivalent to introducing a wide-angle convergent beam illumination. The subsequent elastic scattering of this large 'convergent-angle' beam produces the Kikuchi pattern (see Section 2.7). The geometrical positions of Kikuchi lines can be determined by considering the diffraction of an electron beam that illuminates the crystal with a wide convergent angle, identical to the Kossel pattern. Thus, the Kikuchi pattern is actually the same as the Kossel pattern with respect to the positions of the lines. Both types of lines are thus called K lines.

We first consider a case in which the crystal is illuminated by a plane wave of wavevector $\boldsymbol{K}_0$; then the direction of $\boldsymbol{K}_0$ is varied to cover the entire diffraction angle. For a perfect crystal under the Bragg condition, conservations of energy and momentum require

$$|\boldsymbol{K}_0|^2 = |\boldsymbol{K}|^2, \tag{4.8a}$$

$$\boldsymbol{K} - \boldsymbol{K}_0 = \boldsymbol{g}, \tag{4.8b}$$

where $\boldsymbol{K}$ is the wavevector of the diffracted beam. The combination of the two equations gives the diffraction condition

$$2\boldsymbol{K}_0 \cdot \boldsymbol{g} + g^2 = 0, \tag{4.9a}$$

or

$$\boldsymbol{K}_0 \cdot (-\boldsymbol{g}/2) = (g/2)^2. \tag{4.9b}$$

The geometrical interpretation of (4.9b) gives the following rule for constructing the Brillouin zones. When one point of the reciprocal lattice is chosen as its origin, and the reciprocal-lattice vector $(-\boldsymbol{g})$ is bisected by a perpendicular plane, then any incident wavevector that starts at the origin and ends on the plane will satisfy the

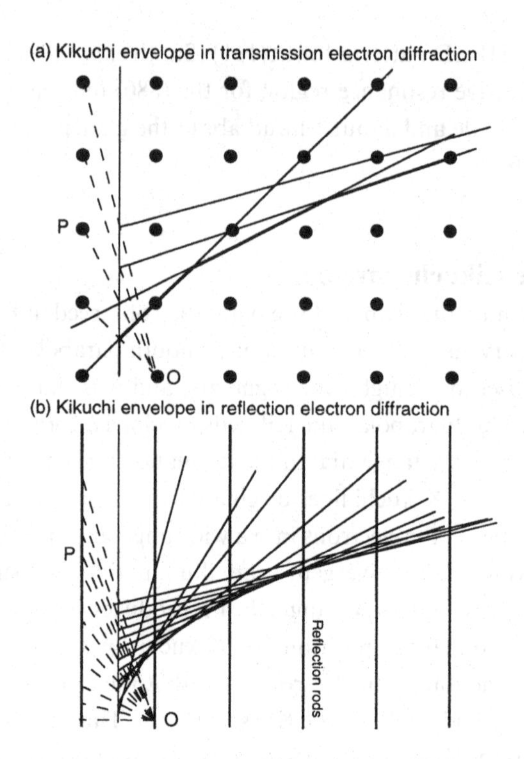

Figure 4.10 Schematic diagrams showing the formation of Kikuchi envelopes in reciprocal space for (a) three-dimensional transmission electron diffraction and (b) two-dimensional reflection electron diffraction. O indicates the incident beam, P is a diffracted beam. The O–P connection is represented by a dashed line. The bisection of the O–P line is the Kikuchi line.

diffracting condition. This procedure is repeated for all the reciprocal-lattice vectors and the obtained planes to form the boundaries of the Brillouin zones, which define the directions in which the incident wave can be diffracted by the periodic scattering lattice. In other words, every point on the Brillouin zone boundaries satisfies the Bragg condition. In comparison with the formation of Kikuchi lines (Section 2.7), the Bragg condition is exactly satisfied when the diffraction beam falls on the Kikuchi lines; thus, the observed Kikuchi lines are the intersections of the Brillouin zone boundaries with the Ewald sphere (Gajdardziska-Josifovska and Cowley, 1991). Therefore, the geometry of Kikuchi lines for high-energy electron diffraction can be constructed using the method for constructing the Brillouin zone boundaries in solid state physics, since the Ewald sphere is very large. For low-energy electrons, however, the K lines are curved due to the finite radius of the Ewald sphere.

Figure 4.10 compares the formation of the Kikuchi envelope in TEM and REM. For a three-dimensional crystal, the discrete Bragg reflection spots result in the straight, 'non-smooth', intensity segments along the envelope. The shape of the envelope can be approximated by a parabola. For two-dimensional surface

Figure 4.11 *A transmission electron diffraction recorded from a thin MgO specimen showing the 'parabolic'-shaped Kikuchi envelopes. The beam energy is 100 keV.*

diffraction, the continuous reflection rod results in a parabolic Kikuchi envelope, and each point on the rod satisfies the Bragg condition. The shape of the envelope is exactly a parabola.

The Kikuchi envelope is a common feature of high-energy electron diffraction, and it should also be observed in transmission electron diffraction of thin foils. A pattern shown in Figure 4.11 was recorded at 100 kV from a thin MgO crystal near the [001] zone axis. Parabola-like Kikuchi envelopes are observed for different orders of reflections. The observed Kikuchi pattern is symmetric about the line as indicated but exhibits excess and deficient intensities due to dynamical scattering. The envelope is composed of straight segments, as expected theoretically (Figure 4.10(a)).

In RHEED, electrons channeling along rows of atoms near the surface may 'see' a potential that is different from the average potential of the crystal, resulting in a slightly larger refraction effect. Hence the continuous parabolas, generated by channeling electrons, are displaced relative to the straight Kikuchi lines, which arise from non-channeled electrons, and there is a gap between the parabolas and the envelopes of the straight lines. This is observed in the RHEED pattern of Pt(111) shown in Figure 4.9(b). The upper curve is the Kikuchi envelope (with straight segments) and the lower one is the resonance parabola (smoothly curved). The width of this gap is a measure of the average depth of the potential well that is sensed by the channeled electrons (Peng *et al.*, 1988) and can be used to estimate the mean inner potential (Yao and Cowley, 1989).

The envelope of the Brillouin zones for a two-dimensional real-space lattice is just the resonance parabola described by Eq. (4.2) (Gajdardziska-Josifovska and Cowley, 1991). In general, the envelopes of the Kikuchi lines are coincident with the resonance parabolas for low-Z materials, but they still do not belong to the same family (Yao and Cowley, 1990). This agrees with the observation that the overlap of the specularly reflected beam with a Kikuchi line results in enhancement only if the relevant part of the Kikuchi line is part of an envelope (Kohra *et al.*, (1962); also seen in Figure 4.2). In RHEED, double splitting of the Bragg and Bragg–Laue reflections can be observed due to the non-refracted effect of the Bragg–Laue beams at the crystal's end exit face. This effect has been used to measure the crystal mean potential (Yamamoto and Spence, 1983). The refraction effect will be introduced in Section 5.5.

4.5 Dynamical calculations of resonance scattering

Dynamical scattering plays an important role in determining the intensity distribution in RHEED patterns. Theoretical calculations usually become difficult and complex whenever multiple-scattering effects are involved. In this section, semi-quantitative calculations using the perpendicular-to-surface multislice theory are presented to illustrate the physical nature of surface resonance and the corresponding experimental evidence.

The PeTSM theory is most useful for REM image and fixed-beam RHEED pattern calculations. The calculation method has been given in Figures 3.8 and 3.10. This theory is now applied to investigate the scattering processes of electrons in the RHEED geometry. Figure 4.12 shows three typical near-zone-axis RHEED patterns of GaAs(110). Under the (440) specular reflection condition (Figure 4.12(a)), the incident beam strikes the surface at such a small angle that the effective penetration into the surface is almost zero. Thus, the (440) reflection is a single-beam 'mirror' reflection that does not suffer strong dynamical scattering, similar to light reflection from a mirror surface. Under the (660) reflection condition (Figure 4.12(b)), strong dynamical diffraction is apparent, and the reflected intensity is maximized. The intensity of the (840) beam is comparable to that of the (660)

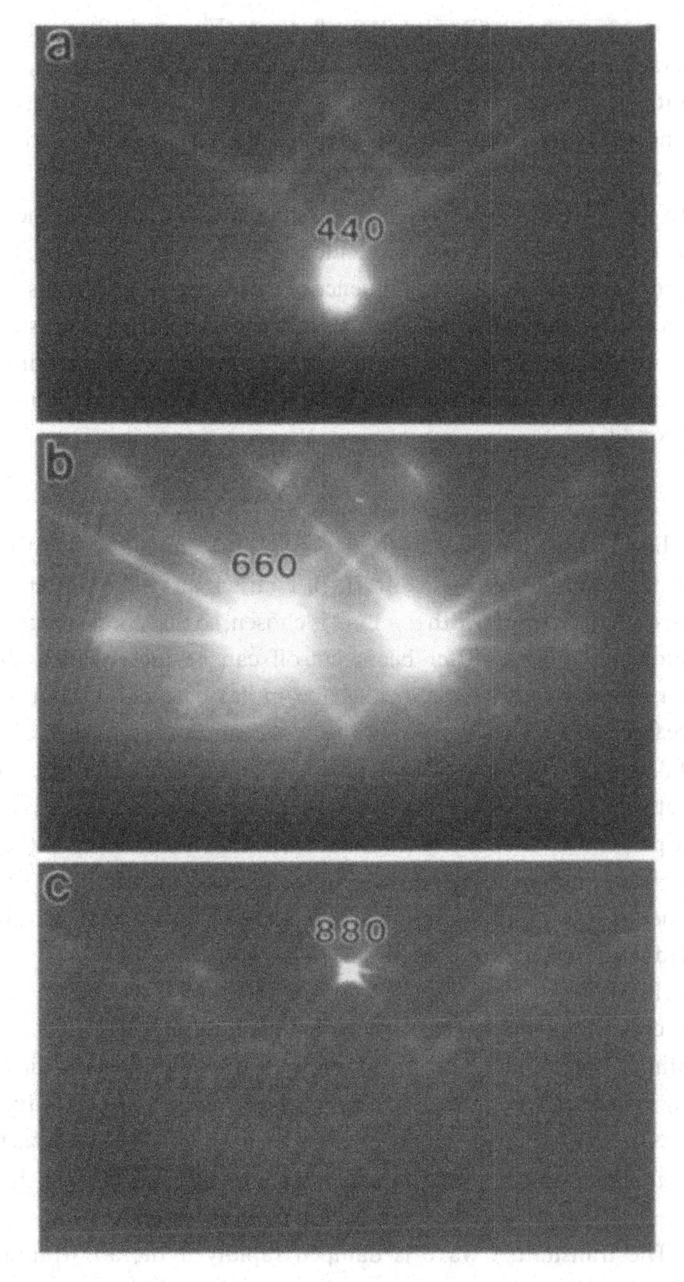

Figure 4.12 *RHEED patterns of GaAs(110) recorded under the resonance condition when (a) (440), (b) (660) and (c) (880) beams are specularly reflected. The beam energy is 120 keV.*

specular beam. For the (880) specular reflection (Figure 4.12(c)), the reflected intensity is rather weak. It is believed that the electron penetration depth into the surface is rather large, effectively decreasing the surface sensitivity of REM. REM images are normally recorded using the (660) or (880) beam in order to reduce the foreshortening factor.

Owing to the RHEED geometry, full dynamical calculations are necessary to describe the reflections of high-energy incident electrons from surfaces. In particular, it is interesting to investigate the influence of surface defects, such as steps, on the propagation of electrons at the surface. This exercise is useful for understanding the contrast of step images in REM. In this section, we intend to examine electron resonance reflection under the (440) (Figure 4.12(a)) and (660) (Figure 4.12(b)) diffracting conditions.

4.5.1 Low-incidence-angle resonance

Our study begins with the resonance at GaAs(110) under the (440) specularly reflected beam condition (see Figure 4.12(a)). To track down the scattering process of electrons, a small beam of width $d = 1$ nm is chosen, so that the propagation of the electrons along the surface after beam cut-off can be investigated. Resonance propagation is best seen after the incident beam has been cut off and there is no interference from the incident wave. The incident beam covers the surface area for a distance of 83 nm ($= d/\theta$, with $d = 1$ nm and $\theta = 12$ mrad) along the direction of the beam azimuth. The intensity distribution of the electrons at the surface is the output in the plane perpendicular to the beam azimuth, so that the build-up of the waves at a crystal surface can be seen by comparing the intensity distributions at different slices. In order to examine the damping of reflected electron intensity, no absorption was included in the calculation.

Figure 4.13 shows the calculated electron intensity distribution at different thicknesses under the (440) specular reflection condition for a perfect GaAs(110) surface without steps. The first three outputs are the intensity distributions representing interference between the incident wave and the reflected wave. The last three outputs occur after the incident wave has been cut off. The wave has been split into three streams, the transmission wave (T) into the bulk crystal, the surface resonance wave (S) propagating along the top monolayer surface and the reflected wave (R). The transmitted wave is damped rapidly if the absorption effect is included in the calculation. The propagation direction of the wave can be easily estimated by measuring transverse and longitudinal traveling distances, and is found to be 17 mrad. This is about 5 mrad larger than the initial assumed incident angle. This difference is the result of the surface refraction effect, as will be discussed in Section 5.5. The surface resonance wave is characteristic of surface resonance and represents the surface sensitivity of REM. The propagation direction of the wave is

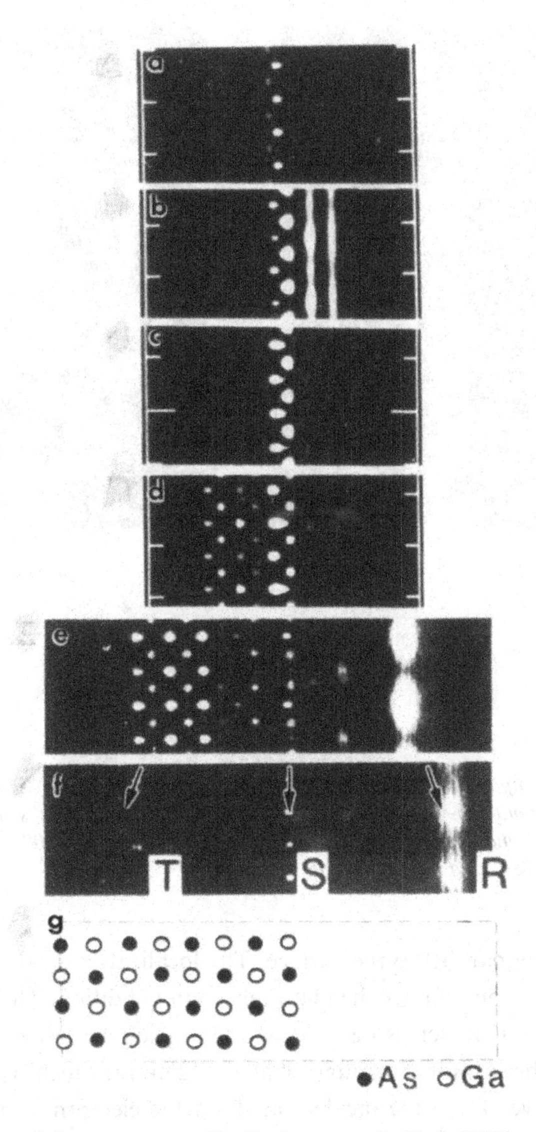

Figure 4.13 *Calculated electron current distributions near the GaAs(110) surface at distances of (a) 22.6, (b) 45.2, (c) 67.8, (d) 90.4, (e) 113 and (f) 135.6 nm along the surface. The beam azimuth is [001], $\theta = 11.9$ mrad, the beam energy is 120 keV, and the beam width is 0.9 nm. In (g) is shown the projected super-cell model used in the calculation. This figure displays the calculated electron intensity distributions, as viewed along the beam direction, near a perfect GaAs(110) surface at different depths along the beam direction. The left-hand side is the crystal, the right-hand side is vacuum, and the interface of the two is the (110) surface.*

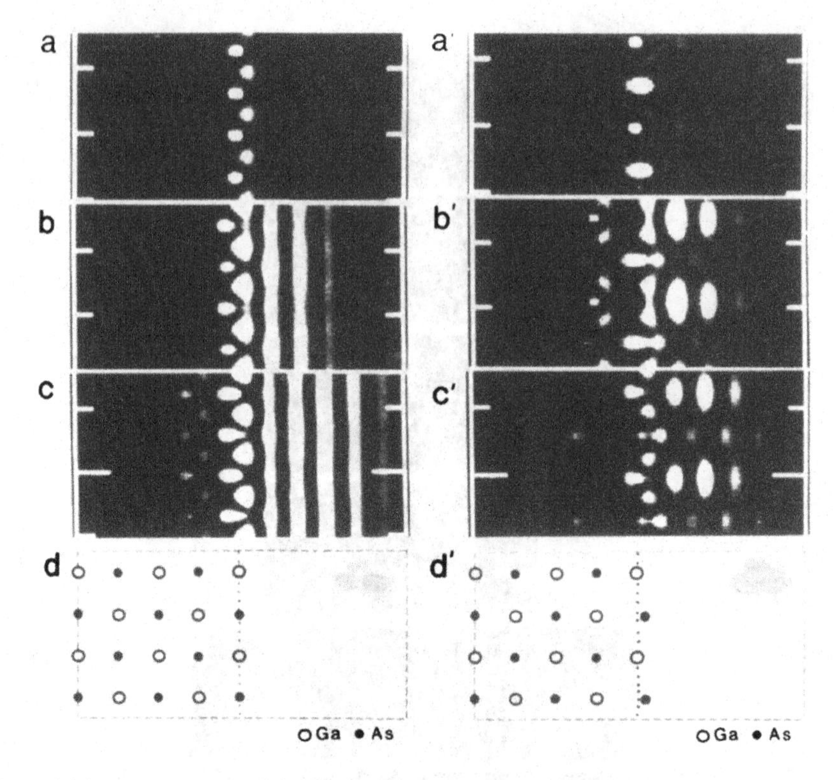

Figure 4.14 A comparison of the electron current distributions near a GaAs(110) surface with (a')–(d') and without (a)–(d) surface reconstruction. The output depths are (a) and (a') 22.6, (b) and (b') 46.2, and (c) and (c') 67.8 nm. The beam azimuth is [001], $\theta = 11.9$ mrad, the beam energy 120 keV, and the beam width is 5.65 nm.

parallel or nearly parallel to the surface. The localization of this wave at the top monolayer occurs only under the (440) reflection condition. The resonance depth slightly increases with increasing incident angle. The reflected wave is propagating outward from the surface at an angle of about 12 mrad, which is about the same as the incident wave. Thus, the incident angle of the electron beam is equal to the reflected angle of the specularly reflected beam under the Bragg reflection condition, similar to that in optics.

In practice, the atomic arrangement on a crystal surface is not always the same as that in the bulk. A clean GaAs(110) surface is reconstructed so that along the [001] direction the top surface layer of Ga is seen to be shifted inward by 0.0094 nm and the As atoms are shifted outward by 0.055 nm (Chadi, 1978). To show the effects of surface reconstruction on the surface resonance reflection, the calculated electron intensity distribution at different thicknesses for a perfect GaAs(110) and a reconstructed GaAs(110) surface is shown in Figure 4.14. The incident wave parameters are exactly the same for the two cases. The fringes seen in the vacuum part are due to the interference of the reflected wave and the incident wave. The channeling current intensity distributions along the atomic columns are perturbed by the surface structural modulation. The distribution of the resonance wave is also critically affected. This calculation has clearly demonstrated the influence of surface atomic structure on the reflected wave. It also has shown that the reflected wave is generated by scattering from the top layer, referred to as 'monolayer resonance' and exhibiting high surface sensitivity. Unfortunately, this low-angle incidence is rarely used for REM imaging because of the strong foreshortening effect.

4.5.2 High-incidence-angle resonance

The (660) specular reflection is commonly used to form REM images of GaAs(110). This beam is strongly excited under the diffracting condition shown in Figure 4.12(b), in which the (660) and (840) reflections are simultaneously excited and the beam azimuth is off the [001] zone axis with a deviation angle of 8.88 mrad. In this section, we will investigate the resonance scattering of reflected electrons under this diffracting condition.

Figure 4.15 shows the calculated electron intensity distributions at a GaAs(110) surface under the specular (660) reflection with an azimuthal deviation angle of 8.88 mrad in the direction parallel to the surface. After cutting off the incident wave, the generated waves are divided into three possible streams. One stream is the transmission wave (T), which propagates into the crystal and will be eventually absorbed. The second stream is the surface resonance wave (S), which propagates along the surface. The last stream is the reflected wave (R), which is in the vacuum and propagates away from the surface (this wave does not appear in Figure 4.15 because of its relatively weak intensity). It is apparent that the surface resonance wave is concentrated in the top two or three atomic layers. This figure shows characteristics similar to those shown in Figure 4.13 except that the surface resonance wave is distributed over about four atomic layers. Therefore, the monolayer resonance occurs only under specific diffracting conditions. The destruction of monolayer resonance may result from two causes. One is the large incidence angle of the beam and the other is the deviation angle of the beam with respect to the [001] zone axis in the plane parallel to the surface, so that the zone axis channeling effect is reduced.

Resonance propagation of electrons at a crystal surface can enhance the reflected intensities. Figure 4.16 shows the calculated total reflectance (i.e., the ratio of the total reflected intensity to the intensity of the entirety of incident electrons) of the electrons from a GaAs(110) surface versus the propagation distance along the beam direction. The incident beam is cut off at the 150th slice. An important feature shown in Figure 4.16 is that the reflected intensity initially increases almost linearly even after the beam has been cut off. This indicates that the resonance propagation of the

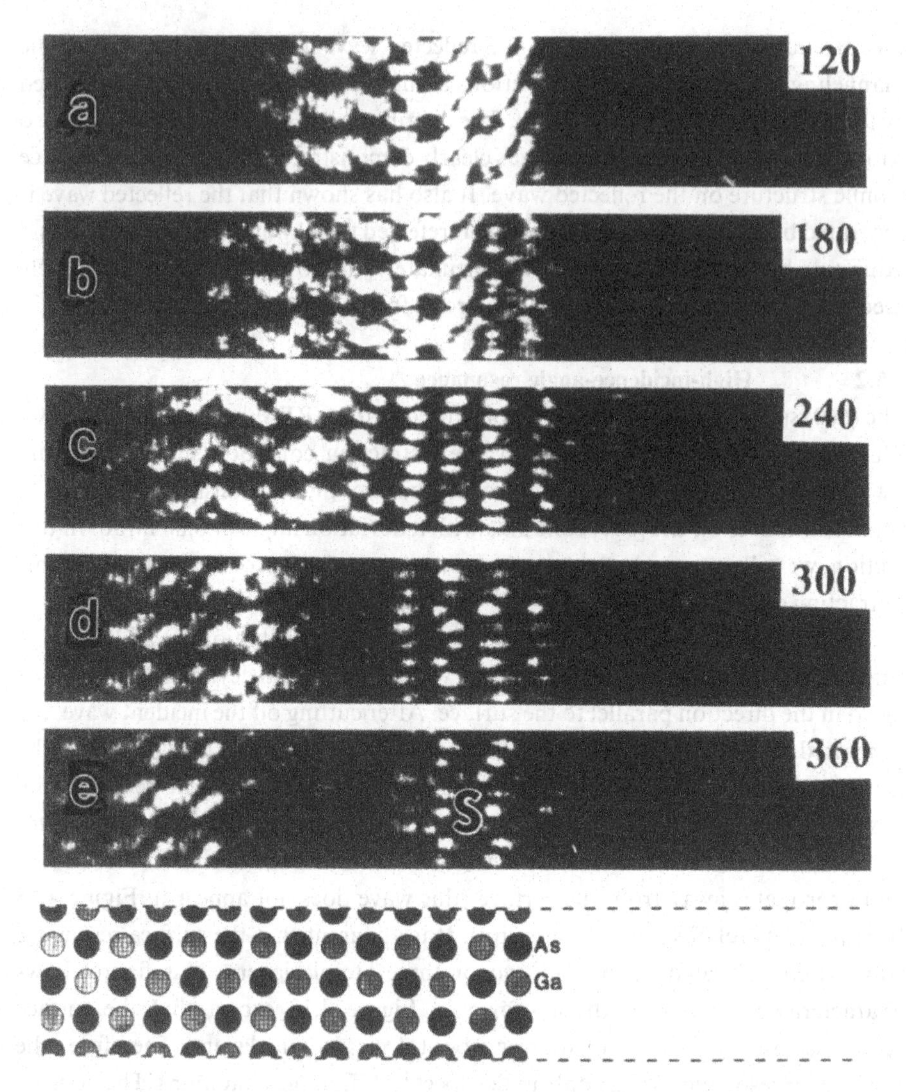

Figure 4.15 *Scattering of high-energy electrons from the surface in RHEED calculated using the perpendicular-to-surface multislice theory. The calculated electron intensity distributions, as viewed along the beam direction, near a perfect GaAs(110) surface at different slice positions (or depth along the beam direction) are displayed. The slice number is indicated at the corner of each intensity output. The projected atomic model is shown at the bottom of the output and the incident beam is directed into the paper. The left-hand side is the crystal, the right-hand side is vacuum, and the interface of the two is the (110) surface. T, S and R indicate the transmitted wave into the bulk, the resonance wave near the surface, and the reflected wave in vacuum, respectively. The beam azimuth is close to [001], the incidence angle is 23.5 mrad and the azimuth deviation angle in the direction parallel to the surface is 8.88 mrad. The incident beam is cut off at the 150th slice. The slice thickness is 0.2827 nm.*

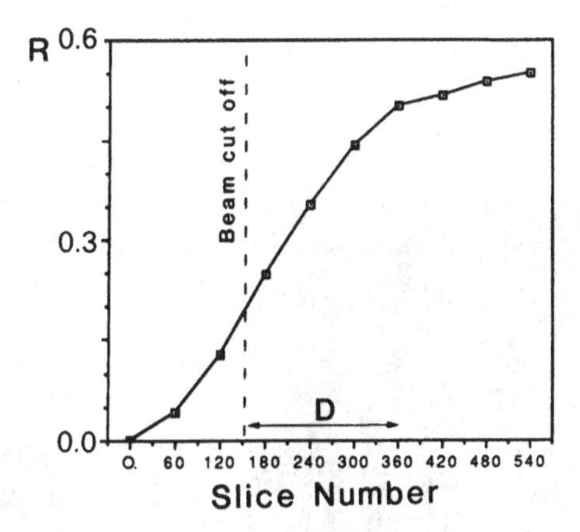

Figure 4.16 *Resonance reflection in RHEED, showing the calculated total reflectance (R) versus the slice number from the results of Figure 4.15. The incident beam illuminates only the surface from the first to the 150th slice, but the total reflected intensity still increases even after the incident beam has been cut off.*

electrons along the surface can generate strong reflection waves. There seems to exist a turning point at which the reflection rate dR/dz decreases significantly. This point is at about the 300th slice. The mean traveling distance of the electrons along the surface is estimated as 50–70 nm. The contribution of the 'direct' reflections, i.e. the immediate reflection of the electrons from the surface, is small, on the order of a third of the total reflected intensity. Most of the incident electrons travel along the surface for some distance before being scattered back into vacuum.

4.5.3 Resonance at a stepped surface

As illustrated in the last section, surface resonance (or channeling) electrons are mainly distributed over the top few atomic layers. It is expected that the propagation of the resonance wave would be greatly affected when it meets a down step (with respect to the beam). Dynamical investigation of electron channeling at stepped crystal surfaces has been reported by Wang (1988a). The interruption of the surface resonance wave by a surface step has been proposed as the origin of double contrast contours observed in REM images (see Section 6.10). This section is designed to summarize the characteristics of surface resonance at stepped surfaces.

Figure 4.17 shows a comparison of the calculated electron intensity distributions at a crystal surface without surface steps (left column), with an up step (middle column) at the 200th slice, and with a down step (right column) at the 200th slice. The incident beam illuminates the first 150 slices. The step occurs at about 12.89 nm after the beam has been cut off along the surface, and it is expected that the step only affects the propagation of the surface resonance wave. The step height is equal to one atomic layer. The electron intensities are output at the same thickness under the

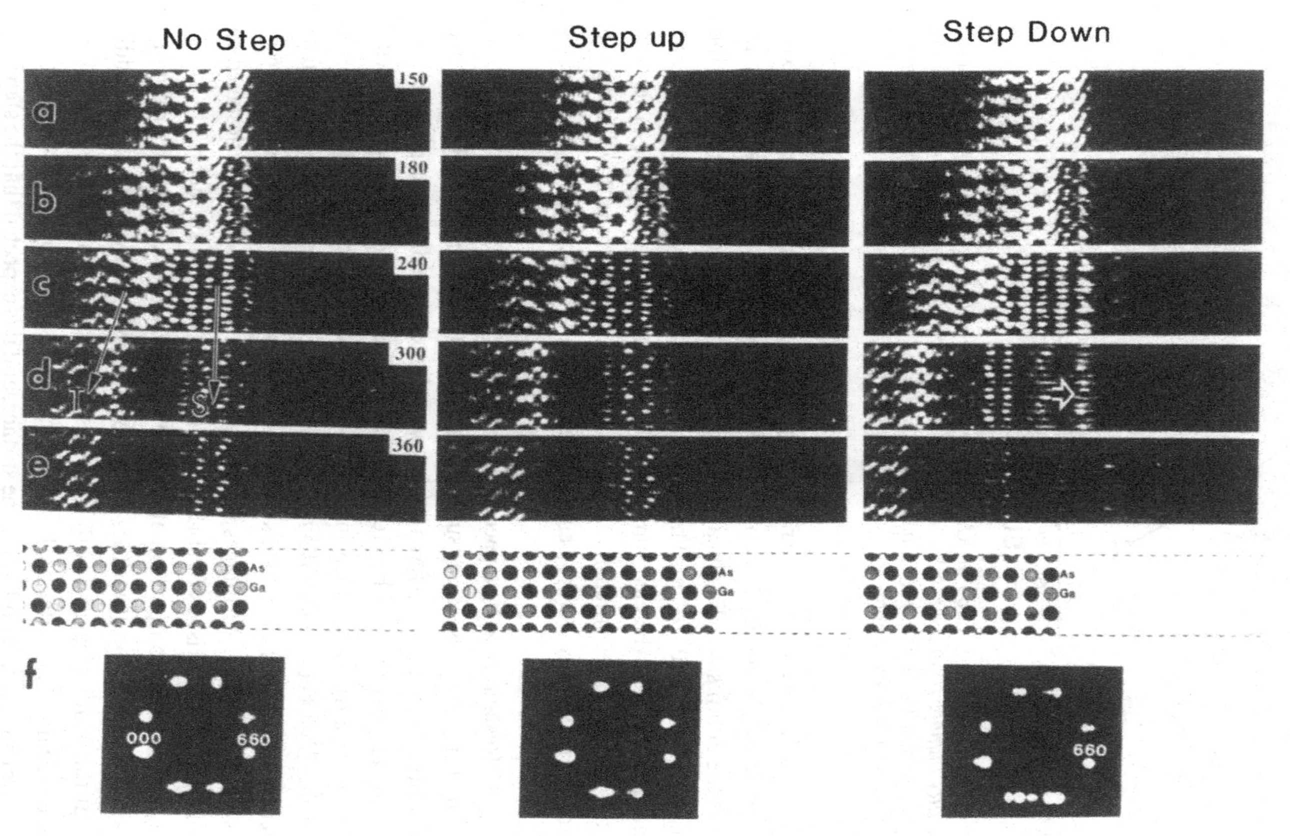

Figure 4.17 *A comparison of the calculated electron intensity distributions for different slice numbers at a GaAs (110) surface without steps (left-hand column), with a step up at the 200th slice (middle column) and with a step down (right-hand column) at the 200th slice. With $\theta = 23.5$ mrad and $\phi = 8.88$ mrad and beam azimuth [001]. The beam cut-off happens at the 150th slice. (f) is the Fourier transform of the electron wave function.*

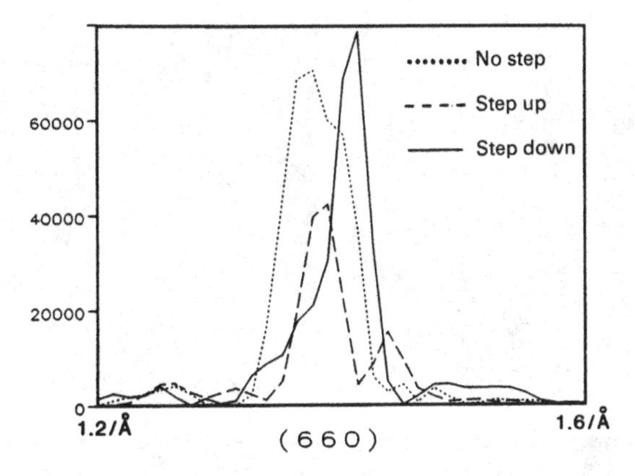

Figure 4.18 *Intensity profiles across the (660) specular reflection spots in the calculated RHEED patterns for the three cases in Figure 4.17. The profiles are calculated according to the wave function output at the 480th slice.*

same incident conditions, so that the perturbation of the resonance wave by the surface steps can be classified. Comparing the intensity output in the left and the middle columns shows that a step up at the 200th slice will not critically affect the propagation of the resonance wave. The resonance wave continues traveling along the original surface terrace from which it was generated. The down step at the 200th slice can effectively interrupt the propagation of the resonance wave along the surface (see Figure 4.17 right-hand columns (c) and (d)). The propagation of the interrupted surface resonance waves in vacuum can be seen in the calculated diffraction patterns in Figure 4.17(f), which introduces some 'extra' spots (Figure 4.17(f) right-hand corner). This may be due to the fact that the propagation directions of the surface resonance waves are perturbed by the potential fields around the step edge. It is possible that the traveling direction of the interrupted wave is different from that of the Bragg scattering.

The steps can significantly affect the reflectance of a surface. Figure 4.18 shows a comparison of the line scan intensities across the (660) specular reflection spot for the three cases as illustrated in Figure 4.17(f). The surface with a down step seems to have the largest reflectance, which is contributed from the interrupted surface resonance wave. The surface with an up step seems to be weak in reflectance, which is due to the blocking effect of the up surface terrace layer with respect to the reflected wave; part of the blocked waves will be scattered back into the crystal by the step edge. Also it is indicated that the positions of the intensity peaks are slightly influenced by the surface steps.

The differences in reflected intensities from a perfect crystal surface and a surface with a down step can be seen by comparing the calculated RHEED patterns, as shown in Figure 4.19. In the case of a perfect flat surface, the reflected intensities are

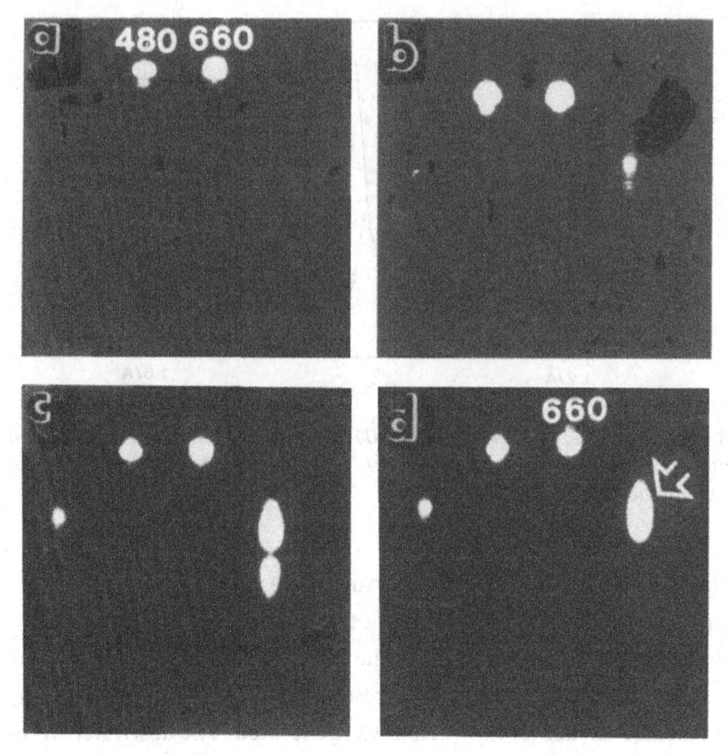

Figure 4.19 *Calculated RHEED patterns of GaAs(110) surfaces for different step-down positions located at (a) ∞, (b) the 200th slice (56.5 nm), (c) the 80th slice (22.6 nm), and (d) the 50th slice (14.1 nm) from the beam cut-off position. The step height is chosen as a monolayer high. The simulation conditions are the same as those for Figure 4.17.*

mainly concentrated at the (660) and the (480) reflections (Figure 4.19(a)). When an atom-high down step is introduced at the 200th slice (56.54 nm) away from the beam cut-off position, besides the regular (480) and (660) reflection beams an 'extra' beam appears in Figure 4.19(b). If the down step is set at the 80th slice (22.62 nm) away from the beam cut-off, then this extra beam becomes remarkably strong and is split into two. The splitting of this beam in the direction perpendicular to the surface shows that this wave comes from a small region of the top few surface layers (Figure 4.19(c)). When the down step is set at the 50th slice (14.14 nm) away from the beam cut-off, the extra beam becomes a single spot and is still strong in intensity (Figure 4.19(d)). The formation of this extra beam can be considered as the propagation of the resonance waves in vacuum after their having been interrupted by the step.

The theoretically predicted results have been observed in RHEED experiments using a VG HB5 STEM. An incident beam 0.5 nm in diameter was chosen in order to study the small-beam reflection effect from a surface. Figure 4.20 shows a series of RHEED patterns made by scanning the beam across a surface down step. When the

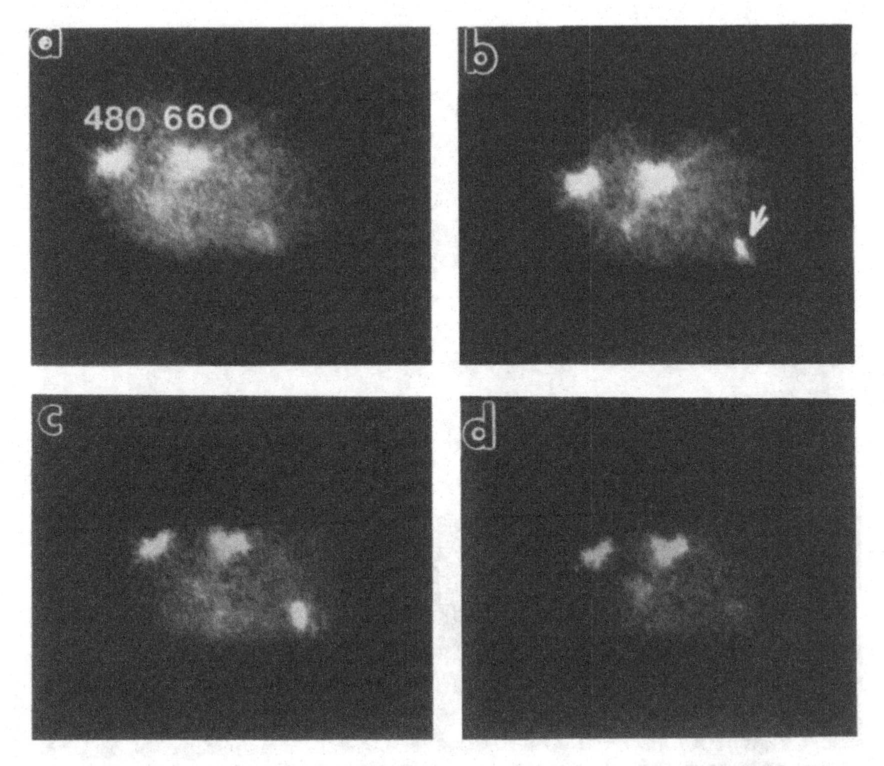

Figure 4.20 *Microdiffraction RHEED patterns with a 0.5 nm beam scanning across a down-step on the GaAs(110) surface. These patterns were obtained when the beam was (a) far away from a step, (b) approaching the step, (c) close to the step, and (d) after passing the step. The reflection located near the surface shadow appears only when the incident beam is near the surface down-step (Wang et al. 1989a).*

beam is far away from a down step (Figure 4.20(a)), there is almost no indication of the existence of the extra beam, in agreement with the prediction in Figure 4.20(a). Only the (660) and (480) reflections are observed. When the beam approaches the step, an extra beam appears near the edge of the surface, as arrowed in Figure 4.20(b). When the beam moves very close to the step, this beam becomes strong (Figure 4.20(c)). This beam disappears suddenly just after the beam crosses the step (Figure 4.20(d)). All these observations agree with the theoretical calculation shown in Figure 4.19.

To confirm that the observed extra beam comes from the surface step, scanning reflection electron microscopy (SREM) images were taken by selecting the (660) reflection and the extra beams, as shown in Figure 4.21. The step structures of the surface can be seen (Figure 4.21(a)), but the up and down steps cannot be identified without much additional effort, as will be described in Section 6.6. In the scanning reflection image (see Section 12.1) taken using the extra beam when the electron probe scans across the surface, very strong bright lines are present. This indicates

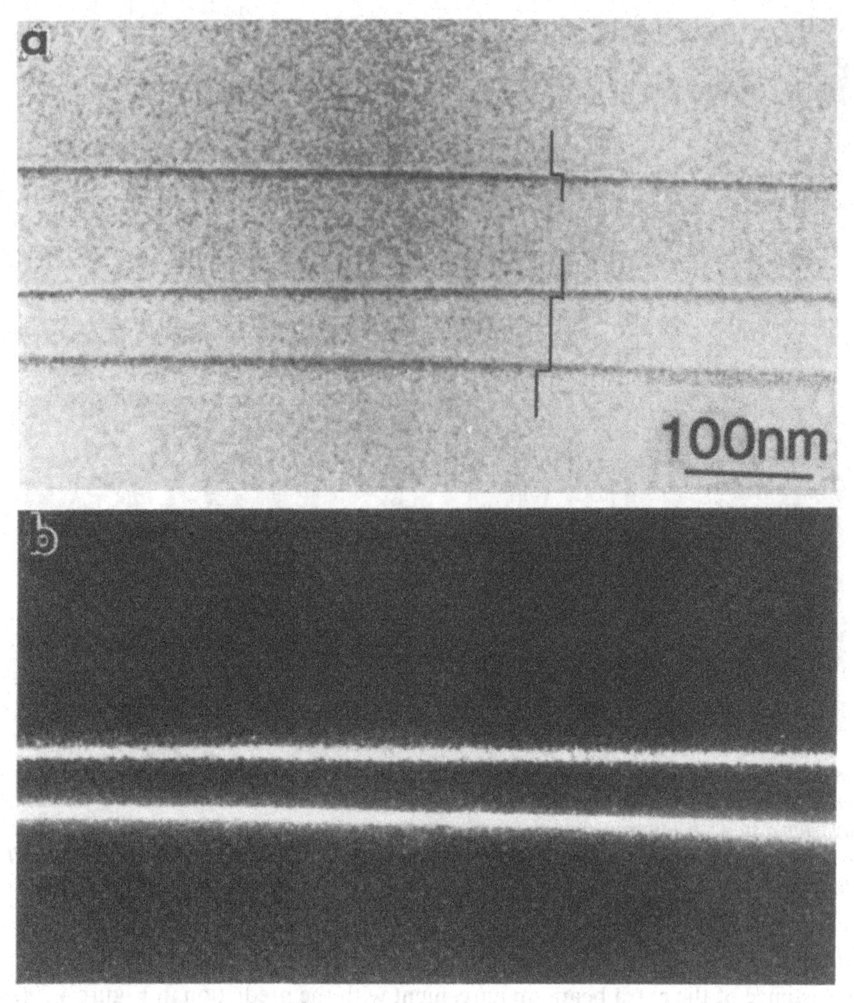

Figure 4.21 *In (a) and (b) are shown the surface dark-field images of GaAs(110) taken by using the (660) and the extra beams, respectively. Both up and down steps are shown in (a), but only down steps are imaged in (b) (Wang et al. 1989a).*

that the observed extra beam comes from the contribution of the surface down steps but not the up steps. The contrast results from the interruption effect of surface resonance waves (or the extra beam) propagating inside the crystal by the down steps (but not the up steps) (Figure 4.21(b)). A step that disappeared in Figure 4.21(b) in comparison with Figure 4.21(a) is an up step; up steps do not interrupt the continuous propagation of the surface resonance waves inside the crystal, and they do not contribute to the intensity of the extra beam, thus showing no contrast in the SREM image (Figure 4.21(b)).

The scattering process of a high-energy electron beam at a crystal surface under

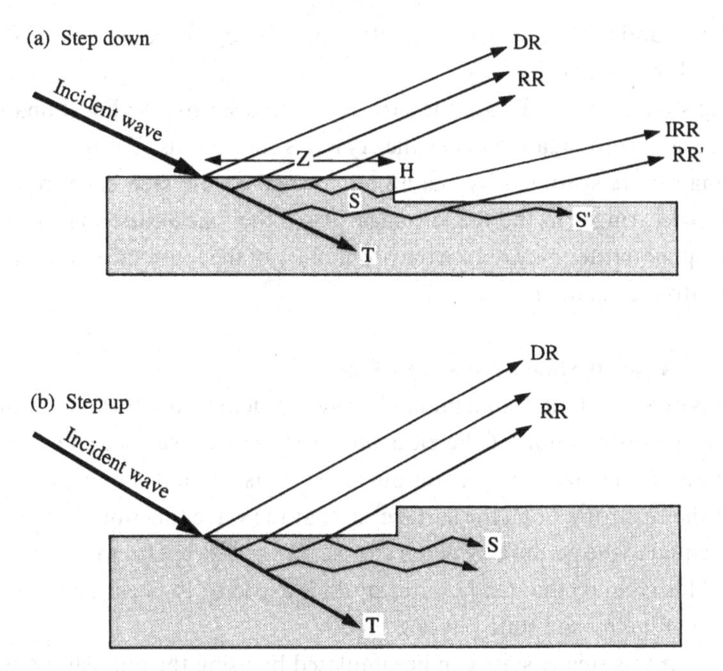

Figure 4.22 *A schematic diagram showing the scattering process of a high-energy electron beam in the RHEED geometry under surface resonance conditions at surfaces with (a) a down and (b) an up step. H is the height of the down step; z is the distance from the beam cut-off point to the position of the step.*

the surface resonance conditions is summarized in Figure 4.22. When an incident beam strikes the surface, it is separated into three streams. One stream is directly reflected from the surface without penetrating into the solid, and is named the directly reflected wave (DR). The second stream is the transmission wave (T). The third stream is the surface resonance wave (S), which travels within a few atomic layers of the surface. The propagation of this wave will be perturbed by the presence of a down step (Figure 4.22(a)). Some of the components of this wave leave the crystal at the step edge, producing a down-step-interrupted resonance reflection (IRR); the remaining components may continue to propagate along the surface, depending on the height of the surface step. The resonance waves localized at different depths from the surface (s and s') may have different propagation directions, so that the resonance reflection produced after reaching the step (RR′) may be different from the other resonance reflections created before reaching the step (RR). The resonance propagation of electrons along the surface before being reflected into vacuum is similar to the Goose–Hanchen effect in optics, and it has been applied to interpret the observed intensity gap between the twin images of a foreign object on the surface in REM (Kambe, 1988). The dynamical calculation of

Lu *et al.* (1991) and the experimental observation of Peng and Cowley (1988b) also support the above conclusion.

For an up step (Figure 4.22(b)), the surface resonance wave is almost unaffected by the step, and its propagation inside the crystal is continuous. The reflected wave, however, may be re-scattered by the atoms located at the step edge, producing possible Fresnel fringes in the REM image. Also, the 'backscattering' of the top layer at the upper terrace may reduce the probability of the resonance wave escaping the crystal into the vacuum.

4.5.4 A steady state wave at a surface

In REM, it is expected that, for an infinitely wide incident beam (this is the situation in practice), the distribution of the electrons at the crystal surface should not be affected by the beam size after the output position is beyond the mean traveling distance of the electrons along the surface, so that the wave function should have a periodicity equal to the periodicity of the atomic unit cell except for a constant phase difference. This is called the steady state, and it is expected to occur on the basis of the Bloch theorem in solid state physics.

To show that this steady state can be simulated by using the multislice theory, a beam of size 8 nm in width was chosen. The beam illuminates a surface area up to 241.7 nm (855 slices for GaAs (880) reflection with $\theta = 33.1$ mrad) along the beam direction, which is much larger than the mean traveling distance of the electron. The electron intensity distributions are output at several successive slices to check the repeatability of the wave fields. Since the atomic unit cell of the GaAs crystal is cut into two slices ($\Delta z = 0.2827$ nm), each of them having different atomic arrangement, it is expected that the intensity distribution of the electrons should repeat every other slice.

Figure 4.23 shows the calculated results. Detailed comparison shows that the intensity distribution at the 400th slice is identical to that at the 402nd slice but not to that at the 401st slice. The intensity distribution at the 401st slice is identical to that at the 403rd slice but not to that at the 402nd slice. Thus the electron intensity distribution does have the periodicity of the atomic unit cell (two slices = 0.5654 nm = the GaAs lattice constant) in the direction of the beam. To confirm the calculation results of Figures 4.23(a)–(d), the intensity distributions at the 600th and 601st slices are shown in Figures 4.23(e) and (f). Compared with the output at the 400th and 401st slices, the identical intensity distributions have been repeated in the 600th and 601st slices respectively. Therefore, these calculation results show that the steady state can be achieved in the calculation by using the multislice theory. The steady state is reached at about the 300th slice, because the steady state can happen only after the detection position is at a distance from the first slice that is longer than the electron mean traveling distance. Thus, the perturbation effect of the sharp window

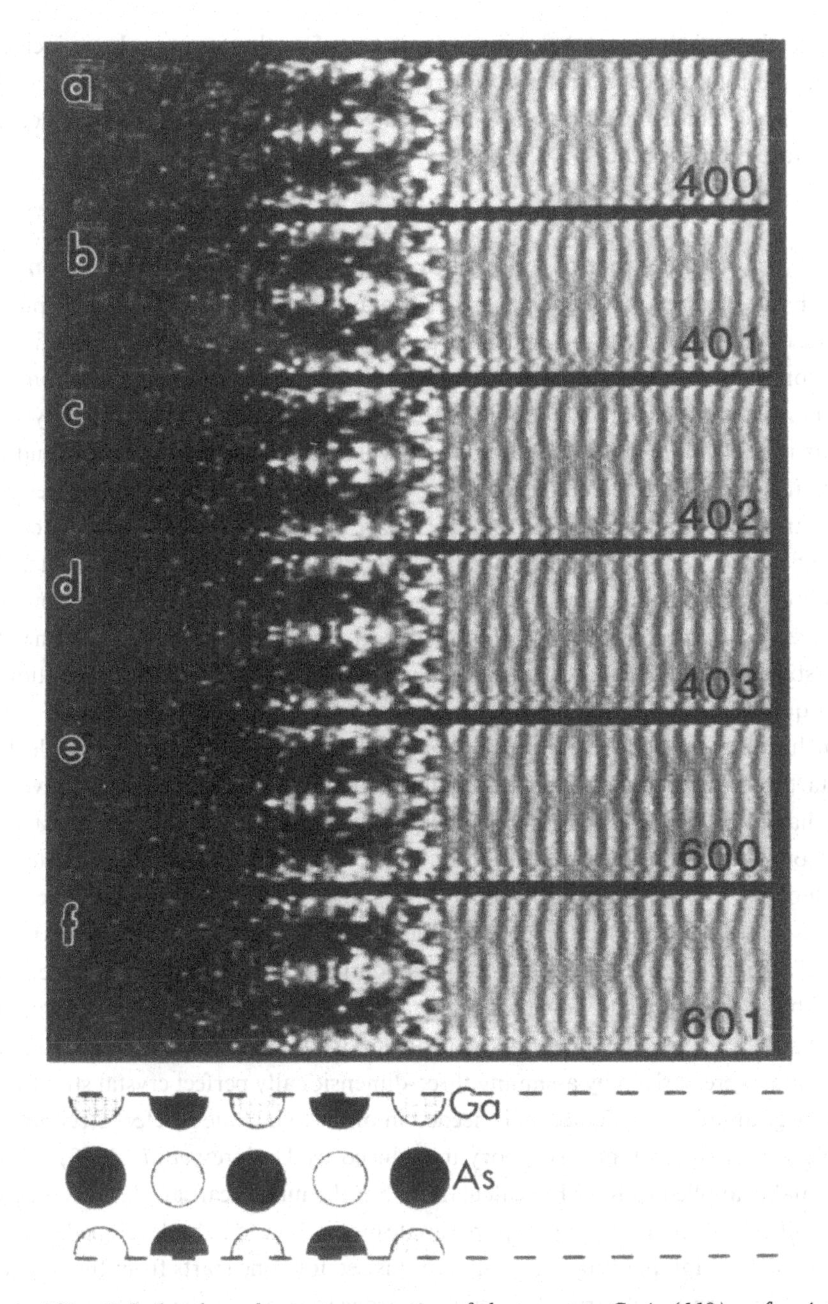

Figure 4.23 *Calculated steady state propagation of electrons at a GaAs (110) surface in the RHEED geometry, with $\theta = 33.1$ mrad and $\theta = 0$. There is no beam cut-off in the output regions.*

edge of the incident wave becomes insignificant. For the case of a large incident beam, the total reflection intensity increases almost linearly after arriving at the steady state. This calculation has demonstrated the great power of perpendicular-to-surface multislice theory.

4.6 The effect of valence excitation in resonance reflection

All of the theories in Chapter 3 and calculations have considered only purely elastically scattered electrons. In RHEED, however, more than 50% of the reflected electrons have lost energies of about 10–40 eV due to valence-band excitations (or plasmon excitations for metals and semiconductors). The spectroscopic analysis of inelastic electrons in the RHEED geometry will be described in Chapters 10 and 11. Here, it is intended to investigate the effect of inelastic valence excitation on electron resonance reflections in RHEED. Unfortunately, it is always difficult to calculate the diffraction behavior of inelastically scattered electrons, especially in RHEED, because the incoherence of the inelastic electrons makes the computation too involved to be performed. For this reason, a simplified theory to examine the inelastic effect in RHEED calculations has been used. Although the calculation is semi-quantitative, a few important physical processes have been illustrated.

Inclusion of inelastic scattering in dynamical diffraction theory, in principle, has to start with the time-dependent Schrödinger equation, with consideration given to the changes in the states of different energies during the interaction of an external electron with the nucleus and electrons in the crystal. This theory was initiated by Yoshioka (1957) and later developed by many authors for the transmission electron microscopy case of an electron penetrating through a crystal (for a review see Wang (1995)). Then a set of coupled equations was derived. Even though several methods have been postulated to solve these equations, concern still remains about the principles, because the many-body problem is involved. Also, these coupled equations were derived by assuming three-dimensionally perfect crystal structures, so a large amount of calculation is needed in order to include a defect structure. A single-particle density matrix theory introduced by Dudarev *et al.* (1993) can be reasonably applied to RHEED calculation, but the numerical calculation could be very difficult. It seems that some approximate methods need to be developed in order to give some semi-quantitative results. In this section, one starts from the physical optics approach for electron diffraction in order to consider the inelastic plasmon scattering in a simple way.

4.6.1 A simplified theory

The perpendicular-to-surface multislice method was first introduced by Cowley and Moodie (1957) based on the physical optics approach. Before we introduce inelastic scattering into the multislice calculation, it is necessary to review the basic equations of this theory. When considering a mono-energetic stream of electrons traveling in the direction of the z axis inside a crystal potential $V(\mathbf{r})$, the kinetic energy of the electrons can be found as

$$E_k = eU_0 - (-eV). \tag{4.10}$$

Then the wavelength of the electrons is

$$\lambda' = \frac{h}{[2m(eU_0 + eV)]^{1/2}} = \frac{\lambda}{(1 + V/U_0)^{1/2}}, \tag{4.11}$$

where λ is the wavelength of the electron in the vacuum. The phase of the electrons, after traveling a distance Δz from point z, will be changed, relative to that in the absence of the field, by an amount $\pi \Delta z V/(\lambda U_0)$ (for $U_0 \gg V$). The modification of the electron wave function due to the potential field is represented by multiplying the wave function by

$$Q(x, y, z) = \exp(\mathrm{i}\sigma \Delta z\, V), \tag{4.12}$$

which is defined as the transmission function of the slice, and $\sigma = \pi/(\lambda U_0)$.

It is important to note that the key step in deriving the multislice theory is to consider the result of the perturbation of the crystal potential V to the electron wavelength. Similar methods can be adopted to consider the perturbation to the electron wavelength due to its energy loss (Wang and Lu, 1988). The position-dependent total energy loss of the electrons due to plasmon excitations will be determined in Chapter 10. The total energy loss of the electron after traveling a distance z is (see Section 10.3)

$$\Delta E(x, z) = \int_0^z \mathrm{d}z \int_0^\infty \mathrm{d}\omega\, \hbar\omega\, \frac{\mathrm{d}^2 P(\omega, x)}{\mathrm{d}\omega\, \mathrm{d}z}, \tag{4.13}$$

where $\mathrm{d}^2P(\omega, x)/\mathrm{d}\omega\,\mathrm{d}z$ is the differential excitation probability per unit distance of the valence band. Here z is a function of x, depending on the scattering trajectory of the electron. The average kinetic energy of the electron is

$$E_k = eU_0 - (-eV) - \Delta E = eU_0 - [-e(V + V_{\mathrm{ef}})], \tag{4.14a}$$

$$V_{\mathrm{ef}} = -\Delta E/e. \tag{4.14b}$$

Therefore, the inelastic valence excitation in the RHEED geometry can be represented by an 'effective' potential, V_{ef}, decreasing the kinetic energy of the incident electrons (i.e., perturbing the wavelength of the electron) and resulting in an extra term in the slice transmission function. In this approach, the modified transmission function (Eq. (4.12)) is replaced by

$$Q' = \exp(i\sigma \Delta z\, V + i\sigma V_{ef} \Delta z), \tag{4.15}$$

where the second term is a phase perturbation function arising from the inelastic energy loss. Since the average energy loss of the electrons is determined by the distance of the electrons from the surface as well as the mean distance that the electron travels along the surface, V_{ef} is actually a positionally dependent function, which is equivalent to an electrostatic force that affects the motion of the electrons near the surface.

This treatment is based on an assumption that all the electrons are equally decreased in energy due to inelastic scattering. This is a qualitative classical physics description and may not be a rigorous quantum mechanical approach, but some useful results can be obtained.

4.6.2 The effect on surface resonance

The following dynamical calculations were made using the phase grating function introduced in Eq. (4.15) in order to show the effect of valence excitation in RHEED. Shown in Figure 4.24 is a set of calculated RHEED intensities across the (660) specular reflection spot of GaAs (110) for different incident angles. The reflected intensity (I) is normalized with respect to the total intensity of the initial incident beam (I_0), so that the magnitudes of I/I_0 can be compared with each other. The reflection intensities calculated by using the effective transmission function Q' (Eq. (4.15)) and using the pure elastic scattering transmission function Q (Eq. (4.12), without absorption) are represented by the dashed and the solid lines, respectively.

At the exact Bragg angle incidence, $\theta = 26.5$ mrad for (660) (Figure 4.24(a)), the intensity calculated by using the total scattering theory is about 40% stronger than that by using the pure elastic scattering theory. When the incident angle is decreased to 24.5 mrad (Figure 4.24(b)), both theories give about the same output except that a slight peak shift (0.8 mrad) is visible.

When the incident angle is changed to 23.5 mrad (Figure 4.24(c)), a large increase in the intensity calculated by using the total scattering theory is seen. This shows that the inelastic scattering can greatly enhance the reflection at some specific angles. It was reported by Wang *et al.* (1989b) that the surface potential trapping resonance occurs only for the 23.5 mrad incident case. This indicates that the incidence at 23.5 mrad corresponds to the strongest resonance excitation of the surface; inelastic scattering then dominates the resonance process.

With the decrease in incident angle (Figure 4.24(d)), the reflection intensity calculated using Q' decreases quite significantly. When the incident angle reaches 21.5 mrad (Figure 4.24(e)), the reflection intensity calculated using Q is much stronger, which is a change in the opposite direction to that for Figure 4.24(c). This suggests that the inelastic scattering and the elastic scattering have their strongest

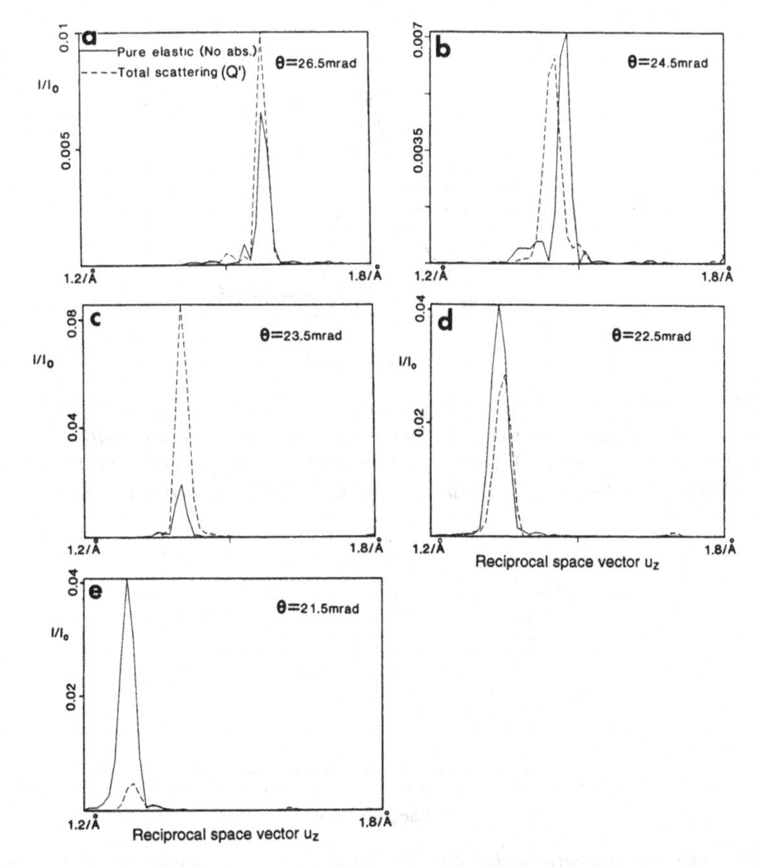

Figure 4.24 *The calculated GaAs (110) RHEED intensity perpendicular to the surface and across the specularly reflected beam as a function of beam incident angle. The solid and the dashed lines represent the results calculated by using the pure elastic scattering theory (without absorption, Q) and the total scattering theory including valence loss (Q'), respectively. The displayed intensity I/I_0 has been normalized with respect to the initial incident beam.*

resonance states at different incident angles. Hence the reflected electrons with and without energy losses are not expected to be distributed evenly in the (660) reflection disk.

The calculated reflectance for the specular reflection spot is plotted in Figure 4.25 for different incident angles according to the results in Figure 4.24. Several features are shown in Figure 4.25. First, the inelastic scattering can quite significantly increase the reflectance of a surface. The maximum observed reflectance is about twice as large as that calculated for pure elastic scattering without absorption (Wang, 1989b). Secondly, a relative shift of the peak positions, between the results calculated by using the total scattering theory and the pure elastic scattering theory, is visible. This shift is a reasonable result of considering the effects of energy loss and the effective potential (Wang, 1989b). It follows from the non-uniform excitation of

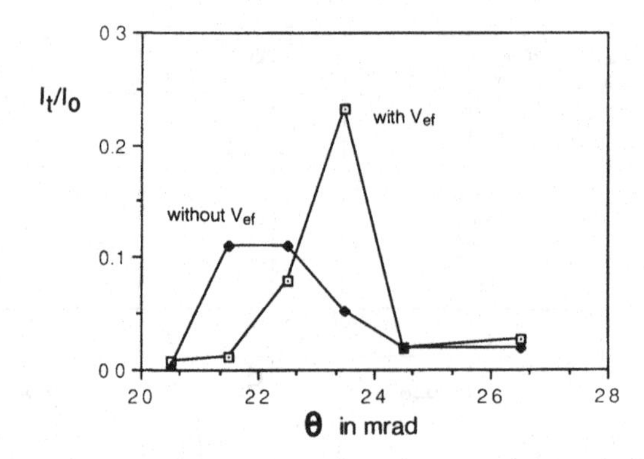

Figure 4.25 *The calculated intensity of the specularly reflected beam of GaAs(110) for 120 keV electrons with and without consideration of the V_{ef} term in the slice phase grating function as a function of incident beam angle. This shows the enhancement of reflected intensity by inelastic scattering (Wang et al. 1989b). The incident beam intensity is assumed to be unity.*

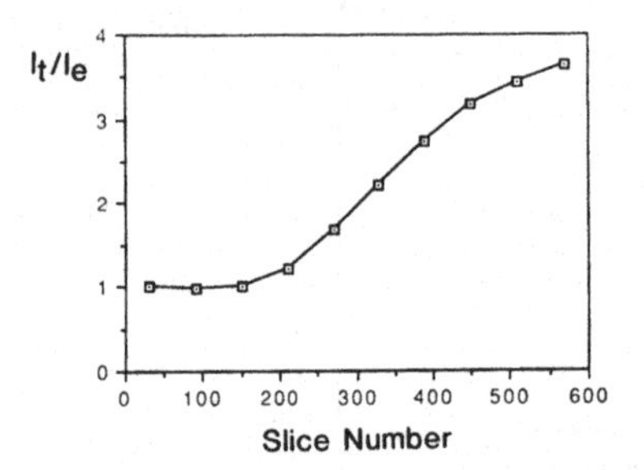

Figure 4.26 *The intensity ratio of the (660) specularly reflected beam calculated by using the perpendicular-to-surface multislice theory using phase grating function Q' (I_t) and Q (I_e), as a function of depth along the surface. The data are from the calculation in Figure 4.24(c).*

the elastic and inelastic scattering in RHEED that the inelastic scattering can change the average direction of propagation of the reflected electrons. The last feature is that about 24% of the incident electrons may be specularly reflected under the surface resonance condition. In off-resonance cases, only about 4–5% of the electrons are specularly reflected. The reflected intensity of the specular (660) spot is expected to increase by a factor of about six when set at the resonance condition compared with when set off resonance.

The multislice calculation makes it possible to trace the build-up process of the reflection wave. Figure 4.26 is a plot of the calculated intensity ratio of the specular reflection, calculated by using $Q'(I_t)$ and $Q(I_e)$ (without absorption), versus the slice positions. When the electrons propagate for short distances (less than 100 slices), the inelastic scattering effect can be neglected. When the propagation distance is close to the electron inelastic mean free path, in the range 70–100 nm, the inelastic scattering becomes significant, resulting in an increase in the ratio I_t/I_e. When the traveling distance is larger than the mean traveling distance D (about 50–70 nm) (Wang *et al.*, 1989a) along which most of the electrons travel before being reflected back into the vacuum, the ratio I_t/I_e begins to attain its steady level.

In RHEED geometry, there are two contributions to the valence excitation (Wang and Egerton, 1989). One contribution is the excitation of the plasmons when the electrons are approaching and departing from the surface in vacuum (direct reflection), which is a fixed term for a fixed incidence angle and does not depend on the propagation status of the electrons inside the crystal (see Section 10.4 for details). Thus, there is always a constant plasmon excitation component due to this term. The other contribution is the excitation of the plasmons when the electrons are resonantly propagating along the surface. The variation of this contribution reflects the resonance propagation of the electrons at the surface. This has been demonstrated by Wang *et al.* (1989b) as a criterion for identifying the occurrence of 'true' resonance, as shown below.

The reflected intensity obtained under the true resonance reflection condition is compatible with that obtained under the false resonance reflection condition but with a more significant amount of the inelastic component (Wang *et al.*, 1989b). This effect introduces some uncertainty in identifying the occurrence of surface resonance purely based on RHEED patterns, but an EELS analysis may help.

Shown in Figure 4.27(a) is a (660) spot obtained with a small objective aperture. Two intensity domains are seen to dominate the (660) disk, separated by about 3–4 mrad and corresponding to the reflection for different incidence angles. The inelastic excitation of the surface can be seen through the REELS spectra acquired from domains A and B, as shown in Figure 4.27(b). The spectra are displayed by normalizing the heights of the zero-loss peak, so that the relative increase in the plasmon-loss part can be considered as the relative increase in the inelastic excitations. In the spectrum acquired from domain A (the dashed line), the ratio of the total inelastically scattered electrons to the elastically scattered electrons is greater by about 50% than that for domain B (the solid line). Also, a stronger multiple inelastic scattering is visible in the tail part of curve A. The results in Figure 4.27(b) can be considered as corresponding to the theoretical predictions shown in Figures 4.24(c) and (e). Then the reflection in domain A is mainly dominated by inelastic resonance scattering; the reflection in domain B is mainly dominated by elastic resonance scattering after subtracting the contribution made by 'direct' reflection. By 'elastic resonance' we mean that the reflection intensity is enhanced but with less inelastic excitations. By 'inelastic resonance' we mean strong excitation of the plasmon losses together with enhancement of the reflection intensity, which is postulated to be due to propagation of the electrons along the surface, and can be called the 'true' surface resonance.

Whether domain A or B is dominated by elastic or inelastic resonance scattering

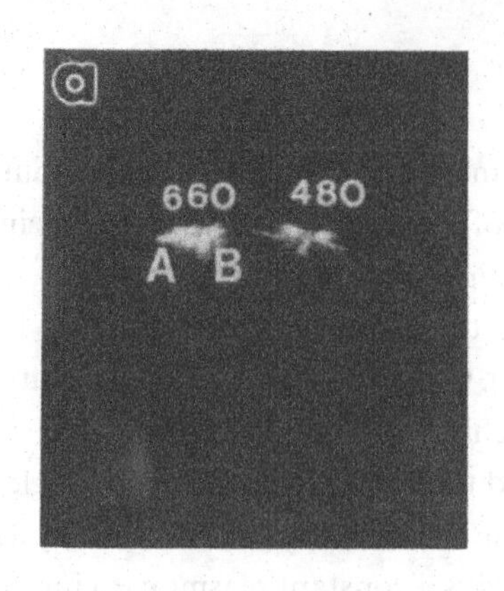

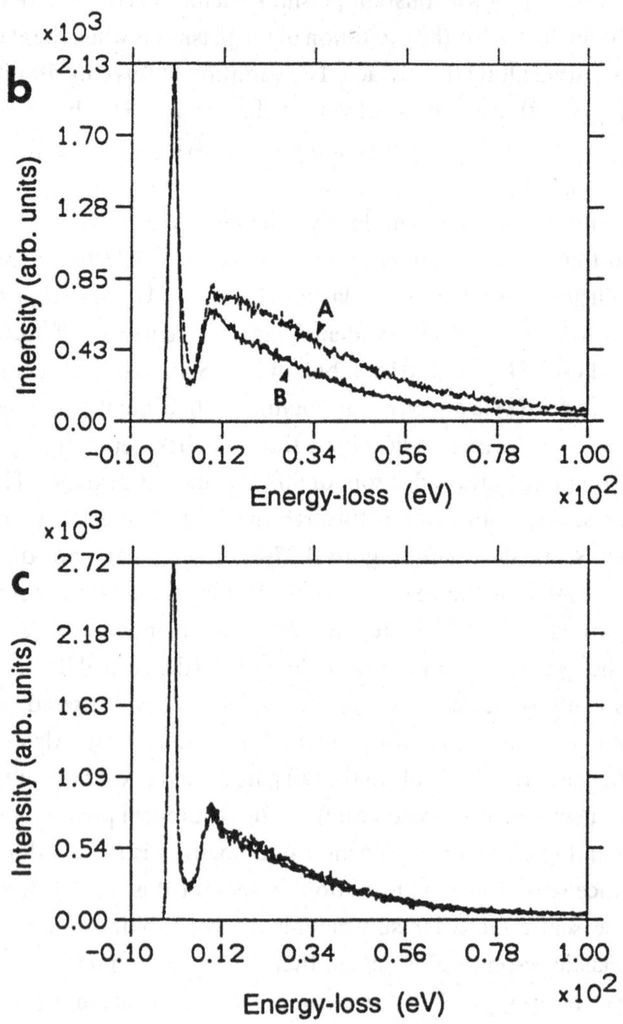

Figure 4.27 *(a) A RHEED pattern from a GaAs (110) surface recorded in a VG HB5 STEM (100 kV). (b) A comparison of the spectra acquired from the domains A and B in (a). (c) A comparison of the spectra acquired from domains A and B but under slightly different resonance conditions.*

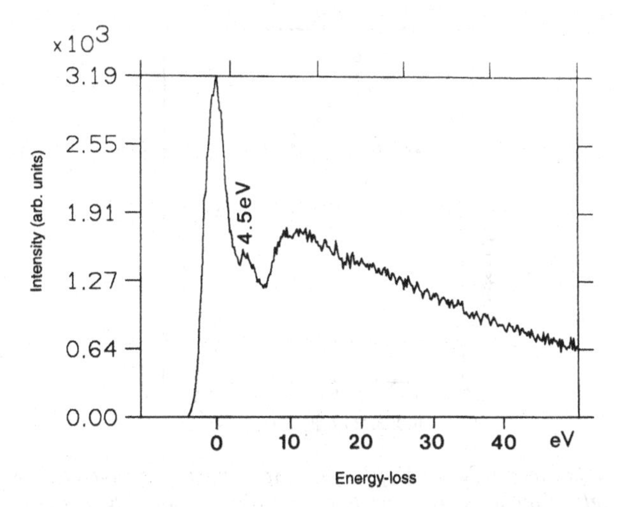

Figure 4.28 *REELS observation of a resonance energy-loss peak of the GaAs(110) surface. The 4.5 eV peak does not appear in the spectrum if there is no resonance (Wang et al. 1989b).*

depends critically on the initial incidence conditions. If the beam is tilted by about 0.5 mrad away from the orientation of Figure 4.27(a) in either direction, then the inelastic excitation behaviors in domains A and B are switched. Also, in some experimental cases, the diffraction conditions appear to be almost exactly the same as those shown in Figure 4.27(a), but the inelastic excitation results are totally different from that shown in Figure 4.27(b). Figure 4.27(c) shows an example of this case. No obvious difference is seen in the spectra acquired from the two domains. These experimental results indicate that, for small changes of incident angle, the diffraction intensity may not change too much but the excitation processes may be totally different.

As shown in Figure 4.27(b), the resonance propagation of the electrons at a crystal surface can significantly increase the inelastic excitations of the surface. It is expected that the resonance energy loss of the electrons could be seen in this process. By resonance energy loss we mean that the extra energy loss of the electrons is due to their resonance propagation along the surface. Figure 4.28 shows a REELS spectrum acquired from the (660) beam under the true resonance reflection. Besides the single (11 eV) and the multiple surface plasmon excitations, an extra peak, located at 4.5 eV, is seen. This peak disappears if the incident condition is slightly (about 0.5 mrad) off resonance. The observed 4.5 eV peak may be attributed to excitation of the surface bound state as described below. According to the bound states model proposed by Marten and Meyer-Ehmsen (1985), the resonance effect can be explained in terms of the electrons diffracting into bound states of a single atomic layer parallel to the crystal surface and channeling along the layer before they are re-diffracted into the vacuum. These states are purely the resonance states

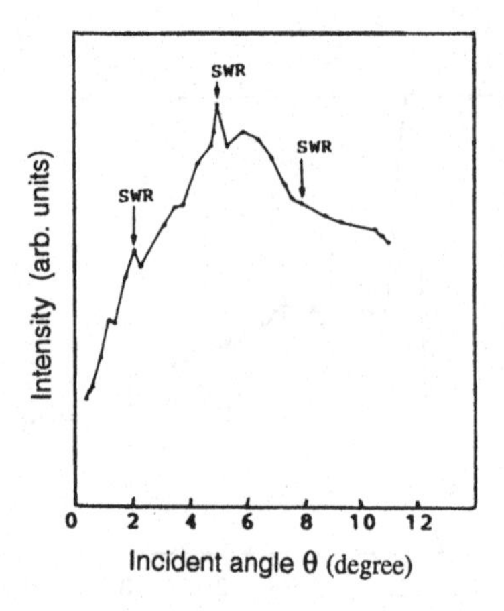

Figure 4.29 *A plot of the Si[2p, V_a, V_a] (V_a = 3s, 3p) Auger spectrum intensity as a function of the glancing angle of the incident electron beam. 'SWR' denotes the position of the surface wave resonance electron-beam incidence. The resonance in the Auger intensity originates from the confinement of the surface electron waves excited under the surface resonance condition. The Auger intensity is divided by the specimen absorption current to avoid variation in the beam current of the primary electrons (Nakayama et al., 1991).*

and have nothing to do with electronic surface states of the crystal. The resonance wave is likely to be confined to the topmost surface layer for relatively low incident energy, such as 19 keV (Mayer-Ehmsen, 1987; Marten and Meyer-Ehmsen, 1988). Auger emission and secondary electron emission measurements from Pt(111) during rocking curve RHEED experiments at 19 keV have shown two bound resonance states (Marten, 1987). Thus, it is inelastic scattering that pushes the channeling electrons out of the bound states, and then the electrons are re-diffracted before being reflected from the surface. Therefore, the occurrence of true resonance is not only an enhancement of the reflected intensity but also the excitation of these bounded states. The observed 4.5 eV peak is evidence for this model. The surface resonance theory of McRae (1979) is also based on the surface resonance states.

4.7 Enhancement of inelastic scattering signals under the resonance condition

Surface resonance refers to the strong beam being accumulated at the top surface layers. These channeled electrons near the surface tend to increase the productivity of secondary and Auger electrons. Since Auger electrons are emitted from a layer thinner than 1 nm near the surface, the measured Auger intensity as a function of incident electron angle can be applied to determine the angles at which the surface resonance wave is excited. Figure 4.29 shows the measured intensity of the Si[2s, V_a, V_a] (V_a = 3s, 3p) Auger spectrum around 90 eV as a function of the glancing angle of the incident electron beam. The Si[2s, V_a, V_a] Auger peaks mainly reflect the valence electronic states of the surface atoms. It is to be noticed that the Auger intensity is enhanced at several specific glancing angles, which correspond to the surface resonance condition. These results are directly confirmed by simultaneously observing the RHEED pattern intensity (Nakayama *et al.*, 1991). Off the resonance condition, the electron beam will penetrate into the bulk and excite the internal Si atoms. Inside the bulk, the Auger electrons are easily scattered by the surrounding internal atoms and cannot escape from the crystal to arrive at the detector, thus decreasing the Auger yields. As can be noticed in Figure 4.29, the Auger-intensity resonance itself is not so strong as the background Auger intensity. This is probably due to the finite localization effect of Auger excitation. For 90 eV Auger peaks, the localization is about 0.5–1 nm. The effect shown in Figure 4.29 may be used to detect impurity elements segregated at the surface (Spence and Kim, 1987). A technique called surface wave excitation Auger electron spectroscopy (SWEAES) has been developed for analyzing composition and electronic states of the top few surface layers (Nakayama *et al.*, 1990) (see Section 12.8).

Similar behavior has been observed by Marten (1987). Figure 4.30 shows the dependences of secondary electron emission yield (top), beam-rocking Auger electron signal (BRAES), RHEED intensity of the specularly reflected beam, and the diffuse scattering background in RHEED on the glancing angle of the electron beam. Figure 4.30 reveals that all the quantities detected are heavily influenced by and only by the surface resonance. Every major structure in the profiles clearly corresponds to a resonance in the rocking curves and, *vice versa*, with each resonance of the specular beam corresponding to a pronounced structure in the other curves. This study clearly shows that surface resonance can be used to improve the surface sensitivity of any technique associated with RHEED.

Besides valence excitations, thermal diffuse scattering (TDS) is another important inelastic process in RHEED. Thermal diffuse scattering refers to non-Bragg scattering of the electron due to a non-periodically perturbed crystal potential resulting from lattice vibration. The kinematical and dynamical theories of TDS in RHEED will be introduced in Chapter 9.

In this chapter, we have thoroughly examined the fundamental characteristics of the surface resonance phenomenon. This is probably the most remarkable effect in RHEED purely from the physics point of view, the study of which is extremely important for understanding the REM image contrast and imaging conditions. The established scattering process of high-energy electrons at crystal surfaces will serve as a fundamental picture for analyzing the spectroscopic data obtained in RHEED geometry (see Chapters 10 and 11).

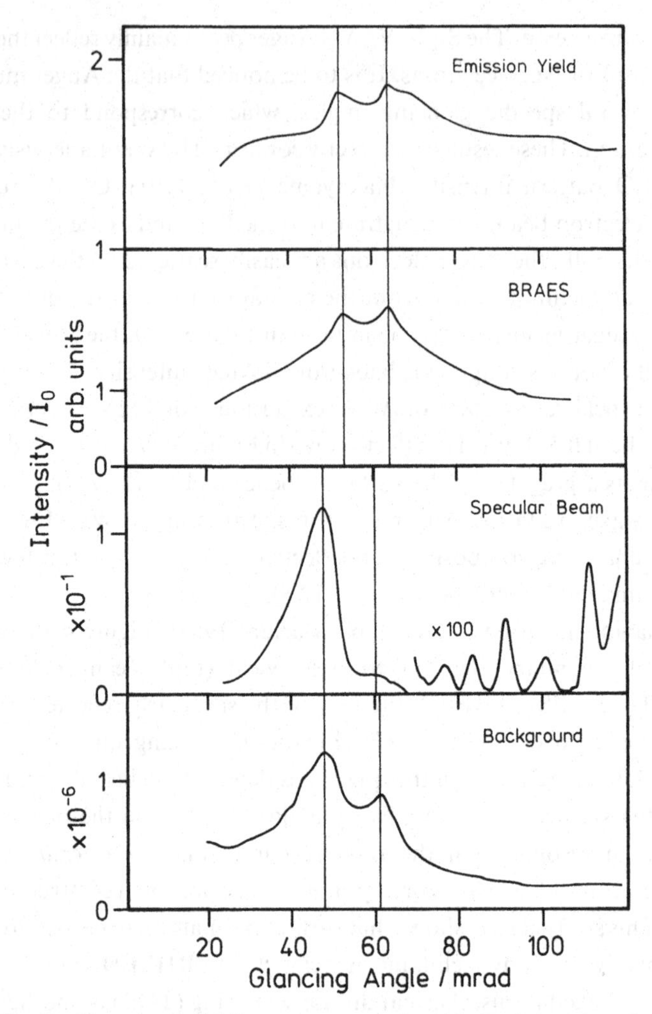

Figure 4.30 *Rocking curves of different quantities for 19 keV electrons incident at a Pt(111) surface. The beam azimuth is* $[2\bar{1}\bar{1}]$. *Positions of resonance maxima are indicated by a vertical line (Marten, 1987).*

PART B

Imaging of reflected electrons

CHAPTER 5

Imaging surfaces in TEM

5.1 Techniques for studying surfaces in TEM

There are two basic modes of TEM operation, namely the bright-field mode, wherein the (000) transmitted beam contributes to the image, and the dark-field imaging mode, in which the (000) beam is excluded. The size of objective aperture in bright-field mode directly determines the information to be emphasized in the final image. When the size is chosen so as to exclude the diffracted beams, one has the configuration that is normally used for low-resolution defect studies, the so-called diffraction contrast. In this case, a crystalline specimen is oriented to excite a particular diffracted beam, or systematic row of reflections. This imaging mode is sensitive to the differences in specimen thickness, distortion of crystal lattices due to defects, strain and bends. High-resolution imaging is usually performed in bright-field mode by including a few Bragg-diffracted beams within the objective aperture. The lattice images are the result of interference between Bragg reflected beams, the so-called phase contrast. In this section, we outline a few techniques that have been extensively developed for studying surfaces in high-resolution TEM (HRTEM) (Cowley, 1986; Smith, 1987).

5.1.1 Imaging using surface-layer reflections

The dark-field imaging technique is favorable for imaging surfaces when the surface layers have different periodicity from that of the bulk of the crystal (Figure 5.1(a)). Then 'superlattice' satellite reflections appear in the diffraction pattern and dark-field imaging with these spots can give high-contrast images of surface structure.

There are two basic mechanisms for generating the superlattice reflections. A sharp termination of the crystal bulk at the surface may form a stacking sequence that is not a complete unit cell. Thus the symmetry of the structure is reduced at the surface, resulting in the appearance of kinematically forbidden reflections in the diffraction pattern. The dark-field images of these forbidden reflections directly reflect the surface structure. Reconstruction of surface layers can also introduce superlattice reflections.

Figure 5.2 shows two dark-field TEM images recorded using the $\frac{1}{3}[4\ \bar{2}\ \bar{2}]$ 'forbidden' reflection reflecting the response of the Si(111) surface to oxidation (Ross and Gibson, 1992). This reflection occurs if the specimen thickness is a non-integral number of unit cells. Since the scattered intensity is a sensitive function of

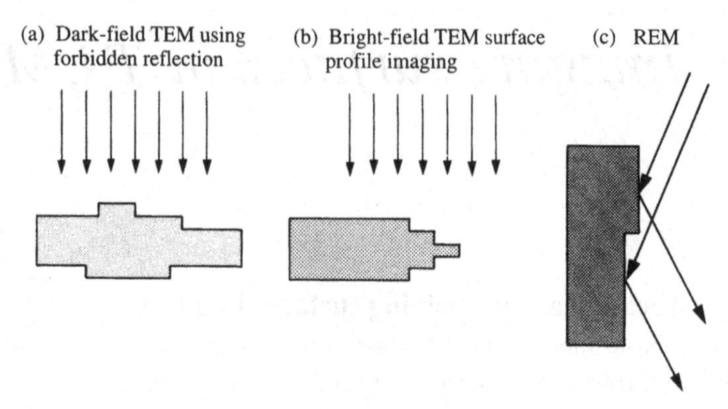

Figure 5.1 *Schematic diagrams showing a few techniques for studying crystal surfaces in TEM (see text).*

the number of atomic layers in the specimen, images formed using electrons scattered into these reflections show steps (on both top and bottom surfaces), an abrupt change in intensity marking step positions (Figure 5.2(a)). The intensity oscillates with a periodicity of one unit cell. Surface steps on the top and bottom surfaces show up as contrast changes. Figure 5.2(b) shows the same area after its having been oxidized to form SiO_2. This real-time *in situ* experiment shows that surface steps do not move noticeably during oxidation of several atomic layers of the silicon specimen (see the step indicated by an arrowhead). This, together with analysis of the changes in appearance of the terraces, demonstrates that oxidation is a terrace-attacking process and suggests that it occurs one monolayer at a time, with each monolayer reacting completely before the next is attacked (Ross *et al.*, 1994).

In general, the weak-beam imaging method has the advantage that, because it relies on the presence of a periodicity at the surface, it is relatively insensitive to the presence of any amorphous layer on the surface. However, this advantages applies only if a very small objective aperture is used to select the spot but exclude the background due to diffuse scattering. Hence the image resolution is limited to some multiple of the surface periodicity. Also, the images are very weak. The exposure times for recording the images are long, and the image resolution and contrast may be affected by specimen drift. The surface reflection can be enhanced by using a suitable diffraction condition. The surface layers are essentially two-dimensional, so that their distribution of scattering power in reciprocal space takes the form of rods perpendicular to the surface plane, where the bulk reflections are more nearly associated with sharp maxima around the reciprocal lattice points that will be strongly excited only when they intersect the surface of the Ewald sphere.

Because the scattering from the surface layers is relatively weak, the surface layer reflections from the bottom face of the thin foil are kinematical, but the surface reflection from the top face of the foil, which will penetrate the entire crystal before

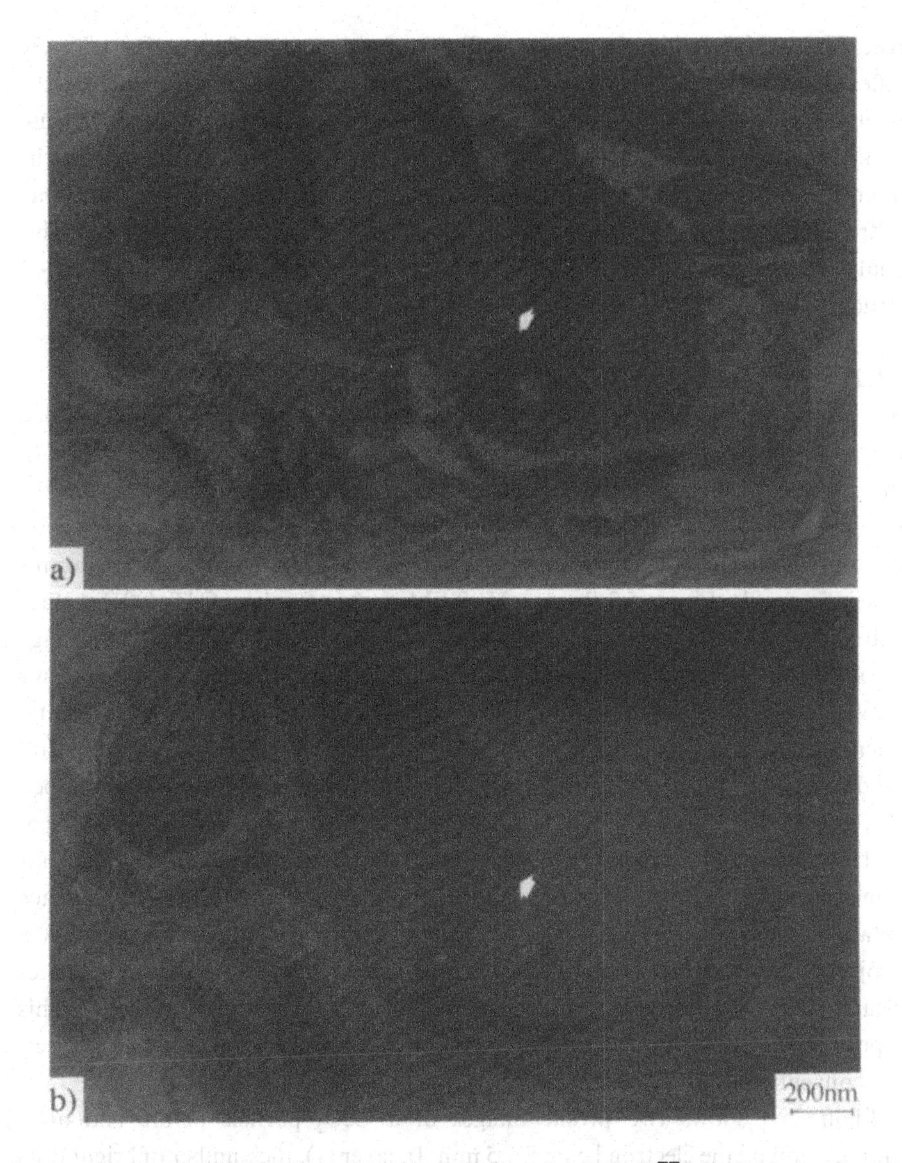

Figure 5.2 *Dark-field TEM images recorded using the $(1/3)[4\bar{2}\bar{2}]$ 'forbidden' reflection showing the step distribution on the Si(111) surface (a) before and (b) after being oxidized. The images were recorded at 200 kV using a modified JEOL 200CX UHV TEM. These real-time* in situ *images provide a clear picture regarding the oxidation process of the silicon {111} surfaces. It also shows the ability of the TEM method to look through buried (amorphous) layers (courtesy of Drs F. M. Ross and J. M. Gibson).*

reaching the diffraction plane, may suffer strong dynamical interaction with the reflections from the bulk crystal lattice. Based on the first-order approximation, the relative intensities of surface reflections could still be reasonably interpreted by use of kinematical diffraction theory if the specimen were thin. On this basis, the intensities of surface layer reflections can be used in the same way as X-ray diffraction intensities for structure analyses to derive relative atomic positions. This analysis has been carried out by Takayanagi *et al.* (1985) for the Si(111) 7×7 structure, to provide the first reliable model for this reconstruction.

5.1.2 Surface profile imaging

A powerful technique has been developed for the study of surface structures, which takes full advantage of the high-resolution capabilities of modern microscopes. In this, a small particle or the edge of a thin crystalline film is so oriented that the incident electron beam is parallel to a surface of the crystal and also parallel to rows of atoms in this surface plane (or zone axis). Then the arrangement of surface atoms is seen in profile. Figure 5.3 shows a HRTEM image of a clean CdTe(001) surface (Lu and Smith, 1991). The beam direction is [110] and $E_0 = 300$ keV. The image resolution of the microscope was about 0.2 nm. It is apparent that the surface atoms moved to new positions due to the formation of the surface. From the experimental micrographs, and based upon comparison with image simulations, a structural model shown in the inset for the (3×1) reconstructed (001) surface was developed (Lu and Smith, 1991). The (3×1) reconstruction involves both the formation of surface dimers and the presence of vacancies at the surface. Every third atomic pair is missing along the $[1\bar{1}0]$ direction, and the remaining two atom pairs at the surface rotate slightly toward each other to form the surface dimers. The image is a projection of the structure on a plane perpendicular to the surface and indicates clearly the distribution in directions perpendicular and parallel to the surface. This type of image can be simulated based on the well-established methods for interpreting conventional HRTEM images.

Figure 5.4 shows two profile images of a CeO_2 particle before and after illumination by the electron beam for 5 min. In layer (1), the number of bright dots increases from four to five. In layer (2), the number of bright dots increases from seven to nine, indicating the migration of unit cells on the CeO_2 surface. It is quite possible that this migration occurs only under the electron beam. This example clearly demonstrates that surface profile imaging is a powerful technique for observing *in situ* surface migration at atomic resolution.

5.1.3 REM of bulk crystal surfaces

The last two sections have shown two TEM techniques for imaging surface structures. The crystal has to be very thin for these techniques to be applicable.

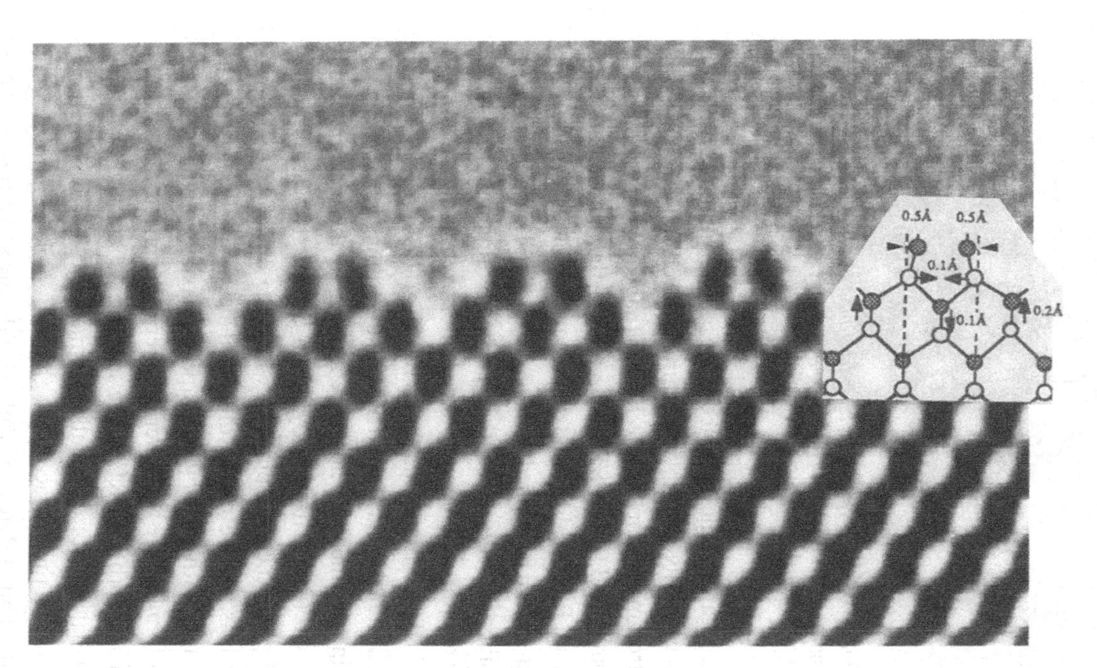

Figure 5.3 *A HRTEM profile image of the CdTe(001) surface in [110] projection showing the (3 × 1) reconstruction. The image was recorded* in situ *at a temperature about 240°C. The inset is a schematic model of the (3 × 1) reconstructed surface, as viewed along the [110] direction. Approximate atom displacements deduced from the experimental images are shown (courtesy of Drs P. Lu and D. J. Smith).*

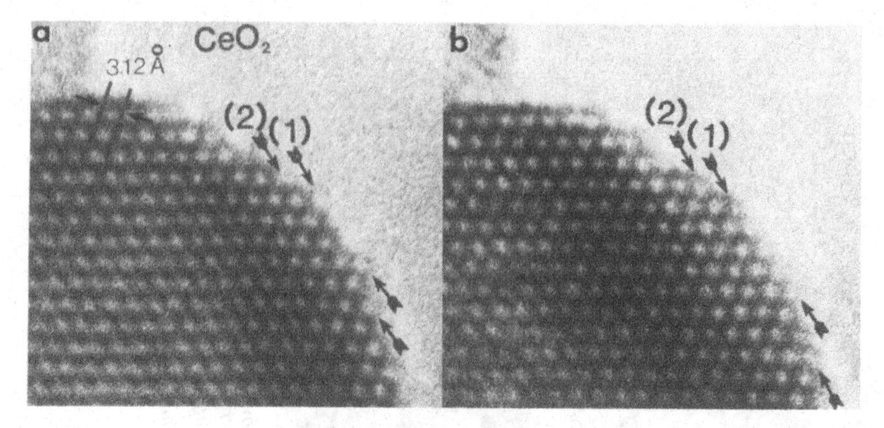

Figure 5.4 Surface profile images of a CeO_2 particle (a) before and (b) after being illuminated by the beam for 5 min, showing the migration of surface atoms at the layers indicated by (1) and (2). The electron beam energy is 200 kV.

Thus, a question may arise regarding the influence of the sample thinning process on the surface structure. It is generally desirable to examine the crystal surface before thinning. Reflection electron microscopy is a technique that can fulfill this task.

REM is usually performed in TEM or STEM because of the requirement for a high-resolution optic system. REM is different from the two methods introduced in Sections 5.1.1 and 5.1.2, and it can be applied to image bulk crystal surfaces in TEM. Thus, the artifacts possibly introduced in the specimen thinning process are minimized. The REM technique is easily introduced after examining the conventional operation modes of a TEM. Figure 5.5 shows ray diagrams corresponding to TEM bright-field (BF), TEM dark-field (DF), REM, and scanning REM (SREM). For TEM imaging, a thin specimen is held perpendicular to the incident beam (Figure 5.5(a)). In TEM, electron scattering by a thin crystal gives a diffraction pattern in the back-focal plane (i.e., the plane in which the objective aperture is located) of the objective lens. If one of the diffracted beams is selected by the objective aperture, preferably with the objective aperture axially aligned and the incident beam tilted so as to minimize the spherical aberration effect, then a dark-field image is formed (Figure 5.5(b)). In REM, the specimen is replaced by a thick bulk crystal and the near-flat surface of a bulk specimen is oriented so that the incident electron beam strikes the surface at a glancing angle (Figure 5.5(c)), resulting in a RHEED pattern on the back-focal plane of the objective aperture. The dark-field (DF) image formed by selecting a reflected beam with the objective aperture is a surface image. Intensity variations in the image may be attributed to variations of the diffraction intensity resulting from changes of crystal structures, surface chemical composition, crystal lattice orientation, or phase contrast arising from a variation of electron path lengths due to surface morphology. An alternative

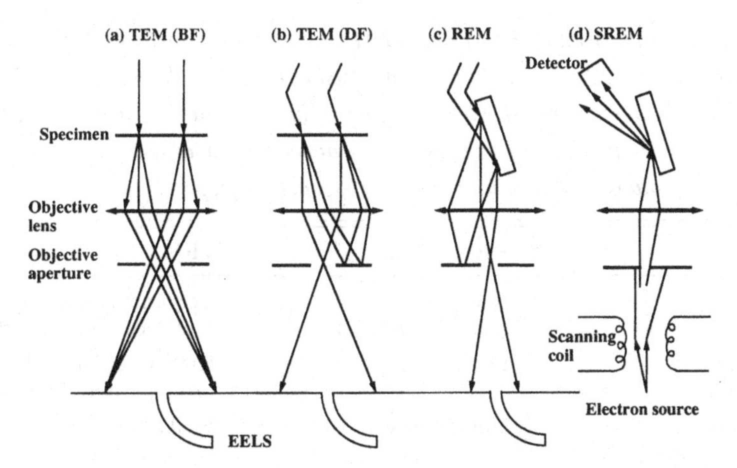

Figure 5.5 *Ray diagrams illustrating the image formation for (a) bright-field TEM with the transmitted and diffracted beams, (b) dark-field TEM with the incident beam tilted and a small objective aperture used to allow only one diffracted beam to contribute to the image, (c) REM, with a geometry similar to (b) but with the diffracted beam produced by the reflection of a bulk crystal surface, and (d) scanning REM (SREM) in STEM, with a small electron probe scanning across a bulk crystal surface and detecting the reflected intensity as a function of scanning position. The deflection of the beam is produced by deflection coils in the microscope.*

to REM is SREM, performed in a dedicated STEM instrument (Figure 5.5(d)), in which the electron beam from a field emission gun is focused to a small probe at the specimen level and scanned over the specimen with the use of deflection coils. In any subsequent deflection plane, a convergent RHEED pattern of the region of the surface illuminated by the beam is formed for each scanning position. Some part (or parts) of the RHEED pattern is (are) detected and the resulting signal is displayed on a cathode ray tube with a scan synchronized with that of the incident beam on the surface. Based on the reciprocity theorem, SREM is equivalent, but not identical, to REM. The REM experiments can be performed in UHV TEMs or conventional TEMs with accelerating voltages ranging from 100 kV to 1 MV. In addition to the imaging capability of REM, the energy-loss distribution of the reflected electrons can be analyzed by electron energy-loss spectroscopy (EELS), so that the electronic and chemical structures of the surfaces can be examined (see Chapters 10 and 11). A comparison of REM and TEM for surface studies is listed in Table 5.1. Since REM and SREM are equivalent in many aspects, the theories described for REM can be adequately applied to SREM.

5.2 Surface preparation techniques

In general, surfaces of single-crystalline materials are most easily studied by REM. The specimen preparation is the most critical step. Listed below are a few routine

Table 5.1. *A comparison of REM and TEM techniques for surface studies. In TEM, surface studies are usually performed using the forbidden reflection (FR) imaging and the surface profile (SP) imaging techniques. In the former case, the forbidden reflections of a perfect crystal may appear in the diffraction pattern if the surfaces are composed of incomplete unit cells, such as steps.*

	REM	TEM
Specimen	Bulk (very flat surface)	Very thin films (not necessarily very flat)
Diffraction	Dynamical	Kinematical for FR imaging Dynamical for SP imaging
Image simulation	Not quite available	Well established
Image		
Image distortion	Foreshortening	None
Resolution	0.33 nm parallel to surface; < 0.1 nm perpendicular to surface	0.12–0.16 nm for SP imaging
Contrast	High	Low for FR imaging
Sensitivity to diffraction conditions	Very sensitive	Sensitive
Subsurface distortion	$\simeq 10^{-4}$ (normal)	$\simeq 10^{-2}$ (lateral)
Sensitivity to surface steps	Very sensitive	Sensitive
Out of phase domains	Easy to see	Easy to see
Signal intensity	Strong	Weak for FR imaging Strong for SP imaging
Chemical analysis	EELS and AES	AES
Valence excitation	Strong	Weak
Thermal diffuse scattering	Strong	Weak

methods for preparing specimens used in non-UHV REM studies. The ultimate goal of specimen preparation is to obtain atomically flat or nearly flat surfaces that are suitable for REM studies. The surface must be flat enough to permit clear imaging of surface features using an electron beam with an incident angle smaller than about 3°. The nominal size of a REM specimen is limited by the size of the TEM specimen holder, which is usually designed for a grid of 3 mm diameter. The typical observable surface area is less than about 1–2 mm^2. There are four different methods for preparing REM specimens: cutting natural or as-grown surfaces, re-crystallization from melting, annealing polished surfaces, and cleaving bulk crystals. Each of these techniques will be introduced below (Hsu, 1992).

5.2.1 Natural or as-grown surfaces

In general, surfaces will be contaminated if they are exposed to air, but the amorphous contamination layer seldom affects the REM observation, because it does not affect the contrast produced by Bragg reflections of the bulk. Useful information has been gathered with REM from the as-grown surfaces of diamond, hematite, sapphire, and III–V semiconductors. Large samples may have to be cut into appropriate sizes with a blade or slow-speed diamond saw. Care should be taken to protect the surface while cutting. Embedding the sample in mounting wax before sawing can prevent damage. It is necessary to clean the surface with organic solvents, such as acetone and alcohol, after cutting. Non-reactive solvents are recommended in order to prevent surface reaction and damage.

5.2.2 Re-crystallization from melting

Very flat surfaces, such as Au(111), Au(100) and Pt(111), can be obtained by melting high-purity gold or platinum thin wire with a torch and then allowing the molten sphere to re-solidify into a single-crystal 'sphere'. A single-crystal sphere can normally be formed if the diameter of the wire is less than about 25 μm. A large temperature gradient across a thicker wire makes it difficult to produce a single-crystalline specimen. An acetylene–oxygen or hydrogen–oxygen torch is normally used. A single-crystal sphere at the end of the wire is formed if the molten tip is quickly cooled. The sphere shows many {111} facets, which are usually atomically flat and are ideal for REM observations. This method, however, has only been successful for a few noble metals.

5.2.3 Annealing polished surfaces

The single-crystal specimen is cut from a large piece. Then the crystal is smoothed first using 600 and 320 μm grinding papers in order to remove mechanical chips and cracks produced by sawing. Then, the faces to be studied are coarsely polished using 6 and 3μm diamond paste on a nylon polishing cloth. The specimen is cleaned before being polished with 1 and 0.25 μm diamond paste on a micro-cloth. 3–5 minutes usually suffices for each stage. Thorough cleaning with freon or acetone between stages is important. The final polishing can be performed using 0.05 μm SiO_2 or Al_2O_3 polishing suspensions, but one has to be sure that the remaining polishing powders are completely removed from the surface before annealing experiments in order to avoid possible surface reaction and contamination.

To prevent solid state reaction, a sample holder in a bridge shape is made using the same material as the specimen, and the specimen is loaded on top of the bridge to avoid its direct contact with any other objects of different chemical compositions. For ceramics, the specimen is annealed in air and the annealing temperature is controlled within the range 1100–1700 °C depending on the melting point of the specimen. It usually takes 10–40 h to smooth the crystal surface for REM observation. A detailed illustration of this preparation method has been given by Crozier *et al.* (1992) and Ndubuisi *et al.* (1992a).

Table 5.2. *Flat surfaces prepared by several methods for REM observations.*

Method	Surfaces
Natural or as-grown surface	CVD diamond; polished natural diamond; and MBE-grown semiconductors.
Re-crystallization from melting	Pt(111) and (100); Au(111) and (100); Cu(110); Ni(111).
Annealing	α-alumina (0001), $(1\bar{1}02)$, $(1\bar{1}00)$ and $(11\bar{2}0)$; MgO(100), (110) and (111); Si(100) and (111); $LaAlO_3(100)$ and (110); $TiO_2(001)$ and (100).
Cleavage	GaAs(110); InP(110); GaP(110); $Ti_2O(110)$; α-alumina$(01\bar{1}1)$ and $(01\bar{1}2)$; MgO(100) and (110).

5.2.4 Cleaving bulk crystals

Cleavage surfaces, such as MgO(100) and GaAs(110), are ideal specimens for REM observations. These surfaces are normally employed to examine some fundamental characteristics of REM, such as step and dislocation contrast. Cleaving can usually be performed with a razor blade or a diamond knife. Scribing one side of the crystal is usually sufficient. The cleavage method is recommended if one is interested in the atomic termination of the surface.

No surface coating is recommended for any REM specimens. The ionic oxides, which are insulators, are usually quite stable under the electron beam. The charging effect 'disappears' when the microscope magnification is greater than 3000 for most Philips TEMs. This is probably due to the small illumination current on the specimen.

Table 5.2 summarizes the surfaces that have been prepared using the methods introduced above for REM observations. In addition, flat surfaces can also be produced by drilling a hole in the middle of a thin silicon foil followed by annealing in UHV (Uchida, 1987). Using this method the (111), (011), (311), (100), (331), (511), (211) and (711) surfaces can be prepared around the inner cylinder on a single silicon slab.

For sensitive surfaces, studies must be carried out in an UHV chamber. There are many techniques for cleaning surfaces to be studied. The most crucial point in surface preparation is to avoid contamination. The selection of surface cleaning method depends on the material and the purpose behind the experiment or film deposition. For small-scale surface studies on certain crystal planes of brittle or easily cleaned materials, *in situ* cleavage in the vacuum chamber is convenient. For large-area substrates such as those encountered in the MBE of semiconductors, a carefully chosen chemical pre-treatment designed to leave a thin uniform layer of protective oxide is employed. The substrate is then heated in vacuum to remove the oxide layer.

In the case of III–V materials in the periodic table, the heating is done in a beam of arsenic molecules in order to maintain a stoichiometric composition. In certain cases, particularly for metals, it is necessary to use ion bombardment with neon or argon ions to remove oxides or carbonaceous products. Argon ion beams of 500–5000 eV are employed with current densities in the range 1–200 $\mu A\ cm^{-2}$. Crystallographic damage can be minimized by employing low angles of incidence and by choosing the ion mass to be approximately equal to the mass of the atom to be removed. Damage is annealed out by thermal treatment with the attendant out-diffusion of trace impurities, taking care not to overheat the sample. Sometimes it may be necessary to perform *in situ* oxidation–reduction cycles by heating the material alternately in a low-pressure atmosphere of oxygen or hydrogen. More detailed surface cleaning procedures are covered extensively in the literature (Pukite *et al.*, 1985; Lewis *et al.*, 1984; Yen *et al.*, 1986, Lee *et al.*, 1986).

5.3 Experimental techniques of REM

5.3.1 Mounting specimens

For REM observations, the surface to be imaged is usually mounted in such a way that the surface is nearly parallel to the optic axis of the microscope. The tilt to orient the surface parallel to the incident beam must be within the tilting range of the specimen holder. There are three usual methods for mounting specimens, as shown in Figure 5.6. Specimens, such as Pt(111), prepared by the re-crystallization method, are mounted on a 3 mm diameter grid hole if the wire is thin (Figure 5.6(a)). The facets on the sphere can be seen under the optical microscope. Specimens prepared by annealing and cleavage can be mounted directly on the specimen holder. The surface to be imaged is placed so that it will be parallel or nearly parallel to the optic axis.

Thin and small specimens can be sandwiched in a folding grid and then loaded in the holder. An anti-twist washer is needed to avoid specimen rotation (Figure 5.6(b)). Very small specimens can be made to adhere on a thin wire using conducting paste.

Moderately small specimens can be attached to a grid cut from a 3 mm diameter grid (Figure 5.6(c)). The size and shape to be cut depends on the size of the specimen. Ordinary conducting paste (Ag paste) is usually added to the epoxy to avoid charging effects. The specimen has to be carefully loaded into the holder. Caution needs to be taken to avoid glue contamination and tweezer scratching on the surface.

There are two rules that one has to keep in mind when loading specimens. No

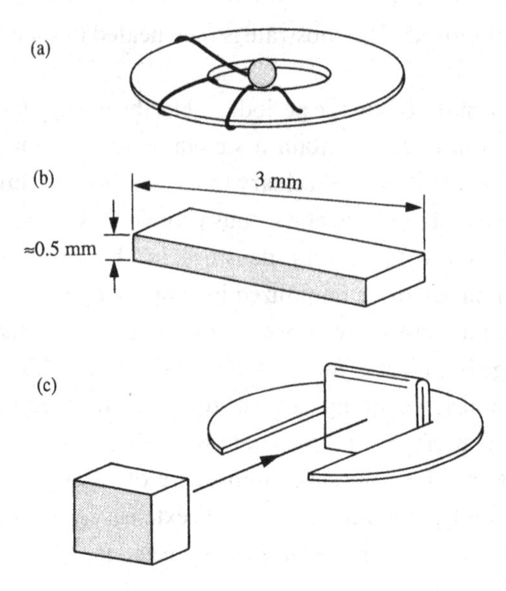

Figure 5.6 Schematic diagrams showing several specimen mounting techniques in REM (Hsu, 1992). The shaded faces are the surfaces to be imaged.

matter what method is used for mounting, it is crucial to be sure that the specimen is not going to fall off in the microscope column. The mounted specimen should be centered at the normal grid position so that the interesting part is close to the focal plane of the objective lens. The other rule is that no contamination or damage should be introduced during specimen loading. The surface of interest must be parallel or nearly parallel to the optic axis. It is preferred that the crystal is oriented in such a way that the incident beam azimuth is close to a low-index zone axis, so that strong Bragg reflections can be generated. This can greatly simplify the specimen tilting process.

5.3.2 Microscope pre-alignment

For general purposes, REM can be performed in conventional TEMs with accelerating voltages 100–1000 kV. Since a REM image is usually obtained by selecting a single Bragg-reflected beam, it is not necessary to choose a TEM with ultra-high resolution. It is important to choose a microscope that permits large-angle tilting. High-voltage microscopes are not recommended in order to reduce the foreshortening factor (see Section 5.4) since the Bragg reflection angle increases with increasing voltage. As will be shown in Section 6.4, a highly coherent emission source is important in order to improve the sensitivity and quality of REM images. Thus, a microscope equipped with a field emission gun (FEG) is highly recommended if one is available, because it can greatly enhance the image contrast and sensitivity.

In RHEED, large numbers of electrons have been inelastically and thermally diffusely scattered, resulting in degraded image contrast. Thus an objective aperture of small size is recommended unless one is interested in the interference fringes of surface superlattice structures. The smaller the objective aperture, the higher the image contrast.

Alignment of the microscope should be carried out as usual for bright-field TEM. An ordinary thin transmission specimen is recommended for performing pre-alignment procedures, because it is most convenient for correcting objective astigmatism. It is important that the transmitted (000) spot is placed at the center of the viewing screen, which defines the optic axis of the microscope. Careful adjustments have to be performed on the voltage center, objective lens current center, condenser astigmatism and objective astigmatism. Then a bulk specimen for REM can be introduced into the microscope.

5.3.3 Forming REM images

After finding the specimen surface under the beam, the specimen should be tilted in bright-field (BF) imaging mode so that both the top and bottom edges of the erect surface can be seen after a slight specimen tilt (Figure 5.7(a)). The magnification of the microscope should be set at 5000 or higher. Then, place the converged beam at the 'edge-on' surface (Figure 5.7(b)) and switch the microscope to diffraction mode (Figure 5.8(a)). Nearly half of the viewing screen shows Kikuchi lines and bands and the other half shows no intensity because of the shadowing of the bulk crystal. Slightly rotate the specimen in order to make the zone axis move towards the optic axis (Figure 5.8(b)). For future reference it is useful to indicate the optic axis on the viewing screen using the beam stop (Figure 5.8(c)). Switch the microscope to the dark-field (DF) tilting mode. Both the specimen tilt and dark-field deflection coils should be used in conjunction in order to get a RHEED pattern. The (000) beam is deflected away from the screen center towards the dark shadow. The specimen should be tilted so that the shadow of the surface moves towards the (000) beam. The ultimate goal of this process is to generate a strong Bragg-reflected spot at the center of the screen (i.e., along the optic axis), which will be used to form REM images (Figure 5.8(d)). The specimen should be tilted in both directions in order to achieve the best diffraction condition. By selecting the reflected beam that falls on the optic axis using the smallest objective aperture and switching to image mode, a REM image is obtained (Figure 5.7(c)). The image can be centered on the screen using both DF image-shift and BF image-shift. It is usually necessary to re-check the diffracting condition after the image has been centered. A slight beam tilt may be necessary in order to compensate the beam deflection.

In REM, the beam convergence can critically affect the image contrast. It is recommended that a less-convergent beam be used, provided that the image

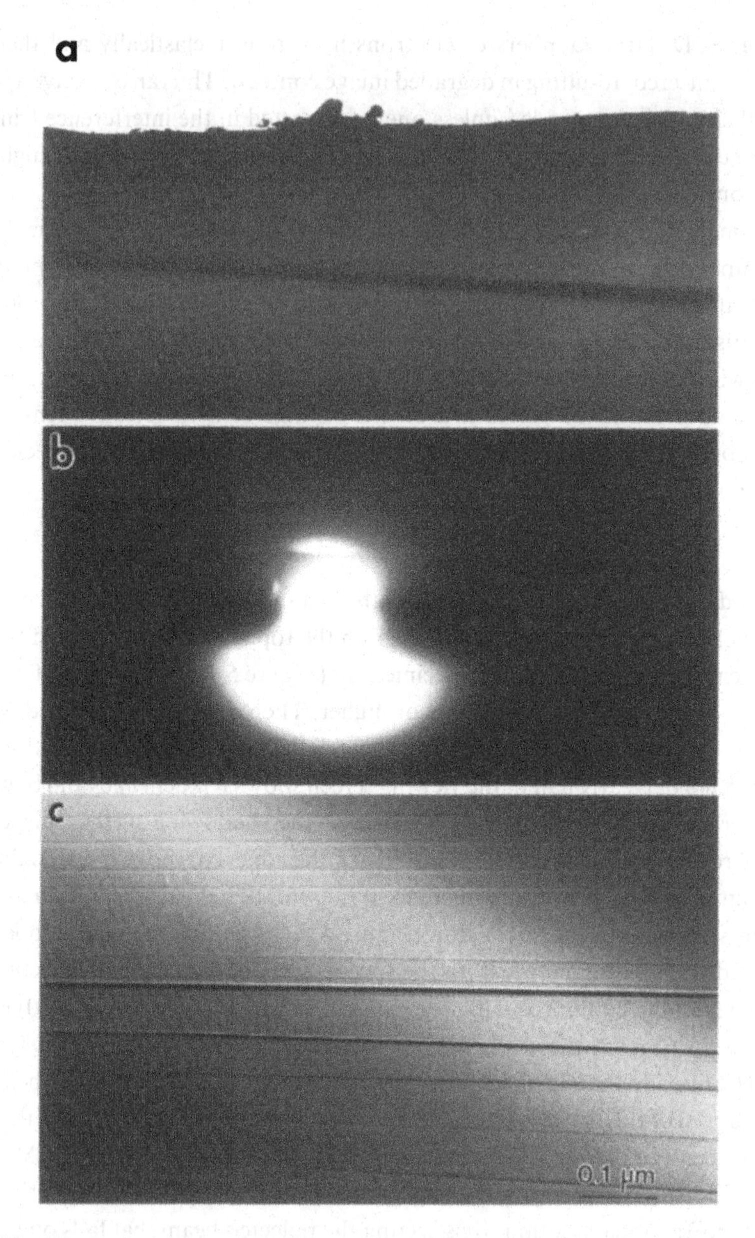

Figure 5.7 Experimental procedures for obtaining REM images in TEM (see text).

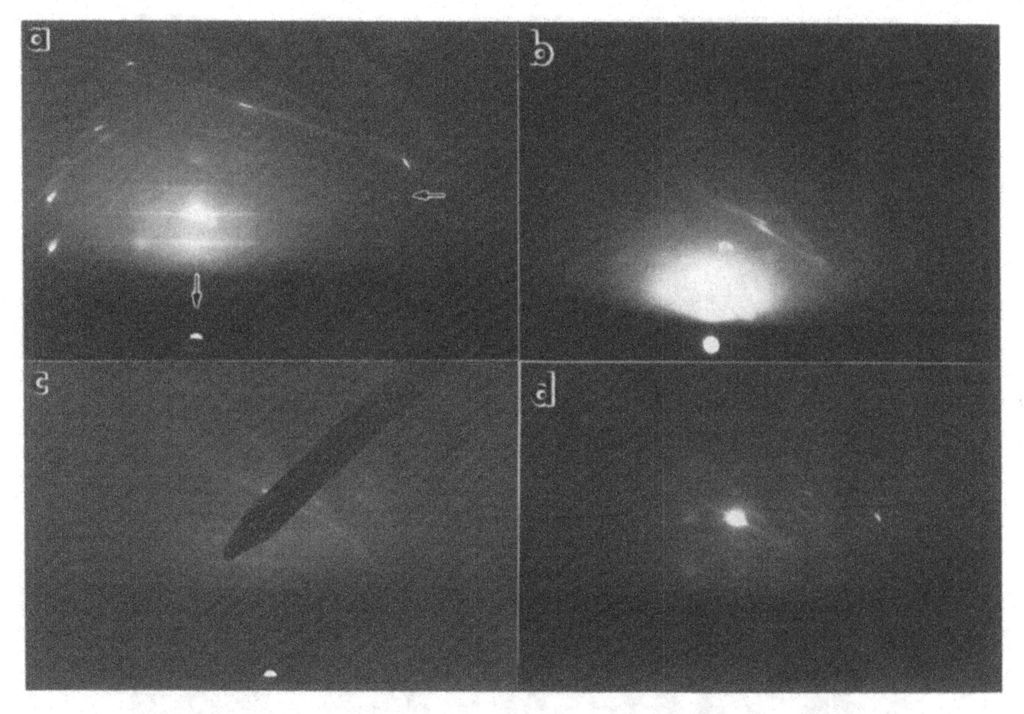

Figure 5.8 *Experimental procedures for obtaining the optimum RHEED pattern for REM imaging (see text).*

intensity is reasonable for photographic recording. The image magnification is 10 000 to 60 000 for recording REM images, and the exposure time is 3–25 s.

5.3.4 Diffraction conditions for REM imaging

The diffraction or resonance condition chosen for REM imaging can have a remarkable effect on the image contrast. The specularly reflected beam can be generated by Bragg reflection but no resonance ($B\bar{R}$) (Figure 5.9(a)), resonance but not Bragg ($R\bar{B}$) (Figure 5.9(b)) and Bragg with resonance (BR) (Figure 5.9(c)). The resonance condition can be either Bragg–parabolic resonance (Figure 5.9(c)) or Bragg–Kikuchi line resonance (Figure 5.9(d)). The former is highly recommended for REM imaging near zone axis, and the latter is useful for REM imaging far from zone axis. The optimum image intensity, contrast, resolution and surface sensitivity are normally achieved under Bragg–resonance reflection conditions.

There are several choices of diffraction conditions under the Bragg–resonance reflection. Figure 5.10 shows several RHEED patterns taken from a MgO(100) surface under different resonance conditions with a beam azimuth near [001]. Figure 5.11 shows corresponding REM images for each case shown in Figure 5.10 Figure 5.10(a) was a common close-to-zone-axis resonance condition for REM imaging.

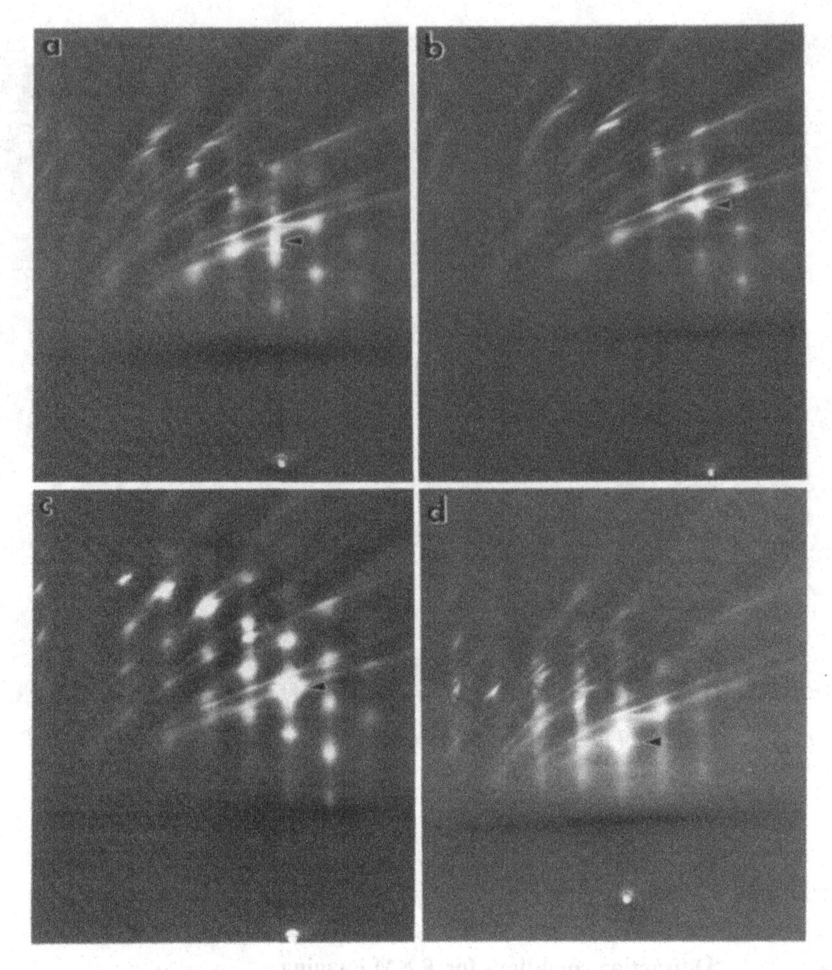

Figure 5.9 RHEED patterns recorded under conditions (a) Bragg but not resonance ($B\bar{R}$), (b) resonance but not Bragg ($R\bar{B}$), (c) Bragg and resonance (BR), and (d) Bragg, parallel to the surface Kikuchi line (see text).

The (400) specularly reflected spot coincides with the (400) and (020) Kikuchi lines; the ($4\bar{2}0$) spot is simultaneously excited. Since strong dynamical effects are involved in this case, and the (400) spot is divided into several intensity sectors by the Kikuchi lines, the image contrast is relatively poor (Figure 5.11(a)). As an alternative to case (a), the ($4\bar{2}0$) reflected spot in Figure 5.10(a) can also be used to record the REM image. The image contrast is poor (Figure 5.11(a′)). Under the on-zone-axis (600) specular reflection condition (Figure 5.10(b)), the REM image contrast is optimized and the image 'distortion' due to off-axis incidence is minimized (Figures 5.11(b_1) and (b_2)). Since the MgO(100) surface was formed by cleavage along [001], the surface steps so formed are parallel to the [001] direction. The strong diffracting

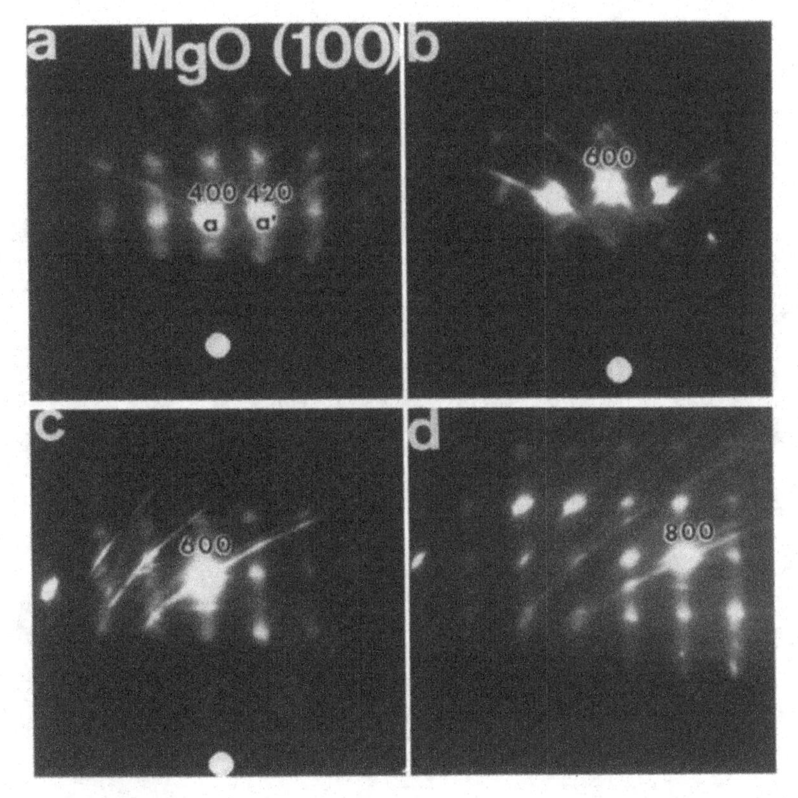

Figure 5.10 RHEED patterns from a MgO(100) surface under different resonance reflection conditions. The beam azimuth B≈[001], the beam energy is 300 keV.

condition in case (b) can be decreased by tilting the incident beam off the zone axis in the direction parallel to the surface, so that only the specularly reflected spot is strongly excited (Figure 5.10(c)). In this case, the (600) spot can be approximately considered as the 'mirror' reflection. The corresponding REM image is 'distorted', but the surface steps can be identified as being of atomic height by reference to a screw dislocation arrowed in Figure 5.11(c). For the (800) specular reflection (Figure 5.10(d)), the REM image is strongly distorted due to a large deviation angle of the incident beam with respect to the [001] zone in the direction parallel to the surface (Figure 5.11(d)).

It is normally recommended that the same surface area is imaged using different orders of Bragg reflections, which is equivalent to viewing the surface from different directions. Under the zone axis resonance in which the deviation angle of the beam azimuth is zero, the electron beam in this case suffers strong dynamical diffraction and the image contrast is weak. The image recorded under this condition can be easily applied to define the direction of the surface steps.

Figure 5.11 *REM images of a cleaved MgO(100) surface taken under the diffracting conditions corresponding to those shown in Figure 5.10 using the reflected spots (see the text for details). Both (b_1) and (b_2) correspond to the diffraction case of Figure 5.10(b). The beam azimuth $B \approx [001]$. The surface can be considered as atomically flat by reference to a step terminated directly at a screw dislocation, as indicated in (c).*

5.3.5 Image recording techniques

The traditional method for image recording uses a photographic plate. It usually takes 5–20 s to complete a single plate recording. For real-time recording of surface dynamic processes, a TV system, which is attached to the bottom of the microscope, is useful for recording time-dependent processes at a rate of 1/30 s per frame. The TV system can be used in conjunction with an image intensifier and a noise-reduction image processor.

For quantitative data analysis, a CCD camera is recommended, which can be used to record weak image signals at single-electron detection sensitivity. The CCD camera has a large dynamical range, and the image intensity can be recorded digitally, so that the observed results can be compared with the calculated results quantitatively. Image processing can be performed later.

The recorded images are normally displayed with the incident beam traveling

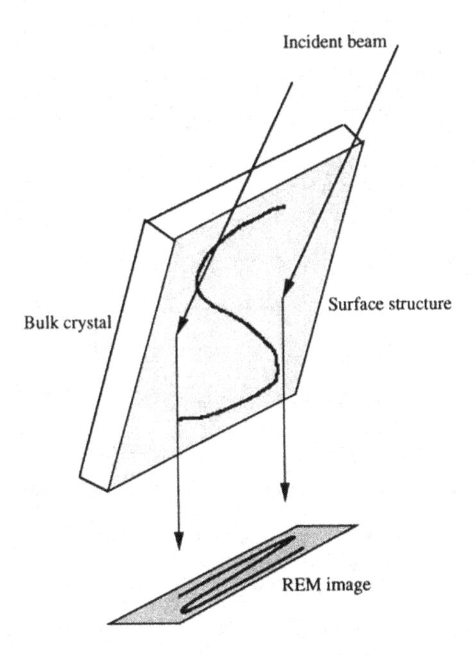

Figure 5.12 A schematic diagram showing the foreshortening effect in RHEED due to glancing angle incidence.

from the top to bottom along the vertical direction in the displayed photograph. The up and down steps are defined with reference to the beam direction.

5.4 Foreshortening effects

The grazing incidence geometry in REM imposes a foreshortening effect in the direction parallel to the beam (Figure 5.12). The foreshortening factor, which is defined as the ratio of the real surface dimension to its apparent dimension in the REM image in the direction along the incident beam, is, to within a first-order approximation, inversely proportional to the glancing angle θ, and varies in the range 20–50. Therefore, surface features are shrunk by a factor of $F = 1/\sin\theta \approx 1/\theta$ in the direction parallel to the incident beam. Surface features separated by a distance less than d_r/θ along the beam direction, where d_r is the resolution limit of REM, would not be resolved. This effect is a major disadvantage of REM. The foreshortening factor decreases with decreasing electron energy, provided that the same order of Bragg reflection is used for REM imaging.

The foreshortening effect in REM limits the spatial resolution along the beam direction. While using a small electron probe, the spatial resolution is actually one-dimensional.

Figure 5.13 shows several typical features observed in REM and their correspon-

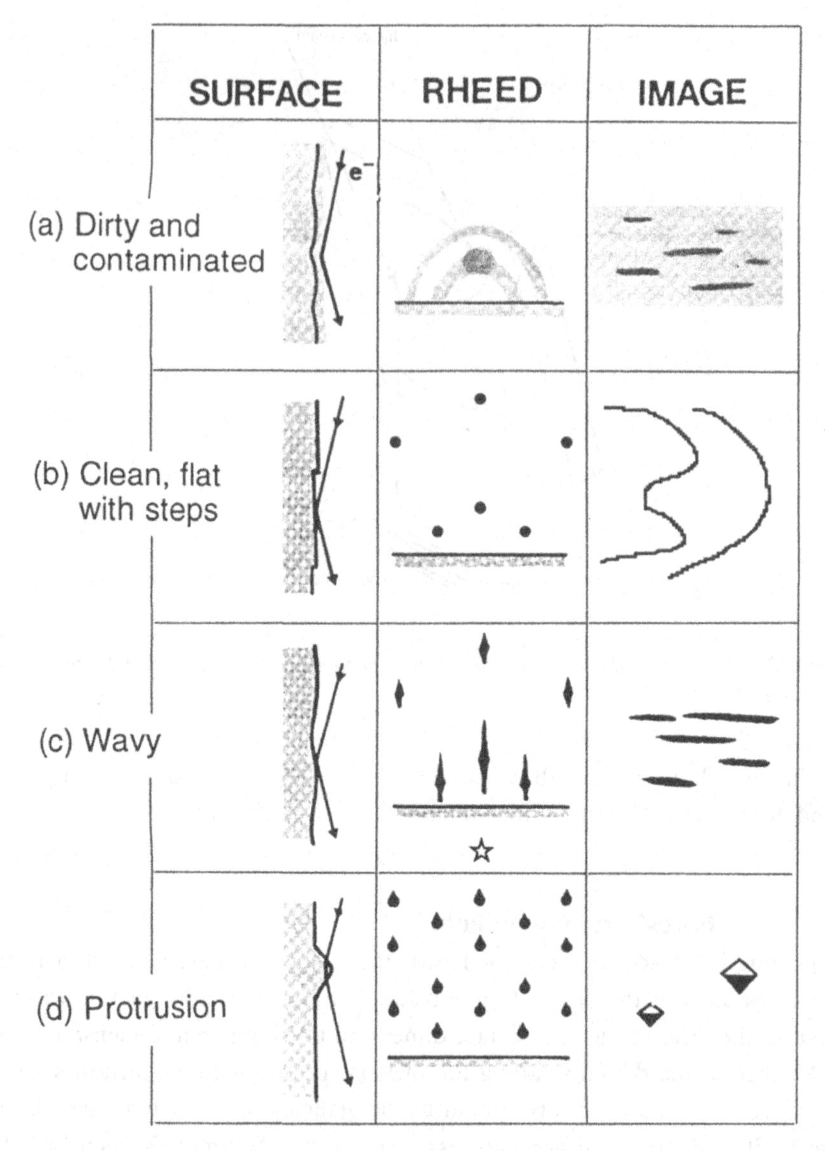

Figure 5.13 *A schematic diagram showing relations between surface structure and the corresponding RHEED patterns and REM images (Yagi, 1993b).*

dences to surface structures. For a contaminated surface, the RHEED pattern is rather diffuse and the Kikuchi lines are almost absent, and the corresponding REM image usually shows poor contrast and the contaminants cover a large portion of the surface. For a clean surface, the RHEED pattern is sharp and the REM image shows clear surface step images. The REM image is dominated by phase contrast (for steps). If the surface has a wavy structure, the RHEED pattern shows streaks due to slight variation of the local orientation, and the REM image is dominated by diffraction contrast so that only the areas that satisfy the Bragg condition show bright contrast. If the surface has many protrusions, many transmitted Bragg reflections will be seen in the RHEED pattern, especially toward a low reflection angle. The REM image will be largely shadowed by the protrusions.

5.5 Surface refraction effects

When an electron beam encounters the surface it experiences an abrupt change in potential (see Figure 2.1), resulting in a change in the de Broglie wavelength of the electron when it enters the solid. The beam traveling angle has to be changed in order to match the boundary condition at the surface, consequently introducing a refraction effect. This effect is usually small but significant at low angles of incidence. This is essentially Snell's law. Inside the crystal, the total energy $E = eU_0[1 + eU_0/(2m_0c^2)] \approx eU_0$ (for $eU_0 \ll m_0c^2$) of the incident electron is related to its kinetic energy E_k by

$$E = E_k - e|\bar{V}_0| = eU_0, \tag{5.1}$$

where $\bar{V}_0$ is the average crystal inner potential. The relationship between $\bar{V}_0$ and the electron scattering factor is given in Appendix B. From Eq. (1.38), the electron wave number inside the crystal is

$$K_I = \frac{1}{h}\left[2m_0e(U_0 + |\bar{V}_0|)\left(1 + \frac{e(U_0 + |\bar{V}_0|)}{2m_0c^2}\right)\right]^{1/2}. \tag{5.2}$$

Thus, the ratio of the wave numbers when the electron is in the crystal and in the vacuum is

$$\frac{K_I}{K_0} = \left[\left(1 + \frac{|\bar{V}_0|}{U_0}\right)\left(1 + \frac{e|\bar{V}_0|}{2m_0c^2 + eU_0}\right)\right]^{1/2}. \tag{5.3}$$

The continuity of the tangential component of the wavevector at the boundary requires

$$K_I \cos\theta_g = K\cos\theta, \tag{5.4}$$

where the beam angle inside the solid is chosen as the Bragg angle θ_g in order to satisfy Bragg's law. Hence, the refractive index of the crystal is defined as

$$n_r = \frac{\mathrm{K}_I}{K_0} = \left[\left(1 + \frac{|\bar{V}_0|}{U_0}\right)\left(1 + \frac{e|\bar{V}_0|}{2m_0c^2 + eU_0}\right)\right]^{1/2} \approx \left(1 + \frac{|\bar{V}_0|}{U_0}\right)^{1/2}, \tag{5.5}$$

where the high-energy approximation was used. This refraction effect becomes particularly important for large-Z materials, such as Au and Pt. The incidence angle

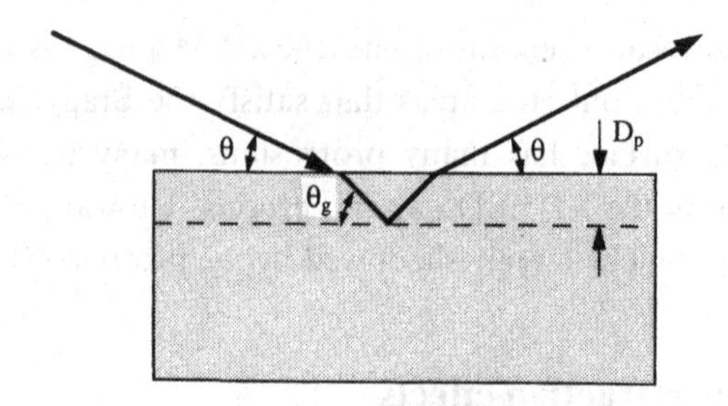

Figure 5.14 *A schematic diagram showing the surface refraction effect in RHEED.*

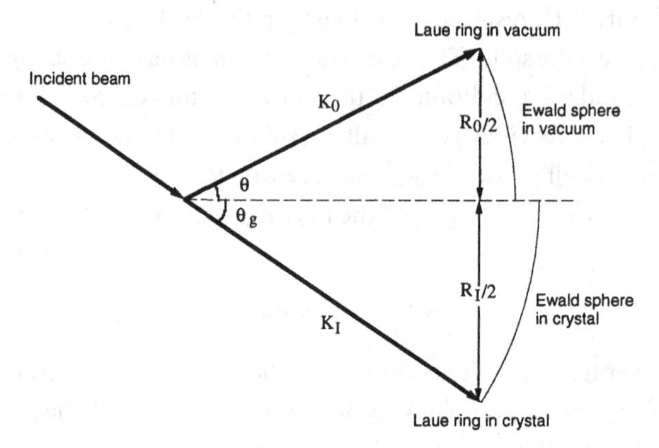

Figure 5.15 *Laue rings of different diameters observed in the RHEED pattern due to the change of Ewald sphere ratio.*

θ, at which the exact inner-crystal Bragg reflection condition is satisfied, is determined by

$$\cos\theta = n_r \cos\theta_g. \tag{5.6}$$

For $|\bar{V}_0| \ll U_0$ and using the small-angle approximation,

$$\theta \approx \left(\theta_g^2 - \frac{|\bar{V}_0|}{U_0}\right)^{1/2}. \tag{5.7}$$

Equation (5.7) means that the initial incident angle θ has to be chosen slightly smaller than the Bragg angle in order to exactly match the inner-crystal Bragg reflection condition. This result is schematically shown in Figure 5.14. The correction factor $|\bar{V}_0|/U_0$ becomes significant for high-Z materials, especially for small-angle Bragg reflections.

The effect of the crystal mean potential is to introduce a slight difference in the radius of the Ewald sphere inside the crystal and in the vacuum (Peng and Cowley, 1987). Figure 5.15 shows the diffraction geometry constructed using the Ewald spheres with different radii. The ratio of the corresponding Laue circle radii is

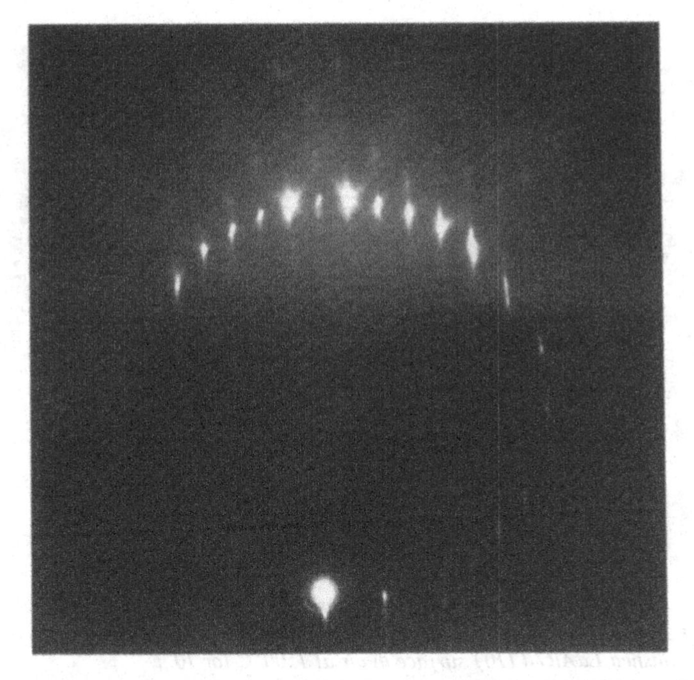

Figure 5.16 *A TRHEED pattern recorded from an annealed $LaAlO_3(110)$ surface showing the difference in sizes of the Laue circles inside and outside the crystal.*

$$\frac{R_\mathrm{I}}{R_0}=\frac{K_\mathrm{I}\sin\theta_g}{K_0\sin\theta}=\frac{\tan\theta_g}{\tan\theta}\approx\frac{\theta_g}{(\theta_g^2-|\bar{V}_0|/U_0)^{1/2}}, \tag{5.8}$$

and the crystal inner potential can be measured according to

$$\bar{V}_0=U_0\theta_g^2\left[1-\left(\frac{R_0}{R_\mathrm{I}}\right)^2\right]. \tag{5.9}$$

In RHEED, the diffraction patterns of both the transmitted and the reflected electrons can be seen simultaneously in a single pattern if the incident electron beam is positioned at the edge of the bulk crystal. This type of pattern is called a *transmission–reflection high-energy electron diffraction* (TRHEED) pattern. Figure 5.16 shows a TRHEED pattern recorded from the $LaAlO_3(110)$ surface. A difference is seen in the ratio of the number of Laue circles inside the crystal and in vacuum, in agreement with Eq. (5.8).

As expected from Eq. (5.7), the refraction effect is more significant for low-angle reflections. Figure 5.17 shows a RHEED pattern and REM image of the $LaAlO_3(110)$ surface, which exhibits many down steps. It is to be noticed that the Kikuchi lines parallel to the surface are doubly split, and the gap between the split components increases with decreasing Bragg angle, consistent with Eq. (5.7). The

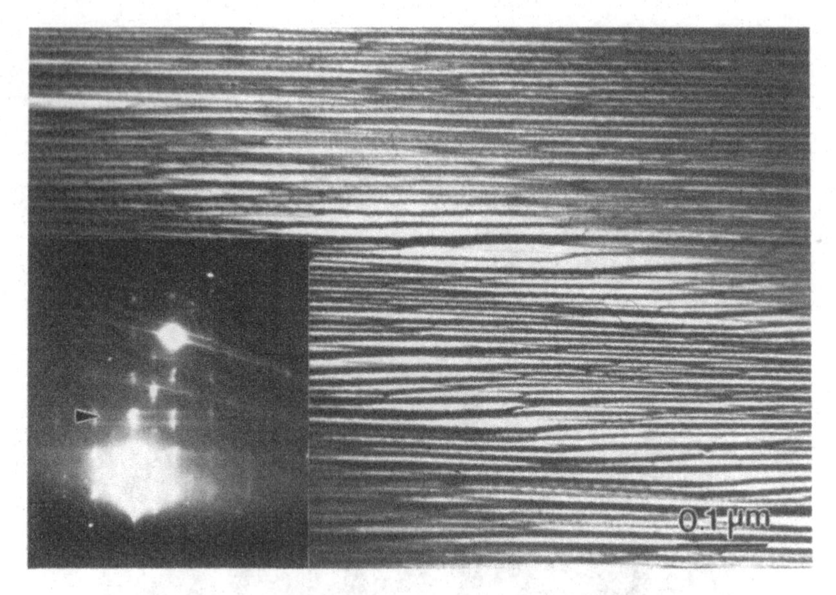

Figure 5.17 *The REM image and RHEED pattern of $LaAlO_3(110)$ showing double splitting of Kikuchi lines parallel to the surface due to the refraction effect. The specimen was prepared by annealing a polished $LaAlO_3(110)$ surface in air at 1500 °C for 10 h.*

double splitting is caused by the refraction effect. Electrons that leave the crystal from the large surface terraces suffer a strong refraction effect. However, the electrons that leave the crystal from the ending edge of the steps are almost unaffected by the refraction effect, because the exit angle is near 90° with respect to the ending faces of the step edges. If the splitting can be measured from the RHEED pattern as $\Delta\theta_{gap}$, then the crystal inner potential can be determined by solving

$$\theta_g - \Delta\theta_{gap} = \left(\theta_g^2 - \frac{|\bar{V}_0|}{U_0}\right)^{1/2}, \tag{5.10}$$

which gives

$$\bar{V}_0 \approx 2U_0\theta_g\Delta\theta_{gap}, \tag{5.11}$$

for $\Delta\theta_{gap} \ll \theta_g$.

The refraction effect can, in principle, be applied to measure the crystal inner potential. The accuracy of this measurement, however, is limited by the quality of the RHEED pattern, the width of Kikuchi lines, and surface flatness. The most accurate method for measuring crystal inner potential is probably the electron holography technique (Gajdardziska-Josifovska, 1994).

The Kikuchi lines located near the surface are also bent due to the surface refraction effect, which is significant at low reflection angles. The amount of bending

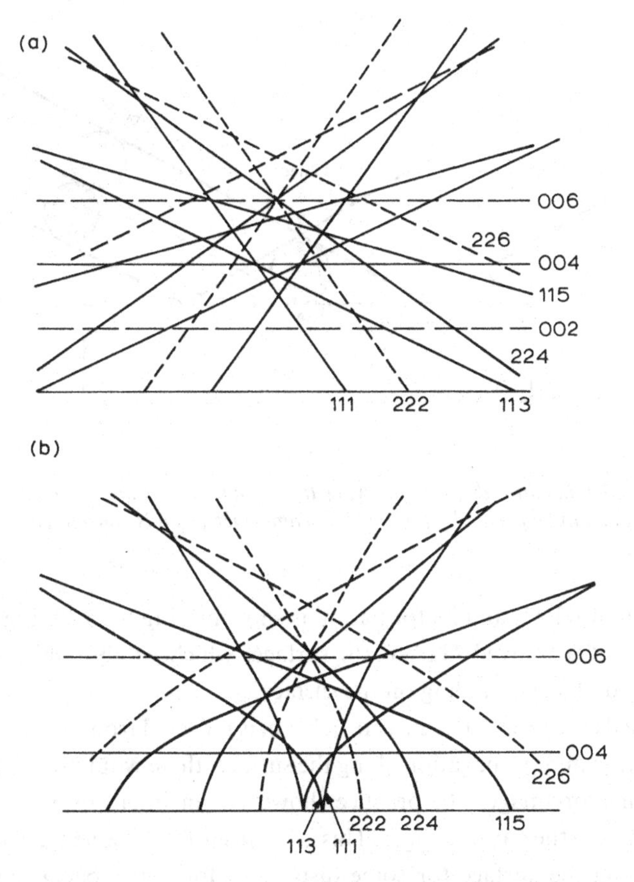

Figure 5.18 *(a) The Kikuchi pattern for the [110] azimuth of GaAs(001) with no refraction correction. The dashed lines refer to weakly allowed Bragg reflections. (b) The same Kikuchi pattern but with refraction estimated for a 12.5 keV energy and an inner potential of 14.5 eV (Dobson, 1987).*

depends on the reflection angle according to Eq. (5.7). This effect is schematically shown in Figure 5.18. In correspondence to this result, the Laue circles observed in Figure 5.16 are actually oval-shaped.

5.6 Mirror images in REM

In REM, an object protruding from the crystal surface shows two images, similarly to the image pair formed with a planar mirror. An erect image is formed by the electrons that are reflected from the surface area before transmitting the object. The inverted mirror image is formed by the electrons that will be reflected by the surface area after transmitting the object. For a perfect flat surface without steps, the two images are identical and inverted with respect to each other. This is the *mirror-image*

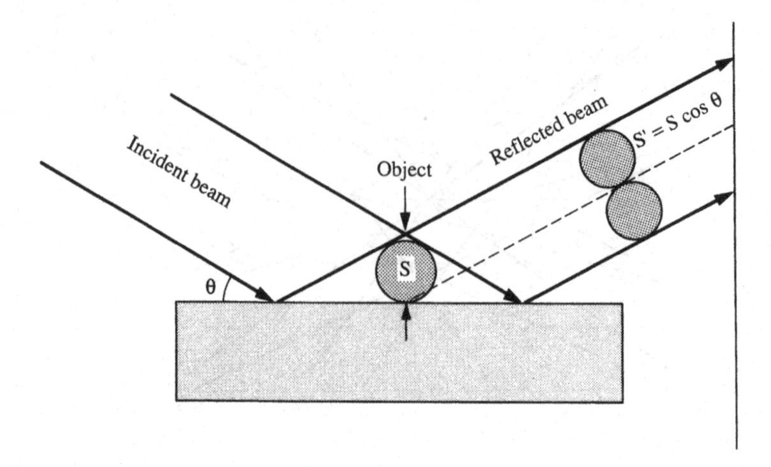

Figure 5.19 *A schematic diagram showing the formation of mirror images in REM. The surface is assumed to be perfectly flat, and no beam penetration is considered.*

phenomenon in REM, and its formation is shown in Figure 5.19. Figure 5.20 is a REM image of a cleaved GaAs(110) surface, which exhibits mirror images of foreign curved objects standing on the surface.

For an ideal case in which the electron beam is reflected from the surface without any penetration and propagation along the surface, there would be no gap between the paired mirror images. In practice, however, an intensity gap is sometimes observed between the paired images. This is because of the resonance propagation of electrons along the surface for some distance before their being reflected back, similar to the Goose–Hanchen effect in optics (Kambe, 1988). The resonance effect causes the beam to propagate inside the crystal and underneath the contact point of the object with the surface; the beam escapes from the surface after passing the object, resulting in an intensity gap between the paired images.

5.7 The surface mis-cut angle and step height

Measuring step height in REM is not always possible, particularly for steps a few atoms in height. The phase contrast produced by atom-high steps depends sensitively on objective lens focus, crystal orientation, beam coherence, resonance and diffraction conditions, and the step height. It is usually difficult to measure the absolute heights of steps directly from the image contrast. In this section, a method proposed by Kim and Hsu (1992) is introduced to measure the surface mis-cut angle and step height for some particular cases.

If the surface plane is the (hkl) reflection plane of the crystal and it is perfectly flat without steps, then the mirror images are identical but inverted with respect to each

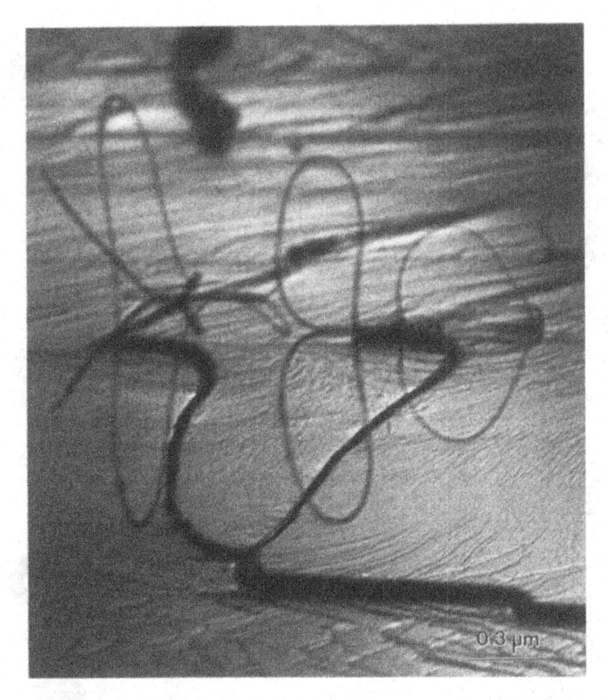

Figure 5.20 *Mirror images of 'foreign' objects standing on a GaAs(110) surface. The beam energy is 120 kV.*

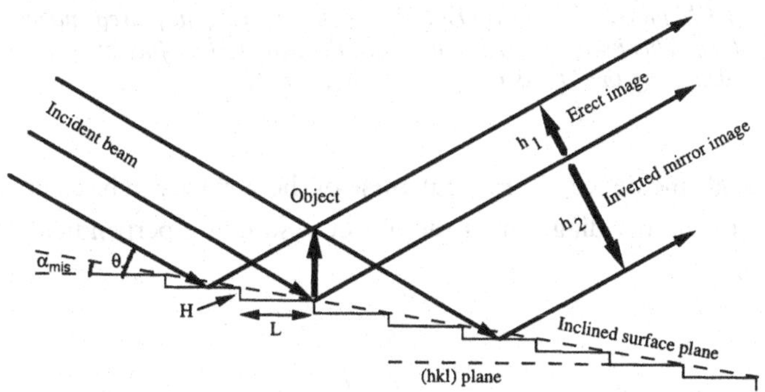

Figure 5.21 *A schematic ray diagram showing the formation of non-symmetric inverted mirror images of an erective object on the surface, where the surface steps are assumed to be of the same height and equally spaced on the surface.*

other (Figure 5.19). However, the images are not symmetric if there are steps on the surface. Figure 5.21 shows a schematic ray diagram for the formation of the paired images. For simplicity, we assume that the one-dimensional surface steps are the same height and are equally spaced in the direction nearly perpendicular to the beam azimuth. The surface plane is assumed to be the (*hkl*) crystal plane for generating the *n*(*hkl*) specularly reflected beam, which is used to record the REM image. The angle between the surface plane and the (*hkl*) crystal plane is defined as the mis-cut angle.

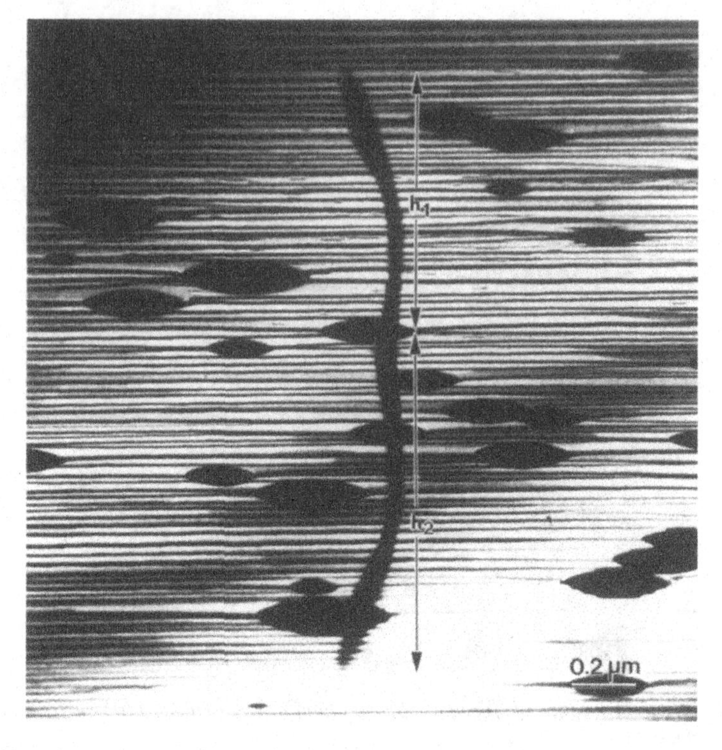

Figure 5.22 A REM image of the $LaAlO_3(100)$ surface showing the paired images of an object erect on the surface. This image is used to measure the height of surface steps (see text). The image was recorded using the (10 0 0) specularly reflected beam.

For grazing angle incidence, the mis-cut angle of the surface can be calculated from the geometry of the ray diagram. If the object is standing perpendicularly on the surface, the result is

$$\alpha_{mis} \approx \theta \frac{h_2 - h_1}{h_2 + h_1}, \tag{5.12}$$

where θ is the glancing incident angle of the beam and can be measured directly from the RHEED pattern, h_1 and h_2 are the heights of the erect and inverted object images, respectively, which can be measured from the REM image. The step height is

$$H \approx L\alpha_{mis}, \tag{5.13}$$

where L is the average distance between two adjacent surface steps in the surface plane. Equations (5.12) and (5.13) can be applied to measure the mis-cut angle of the surface and the step height.

Figure 5.22 shows a REM image of an object on the $LaAlO_3(100)$ surface. The paired images are similar to paired 'bird wings'. The inverted image is longer (or

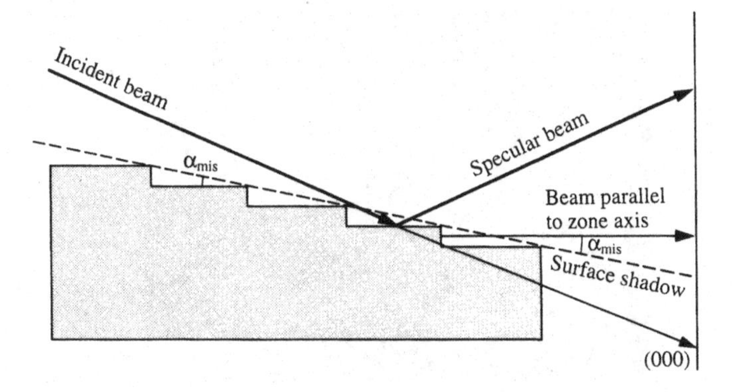

Figure 5.23 *A method for determining the surface mis-cut angle.*

shorter) than the erect image, indicating that the steps are down (or up) steps. The mis-cut angle of the surface is determined as $\alpha_{mis} \approx 0.11 \pm 0.01°$. The average step height is thus $H = 1.1 \pm 0.1$ nm. It must be pointed out that the measurement only gives an average value of step heights.

We now need to consider the restriction of the mirror-imaging technique for measuring step height. In Eq. (5.12), since $|h_2 - h_1| < h_2 + h_1$, the maximum mis-cut angle that can be measured using this technique is $\alpha_{mis} \approx \theta$, the beam incident angle. The maximum step height that can be measured by the technique is $H_{max} = L\theta$.

It must be pointed out, however, that Eq. (5.12) was derived based on the assumption that the object is standing on the surface vertically. A slight inclination of the object can introduce additional complexity into the equation (Kim and Hsu, 1992).

The method presented above can be applied to measure the mis-cut angle of the surface if the steps are clearly resolved in REM images. For surfaces having many densely distributed steps, the situation can be complex. Here we introduce an alternative method, which can be applied to measure the mis-cut angle of surfaces.

Figure 5.23 shows a schematic diagram of a mis-cut surface. When the incident beam strikes the surface, part of the diffraction pattern is shadowed by the crystal. Beams coming from the end faces of surface steps project onto the screen. The mis-cut angle is just the angle between the surface shadow and the center of the zone axis. Figure 5.24(a) is a RHEED pattern of the $LaAlO_3(110)$ surface, in correspondence to the situation shown in Figure 5.23. The surface mis-cut angle is directly measured from the RHEED pattern as 2°. This technique permits the measurements of mis-cut angles larger than θ.

The beam circled in Figure 5.24(a) is propagating parallel to the surface terraces. The imaging of this beam directly lights up the step edges, the step height being the distance between two adjacent bright lines in the image. The step height measured

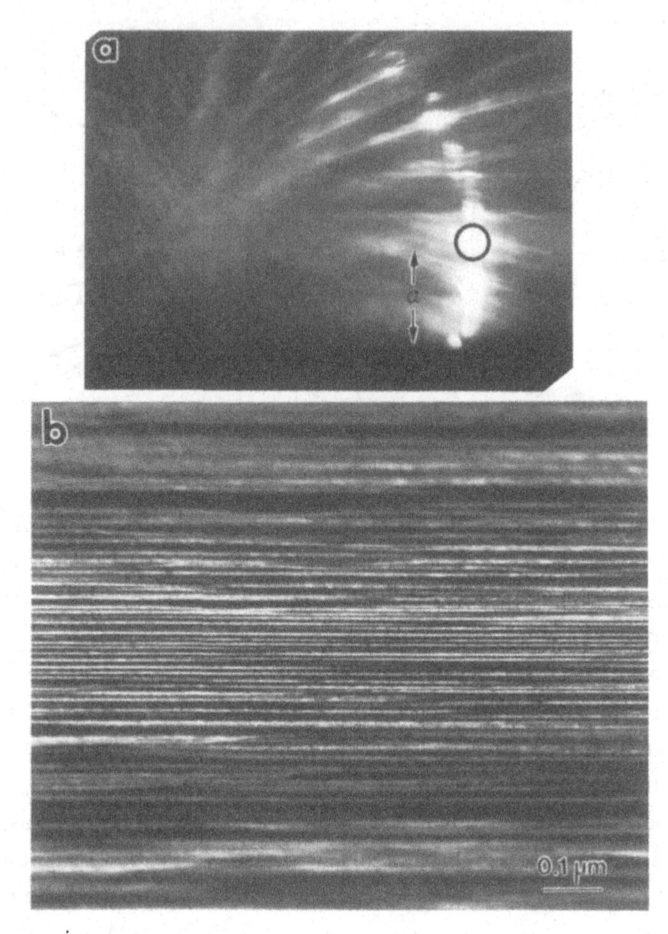

Figure 5.24 (a) The RHEED pattern of a mis-cut $LaAlO_3(110)$ surface, from which the mis-cut angle can be directly measured. (b) The REM image recorded using the circled beam as indicated in (a) for measuring step height. The beam energy is 120 keV.

from this image does not suffer any foreshortening effect. The result indicates that the observed step height is approximately 9 nm. This method, unfortunately, is only suitable for measuring the height of higher steps.

5.8 Determining surface orientations

In RHEED, the diffraction pattern usually appears in one half space with the other half blocked by the crystal edge. It can be difficult to determine the orientation of a high-index surface, especially in the case of hexagonal crystal structures (Wang and Bentley, 1992). The following method is useful for determining surface orientation and beam azimuth. Instead of using the RHEED pattern directly (Figure 5.25(a)), one can tilt the specimen away from the RHEED geometry and gradually deflect the

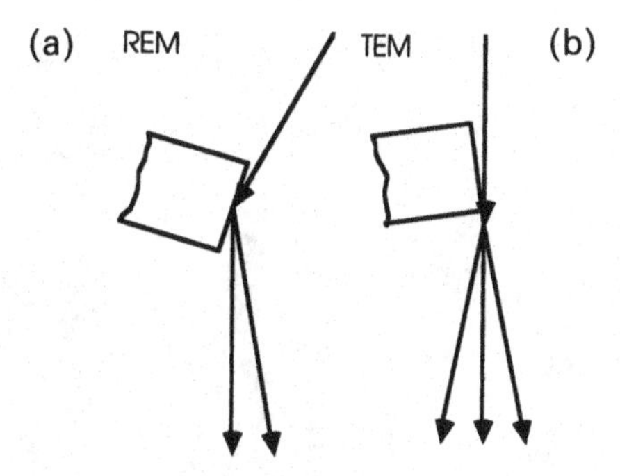

Figure 5.25 *Schematic diagrams showing a method for determining surface orientation using the transmission electron diffraction technique.*

electron beam from dark-field back to bright-field mode such that the beam penetrates a thin edge of the bulk crystal to get a transmission high-energy electron diffraction (THEED) pattern (Figure 5.25(b)). However, caution must be exercised to confirm that the transmission pattern comes from the edge of the bulk crystal rather than from adhered fragments close to the edge. This can be done easily, either by translating the specimen parallel to the crystal edge for a long distance and observing any change of the pattern, or by tilting the crystal from RHEED to THEED geometry and observing the continuous change in the diffraction pattern. From the pattern in Figure 5.26, the cleaved sapphire surface is found to be $(01\bar{1}2)$; the reciprocal-lattice direction parallel to the surface and normal to the beam is $[10\bar{1}\bar{2}]$; the corresponding beam azimuth is close to $\boldsymbol{B}=(0,a^*,2c^*)\times(a^*,0,-2c^*)$, where a^* and c^* are the unit cell constants of sapphire in reciprocal space. This yields $\boldsymbol{B}=[2a,-2a,c]=[2\bar{2}1]$ (or $[2\bar{2}01]$). This method has been found to be very useful for determining the indexes of unknown surfaces.

5.9 Determining step directions

Determining surface facets and crystallography is an important subject in surface science. This task can be performed in REM by examining the surface from several beam directions. However, the foreshortening effect in REM can greatly enlarge or reduce the angle between two steps. In this section, we outline a method that directly links the direction of a surface step observed in the REM image with its true direction on the surface.

For convenience of derivation, a three-dimensional coordinate system is defined to correlate the orientation of the surface feature with respect to its REM image (Figure 5.27(a)). The Cartesian coordinate system is designed according to the following rules. (1) The coordinate system and the relative rotation are right-handed. (2) The electron beam strikes the surface along the positive direction of the y

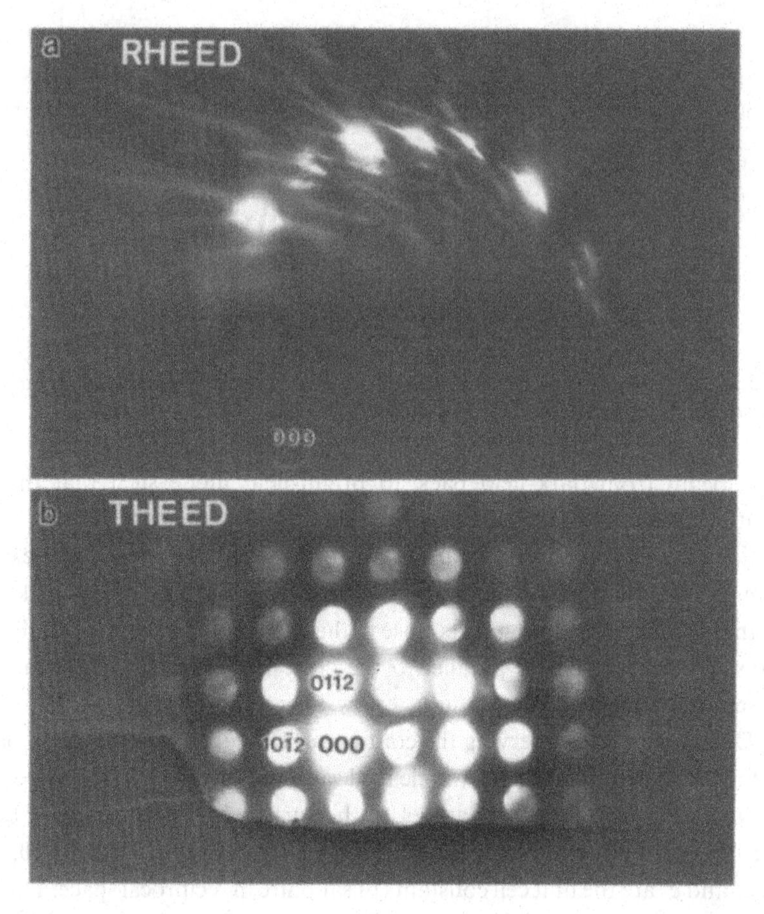

Figure 5.26 *In (a) and (b) are shown RHEED and THEED patterns, respectively, recorded from crystals of α-alumina of the same bulk to index the surface normal direction and incident beam azimuth.*

axis. (3) The normal direction of the surface is the z axis. (4) The REM image is a view of the surface looking along the electron beam, because the reflected beam is perpendicular to the REM image plane.

In Figure 5.27(a), the electron beam travels along a trajectory A–O–B. The projection of this trajectory on the surface plane is a line along A′–O–B′. The dashed line is the direction of the crystal zone axis, which is parallel to the y axis. The orientations of surface features are given in terms of $\boldsymbol{r}=(x, y, z)$. The y axis is conveniently chosen as the direction of the crystal zone axis near which the REM image was recorded. A REM image is recorded when the incident beam is at an azimuth angle θ, the angle between OB and OB′, and a deviation angle φ, the angle between OB′ and OC. These two angles are equivalent to introducing two separate rotations in the coordinate system. The transformation from the $\boldsymbol{r}$ system to the $\boldsymbol{r}'$

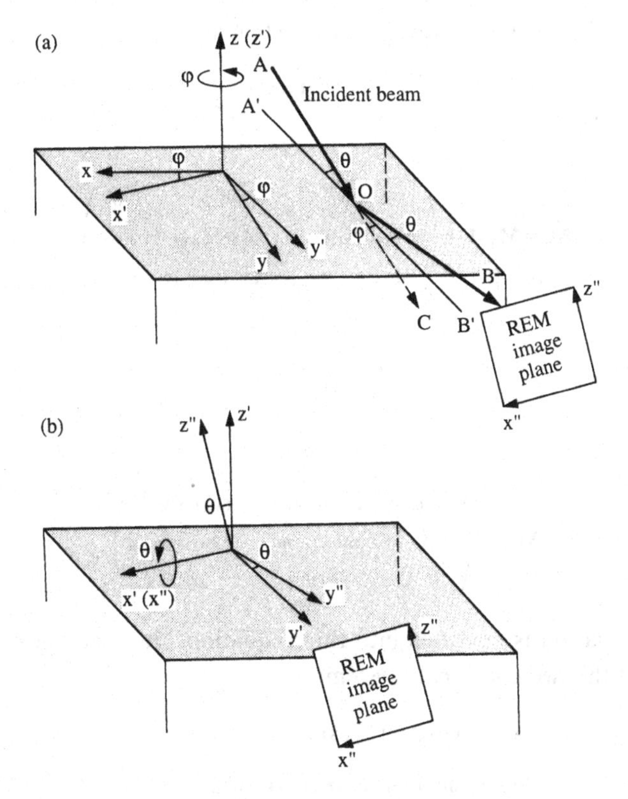

Figure 5.27 Coordinate system rotations defined by two Euler-angles in order to correlate the step orientation observed in the REM image with its true crystallographic orientation on the surface.

system is a rotation around the z-axis for an angle φ, and the corresponding rotation matrix is

$$\mathbf{M}_1 = \begin{pmatrix} \cos\varphi & \sin\varphi & 0 \\ -\sin\varphi & \cos\varphi & 0 \\ 0 & 0 & 1 \end{pmatrix}. \tag{5.14}$$

The y' axis is parallel to A′–O′–B′, the projection of the electron trajectory on the surface. The second transformation is a rotation of θ around the x' axis, so that the y'' axis is parallel to the direction of the reflected electron beam O–B. The rotation matrix is

$$\mathbf{M}_2 = \begin{pmatrix} 1 & 0 & 0 \\ 0 & \cos\theta & \sin\theta \\ 0 & -\sin\theta & \cos\theta \end{pmatrix}. \tag{5.15}$$

Therefore, the $\boldsymbol{r}''$ system is related to the $\boldsymbol{r}$ system by

$$\boldsymbol{r}'' = \mathbf{M}_t \boldsymbol{r}, \tag{5.16}$$

where

$$\mathbf{M}_t = \mathbf{M}_2\mathbf{M}_1 = \begin{pmatrix} \cos\varphi & \sin\varphi & 0 \\ -\cos\theta\sin\varphi & \cos\theta\cos\varphi & \sin\theta \\ \sin\theta\sin\varphi & -\sin\theta\cos\varphi & \cos\theta \end{pmatrix}, \tag{5.17}$$

or

$$\boldsymbol{r} = \mathbf{M}_t^{-1}\boldsymbol{r}'', \tag{5.18}$$

where

$$\mathbf{M}_t^{-1} = \begin{pmatrix} \cos\varphi & -\cos\theta\sin\varphi & \sin\theta\sin\varphi \\ \sin\varphi & \cos\theta\cos\varphi & -\sin\theta\cos\varphi \\ 0 & \sin\theta & \cos\theta \end{pmatrix}. \tag{5.19}$$

The matrix equation is rewritten into three equations that determine the relative orientation of the two coordinate systems:

$$x = x''\cos\varphi - y''\cos\theta\sin\varphi + z''\sin\theta\sin\varphi, \tag{5.20a}$$

$$y = x''\sin\varphi + y''\cos\theta\cos\varphi - z''\sin\theta\cos\varphi, \tag{5.20b}$$

$$z = y''\sin\theta + z''\cos\theta. \tag{5.20c}$$

The REM image is a projection of $\boldsymbol{r}''$ on the x''–z'' plane. For surface steps that are confined in the x–y plane, i.e. $z = 0$, Eq. (5.20c) yields $y'' = -z''\cot\theta$. Substituting this relation into (5.20a) and (5.20b) yields

$$x = x''\cos\varphi + z''\frac{\sin\varphi}{\sin\theta}, \tag{5.21a}$$

$$y = x''\sin\varphi - z''\frac{\cos\varphi}{\sin\theta}, \tag{5.21b}$$

where the factor $1/\sin\theta$ accounts for the foreshortening effect along the beam direction (i.e., the z'' axis in the REM image). Since θ and φ can be directly measured from the RHEED pattern, a feature observed in the REM image can be directly related to its true orientation on the surface. If the projected beam direction on the surface can be recognized in the REM image, then the x'' axis is the direction perpendicular to the beam direction. It is generally recommended that the surface step orientations are most easily determined if the deviation angle $\varphi = 0$.

From (5.21a) and (5.21b), the angle Θ_s between two intersecting steps of directions (x_1, y_1) and (x_2, y_2), respectively, in the crystal system is

$$\cos\Theta_s = \frac{x_1x_2 + y_1y_2}{(x_1^2 + y_1^2)^{1/2}(x_2^2 + y_2^2)^{1/2}} = \frac{x_1''x_2'' + z_1''z_2''/\sin^2\theta}{(x_1''^2 + z_1''^2/\sin^2\theta)^{1/2}(x_2''^2 + z_2''^2/\sin^2\theta)^{1/2}}, \tag{5.22}$$

independent of the deviation angle φ. We now introduce a 'stretching' transformation in order to eliminate the foreshortening factor $1/\sin\theta$ along the z'' direction, which is

$$X = x, \tag{5.23a}$$

$$Z = z''/\sin\theta. \tag{5.23b}$$

Equation (5.22) is converted into

$$\cos\Theta_s = \frac{X_1X_2 + Z_1Z_2}{(X_1^2 + Z_1^2)^{1/2}(X_2^2 + Z_2^2)^{1/2}} = \cos\Theta_t, \tag{5.24}$$

which means that the angle Θ_t between two surface steps measured in the (X, Z) system is exactly the same as the angle Θ_s between the two steps measured in the (x, y) crystal system. Therefore, the true angle between two steps can be directly determined from the REM image following the procedures listed below.

(1) Determine the projected incident beam direction in the surface plane, which is the z'' axis; the axis perpendicular to z'' is x''.
(2) Determine the incident angle θ from the RHEED pattern.
(3) Apply the stretching transformation given by Eq. (5.23) to the (x'', z'') system.
(4) Measure the angle between the two steps.

In this chapter, we have introduced the techniques that have been applied to image surface structures in TEM. It appears that REM is the only technique in TEM that can be applied to study bulk crystal surfaces. Details have been given regarding specimen preparation and mounting techniques. Practical operations for obtaining REM images have been introduced. Several important effects of reflected electron imaging and their possible applications for surface studies have been described. Techniques for determining surface orientations and mis-cut angles have been introduced.

CHAPTER 6

Contrast mechanisms of reflected electron imaging

Image calculations for REM are usually difficult because it is necessary to incorporate surface defects, such as steps and dislocations. This, in principle, can be done with the PeTSM theory; but, in practice, the huge amount of computation required and the sensitivity of REM image contrast to focus, beam convergence and diffracting conditions make the calculations rather involved and difficult compared with experimental observations. Attempts have been made to simulate surface step images (Peng and Cowley, 1986; Ma and Marks, 1990), but the results are still not very satisfactory. For this reason, in this chapter, the contrast mechanisms of REM imaging are described based on simplified models, in order to illustrate the physical concepts.

There are four basic contrast mechanisms in REM. Phase contrast (or Fresnel contrast) is produced by the path-length difference (or phase shift) of electron waves due to the change in surface morphology. This contrast dominates the images of atom-high surface steps. High-resolution information can be obtained from the phase contrast mechanism. Diffraction contrast (or Bragg contrast) is produced by the variation of local Bragg reflection angles due to lattice distortion from dislocations, local strain and crystal boundaries. Compositional contrast is produced by a variation in scattering power of different elements. This type of contrast appears on different surface domains or regions, depending on the local composition, and usually has relatively lower resolution. Finally, geometrical (or morphology) contrast is produced by variation in surface geometry, such as large surface steps and facets. For atomic level images, we mainly discuss the phase and diffraction contrast in the following sections. If the surface shows magnetic domains, then magnetic contrast could also be observed using REM (Wang and Spence, 1990).

6.1 Phase contrast

In REM, a common feature is the observation of atom-high surface steps. The image contrast of one-atom-high surface steps is primarily determined by the interference properties of the waves reflected from the upper and lower surface terraces near the step. For simplification, a surface step is assumed to be sharp, without relaxation. The resonance propagation of the electrons along the surface is not considered, but

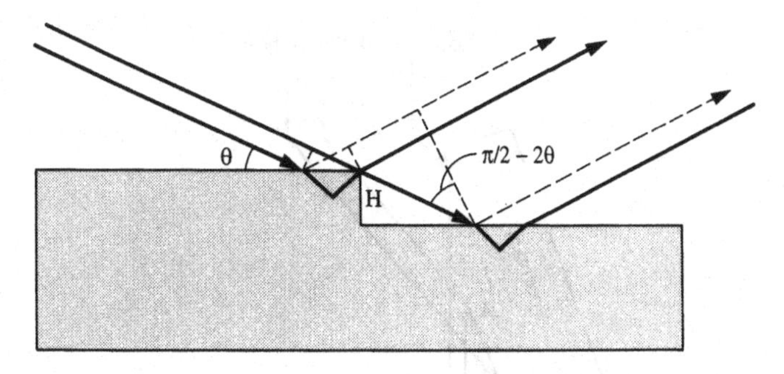

Figure 6.1 *A schematic ray diagram for calculating the electron phase jump at a surface step.*

the finite penetration of the electron is included (Figure 6.1). Owing to the refraction effect, the incident angle θ of the electron beam is related to the Bragg angle θ_g by Eq. (5.7). The path-length difference of the electrons reflected from the top and bottom terrace is

$$\Delta L=\frac{H}{\sin\theta}-\frac{H}{\sin\theta}\sin\left(\frac{\pi}{2}-2\theta\right)=2H\sin\theta, \tag{6.1}$$

where H is the step height. Using the Bragg law $2d_g\theta_g=n\lambda$ for $\theta_g \ll 1$, the phase shift of the reflected wave is (Bleloch *et al.*, 1989a)

$$\begin{aligned}\phi_p&=\frac{2\pi}{\lambda}\Delta L=\frac{4\pi H}{\lambda}\left(\theta_g^2-\frac{\bar{V}_0}{U_0}\right)^{1/2}\\&=\frac{2\pi H}{d_g}\left(n^2-\frac{8m_e d_g^2}{h^2}\bar{V}_0\right)^{1/2}\approx\frac{2\pi nH}{d_g}-\frac{2m_e H d_g \bar{V}_0}{n\pi\hbar^2},\end{aligned} \tag{6.2}$$

where d_g is the interplanar distance between the surface atomic planes, m_e is the electron mass including relativistic correction, and n indicates the nth-order reflection. The phase jump ϕ_p can be very large, even for atom-high steps. For a 100 keV electron, $H=0.1$ nm, $\theta=25$ mrad, and $\phi_p=1.35$.

Under the phase object approximation, the wave function of the reflected electrons before passing through the objective lens (i.e., in the x plane) can be written as

$$q(x)=\begin{cases}1 & \text{for } x<0\\ \exp(i\phi_p) & \text{for } x\geq 0.\end{cases} \tag{6.3}$$

Equation (6.3) can be conveniently rewritten as

$$q(x)=\tfrac{1}{2}[\exp(i\phi_p)+1]+\tfrac{1}{2}[\exp(i\phi_p)-1]p(x), \tag{6.4}$$

where $p(x)$ is a switch function defined as

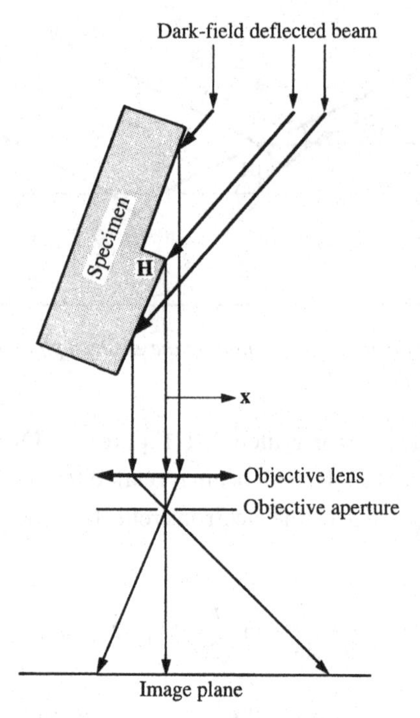

Figure 6.2 *A ray diagram showing the image formation of a surface step under the phase object approximation in REM.*

$$p(x)=\begin{cases}-1 & \text{for } x<0\\ 1 & \text{for } x\geq 0.\end{cases} \tag{6.5}$$

We now use Abbe's imaging theory to calculate the contrast of surface steps in REM. Figure 6.2 shows the ray diagram for forming REM images. For simplification, we treat a step image as a one-dimensional case. In the back focal plane (i.e., at the position of the objective aperture), the diffraction amplitude is a Fourier transform of $q(x)$ multiplied by $T_{\text{obj}}=A_{\text{obj}}(u)\exp[i\chi(u)]$ due to defocus and aberration,

$$\psi(u)=\mathrm{FT}\,(q(x))T_{\text{obj}}(u). \tag{6.6}$$

On converting the diffraction amplitude back to real space by the inverse Fourier transform, the intensity distribution of a surface step in the REM image is (Cowley and Peng, 1985)

$$\begin{aligned}I(x)&=|q(x)\otimes \mathrm{FT}^{-1}[T_{\text{obj}}(u)]|^2\\ &=\tfrac{1}{2}(1+\cos\phi_{\text{p}})-\sin\phi_{\text{p}}p(x)\otimes \mathrm{Im}\{\mathrm{FT}^{-1}[T_{\text{obj}}(u)]\}+\tfrac{1}{2}(1-\cos\phi_{\text{p}})|p(x)\otimes \mathrm{FT}^{-1}[T_{\text{obj}}(u)]|^2.\end{aligned} \tag{6.7}$$

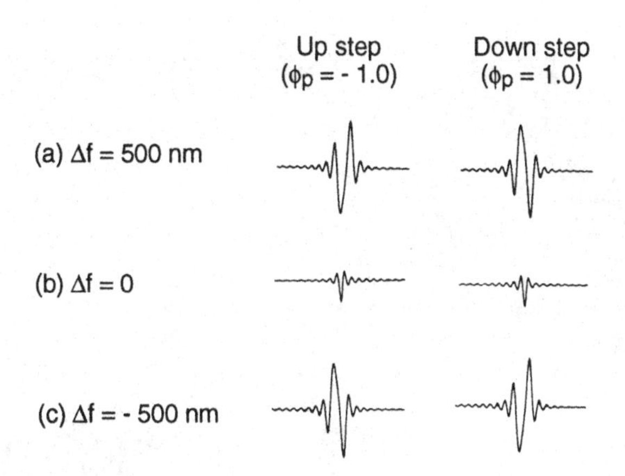

Figure 6.3 *Calculated intensity profiles for REM images of up and down surface steps, under the phase object approximation. The phase shift* $|\phi_p| = 1$ *(Cowley and Peng, 1985).*

If the step height H is a multiple of the interplanar distance d_g, then Eq. (6.2) becomes

$$\phi_p = 2\pi n' - \delta\phi_p, \tag{6.8a}$$

where

$$\delta\phi_p = \frac{2m_e H d_g \bar{V}_0}{n\pi\hbar^2}, \tag{6.8b}$$

and n' is an integer. To see the physical meaning of Eq. (6.7), we make the approximation $\delta\phi_p \ll 1$, so that

$$I(x) \approx 1 - \sin(\delta\phi_p)\, p(x) \otimes s(x), \tag{6.8c}$$

where $s(x) = \text{Im}\{\text{FT}^{-1}[T_{\text{obj}}(u)]\}$. Since $p(x)$ is anti-symmetric and $s(x)$ is symmetric, this gives an anti-symmetric, black–white contrast, which is zero for $\delta\phi_p = 0$ when the exact Bragg reflection condition is satisfied (assuming that $\bar{V}_0 = 0$) and reverses sign with the sign of $\delta\phi_p$ (step up or down) or with the sign of the defocus Δf (which reverses the sign of $s(x)$). These are the basic features of phase contrast imaging.

Figure 6.3 shows the calculated results using Eq. (6.7) for a down step ($\phi_p = 1$) and an up step ($\phi_p = -1$). Several features are predicted. First, the down and up steps show contrast reversal at both under and over focus conditions. Second, the step contrast is reversed when the objective lens focus changes from over to under focus. Third, the minimum contrast is seen under the in-focus condition. These predicted results have been observed experimentally (see for example Figure 4.3(a)). Finally, residual Fresnel fringes are seen near a surface step.

Figure 6.4(b) shows a REM image of the GaAs(110) surface recorded using a

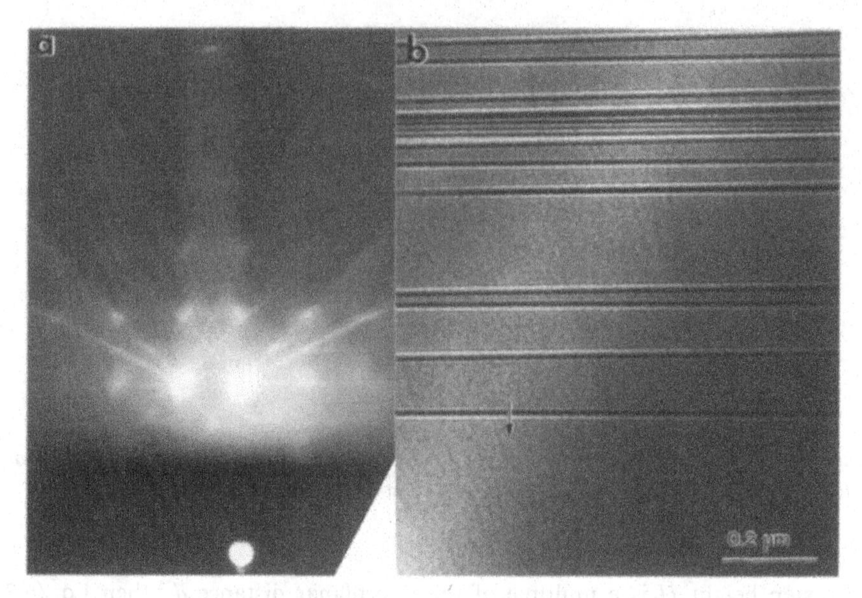

Figure 6.4 *(a) A RHEED pattern and (b) the corresponding REM image recorded from GaAs(110) using a FEG TEM. The beam energy is 100 keV.*

FEG TEM. The fine fringes appearing in the image (indicated by an arrowhead) are in good agreement with the calculated intensity profile for a down step under the under focus condition (Figure 6.3(c)). Agreement is seen even for weak residual fringes.

Contrast of surface steps has also been calculated using the perpendicular-to-surface multislice theory (Peng and Cowley, 1986) and the parallel-to-surface multislice theory (McCoy and Maksym, 1993 and 1994). In both cases, a truncated super cell is used, which is designed as half crystal and half vacuum and contains a step. The entire space is occupied by the super cells.

From Eq. (6.8), the effect of a phase shift at a step in the REM image is to modify the image intensity by a constant factor $\sin(\delta\phi_p)$. Thus the contrast profile of a surface step is expected to be independent of the phase shift at the step, which means that there is no change in intensity oscillation near a step with the change in step height. As shown in Eq. (6.2), on the other hand, the phase at the step depends on the order of the reflection to be used for REM imaging. Therefore, it is expected that the contrast profile of steps should not be significantly affected in the images recorded using different orders of Bragg reflections, provided that there is no change in electron optical configuration (such as focus).

Figures 6.5(a)–(c) are REM images of Pt(111) taken from the same surface area using the (444), (555) and (666) specular reflections as shown in Figures 6.5(d)–(f), respectively. Apart from some image 'distortion' in Figures 6.5(a)–(c) due to

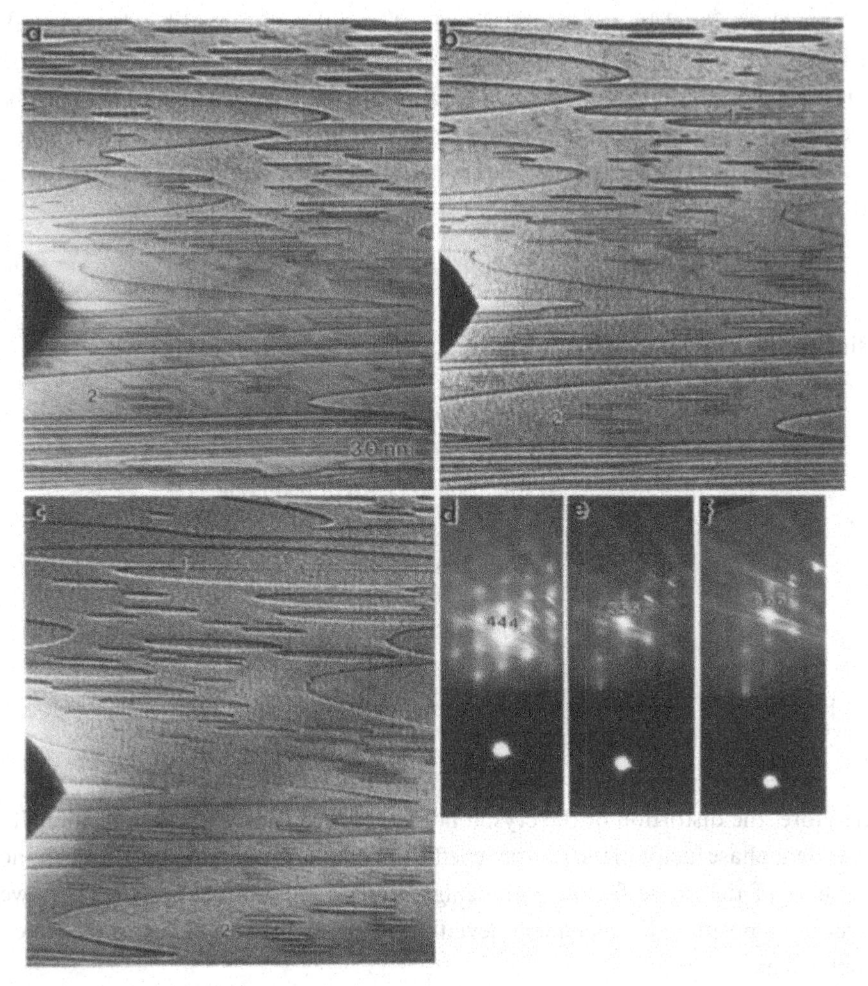

Figure 6.5 (a)–(c) REM images of Pt(111) recorded using a FEG TEM under the diffracting conditions of (d) (444), (e) (555) and (f) (660) specularly reflected beams, respectively. The beam energy is 100 keV.

specimen rotation in order to achieve the desired resonance conditions in the experiments, no dramatic contrast change is seen for the same step. Also, there is no significant change in the distance between the black–white Fresnel fringes associated with the step. This is consistent with the theoretical prediction.

6.2 Diffraction contrast

Diffraction contrast usually refers to the contrast as the result of variation of diffraction condition across the specimen, and it is normally observed in crystals containing defects, imperfections and strain. We first consider the perturbation of the crystal potential by the crystal defects. Then, the perturbed local reciprocal-lattice vector, which is important for determining the local Bragg reflection condition, is examined. For a general case, the displacement of atoms may be described by a displacement vector $\boldsymbol{R}(\boldsymbol{r})$, which depends on the position of the atom. Thus the potential distribution in the crystal, under the rigid-ion approximation, is written as

$$U(\boldsymbol{r})=\sum_n\sum_\alpha U_\alpha(\boldsymbol{r}-\boldsymbol{R}_n-\boldsymbol{r}_\alpha-\boldsymbol{R}(\boldsymbol{r})). \tag{6.9}$$

Using the inverse Fourier transform of U_α, the crystal potential is rewritten as

$$\begin{aligned}U(\boldsymbol{r})&=\sum_n\sum_\alpha\int d\boldsymbol{u}\,U_\alpha(\boldsymbol{u})\exp[2\pi i\boldsymbol{u}\cdot(\boldsymbol{r}-\boldsymbol{R}_n-\boldsymbol{r}_\alpha-\boldsymbol{R}(\boldsymbol{r}))]\\&=\sum_g\left(\sum_\alpha U_\alpha(\boldsymbol{g})\exp[-2\pi i\boldsymbol{g}\cdot(\boldsymbol{r}_\alpha+\boldsymbol{R}(\boldsymbol{r}))]\right)\exp(2\pi i\boldsymbol{g}\cdot\boldsymbol{r})\\&=\sum_g\left(\sum_\alpha[U_\alpha(\boldsymbol{g})\exp(-2\pi i\boldsymbol{g}\cdot\boldsymbol{r}_\alpha)]\right)\exp[2\pi i(\boldsymbol{g}\cdot\boldsymbol{r}-\boldsymbol{g}\cdot\boldsymbol{R}(\boldsymbol{r}))]\\&=\sum_g U_g\exp[2\pi i(\boldsymbol{g}\cdot\boldsymbol{r}-\boldsymbol{g}\cdot\boldsymbol{R}(\boldsymbol{r}))].\end{aligned} \tag{6.10}$$

The Fourier coefficient of the modified potential is

$$U_g(\boldsymbol{r})=U_g\exp(-2\pi i\boldsymbol{g}\cdot\boldsymbol{R}(\boldsymbol{r})). \tag{6.11}$$

Therefore, the distortion of the crystal lattice by defects introduces a positionally dependent phase factor in the Fourier coefficient of the crystal potential. To examine the effect of the phase factor $\exp(-2\pi i\boldsymbol{g}\cdot\boldsymbol{R}(\boldsymbol{r}))$ on the diffraction condition, we introduce a positionally dependent deviation reciprocal-lattice vector $\Delta\boldsymbol{g}(\boldsymbol{r})$, which is defined as

$$-\boldsymbol{g}\cdot\boldsymbol{R}(\boldsymbol{r})=\Delta\boldsymbol{g}(\boldsymbol{r})\cdot\boldsymbol{r}, \tag{6.12}$$

so Eq. (6.10) is thus rewritten as

$$U(\boldsymbol{r})=\sum_g U_g\exp[2\pi i(\boldsymbol{g}+\Delta\boldsymbol{g})\cdot\boldsymbol{r}]. \tag{6.13}$$

Therefore, the local reciprocal-lattice vector is $\boldsymbol{g}^{(loc)}=\boldsymbol{g}+\Delta\boldsymbol{g}$, which means that the diffraction vector $\boldsymbol{g}^{(loc)}$ changes locally in orientation and length in the vicinity of the defect, so that the local Bragg angle, determined by $(2/g^{(loc)})\sin\theta_g^{(loc)}=n\lambda$, or equivalently $\sin\theta_g^{(loc)}=n\lambda g^{(loc)}/2$, changes across the defected region. In REM, if the objective aperture is set at a fixed angle $\theta^{(obj)}$, then the observed reflected intensity will rapidly decrease with the increase of $\Delta\theta=|\theta^{(obj)}-\theta_g^{(loc)}|$. This is the diffraction contrast observed in REM, as schematically shown in Figure 6.6.

The above discussion was made based on the column approximation (Hirsch *et*

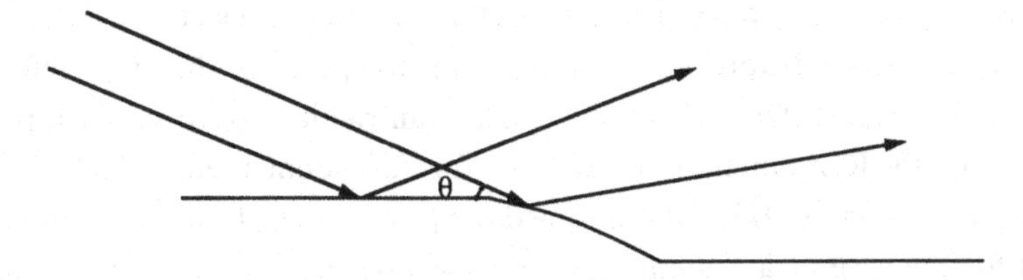

Figure 6.6 A schematic diagram showing the diffraction contrast mechanism for imaging surface emergent dislocations (see text).

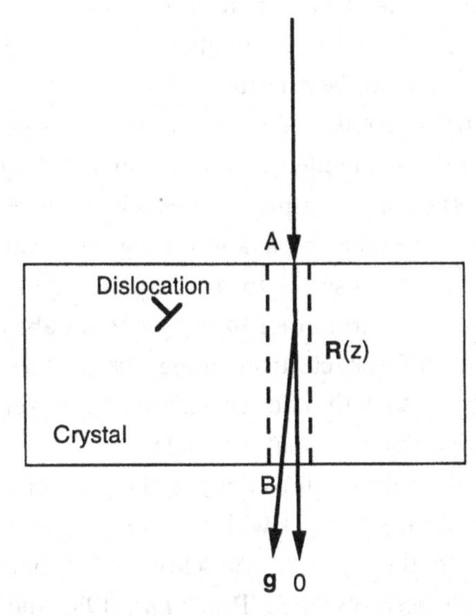

Figure 6.7 The column approximation in TEM for calculating diffraction contrast images.

al., 1977), as schematically shown in Figure 6.7 for TEM, in which the diffraction intensity is purely determined by the local Bragg condition, independent of the subsequent scattering. For a thin foil, a dislocation is situated at D inside the foil. An electron wave is incident on the top surface. The dislocation line causes displacement of an atom in the column along AB from its true position by an amount $\boldsymbol{R}$, which depends on its distance z from the upper surface. The column approximation amounts to assuming that the electron wave function at B is the same as that at the lower surface of a crystal of infinite lateral extension, with the same displacement $\boldsymbol{R}(z)$, depending only on z and not on the position of the column. Such a crystal can be considered as an assembly of thin crystal rods, each perfect, but displaced and rotated relative to each other. The wave function at the exit face of each column (or rod) depends only on the scattering of that column, so that there is no interaction between columns. The width of the column is about 2 nm in TEM. This approximation may not be valid at dislocation cores, where rapid variation of $\boldsymbol{R}$ is possible.

In REM, a surface emergent dislocation shows a long-range contrast tail as the result of crystal lattice distortion, which consequently perturbs the Bragg reflection intensity of the local surface. This is the result of diffraction contrast. An approach to dealing with the REM image interpretation for surface emergent dislocations was first made by Shuman (1977), who outlined the principles of Bloch wave matching involved for the case of a stacking fault intersecting the surface. For the case of dislocations with strain fields, the theory is simplified by assuming that the reflected amplitudes at each point on the distorted crystal surface are identical to those from a perfect crystal with the same local orientation. This came from the column approximation described above, but is modified in the way that the columns are arranged to be perpendicular to the external surface and the electron waves do not couple between columns. The reflected intensity from each column is determined only by the deviation of the local reflection angle from the Bragg angle (Peng *et al.*, 1989). The diameter of the column is probably much larger than that in the TEM case so that the electron resonance propagation along the surface does not affect the final result. This approximation seems to be adequate due to the foreshortening effect and the moderate resolution in REM images (Osakabe *et al.*, 1981a). If the mean traveling distance of the electron along the surface is 100 nm, for a foreshortening factor of 40, then the effective column size observed in REM images is $100/40 = 2.5$ nm, compatible with that for TEM.

If one is interested only in diffraction contrast, then this model can be simplified into an optic mirror reflection model without consideration being given to the electron penetration into the surface, and reasonably good comparisons with experimental results have been obtained (Peng *et al.*, 1987 and 1989).

Consider a surface emergent screw dislocation (Figure 6.8). When the incident beam is exactly at the Bragg condition, any deviation from the perfectly planar surface arising from the local strain field will weaken the reflected beam intensity relative to the unstrained perfect surface far away from the dislocation core. Owing to the severe foreshortening in the beam direction, the image of a dislocation then appears as a dark line. When the incident beam deviates from the Bragg angle, on one side of the dislocation core the local angle of incidence is always less than that for the unstrained regions and the image of the dislocation will then appear as a dark wing (the left-hand side of the core in Figure 6.8). On the other side of the dislocation core, the local incident angle is always larger than that of the unstrained regions, so that some of the reflected electrons satisfy the Bragg reflection condition, resulting in a bright contrast wing in the image (the right-hand side of the core in Figure 6.8).

If the incident beam direction is slightly changed so that the local Bragg condition is perturbed, then a contrast reversal between the two wings can be seen. Figure 6.9 shows two REM images of the same dislocation when the reflected beam is positioned at the left- and right-hand sides of the dislocation. The variation of local

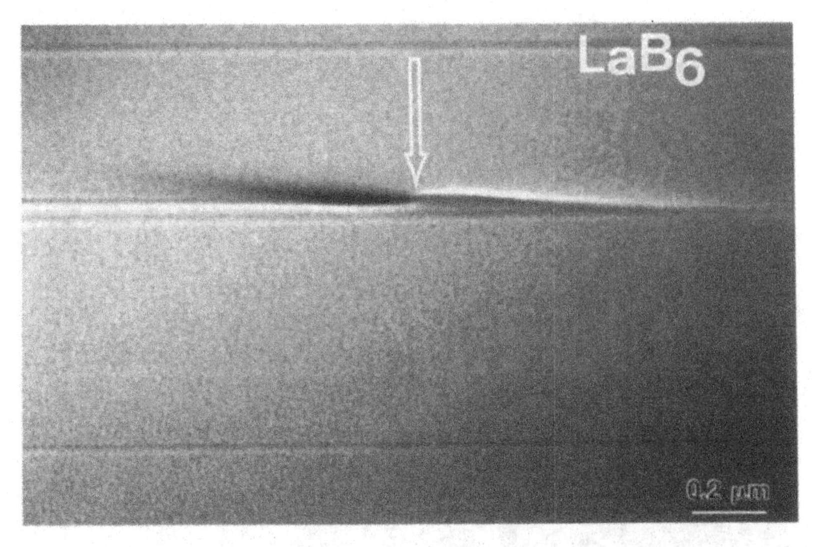

Figure 6.8 A REM image of a surface emergent screw dislocation on a GaAs(110) surface recorded with a LaB_6 filament TEM at 100 kV, showing symmetric dark- and bright-contrast tails due to the effect of diffraction contrast.

reflection angle leads to the reversal of dislocation contrast. The diffraction contrast in REM imaging is sensitive to the surface strain generated by defects (Takeguchi *et al.*, 1992). Compared with the dislocation image shown in Figure 6.8, the asymmetric contrast of the dislocation shown in Figure 6.9 is due to the inclined Burgers vector with respect to the surface, which is likely to be (a/2)[101] at 60° to the surface normal. The dislocation shown in Figure 6.8 is of pure screw type with Burgers vector (a/2)[110], perpendicular to the surface. Quantitative analysis of dislocation contrast in REM is useful for determining the Burgers vector (Peng *et al.*, 1989).

In REM, changing focus is equivalent to shifting the position of the beam cross-over across the surface, which is equivalent to changing the beam reflection angle, leading to possible contrast reversal. Figures 6.10(a), (b) and (c) show REM images of two paired dislocations recorded at under-, in- and over-focus conditions, respectively. It is apparent that the dislocation contrasts in Figures 6.10(a) and (c) are reversed. The contrast of the two paired dislocations is mutually reversed under any focus condition, due to the opposite directions of the Burgers vectors.

Based on the theory presented in the last section, phase contrast is minimized if the step height is a multiple of the interplanar distance under the exact Bragg condition. However, contrast is always observed at surface steps, for two possible reasons. One reason is that the phase shift at a step can never be $2\pi n'$ because of the significant effect of the crystal inner potential (see Eq. (6.2)). The other reason is diffraction contrast. Atomic relaxation near step edges can influence the local diffraction condition. Diffraction (or scattering) at a step edge can significantly affect the

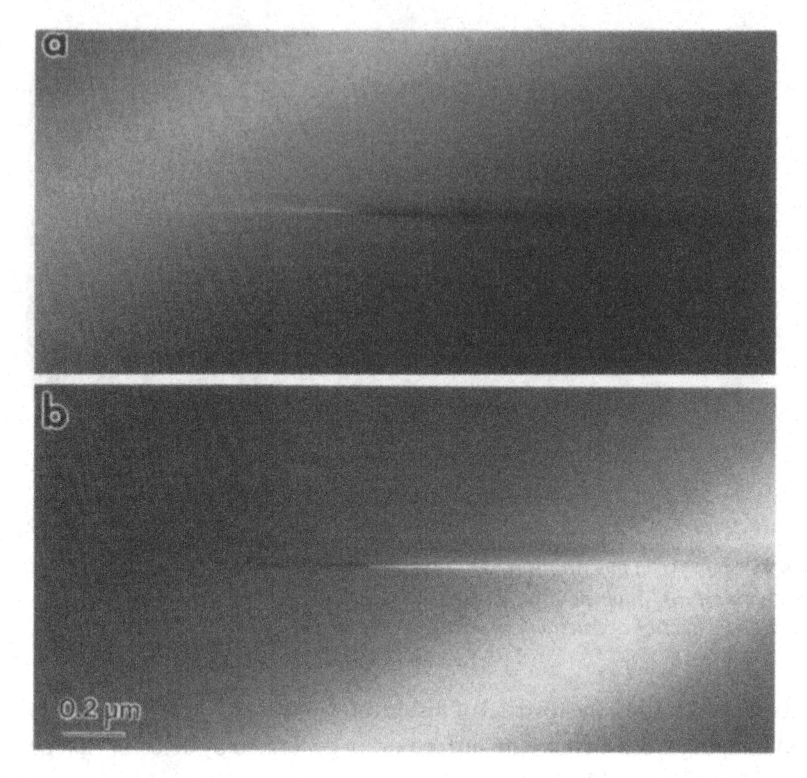

Figure 6.9 *REM images of a surface emergent dislocation on GaAs(110) showing contrast reversal when the incident beam is positioned on the left- and right-hand sides of the dislocation. The dislocation contrast, however, is not as symmetric as that shown in Figure 6.8.*

propagation direction of the beams reflected from the surface terrace, resulting in diffraction contrast at surface steps. Figure 6.11 shows a REM image recorded from Pt(111). Surface steps are clearly resolved at the surface, and most of the steps show normal contrast. The steps located at the top of the image with dark background appear bright due to diffraction contrast.

Based on the contrast characteristics of REM images, a rule for identifying up and down steps (with respect to the beam direction) has been established, as shown in Figure 6.12. Under the exact Bragg condition, the step contrast is dominated by phase contrast, thus the calculations given in Section 6.1 can qualitatively describe the image contrast. For the cases of off-Bragg reflections, the diffraction effects could be important if there is step relaxation.

Before closing this section, it is necessary to point out the complexity in analyzing REM image contrast. In RHEED, the excitation of surface resonance critically depends on diffraction conditions. It has been found that the resonance occurs only within an angular width smaller than about 2 mrad. The $LaAlO_3(100)$ reflection, which exhibits {100} twin structure with twin angle about 0.12°, can be used to illustrate the problem, as shown in Figure 6.13. The entire surface is illuminated by a

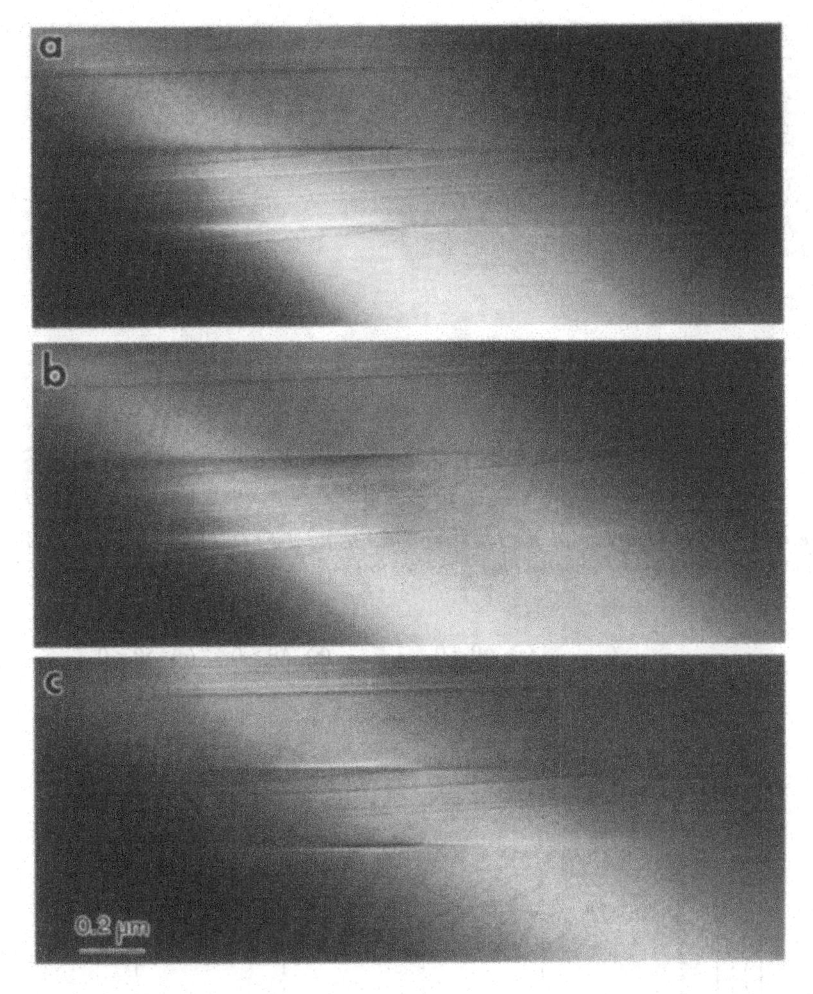

Figure 6.10 *REM images of two paired dislocations on GaAs(110) showing the dependence of dislocation contrast on the image focus (see text).*

convergent beam, so that the incident direction of the beam varies continuously across the surface. The areas showing bright contrast are the twin grains, which are oriented under the resonance condition. However, it is interesting to note the reversal of contrast within the same twin grain on going from the left-hand side to the right-hand side. This effect is entirely due to the variation in the beam illumination condition across the surface. Therefore, caution must be exercised when analyzing REM image contrast for sophisticated purposes.

6.3 Spatial incoherence in REM imaging

The schematic model shown in Figure 6.1 is based on the assumption that the illumination source is highly coherent and the beam illumination is ideally parallel

Figure 6.11 A REM image of Pt(111) recorded using a FEG TEM, showing diffraction contrast at surface steps. The beam energy is 100 keV.

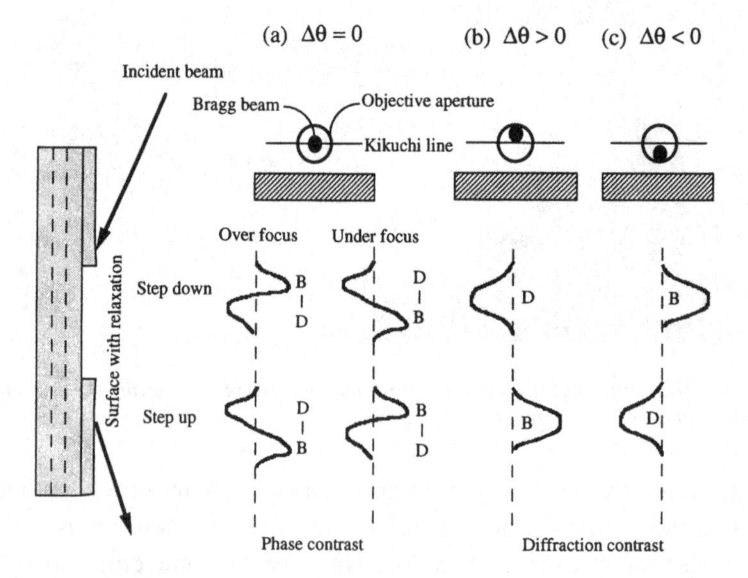

Figure 6.12 A schematic illustration of bright (B) and dark (D) contrast characteristics of surface steps under different imaging conditions. Note that 'bright–dark' means bright–dark Fresnel fringes along the beam direction. The down or up step is defined with respect to the beam direction. The general features of steps due to diffraction contrast are also shown. The horizontal line indicates a Kikuchi line. The Bragg condition is met if the reflected beam is coincident with the Kikuchi line.

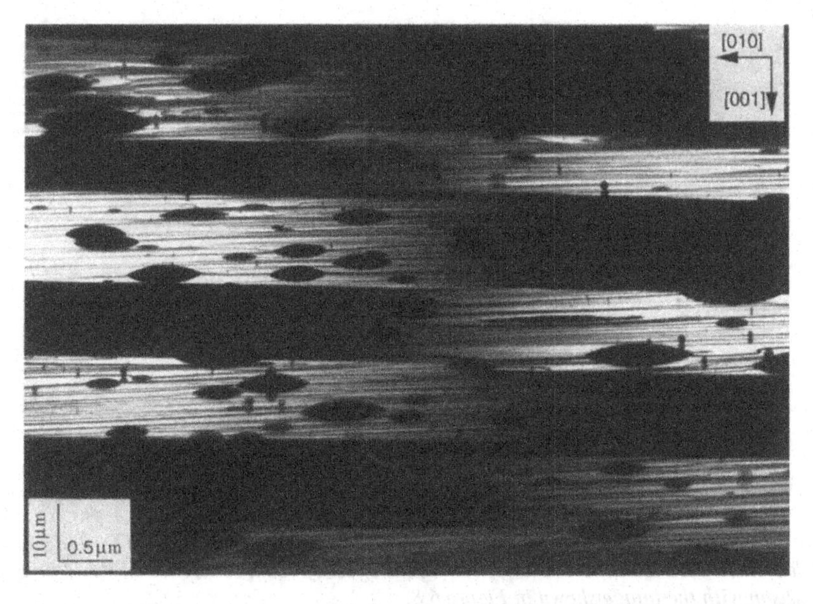

Figure 6.13 *A REM image, recorded from a twinned $LaAlO_3(100)$ surface under the diffracting condition given in Figure 6.5(a), shows contrast reversal due to the variation of diffracting condition across the illumination area. The bright- and dark-contrast bands are the crystals that form the (001) twins. The beam azimuth is [001].*

so that the phase jump at a step would be preserved for a long distance on the surface terraces on both sides. In practice, it has been suggested that the effective angular range of the illumination in REM might be limited to a range $\Delta\theta_g$ by the width of the Bragg reflection and that, because of the finite penetration depth D_p (or equivalently the shape factor in kinematical scattering theory), the phase jump at a step would take place over a finite distance (Howie *et al.*, 1992). The ratio of the Bragg angle with the width of the Bragg beam is

$$\frac{\theta_g}{\Delta\theta_g}=\frac{D_p}{d_g}. \tag{6.14}$$

The corresponding spatial coherence distance can be estimated by

$$X_c=\frac{\lambda}{\Delta\theta_g}=\frac{\lambda D_p}{d_g\theta_g}=2D_p, \tag{6.15}$$

where Bragg's law $2d_g\theta_g=\lambda$ was used. This expression shows that, unless the illumination has an angular spread $\Delta\theta<\Delta\theta_g$, the angular filtering effect of the Bragg reflection will automatically determine a coherence distance (X_c). Since the electron penetration depth is typically about 1 nm, the coherence length is 2 nm. The corresonding distance in the REM images is $X_c/\theta=80$ nm if the foreshortening effect

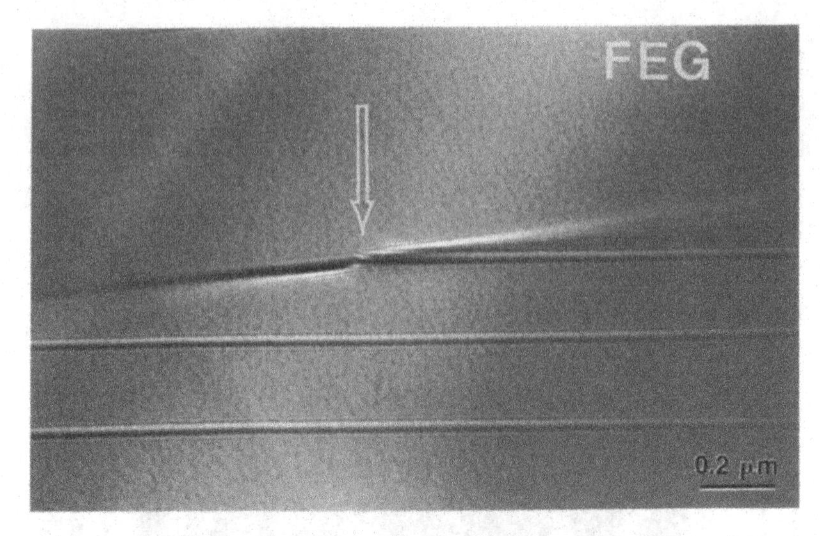

Figure 6.14 *A REM image of a surface emergent screw dislocation on a GaAs(110) surface recorded with a FEG TEM at 100 kV, showing the effect of beam coherence on REM images in comparison with the image shown in Figure 6.8.*

is considered, where $\theta = 25$ mrad. The occurrence of phase contrast, but only limited phase contrast, in typical REM images of steps can thus be explained.

It is thus expected that with more parallel illumination ($\Delta\theta < 1$ mrad) or at least more coherent illumination, multiple fringes would be observed at steps due to the increased spatial coherence. The REM image shown in Figure 6.4(b) was recorded using a FEG TEM. The residual fringes are observed at each step. It is thus suggested that parallel or nearly parallel beams should be used for REM imaging in order to enhance phase contrast.

6.4 Source coherence and surface sensitivity

A screw dislocation with its Burgers vector normal to the surface appears with symmetric dark–bright contrast wings in REM images. The strain field variation around the core of the dislocation is expected to create strong phase contrast, but this effect has not been observed in typical REM images recorded with the use of a LaB_6 filament (Figure 6.8). However, detailed phase contrast (fine fringes) can be seen near the dislocation core in a REM image recorded on an instrument with a field emission gun (FEG), as shown in Figure 6.14. Non-symmetric fringes extend to about 200 nm from the dislocation core. At larger distances, the phase contrast effect decreases and diffraction contrast becomes dominant, producing a long tail that extends to more than 1 μm. Therefore, the REM images are severely affected by the incoherence effect. This does not arise from the strong inelastic scattering (or

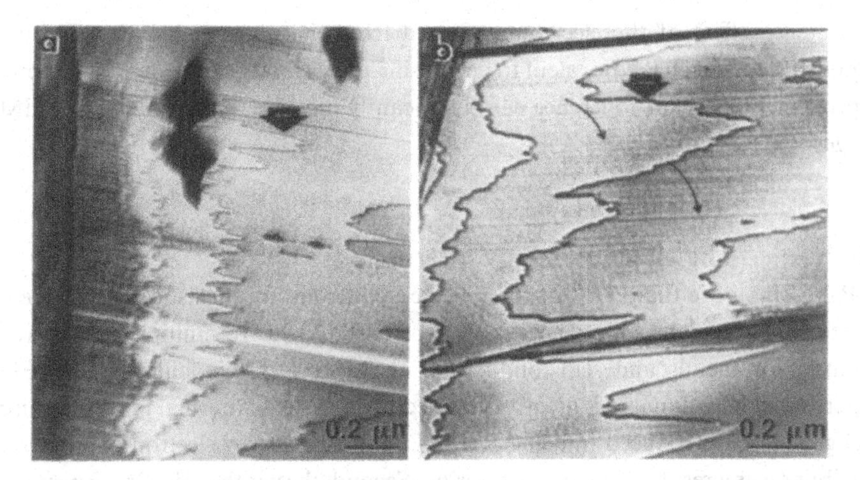

Figure 6.15 *In (a) and (b) are shown REM images recorded using a LaB_6 filament and FEG, respectively, from the same α-alumina (01$\bar{1}$2) surface. Note the improvement in REM sensitivity with increasing electron source coherence. The beam energy is 100 kV.*

energy spread), which is often present regardless of whether a LaB_6 gun or a field emission gun is used. Instead, this incoherent effect originates from the effective angular spread in the illumination, which is frequently determined by the width of the Bragg reflection.

The improvement of source coherence not only can enhance the phase contrast in REM images but also can improve the image sensitivity. Figure 6.15 shows a comparison of the REM images taken from an α-Al_2O_3 $(01\bar{1}2)$ cleavage surface using a LaB_6 source and a FEG respectively under the same diffracting condition. Large surface steps can be seen at the arrowheads in both Figures 6.15(a) and (b). The smaller surface steps, however, can only be clearly identified in the image taken using a FEG, as indicated in Figure 6.15(b). The heights of these small steps are believed to be less than 0.08 nm. The absence of small steps in the REM image in Figure 6.15(a) indicates clearly that the sensitivity of REM imaging is related to beam coherence. In addition, relatively fewer residual Fresnel fringes are observed in Figure 6.15(a).

The sensitivity to the steps of heights on the order of 0.08 nm of REM can be understood from simple phase contrast theory. The contrast of a surface step is determined by the phase difference of the electron waves reflected from the upper and lower surface terraces around the step. If $\theta \approx 25$ mrad, $\lambda = 0.0037$ nm for 100 kV and $H = 0.02$ nm, then Eq. (6.2) yields $\phi_p \approx 0.54\pi$. This is a large phase shift that can create visible Fresnel fringes in REM images. Therefore, the interference of the waves reflected from the upper and lower surface regions around a step gives strong contrast, resulting in the observation of small surface steps of heights less than about 0.08 nm. This simple calculation applies only to a perfectly coherent source, such as a FEG. However, for a LaB_6 gun, the partially coherent beam would limit the interference effect of the electrons reflected from the upper and lower surfaces around a step, which is equivalent to decreasing the visibility of the step. Therefore, improvement in beam coherence can significantly increase the sensitivity of REM images.

6.5 The effect of energy filtering

In RHEED, more than 50% of the reflected electrons are inelastically scattered. Most of the electrons suffer energy losses of 5–50 eV due to multiple valence (or plasmon for metals and semiconductors) excitations. The quality of RHEED patterns and REM images is greatly modified by inelastic scattering. First, a diffuse background is introduced into the RHEED patterns due to electron angular redistribution as a result of inelastic scattering. Second, the width of the Bragg peaks is increased due to the convolution of the inelastic scattering's angular function. Third, since the inelastically scattered electrons are incoherent, the phase contrast from surface steps is reduced. Finally, chromatic aberration introduces a focus-spreading, which reduces the image resolution and contrast. Therefore, it is desirable that the inelastic component be filtered in RHEED and REM, so that quantitative data analysis is feasible using available elastic scattering RHEED theories.

There are two methods for performing energy filtering in TEM. One method uses a Castaing–Henry filter (Castaing and Henry, 1962). The Castaing–Henry filter consists of two 90° magnetic prisms and a retarding electric field. The filter is located between the objective lens and the intermediate lens. The electrons are sent to a 90° electromagnetic sector, and then they are reflected by an electrostatic mirror. Electrons of different energies are dispersed. The second 90° prism deflects the electrons back onto the optic axis again. A slit is placed before the intermediate lens and selects electrons with specific energy losses. Then the energy-filtered diffraction pattern or image is formed by the intermediate lens and projection lens as in conventional optic systems. The Castaing–Henry filter is unsuitable for primary beam voltages greater than about 100 kV. The Ω-filter introduced by Senoussi *et al.* (1971) is a better choice for higher energy electrons. This energy filtering can only be performed in a specially built TEM. A detailed introduction to this energy-filtering system and its application has been given by Reimer *et al.* (1988 and 1990).

The other energy-filtering method uses the parallel-detection EELS system (Shuman *et al.*, 1986; Krivanek *et al.*, 1991). The electrons are dispersed by the magnetic sectors in the EELS spectrometer, and they are focused on the energy–momentum (E–q) plane, at which an energy-selecting slit is introduced. Then a set of lenses are arranged to re-disperse the electrons and form the image (or diffraction pattern) using the energy-filtered electrons. The final image/diffraction pattern is

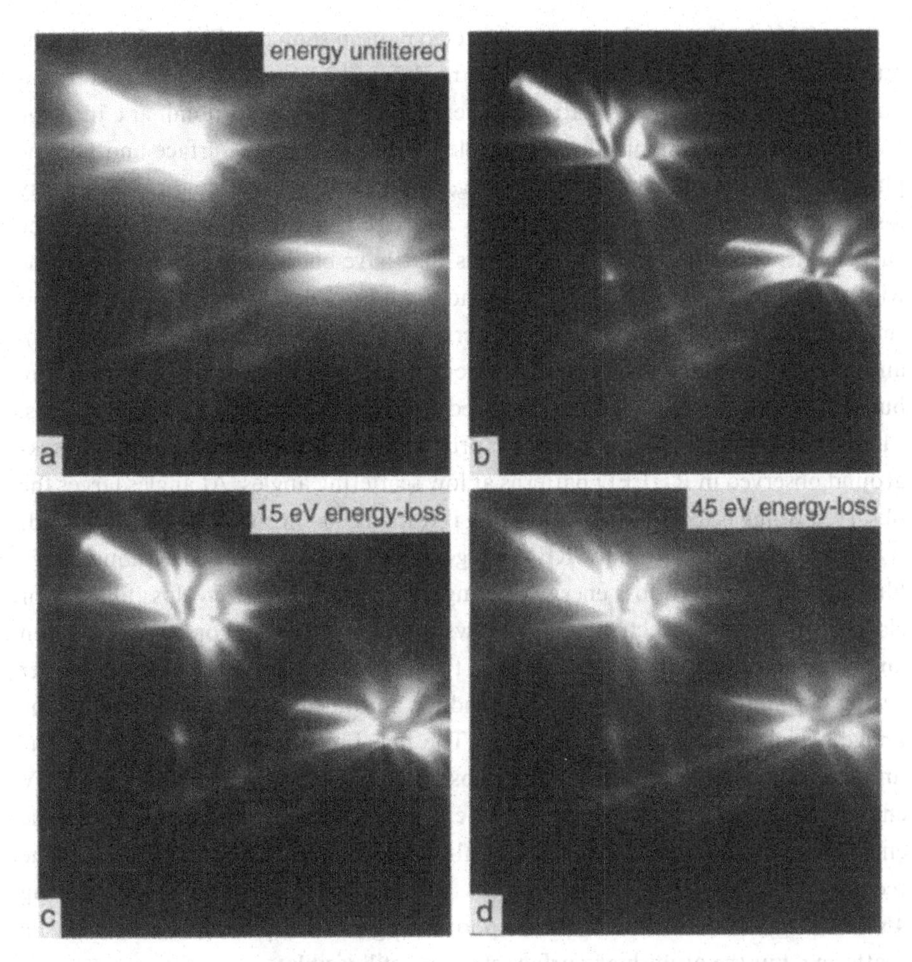

Figure 6.16 *Convergent-beam RHEED patterns of GaAs(110) recorded (a) without energy-filtering (i.e. the entirety of the electrons), (b) zero-loss energy-filtered (i.e., elastic electrons), (c) 15 eV energy-loss-filtered and (d) 45 eV energy-loss-filtered electrons. The primary beam energy is 300 keV, and the energy-selecting window width is 10 eV.*

recorded digitally using a CCD camera. This energy-filtering system can be added to any existing TEMs without any modification to the existing electron optics. The disadvantage of this system is that the high-energy-loss filtering is limited by weak intensity due to the finite size of the EELS entrance aperture, typically about 3 mm in diameter. This energy-filtering technique has been applied to obtain chemical images of thin foils in TEM. A spatial resolution better than 0.4 nm has been achieved (Wang and Shapiro, 1995d).

Figure 6.16 shows a group of RHEED patterns recorded by selecting the electrons with different energy losses, under identical diffraction and electron microscope conditions. In comparison with the pattern recorded without using the energy filter

(Figure 6.16(a)), the zero-loss energy-selected pattern shown in Figure 6.16 exhibits much better quality. It is apparent that the diffuse background has been substantially reduced and all the features have become clear. The dominant inelastic scattering processes involved are multiple valence (including surface and volume plasmons) excitation, phonon scattering, and electrons with continuous energy losses (due to *Bremsstrahlung* radiation and electron Compton scattering). The energy filter has eliminated the electrons that have energy losses larger than the width of the energy window, but the phonon-scattered electrons remain due to their small energy losses (<0.1 eV). The patterns recorded using 15 eV (Figure 6.16(c)) and 45 eV (Figure 6.16(d)) energy-loss electrons also show fine details in the pattern, but the diffuse background is enhanced with increasing electron energy loss. Therefore, multiple valence losses are primarily responsible for the diffuse background observed in RHEED patterns at low scattering angles. At angles larger the observed Bragg peaks, phonon scattering is responsible for the diffuse background.

Figures 6.17(a) and (b) show REM images of GaAs(110) recorded without energy filtering and with zero-loss energy filtering, respectively. The zero-loss filtered (or elastically scattered) electron image shows significantly better contrast, resolution and structure sensitivity. This is because the inelastically scattered electrons suffer strong chromatic aberration and so produce an out-of-focus background in the image, which is removed by the filter. The fine Fresnel fringes not seen in the unfiltered image are resolved in the zero-loss filtered image. The contrast of the 15 eV energy-loss filtered image (Figure 6.17(c)) shows better contrast than does the energy unfiltered image (Figure 6.17(a)), although it is slightly poorer than the image recorded using zero-loss filtered electrons. The contrast in the image recorded using 45 eV energy-loss electrons is rather poor owing to multiple inelastic and elastic scattering, but the atom-high surface steps are still visible.

In summary, energy filtering dramatically improves the quality of RHEED patterns. The contrast, resolution and sensitivity of REM images are greatly improved in the zero-loss filtered images. Therefore, a high-coherence electron source and an electron energy-filtering system are the two key factors that are required in order to improve the quality of REM imaging.

6.6 Determining the nature of surface steps and dislocations

6.6.1 Step height

As described in Section 6.1, images of surface steps are dominated by phase contrast. It is thus difficult to identify step height purely based on interference Fresnel fringes. In general, the following methods could be used to determine surface step height.

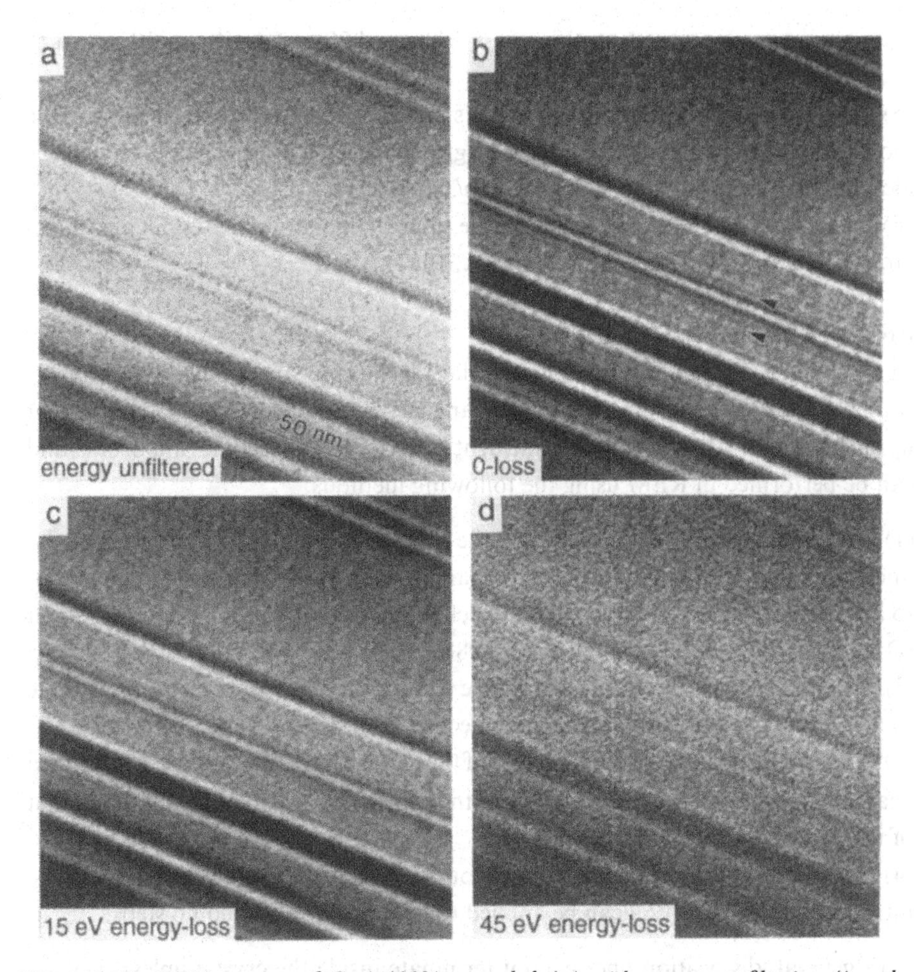

Figure 6.17 *REM images of GaAs(110) recorded (a) without energy-filtering (i.e. the entirety of the electrons), (b) zero-loss energy-filtered (i.e., elastic electrons), (c) 15 eV energy-loss-filtered and (d) 45 eV energy-loss-filtered electrons. The primary beam energy is 300 keV, and the energy-selecting window width is 10 eV.*

(1) If the step ends at a dislocation, it is then possible to conclude from the contrast of the step and its strain field whether or not the step is a one-atom-high step, provided that the dislocation system in the crystal is known (Hsu *et al.*, 1984).
(2) If the macroscopic mis-cut of the surface from a main zone axis is known, or measurable, and the steps are of uniform contrast and their distribution is periodical, then the step height can be calculated using the mis-cut angle and the lateral distance between steps.
(3) By measuring the phase shift at a surface step using electron holography, the step height can be precisely determined (see Section 12.4).
(4) By referring to theoretical calculations, the images formed by the surface

resonance wave (i.e., the 'extra' beam in Figures 4.20 and 4.21) could be used to determine step height (Wang, 1989b).
(5) When evaporation or growth on a stepped surface is observed *in situ*, the movement of steps indicates monolayer growth or evaporation. The step can be identified as one atom high (Osakabe *et al.*, 1980).
(6) In some cases, it is possible to observe steps of one-atom, two-atom and three-atom height simultaneously (Ogawa *et al.*, 1987).

6.6.2 Down and up steps

In crystal growth, the observation of surface island-type or vacancy-type terraces can help to explain the film growth mechanism. Thus the identification of down or up surface steps with respect to the incident beam direction becomes important, and can be performed in REM using the following methods.

(1) The **contrast method**. By changing the objective lens focus from over focus to under focus, the contrast of an up step changes from dark–bright (along the beam) to bright–dark, and a down step contrast changes from bright–dark to dark–bright (Osakabe *et al.*, 1981a; Hsu and Peng, 1987b).
(2) **Electron holography**. By measuring the phase shift of a surface step, it is possible to determine the nature of the step (see Section 12.4).
(3) **The SREM imaging method**. Images of the surface resonance wave (i.e., the extra beam in Figure 4.18) can directly provide the distribution of down steps but not that of up steps.
(4) **Secondary electron imaging**. The secondary electron imaging technique to be introduced in Section 12.2 can be applied to determine down and up steps.

In general, dislocation lines cannot terminate inside the crystal unless they are loops. Thus, the distribution of dislocations on the surface may give a direct measure of dislocation density inside the bulk crystal. The Burgers vector of a surface emergent dislocation could be determined in REM by either comparing a simulated REM image contrast with an observed dislocation image (Peng *et al.*, 1989) or using electron holography (see Section 12.4).

6.7 REM image resolution

The REM image resolution is limited not only by the optical system of the electron microscope but also by the glancing angle scattering geometry. In general, surface lattice images may not be formed by selecting two Bragg reflections (not superlattice reflections) as in TEM imaging for the following reasons (Yagi, 1987). First, two beams have different exit angles and different foreshortening factors in REM. In this case, the obtained lattice fringes or interference fringes do not have one-to-one

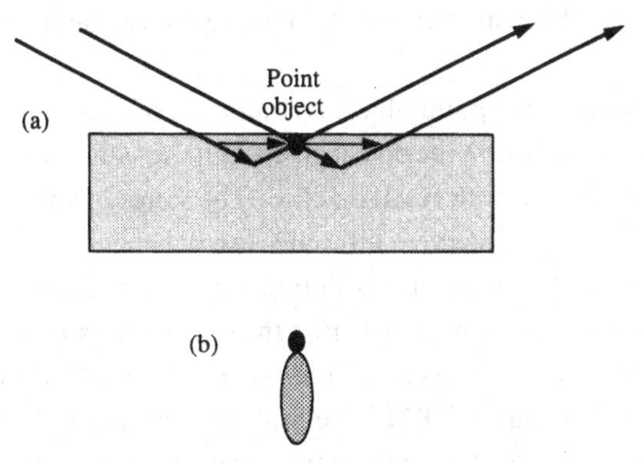

Figure 6.18 *(a) Electron beam paths before and after a point object in a crystal surface. (b) A possible form of the resulting REM image.*

correspondence with real lattices on the surface. Second, REM images are in focus only along lines perpendicular to the beam and the obtained fringes suffer large defocus and beam divergence effects. Therefore, fringes are seen only over a narrow range of defocus. Third, the lattice fringes can be seen reasonably well only if the surface lattice is nearly perfectly straight for a distance of more than about 10 nm due to the foreshortening effect. Furthermore, owing to the forefield of the objective lens, the incidence angle changes slightly on the surface along the beam direction and the phases and intensities of the reflected beams change from place to place. All of the above factors do not exist in high-resolution TEM imaging, so that REM resolution is lower than that of TEM under equivalent electron optical conditions.

The point-to-point image resolution in TEM can be measured from the optical diffractogram of a thin amorphous film, such as Ge or C. In REM it is not easy to find a test specimen that may be used to determine the experimental resolution limit, because the image resolution is greatly affected by the scattering process of electrons at the surface. For easy discussion we propose an ideal case of a point object, such as an isolated single very heavy atom replacing a lighter atom of the top surface layer (Cowley and Liu, 1993). Consider the resonance scattering illustrated in Chapter 4; the ray diagram for image formation is given in Figure 6.18.

There are several ways that the incident wave may be scattered. The wave may be scattered directly by the object into the vacuum without any penetration into the surface. This component is rather small in practice, especially for high-energy electrons. Larger contributions to the image come from beams that are strongly diffracted in the crystal (i.e., Bragg reflection), channeled along the surface layers of atoms and then scattered through small angles by the object (i.e., the surface resonance beam). Further contributions to the image arise from the ray paths through the crystal to the right of the object. Waves transmitted through the object differ in amplitude and phase from those transmitted through other surface atoms and, hence, they affect the image intensities for waves subsequently diffracted by the crystal.

Therefore, the image of the point object is composed of two components. The ray paths through the crystal before the object give much the same effect as for an object on the exit face of a rather thick crystal in TEM. The range of angles of incidence of the waves reaching the object has very little effect on the imaging process and, hence, the object may be imaged with the full resolution of the microscope, typically 0.2 nm, if elastic scattering alone is considered. For those ray paths passing through the crystal after the object, the effect is much the same as for an object on the entrance face of a rather thick crystal in TEM. The multiple scattering of the waves in the crystal results in uncertainty about where the waves originate and, hence, gives an apparent spreading of the object. The observed image of the object may therefore be as suggested in Figure 6.18(b). A sharp image of the object with resolution about 0.2 nm is accompanied by a diffuse tail or 'shadow', which may have contrast of the same or opposite sign and dimensions 1–2 nm. Therefore, REM image resolution is greatly limited both by the scattering geometry and by multiple electron scattering regardless of the optic system.

The above discussion has outlined the effect of multiple scattering on REM resolution. However, this limitation affects primarily the resolution along the beam direction (called 'longitudinal' image resolution). The resolution perpendicular to the beam, the 'transverse' image resolution, however, is mainly determined by the optic system.

To give an optimum estimation of the transverse image resolution, we now consider only the limitation of the optical system. In REM, since more than half of the electrons have suffered one or more plasmon energy losses, the chromatic aberration limits the image resolution. For axial rays the least resolvable distance d_r is limited by spherical and chromatic aberrations as well as the diffraction limitation, and it is given by incoherent imaging theory as (Booker, 1970)

$$d_r^2 = C_s^2\alpha^6 + C_c^2\alpha^2\left(\frac{\Delta E}{E}\right)^2 + \frac{0.61^2\lambda^2}{\alpha^2}, \tag{6.16}$$

where α is the semi-angle of the objective aperture and ΔE is the average energy loss of electrons used for imaging. The first and second terms are due to spherical and chromatic aberrations, respectively. The last term is the usual Rayleigh criterion in optics. In REM, the size of the objective aperture is generally small in order to enhance image contrast. Thus, by neglecting the first term in Eq. (6.16), the optimum choice of objective aperture size is

$$\alpha_0^2 \approx \frac{0.61\lambda}{C_c}\frac{E}{\Delta E}, \tag{6.17}$$

and the corresponding resolution limit is

$$d_r \approx \left(\frac{1.22 C_c \lambda \Delta E}{E}\right)^{1/2}. \tag{6.18}$$

For $C_c = 2$ mm and $\Delta E = 10$ eV, $d_r \approx 0.82$ nm for $E = 120$ keV and 0.15 nm for $E = 1$ MeV. These values are in good agreement with the experimentally obtained resolutions of 0.9 nm at 120 keV (Hsu, 1983) and 0.33 nm at 1 MeV (Koike *et al.*, 1989). The resolution given by Eq. (6.18) is an optimum estimation of the transverse image resolution.

As shown in the last section, the image resolution can be significantly improved using the energy filter, which removes the electrons with energy losses. Thus, the chromatic aberration term disappears in (6.16). For the zero-loss energy-filtered REM images, the resolution is determined by

$$d_r^2 = C_s^2 \alpha^6 + \frac{0.61^2 \lambda^2}{\alpha^2}. \tag{6.19}$$

The best image resolution is

$$d_r = 0.61^{3/4}(3^{-3/4} + 3^{1/4})^{1/2} \lambda^{3/4} C_s^{1/4} \tag{6.20a}$$

for

$$\alpha^4 = 0.61\lambda/(3^{1/2} C^s). \tag{6.20b}$$

For 100 kV electrons and $C_s = 1.2$ mm, $d_r = 0.6$ nm.

6.8 High-resolution REM and Fourier imaging

6.8.1 Imaging a reconstructed layer

Although the REM image resolution is limited by various factors, it is still possible to form lattice fringes using superlattice-reflected beams, and this has been shown to be an effective method for imaging the periodicity of surface reconstruction (Tanishiro and Takayanagi, 1989; Hsu and Kim, 1991), which is performed by enclosing the specularly reflected beam and a few superlattice-reflected beams by an objective aperture of small size; the interference among these beams exhibits the periodicity of surface reconstruction in the direction perpendicular to the incident beam. The movement of surface steps on Si(111) has been recorded in real time using this technique (Murooka *et al.*, 1992). Figure 6.19 shows a series of high-resolution REM images of the Si(111) 7×7 surface during gold deposition at 923 K. These images were recorded at 200 kV in an UHV electron microscope (JEM 2000FXV with $C_s = 0.7$ mm and $C_c = 1.2$ mm). The growth of the 5×2 structure is the result of

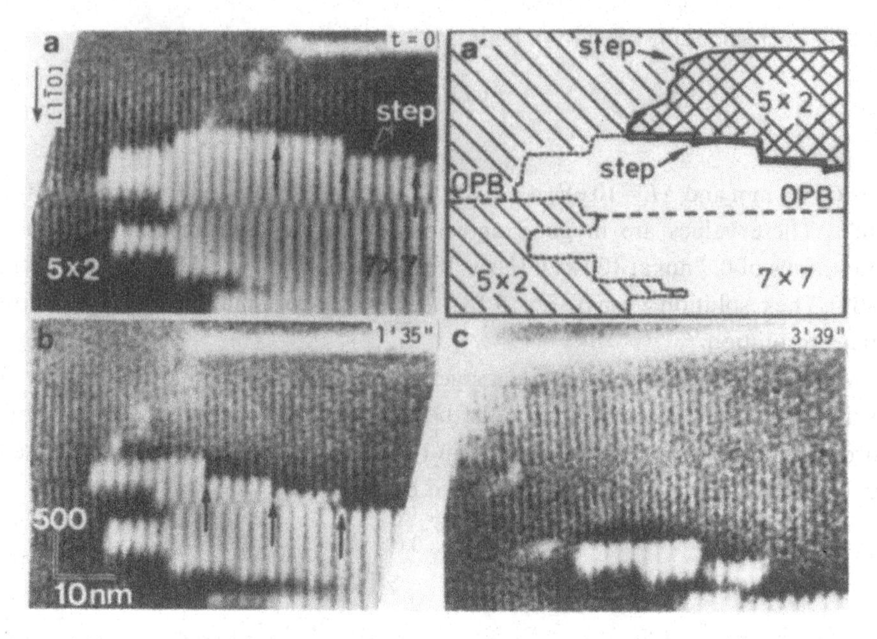

Figure 6.19 *A series of high-resolution REM images of the Si(111) 7 × 7 surface (bright contrast) recorded at (a) t = 0, (b) t = 1 min 35 s, and (c) t = 3 min 39 s during transition to the 5 × 2 structure (gray contrast) caused by the adsorption of gold. (a′) Illustrates the domains, a step and OPBs seen in (a). The areas of the 5 × 2 structure of the lower and higher terraces are hatched and cross-hatched, respectively. The step between the 7 × 7 and the 5 × 2 surface took the shape of a staircase of 5 × 7a = 35a as indicated by vertical arrows in (a) and (b), where a = 0.384 nm is the unit mesh of the 1 × 1 surface (courtesy of Drs Tanishiro and Takayanagi, 1989).*

gold deposition. Besides the observed 1.7 nm 5 × 2 and 2.3 nm 7 × 7 superlattice fringes, out-phase boundaries (OPB) are also seen. The appearance of an OPB is indicated by a sharp shift of the fringes, different from the Fourier image to be shown in the next section. With increasing deposition time, the entire surface gets almost completely covered by the 5 × 2 structure (Figure 6.19(c)). This type of image together with the RHEED pattern can be recorded in real time, revealing the *in situ* structural evolution of the surface.

6.8.2 Fourier images

A more careful examination of Figure 6.19(c) shows that the 5 × 2 fringes in the upper part are shifted by a half period with respect to fringes in the bottom part. This shift occurs slowly within a certain distance across the surface and is different from the sharp phase shift introduced by the out-phase boundary. This effect is a feature of the Fourier image of a periodic object (Cowley, 1981) and it is often seen in REM. We now introduce the nature of the Fourier image.

For simplification, we consider a flat surface without steps and treat the problem as one-dimensional. The reflected wave can be written as a superposition of the plane waves

$$q(x) = \sum_{g_s} F_{g_s} \exp(2\pi i g_s x), \tag{6.21}$$

where g_s refer to the reciprocal-lattice vectors of the superlattice reflections. From Abbe's imaging theory, the diffraction amplitude is a Fourier transform of $q(x)$,

$$\Psi(u) = \sum_{g_s} F_{g_s} \delta(u - g_s). \tag{6.22}$$

Applying the contrast transfer function of the optic system and taking into account the size of the objective aperture, the wave observed in the image plane is

$$\Psi(x) = \text{FT}^{-1}\left(\sum_{g_s} F_{g_s} \delta(u - g_s) T_{\text{obj}}(u)\right) = \sum_{g_s} F_{g_s} T_{\text{obj}}(g_s) \exp(2\pi i g_s x). \tag{6.23}$$

Since the size of the objective aperture is small, we can ignore the spherical aberration effect, so that T_{obj} can be approximated by

$$T_{\text{obj}}(u) \approx A_{\text{obj}}(u) \exp(\pi i \Delta f \lambda u^2), \tag{6.24}$$

and Eq. (6.23) becomes

$$\Psi(x) = \sum_{g_s} F_{g_s} A_{\text{obj}}(g_s) \exp(\pi i \Delta f \lambda g_s^2) \exp(2\pi i g_s x). \tag{6.25}$$

For a special case in which the defocus value $\Delta f = nA_s^2/\lambda$, where $n = 2m + 1$ is an odd integer and A_s is the periodicity of the superlattice structure ($g_s A_s = l =$ integer), the phase factor $\exp(\pi i \Delta f \lambda g_s^2) = \exp(\pi i g_s A_s)$ because l is odd if nl^2 is odd, thus

$$\Psi(x) = \sum_{g_s} F_{g_s} A_{\text{obj}}(g_s) \exp[2\pi i g_s (x + A_s/2)], \tag{6.26}$$

and the image intensity distribution $|\Psi(x)|^2$ is exactly the same as for $\Delta f = 0$ except that it is translated by half the periodicity. In REM images, the objective lens focus changes continuously along the beam direction, so that at some surface area the condition $\Delta f = nA_s^2/\lambda$ is satisfied, resulting in half-period-shifted image contrast, as observed in Figure 6.19.

For the special case in which the defocus value is $\Delta f = 2nA_s^2/\lambda$, the wave function

$$\Psi(x) = \sum_{g_s} F_{g_s} A_{\text{obj}}(g_s) \exp(2\pi i g_s x), \tag{6.27}$$

which is again periodic but without translation. This means that the image contrast really reflects the periodicity of the surface superlattice.

It must be pointed out that the appearance of superlattice reflections in RHEED may not necessarily indicate the occurrence of surface reconstruction, since

Figure 6.20 *High-resolution REM images of the $LaAlO_3(100)$ surface, showing the interference fringes between (10 0 0) and the 5 × 5 superlattice reflected beams. In (b) and (c) are shown enlargements of the areas b and c indicated in (a), respectively.*

superlattice reflections can be generated by new phases that are grown on the surface. In this case, the HR-REM technique can be effectively used to identify the distribution of surface reconstructed areas. Figure 6.20 shows a group of HR-REM images of the $LaAlO_3(100)$ surface. The RHEED pattern of the surface was shown in Figure 2.3. HR-REM images are recorded by selecting the (10 0 0) and superlattice-reflected beams using a larger objective aperture. Figures 6.20(b) and (c) are enlargements of the areas b and c, respectively, indicated in Figure 6.20(a). The 1.89 nm fringes are produced by the interference of the first-order superlattice

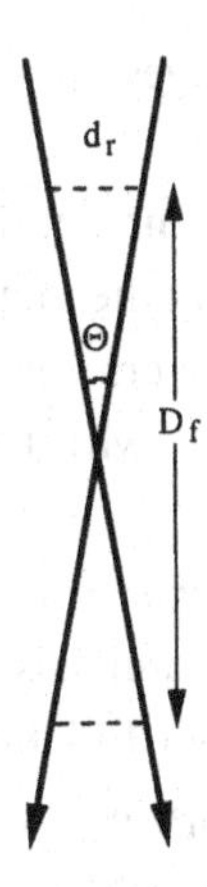

Figure 6.21 *A schematic ray diagram for defining the depth of field in TEM.*

reflection with the (10 0 0) beam. Fringes separated by 0.95 nm result from the interference of the second superlattice reflection with the (10 0 0) beam. The unreconstructed regions are in brighter contrast and do not display fringes. The fringes can only be seen in the near focus region due to the smaller depth of field for larger objective aperture. This example demonstrates the application of REM for imaging surface reconstruction.

6.9 Depth of field and depth of focus

In TEM, focal depth is not a question because the thickness of the foil is less than about 200 nm. The entire illumination area is assumed to be in-focus and the focal depth is much larger than the foil thickness. In REM, however, the glancing incidence geometry of the beam covers a large surface area that is almost parallel to the incident beam. If the beam diameter is 100 nm and incident angle $\theta = 25$ mrad, the illuminated surface area is about 4 μm along the beam direction. Thus, the depth of field is important in REM. We now use a simple model to show which factors affect the depth of field.

The amount by which the objective lens may be out of focus on either side of the object plane before blurring becomes comparable to the attained resolution d_r is called the depth of field D_f. D_f is the distance between two marginal planes on either side of the object plane. D_f is related to the objective aperture angle $\Theta(=2\alpha)$ and image resolution d_r by $D_f = 2d_r/\Theta$ (see Figure 6.21). In REM, for the smallest objective aperture, $\Theta \approx 2$ mrad, $D_f = 1$ μm for $d_r = 1$ nm and $D_f = 2$ μm for $d_r = 2$ nm. If a large aperture is used to record lattice images of surface reconstruction, $\Theta \approx 6$ mrad, then $D_f = 333$ nm for $d_r = 1$ nm. If the foreshortening effect is considered, then the width of the in-focus area in the REM image is $D_f\theta \approx 8.3$ nm for $\theta = 25$ mrad. This is the reason that the interference fringes are only observed at near focus positions (see Figure 6.20).

For low-resolution REM imaging, the depth of field can be very large if a small objective aperture is used for recording the image. Figure 6.22 shows a REM image of $LaAlO_3(100)$. The in-focus area covers a large portion of the image. The bright and dark contrast areas are the surfaces with 1×1 (i.e., no reconstruction) and 5×5 reconstruction, respectively.

The depth of field may also be increased with the use of a FEG because of small beam convergence. Figure 6.23 is a comparison of two REM images of the same surface area of Pt(111) recorded using FEG and LaB_6 sources under similar optical configurations. In addition to the improvement in image contrast and resolution, the image recorded with the FEG also shows a larger depth of field.

The distance in the final image space corresponding to D_f in object space is called the depth of focus. Under the thin lens approximation, the distances of the object and its image from the lens are related by (see the ray diagram shown in Figure 1.11)

$$\frac{1}{f}=\frac{1}{R}+\frac{1}{d_i}, \tag{6.28}$$

where $d_i = f + R'$, with R' being the distance from the diffraction plane to the image plane (see Figure 1.11). Since the focal length f is fixed,

$$\frac{\Delta d_o}{d_o^2}+\frac{\Delta d_i}{d_i^2}=0. \tag{6.29}$$

Since magnification $M = |d_i|/|d_0|$ and $\Delta d_o = D_f$, the depth of focus is

$$D_i = |\Delta d_i| = M^2 D_f. \tag{6.30}$$

Owing to the high magnification employed in REM, the depth of focus is for all practical purposes infinite. If the magnification used for REM is $M = 10000$, the depth of focus in the image plane is $D_i = 33.3$ m for $D_f = 333$ nm. This distance is huge in comparison with the size of the photographic recording system of the microscopy. Thus a clear image can be recorded if the film plate is placed at any position near the viewing screen.

6.10 Double images of surface steps

In REM, the contrast of step images depends sensitively on diffracting conditions. A single-atom-high surface step can show double images (i.e., the step is imaged twice) (Uchida and Lehmpfuhl, 1987). Figures 6.24(c) and (d) show two REM images of Pt(111) recorded under diffraction conditions shown in Figures 6.24(a) and (b), respectively. In Figure 6.24(a), the specularly reflected beam intersects with the

Figure 6.22 *A REM image of $LaAlO_3(100)$ showing the depth of field in REM.*

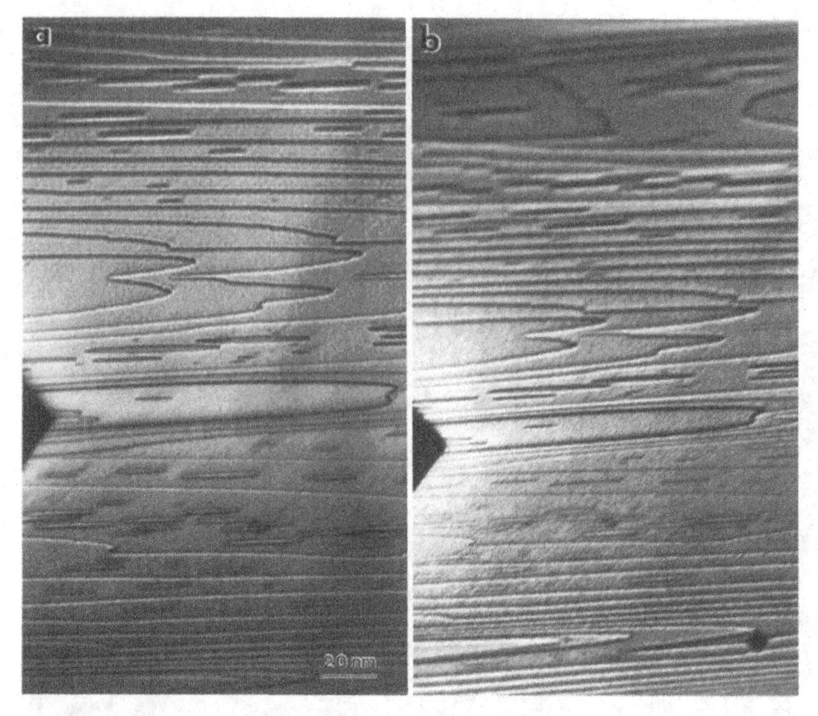

Figure 6.23 *REM images of Pt(111) recorded using a TEM equipped with (a) FEG and (b) LaB_6, showing the improvement of depth of field using a FEG TEM. The beam energy is 100 keV.*

resonance parabola (the lower one indicated by an arrowhead), and the corresponding image (Figure 6.24(c)) shows the normal single-step contrast. However, when the specularly reflected beam is tilted to intersect with the Kikuchi envelope (indicated by an arrow head in Figure 6.24(b)), a double image of surface steps is shown (Figure 6.24(d)). The doubling in contrast is preserved even when the surface is coated by a layer of carbon contaminant, suggesting that the modification of the surface potential is irrelevant to the formation of the abnormal contrast. This double-image effect is observed for both up and down steps.

In general, the double-step image can only be seen under the diffraction condition shown in Figure 6.24(b), in which the beam used for imaging coincides with an intersection of the Kikuchi lines running parallel to and inclined to the crystal surface (Yao and Cowley, 1992). As can be noticed in Figure 6.24(b), the specularly reflected beam intersects with both the resonance parabola (the lower parabola) and the Kikuchi envelope (the upper parabola) due to slight convergence of the incident beam. Thus the image is formed by two components of electrons, which are propagating with slightly different angles. Therefore, a small displacement in the direction parallel to the incident beam is produced in the image, resulting in the double contrast of a single surface step. Under the diffraction condition shown in

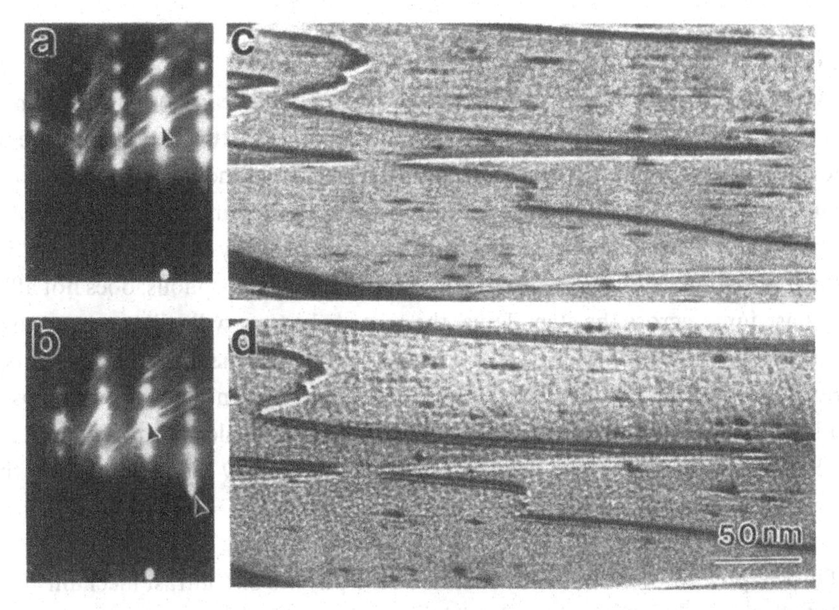

Figure 6.24 In (a) and (b) are shown RHEED patterns recorded from a Pt(111) surface with slightly different diffraction conditions: (c) and (d) are REM images recorded using the (666) specularly reflected beam under the conditions shown in (a) and (b), respectively (courtesy of Dr N. Yao). In (b), the surface resonance wave, propagating parallel to the surface, is strong because it is not re-diffracted to contribute to the specular beam (666) under the non-resonance condition. In (a), however, the surface resonance wave is not seen because it is totally diffracted to contribute to the (666) beam under the resonance condition.

Figure 6.24(a), however, the reflected beam is strongly excited only when it falls on the resonance parabola. Thus, the image is formed primarily by the surface resonance electrons, resulting in the disappearance of the double line contrast. If one examines Figure 6.24(c) carefully, then it will be seen that a weak contrast still shows up besides the main step image, which results from the weak contribution of the electrons coinciding with the Kikuchi envelope.

Dynamical calculations for stepped surfaces suggest that the double images could be formed by the two beams propagating along slightly different directions, one being the directly reflected wave and the other being the surface-resonance-reflected wave (Wang, 1988a). A simple interpretation could be outlined based on the splitting of the resonance parabola from the Kikuchi envelope as shown in Figure 4.9(b). It is possible to excite two surface resonance waves propagating in slightly different directions. If both the reflected beams were selected by the objective aperture for surface imaging, then it would be possible to obtain double images of a single step. A recent dynamically calculated result of Anstis (1994) has shown reasonable agreement with experimental observations.

6.11 Surface contamination

REM images are normally recorded using Bragg-reflected beam(s). Thus, the image contrast is directly affected by the periodically arranged lattices. If a thin layer of amorphous contaminant is present, then the phase shift of the electron beam across the upper and lower terraces of a surface is unaffected, because the image is formed by the Bragg-reflected electrons, which are generated only by the crystal lattice. The surface contaminated layer, as long as it is uniform and amorphous, does not affect the phase jump across the step. Thus, the recorded image will still preserve phase contrast, showing atom-high steps. This is probably the key that permits a conventional non-UHV TEM to be applied to image atom-high surface steps. A similar argument applies to diffraction contrast images of dislocations. Therefore, in conclusion, REM image resolution is almost unaffected by the presence of a thin surface contaminant.

This chapter has systematically described the physics of contrast mechanisms of REM. Experimental factors, such as beam coherence and energy-spreading and diffracting conditions, which significantly influence the resolution and contrast of REM images, have been examined in detail. Methods for determining step height and the nature of dislocations have been summarized. This chapter together with Chapter 5 has dealt with the fundamentals of REM.

CHAPTER 7

Applications of UHV REM

As a surface-sensitive technique, REM provides a real-space image of surface structures accompanied by crystallographic information from RHEED. This method has been extensively applied to image surface steps and related surface phase transformations. *In situ* surface dynamical processes that have been studied by REM include sublimation (or step movement), structure transformation, atom/molecule adsorption, epitaxial thin film growth, surface oxidation, surface–gas reaction, electric-current-induced surface phenomena, and ion sputtering and annealing. In this section, the surfaces that have been studied by REM in ultra-high-vacuum (UHV) TEMs are summarized. A more complete review regarding UHV REM has been given by Yagi (1993a).

7.1 UHV microscopes and specimen cleaning

For surface studies, UHV is mandatory in order to maintain clean surfaces. The need for the highest possible vacuum can be seen by recalling the kinetic theory of gases. The arrival rate of molecules at a unit surface area can be calculated from Maxwell's distribution function and the result is

$$\frac{dn_a}{dt}=\frac{n_m v_m}{4}, \tag{7.1}$$

where n_m is the number density of molecules at pressure P and temperature T,

$$n_m = P/(k_B T), \tag{7.2}$$

and v_m is the average velocity of molecules of mass M

$$v_m=\left(\frac{8k_B T}{\pi M}\right)^{1/2}, \tag{7.3}$$

with k_B Boltzmann's constant. Thus, the arrival rate is

$$\frac{dn_a}{dt}=\frac{P}{(2\pi M k_B T)^{1/2}}. \tag{7.4}$$

For example, if we substitute the number for nitrogen ($M = 28 \times 1.66 \times 10^{-27}$ kg) at 300 K and a pressure of 10^{-6} Torr $= 1.33 \times 10^{-6}$ mbar $= 1.33 \times 10^{-4}$ N m^{-2}, then we find from Eq. (7.4) that $dn_a/dt = 3.8 \times 10^{18}$ m^{-2} s^{-1}. This corresponds to approximately half a monolayer per second. This means that the surface will be immediately covered by some adsorbed molecules once it is exposed to the vacuum. Even an improvement in the pressure by four orders of magnitude to 10^{-10} Torr (or mbar) will give a gas bombardment rate of 10^{-4} monolayer per second.

Therefore, surface studies must be performed at pressures better than 10^{-10} Torr in order to give enough time for data acquisition from clean surfaces. The pressure in a conventional TEM, however, is about 10^{-7}–10^{-8} Torr. For sensitive surfaces, an UHV microscope and a specimen preparation chamber must be available. The entirety of the specimen preparation procedures, such as sputtering, heat cleaning, deposition, cleavage and ion sputtering, have to be performed under UHV conditions. RHEED and Auger electron spectrometers are usually installed in the specimen chamber in order to examine the surface before its introduction into the microscope. An *in situ* experiment is performed when the specimen is loaded on the microscope stage inside the column. *In situ* techniques include use of a specimen temperature control stage for studying surface phase transitions and sublimation, deposition for studying surface adsorption and thin film growth, use of a gas inlet for surface reaction experiments, and application of DC through the specimen to study current effects. The specimen holder must have double-tilting capability and a large capacity so that specimens of thickness about 0.5 mm can fit in. A heating furnace needs to be built in order to observe *in situ* phenomena.

While an UHV environment guarantees that a surface should not be influenced by the adsorption of ambient atoms and molecules on a time scale of a few hours, an additional requirement for studying properties of clean surfaces is to be able to clean them in the vacuum system and then reconstruct an ordered surface. The commonly used techniques for *in situ* cleaning are cleavage, heating, ion bombardment and chemical processing. The cleavage method is limited mostly to oxides and semiconductor specimens. It may be that a cleaved surface presents a structure different from that obtained by heating to allow the surface to reach equilibrium. Heating a surface, like heating the vacuum chamber, can lead to desorption of adsorbed species. This method of cleaning has been most used for tungsten and similar materials with high melting points, for which the surface oxides are flashed off below the melting points. However, the method is not effective for removing carbon contamination. The use of Ar ion bombardment of a surface to remove layers of the surface by sputtering is the most widely used, particularly for metal surfaces. The technique is effective in removal of many atomic layers of a surface. Even if an impurity species is far less effectively sputtered than the substrate, it can be removed eventually. One disadvantage of ion bombardment is that the surface is left in a heavily damaged state, usually with embedded Ar atoms, so that the surface must be annealed to restore the order. It has to be pointed out that impurity segregation at the surface can be produced by heating. *In situ* chemical cleaning involves introduction of gases into the vacuum system at low pressures (10^{-6} Torr or less), which react with the impurities on a surface to produce weakly bound species that can be thermally desorbed. It is most widely used for removal of carbon from refractory metals such as tungsten. Exposure of such a surface to O_2 at elevated temperatures leads to removal of C as desorbed CO, leaving an oxidized surface, which can then be cleaned by heating alone.

7.2 *In situ* reconstruction on clean surfaces

An important application of UHV REM is to directly image surface reconstruction. An outstanding example is the observation of the phase transition between 7 × 7 and 1 × 1 structures on clean Si(111) surfaces (Osakabe *et al.*, 1981b; Latyshev *et al.*, 1991), as shown in Figure 7.1. The Si(111) surface is an ideal sample for REM imaging in UHV, and it is easy to prepare and clean. The surface shows many wavy atom-high surface steps. Many *in situ* deposition experiments have been performed on Si(111) (Yagi, 1993a and b). The Si(111) surface is covered by the 1 × 1 structure at temperatures above $T_c = 1083$ K. Below this temperature, the surface starts to exhibit the 7 × 7 structure. The formation of the 7 × 7 structure is characterized by the superlattice reflections appearing in the RHEED pattern. Since more electrons are scattered out of the bulk reflection, which will be used to form the REM image, the surface areas exhibiting 7 × 7 structure show darker contrast in the image. More importantly, the formation of the 7 × 7 structure is clearly associated with one-atom-high surface steps. This information cannot be easily provided by other surface-sensitive techniques, such as LEED. Thus, the step edges are nuclear sites for forming the 7 × 7 structure. The same surface terrace can exhibit both 1 × 1 and 7 × 7 structure, and there are no steps to isolate them.

In situ surface step movement on Si{111} has been reported by Alfonso *et al.* (1992). The thermal fluctuation of isolated steps at 900 °C was measured and the result was applied to deduce the step stiffness.

Observations of surface structure evolution at early stages of thin film growth is an important application of REM. The REM images can be recorded in real-time while the deposition is conducted continuously, providing an *in situ* technique. In general, the surface temperature has to be controlled to be within a certain range in order to stimulate the film growth. Heating of the specimen can remarkably change the behavior of surface reconstruction, eventually affecting the film growth. Under the stimulation of heating, a surface phase transition between 2 × 1 and 1 × 2 structures has been observed on Si(001) (Kahata and Yagi, 1989a). Heating can also reduce metal particles from an InSb(111) surface (Nakada *et al.*, 1989). After being heated to 693 K, an Sb-stabilized 2 × 2 surface transforms into an In-stabilized 3 × 1

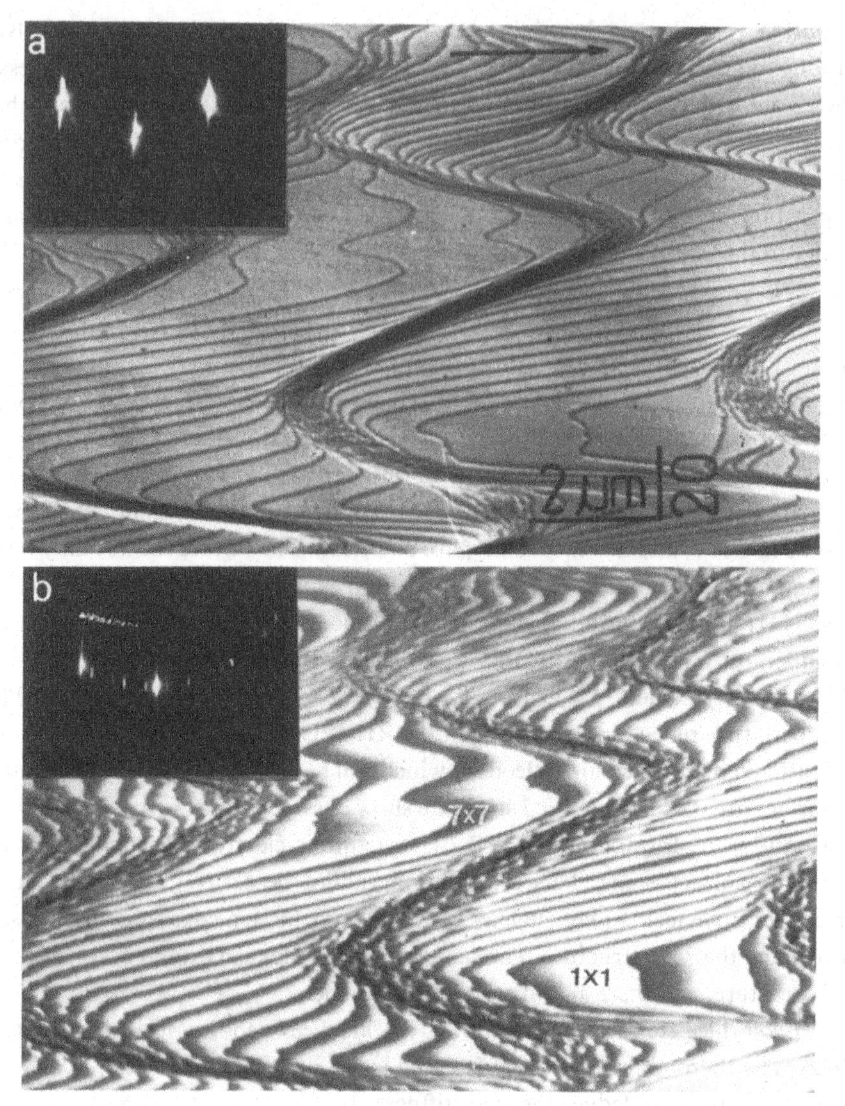

Figure 7.1 *Reflected electron images and diffraction patterns of the Si(111) surface during the 1 × 1 to 7 × 7 transition. In (a) is shown the image of the 1 × 1 surface reconstruction above the critical temperature $T_c = 1083$ K. In (b) is shown the coexistence of the 1 × 1 (bright areas) and the 7 × 7 (dark contrast) surface structures. The transition to the 7 × 7 reconstruction is clearly associated with one-atom-high surface steps (courtesy of Dr Latyshev et al., 1991).*

surface, resulting in the formation of In particles; and an Sb-stabilized 2 × 6 surface transforms into an In-stabilized 2 × 2 surface at 573 K.

Reconstructions of crystal surfaces have been observed on clean metal surfaces, such as Au(111) 23 × 1 (Uchida, 1987), Au(100) 5 × 20 (Uchida, 1987) and Pt(100) 5 × 20 (Lehmpfuhl and Uchida, 1988). Gold atom deposition on Pt(111) surfaces

Table 7.1. *Surface reconstructions observed by* in situ *REM after atom depositions.*

Structure	References
Si(111) 5 × 1–Au	Osakabe *et al.* (1980), Krasilnikov *et al.* (1992)
Si(111) 6 × 6–Au	Osakabe *et al.* (1980)
Si(111)–Cu	Ishizuka *et al.* (1986)
Si(001) $\sqrt{26} \times 3$–Au	Yagi *et al.* (1980)
Si(111) 7 × 7–1 × 1–Au	Tanishiro *et al.* (1981 and 1986)
Si(111) 7 × 7–Au	Tanishiro *et al.* (1981 and 1990)
Si(111)–Au	Latyshev *et al.* (1992)
Si(111) 5 × 2–Au	Tanishiro and Takayanagi (1989), Tanishiro *et al.* (1990)
Si(111) $\sqrt{3} \times \sqrt{3}$–Au	Tanishiro and Takayanagi (1989)
Si(111) 7 × 7–Ag	Osakabe *et al.* (1980)
Si(111) 7 × 7–Ge	Kajiyama *et al.* (1986)
Si(111) 7 × 7–Cu	Ishizuka *et al.* (1986)
Si(111) 7 × 7–1 × 1–Si	Latyshev *et al.* (1989b)
Si(111) $\sqrt{3} \times \sqrt{}$–Ag	Tanishiro *et al.* (1991a)
Si(111) 7 × 7–Pb	Tanishiro *et al.* (1991b)
Si(111) 7 × 7–Sn	Iwanari and Takayanagi (1991)
Si(111) 7 × 7–Ge	Krasilnikov *et al.* (1992)
Si(111) 5 × 5–Ge	Krasilnikov *et al.* (1992)
Si(111) 1 × 1–Ge ⇔ 7 × 7–Ge	Krasilnikov *et al.* (1992)
Si(111) $\sqrt{3} \times \sqrt{3}$–(Cu, Au)	Yagi *et al.* (1992); Homma *et al.* (1991a)
Si(111) 7 × 7–(Cu, Au)	Yagi *et al.* (1992); Homma *et al.* (1991a)
Si(5 5 12)	Suzuki *et al.* (1994)
InP(110), P sublimation	Gajdardziska-Josifovska and Smith (1994)
Pt(111)–Au	Yagi *et al.* (1987)
Au(100)	Wang N. *et al.* (1993)

has shown monolayer nuclei around Pt surface steps (Ogawa *et al.*, 1986; Yagi *et al.*, 1987).

7.3 Surface atom deposition and nucleation processes

Another important application of UHV REM is *in situ* imaging of the surface structure transition during atom deposition. This is particularly useful during MBE growth. Various atom depositions have been studied in cases of metal-on-semiconductor, semiconductor-on-semiconductor, and metal-on-metal systems. Table 7.1 gives a summary of the reconstructed surfaces due to atom depositions. In general, RHEED patterns provide crystallographic structures of the top few surface layers, and REM directly gives the distribution of deposited atoms and its relation with surface steps.

Figure 7.2 shows REM images of a Si(111) surface before and after deposition of 0.7 monolayer of Si at $T = 843$ K. It can be seen that Si deposition results in the

Figure 7.2 *REM images of a clean Si(111) surface (a) before and (b) after the deposition of 0.7 Si monolayer at T = 843 K, showing the distribution of the adsorbed Si atoms (courtesy of Latyshev et al., 1992).*

development of black–white spotted contrast in the images, caused by two-dimensional islands (Latyshev *et al.*, 1989a). Near the one-atom-high steps, there are areas that are free from islands. Their width increases with temperature and is independent of the distance between the steps. Thus, a one-atom-high step is an effective sink for adatoms within a surface area whose width is determined by the adatom diffusion length. This effect is expected from the consideration of surface

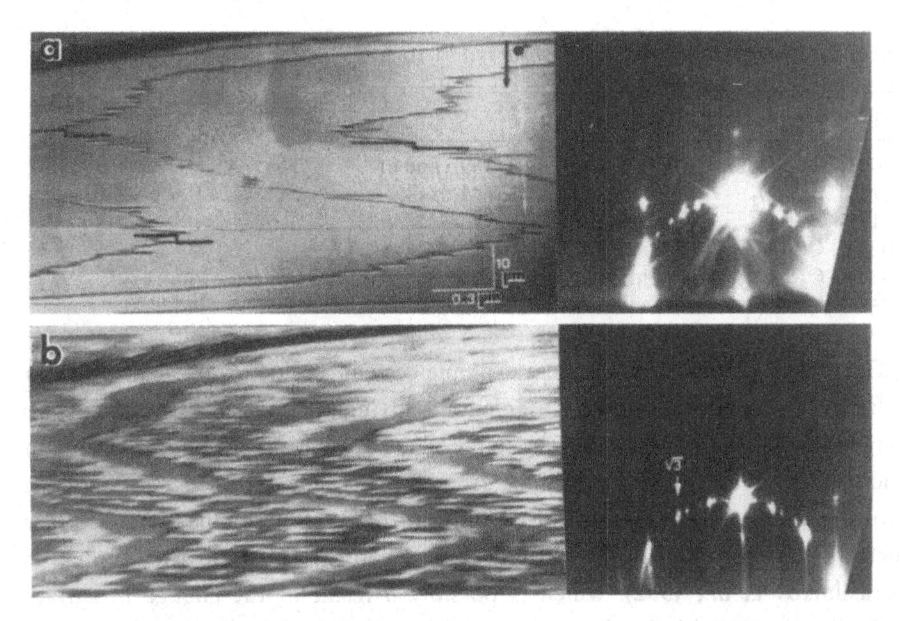

Figure 7.3 REM images and the corresponding RHEED patterns of a Si(111) surface recorded (a) before and (b) after simultaneous deposition of Au and Cu for a thickness of a half monolayer each. Bands of dark regions formed on higher side terraces of steps and dark patches on the terraces are $\sqrt{3} \times \sqrt{3}$ structures formed by co-deposition. Bright regions are the 7×7 structure (courtesy of Yagi et al., 1992).

energy. The temperature-dependence of the widths of the areas free from islands allows one to estimate the surface diffusion activation energy of adatoms as 1.3 ± 0.2 eV (Latyshev *et al.*, 1992).

In two-dimensional alloy adsorbates, the surface structure depends on the atomic ratio of the two elements (Yagi *et al.*, 1992; Homma *et al.*, 1991a). Figure 7.3 shows REM images of a Si(111) surface taken before and after simultaneous deposition of Cu and Au at about 1070 K. The atomic ratio of Cu to Au was controlled to be 1 : 1, and the total amount of deposition was one monolayer. In Figure 7.3(a) the up and down steps are indicated. In Figure 7.3(b), dark regions are seen to form along the steps but on the higher side of the steps, corresponding to the formation of a $\sqrt{3} \times \sqrt{3}$ structure. On the left-hand side of the dark regions, bright regions are seen and covered by the 7×7 structure. Mixtures of bright and dark regions are seen along the lower side of the steps. These observations have shown clearly the importance of surface steps for thin film growth. As illustrated in Chapter 6, REM is a sensitive technique, which can easily show the step distribution on surfaces.

In thin film growth, the amount of atom deposition is normally given in monolayers. The grown layers are given based on the average growth thickness and this does not reflect the uniformity of the growth. Surface islands are usually formed during initial growth due to their lower surface energy.

Table 7.2. *Surface–gas reactions observed by* in situ *REM*.

Structure	References
Si(001) 2×1–O_2	Kahata and Yagi (1989b and c)
Si(111) 7×7–O_2	Shimizu *et al.* (1985) Murooka *et al.* (1992)
Si(111) 7×7–H	Ohse and Yagi (1989)
Pt(111)–CO	Uchida *et al.* (1990)
Pt(111)–O_2	Uchida *et al.* (1990)

Growth of Ge on Si is a typical example of semiconductor-on-semiconductor MBE growth. This is a typical example for studying the interface structure of two semiconductor materials. *In situ* imaging of the growth process is important for understanding the phase transitions from Si (1×1)Ge to Si (7×7)Ge. REM provides an easy way of monitoring the diffusion of Ge atoms on the surface and the surface step clustering during the Si (7×7)Ge to Si (1×1)Ge transformation (Krasilnikov *et al.*, 1992), which is possibly a result of the change in surface activation energy with the change in surface atomic structures.

7.4 Surface–gas reactions

Studies of solid–gas reactions are important in surface science in order to understand the adsorption, reaction and diffusion of gas molecules on the surface. REM has provided an effective method for examining the *in situ* process of surface reaction. Table 7.2 gives a summary of REM studies of gas–solid reactions. By using a gas inlet device in an UHV REM, the initial stage of surface–gas reactions has been imaged by examining the movement of surface steps, changes in image contrast, and formation of one-atom-high surface hollows. Figure 7.4 shows a series of REM images of a Si(001) surface during exposure to oxygen at 1070 K. The surface is atomically flat with some atom-high steps. The boundaries between 1×2 and 2×1 reconstructed areas have shifted as the result of surface oxidation. Circular hollows are nucleated at the middle of the 2×1 reconstructed surface, which is produced as the result of the following reaction (Kahata and Yagi, 1989b and c),

$$\tfrac{1}{2}O_2 + Si \rightarrow SiO\uparrow, \tag{7.5}$$

where the SiO vaporizes and exposes the underlying Si atoms. The surface oxidation of Si(111) results in the movement of surface steps (Murooka *et al.*, 1992).

Gas–solid reactions are important in studies of catalysis. Catalytic CO oxidation on a Pt single-crystal sphere has been studied by REM (Uchida *et al.*, 1990). Images of surface structures before and after the chemical reaction revealed that Pt(100) is much more strongly attacked by the CO gas than is the Pt(111) surface. Similar

Figure 7.4 *A series of REM images of a Si(001) surface taken during exposure to oxygen, showing the gas–solid reaction. The beam azimuth is [110]. Bright-contrast hollows are formed as the result of initial surface oxidation (courtesy of K. Yagi).*

results have been observed in the Pt-assisted NO + CO reaction (Uchida *et al.*, 1992a, b). The difference could be ascribed to the existence of an adsorbate-induced phase transition on the Pt(100) surface, which enhances the mass transport of Pt atoms under the reaction conditions. Surprisingly, the surface inhomogeneities, such as steps, play a minor role in the roughening process of Pt(100) due to chemical reaction. Reaction of hydrogen gas with Si(111) has been observed by Ohse and Yagi (1989).

7.5 Surface electromigration

In REM, specimens are usually heated either by a furnace cup or by passing an electric current through the sample. Different surface phenomenon may be produced by the two methods because the electric and magnetic forces are present in the latter but not the former. Surface atom migration as a result of electric current is an active area of REM studies.

Direct imaging of surface structures when an electric current flows through the bulk is important for understanding the surface reconstruction process under electric heating. The REM image has shown clearly how the surface dynamic process occurs by changing the current direction with respect to the crystallographic directions (for a review see Latyshev *et al.* (1992) and Yagi *et al.* (1992)). Figure 7.5 shows two REM images of the Si(001) surface obtained after applying a direct

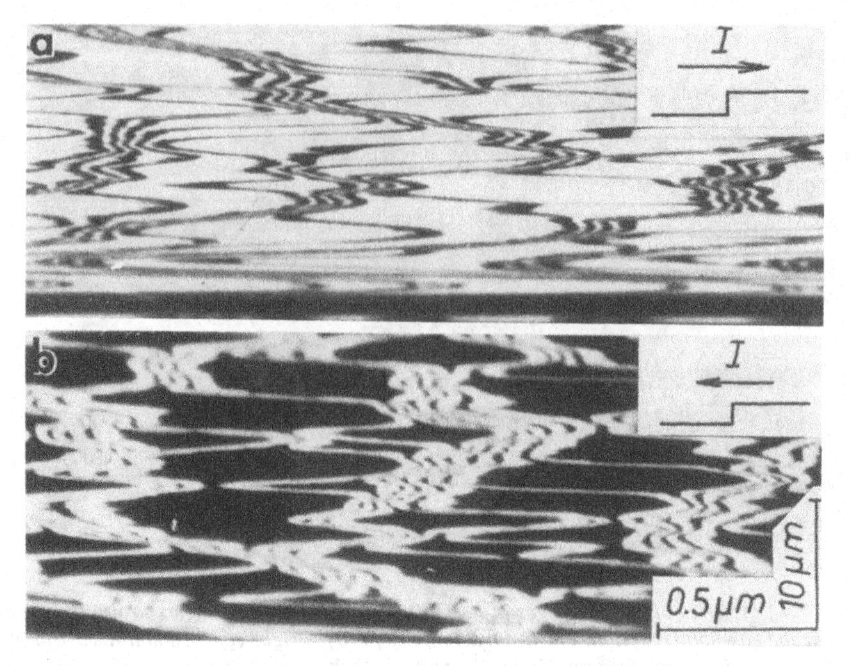

Figure 7.5 *REM images of the Si(001) surfaces after annealing at T = 1413 K for 2 min, showing the transition of surface structure when the electric current direction is reversed (courtesy of Dr L. V. Litvin et al., 1991).*

current along different directions with respect to the surface step terraces. The 2×1 terraces (bright contrast) expand when the electric current is in the step up direction (Figure 7.5(a)), and the 1×2 terraces (dark contrast) expand when the electric current is in the step down direction (Figure 7.5(b)).

The bunching of surface steps is often seen under the DC driving force. The criterion for formation of step bunching is that the probability of adatom incorporation into a step on the surface must be higher from the lower terrace of the step than from the upper terrace (Schwoebel and Shipsey, 1966). The step bunching on Si(100) occurs only when the DC direction is antiparallel to the step down direction (Yagi, 1993a).

Table 7.3 summarizes the reconstruction processes of clean surfaces and surfaces with adatoms under electric current. It is interesting that the surface structure and the direction of the adatom diffusion depend sensitively on the direction of the applied electric current. This is possibly due to the anisotropic properties of these surfaces. These observations are important in determining the growth of surface films in MBE if the substrate is heated by electric current. The migration of surface atoms is driven not only by thermal diffusion but also by electric and magnetic forces. It has been suggested that the observed electromigration is different from conventional electromigration with respect to the direction of the mass transport

Table 7.3. *The REM observations of electromigrations on clean surfaces and surfaces with adatoms.*

Structure	References
Si(001) $2 \times 1 \Leftrightarrow 1 \times 2$ via current direction	Latyshev *et al.* (1988), Kahata and Yagi (1989a), Litvin *et al.* (1991), Ichikawa and Doi (1990)
Si(001) step bunching via current direction	Latyshev *et al.* (1992)
Si(111) $1 \times 1 \Leftrightarrow 7 \times 7$ via current direction	Yamanaka *et al.* (1990), Latyshev *et al.* (1989b), Yamaguchi and Yagi (1993)
Si(111) step bunching via current direction	Yamanaka *et al.* (1990), Latyshev *et al.* (1989b)
Si(111) 7×7–In $\rightarrow$ Si(111) 4×1–In	Yamanaka *et al.* (1989)
Si(111) 7×7–Au $\rightarrow$ Si(111) 5×1–Au	Yamanaka *et al.* (1989)
Si(111) 7×7–Cu	Yamanaka and Yagi (1991)
Si(001) 2×1–Au $\rightarrow$ Si(001) c2×18–Au	Yamanaka *et al.* (1992)
Si(001) 2×1–Si	Ichikawa and Doi (1990)
Si(111) step bunching	Latyshev *et al.* (1994)
Si(111)–Ga with Si	Nakahara and Ichikawa (1993)

and the emerging morphology (Yasunaga and Natori, 1992). The surface electromigration is possibly driven by either DC or electric field. Yasunaga and Natori (1992) appear to conclude that the electric field is the driving force. A review has been given recently by Vook (1994), who compared the electromigration phenomena observed in bulk solid solution and on surfaces. Although many remarkable surface phenomena have been observed under DC heating, a clear physical model is not yet available for interpreting the observed results.

7.6 Surface ion bombardment

Examination of surface degradation, amorphization, and recrystallization is important for understanding the mechanisms of damage of crystal surfaces by charged particles or atoms. REM has also been applied to image *in situ* the noble gas ion bombardment of Si(111) surfaces (Claverie *et al.*, 1992), which is a technique normally used to clean surfaces. The experimental results have shown that it is difficult to re-grow a defect-free material after amorphization by noble gas bombardment. At high substrate temperatures, the surface shows monocrystallized but atomically rough surfaces.

REM imaging of Pt(111) and Si(111) surfaces bombarded by argon atoms has been reported by Ogawa *et al.* (1988). Their results showed that low-temperature bombardment produces defects on surfaces, and high-temperature annealing is required to get atomically flat surfaces for long-time bombardment. Also, the movement of surface steps is primarily the result of sputtering effects.

7.7 Surface activation energy

REM not only gives a real-space image of crystal surfaces but also can be applied to measure surface activation energy, as described by Uchida and Lehmpfuhl (1991). Recrystallized Pt(111) surfaces, after being cleaned by Ar ion bombardment, were quenched from different temperatures; and the area covered by small vacancy-type terraces was measured. The concentration (C/C_0) of the ad-vacancies is then given by the ratio of this area (C) to the total surface (C_0) covered by the loops. If the model for the massive crystal vacancies is assumed valid for surfaces, then the vacancy activation energy is given by the slope of the $\ln(C/C_0)$–$1/T$ plot. The calculated result is $E=0.62\pm0.06$ eV, which is considered to be the formation energy of an ad-vacancy on the Pt(111) surface.

This chapter has systematically reviewed the applications of REM in UHV TEMs for surface studies. Numerous examples have been shown to demonstrate the usefulness of the REM technique for revealing clean surface structural information and associated phase transformations. REM is particularly powerful for *in situ* imaging of initial nucleation and growth of thin films on a variety of substrates. The temperature and heating current effects can be examined at high resolution. It is anticipated that REM will be one of the key techniques for real-space imaging of *in situ* surface processes in thin film growth.

CHAPTER 8

Applications of non-UHV REM

Although limited by the popularity of UHV TEMs, many surface studies have been performed under non-UHV conditions, and useful atomic-level structure can be provided by REM. The REM technique was actually first explored in non-UHV TEMs. Large amounts of REM imaging have been performed in conventional TEMs for single-crystal surfaces of stable structures. In this chapter, we will review almost all types of surfaces that have been studied with REM.

8.1 Steps and dislocations on metal surfaces

Shown in Figure 8.1 is a REM image of a Pt(111) surface exhibiting many steps. Besides the numerous atom-high surface steps, an important application of REM is to image the surface emergent dislocations. Screw (s) and edge (e) dislocations are observed (Lehmpfuhl and Uchida, 1990). It is clear that a surface step is directly terminated at a screw dislocation in Figure 8.1(a), the height of which is related to the Burgers vector by $H=\hat{z}\cdot \boldsymbol{b}_B$. A sequence of equally separated edge dislocations is observed, and they produce a low-angle boundary in the bulk (Figure 8.1(b)). Their strain fields are weaker than that from a screw dislocation. The two neighboring crystal grains have slightly different orientations as can be seen from the different image intensities. The angle between the two grains, determined from the separation of the edge dislocations and the Burgers vector (1/2) [110], is 0.64 mrad (or 0.0367°). Such a small rotation may not be detected by any other surface imaging techniques. The direction of the grain boundary is $[11\bar{2}]$, perpendicular to the direction of the incident beam, which is close to $[1\bar{1}0]$. Atomic steps crossing the low-angle grain boundary do not show any change. This is a clear example of the sensitivity of REM image contrast to the crystal lattice distortion.

REM has mainly been applied to study surfaces that are stable in air. Table 8.1 gives a summary of the metal surfaces that have been imaged by REM under non-UHV conditions.

8.2 Steps on semiconductor surfaces

Surfaces of semiconductors have also been imaged with REM. Cleaved semiconductor surfaces, such as GaAs(110) and InP(110), exhibit atomic flatness, and have

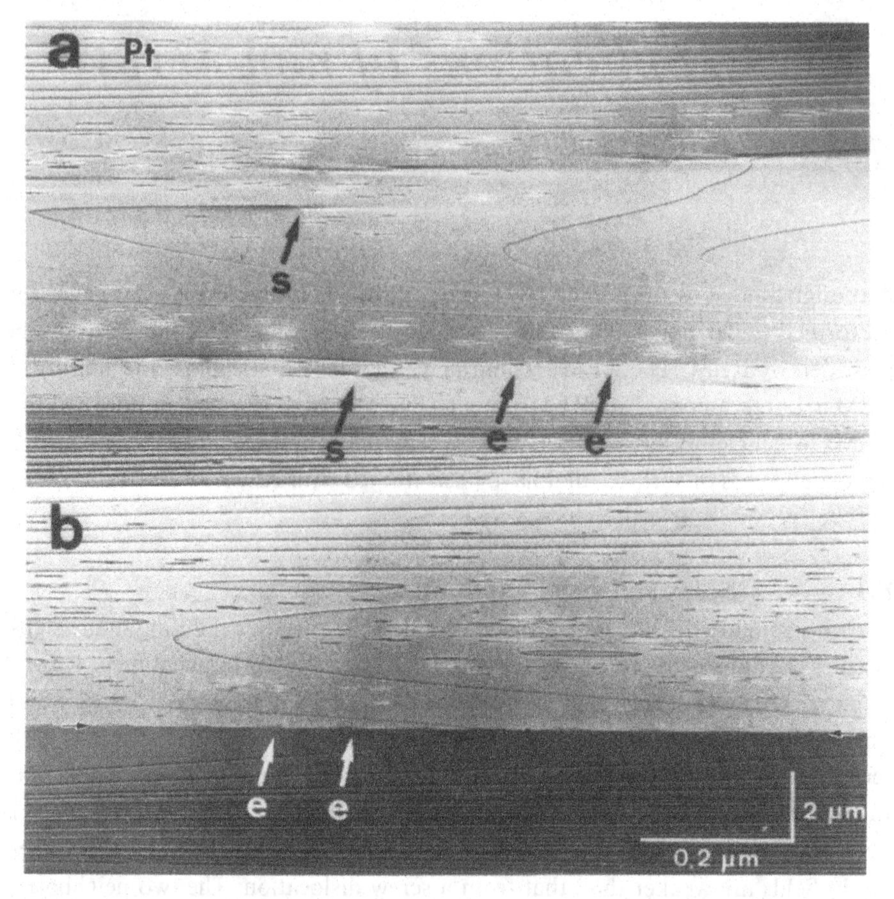

Figure 8.1 *REM images of the Pt(111) surface showing screw dislocations (s), edge dislocations (e) and steps. The step height is no more than the magnitude of the Burgers vector of the screw dislocation. A small-angle grain boundary is formed by a row of edge dislocations in (b) (courtesy of Drs G. Lehmpfuhl and Y. Uchida, 1990).*

been used as test specimens for fundamental REM studies such as dislocation contrast and surface-step phase contrast. Figure 8.2 shows a group of REM images recorded from a cleaved GaAs(100) surface. Many surface islands and bands are formed by cleavage (Figure 8.2(a)). Bunching of surface steps is also observed (Figure 8.2(b)). A more interesting phenomenon is seen in Figure 8.2(c), in which two types of surface terrace with bright and gray contrast, respectively, are observed. From the structure of GaAs, the (100) surface is terminated initially with either Ga or As when the cleavage surface is formed. Thus, the surface terraces showing brighter contrast may be terminated with As because of stronger atomic scattering power in comparison with the surface terraces terminated with Ga, provided that there is no beam damage at the surface. Finally, steps running nearly

Table 8.1. *Metal surfaces studied by REM*

Surface	References	Remarks
Pt(111) and Au(111)	Hsu (1983)	Steps, 0.9 nm image resolution
	Hsu and Cowley (1983)	Steps, dislocations and slip traces
	Uchida *et al.* (1984a)	Steps and surface treatments
	Ogawa *et al.* (1987)	Steps one, two and three atoms high, and surface sublimation at 1550 K
	Wang N. *et al.* (1992)	Surface reconstruction
Pt(111) and Pt(100)	Canullo *et al.* (1987)	Steps
	Lehmpfuhl and Uchida (1990)	Surface facets, steps, reconstructed surface domains, and a low-angle grain boundary (<0.64 mrad) formed by edge dislocations
Ni(111)	Wang and Bentley (1991c)	Surface terraces
Cu bicrystal	Li *et al.* (1990)	Index facets
Cu(110)	Liu and Cowley (1989)	Facets
Cu(100)	Milne (1990)	Oxide islands
Ag(111), (100) and (110)	Uchida (1987)	Atom steps
$Cu_3Au(111)$	Huang *et al.* (1994)	Surface morphology
Au–Ag alloy	Uchida *et al.* (1994)	Surface reconstruction

parallel to [011] and [01$\bar{1}$] are seen (Figure 8.2(d)), indicating that ⟨011⟩ steps are lower energy surface steps. Table 8.2 gives a summary of semiconductor surfaces that have been studied with REM.

8.3 Ceramics surfaces

Ceramics, such as α-alumina and MgO, are important substrates for MBE growth of superconductor and semiconductor thin films. Imaging of ceramic surface microstructures can provide useful information about possible nucleation sites and interfacial dislocations. In general, there are two methods for getting flat ceramic surfaces: annealing polished or chemically etched surfaces, and cleaving bulk crystals. We now use $\alpha\text{-}Al_2O_3$ to show the surface structural information that can be provided by REM.

$\alpha\text{-}Al_2O_3$ has a rhombohedral structure with space group $R\bar{3}c$ and six formula units in the commonly used hexagonal unit cell. Two-thirds of the octahedral positions within the hexagonal anion sublattice in corundum are occupied by aluminum ions; the remaining third are vacant. Thus the arrangement of aluminum cations in successive overlying layers appears displaced. This motif repeats every three layers

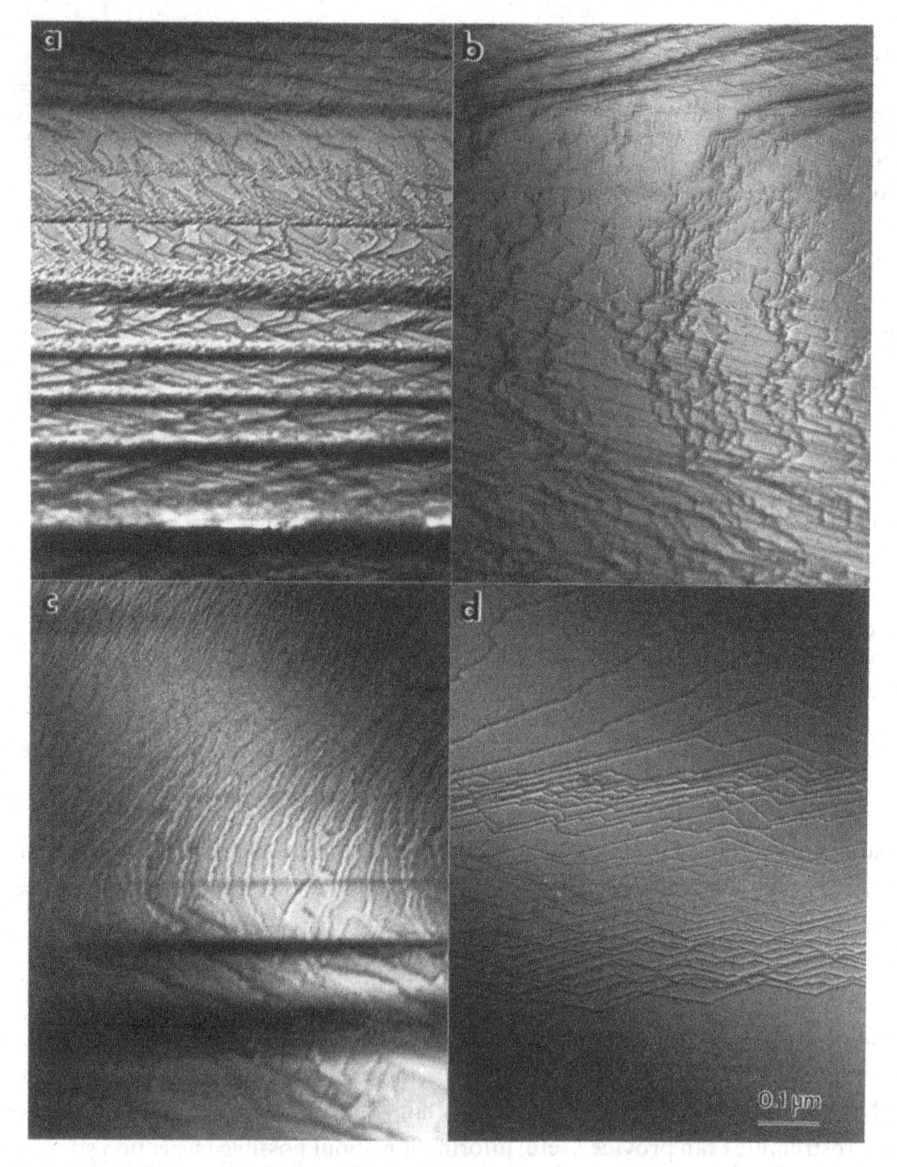

Figure 8.2 *REM images of a cleaved GaAs(100) surface showing various surface features (see text). The beam energy is 120 keV. The beam azimuth is [001].*

along [0001]. The arrangement of oxygen anions is repeated every other layer along [0001]. Thus the complete structure of corundum repeats after every six pairs of oxygen and aluminum layers along [0001]. The atomic structures and corresponding possible dislocation Burgers vectors in the basal plane have been described in detail by Kronberg (1957). REM is particularly sensitive to atomic structural modifications in the direction normal to the surface. The $(01\bar{1}2)$ surface is the lowest energy

Table 8.2. *Semiconductor surfaces that have been studied by non-UHV REM.*

Surface	References	Remarks
GaAs(110)	Hsu *et al.* (1984) Shimizu and Muto (1987)	Cleavage face, steps and dislocations As-grown surfaces, large steps
GaAs(100)	Hsu *et al.* (1990) Wang (unpublished)	As-grown surfaces Cleavage face
GaP(110)	Yamamoto and Spence (1983)	Cleaved face, steps and dislocations
InP(110)	Wang (1988b)	Cleaved face, steps and screw dislocations
$GaAs/Al_{0.39}Ga_{0.61}As$	Hsu (1985)	Cross-section of multilayer films
$Al_xGa_{1-x}As/GaAs$	Yamamoto and Muto (1984)	Superlattices
GaInAs/InP	Hsu *et al.* (1991)	As-grown quantum wells

surface of α-alumina and has been used as a substrate for silicon-on-sapphire semiconductor devices. For the $(01\bar{1}2)$ surface, the aluminum and oxygen ions are located in different layers parallel to the surface plane (Wang, 1992b), so that the surface can be terminated with either an aluminum layer or an oxygen layer depending on where the bonds break. This results in many atomic-height steps on the surface.

Complex fracture mechanisms are involved in forming a cleavage surface. In Figure 8.3, both zig-zag and closely-spaced surface steps distributed around a plateau-like island are imaged. The height of the island in Figure 8.3(a) can be determined from the width of the dark shadow as about 0.1 μm (note that the width of the shadow is not affected by the foreshortening effect). The surface can show two different contrast effects sharply separated by steps; one is bright and the other is dark (Figure 8.3(b)). This contrast effect has been interpreted based on electron beam radiation damage to the surface domains terminated with different atomic layers. The analysis found that the bright and dark contrast domains are initially terminated with aluminum and oxygen ions, respectively (see Section 8.5).

As well as zig-zag ledges, vacancy-type and island-type terraces are also present, as shown in Figure 8.4(a). The small steps indicated with arrows are less than one atom high (Wang, 1992b). As shown in Figure 8.4(b), strain contrast can appear in REM images. Some of the surface regions exhibit many steps, and some of the regions are almost free of steps. The formation of these dense arrangements of surface steps may indicate accumulation of stress at this area when the surface was formed. The zig-zag shaped terrace ledges are actually slowly curving steps, if the foreshortening effect is considered in the beam direction (or the vertical direction in the displayed image). Since the cleaved surface may not be the lowest surface energy state, annealing should remove the steps distributed on the flat surface regions and

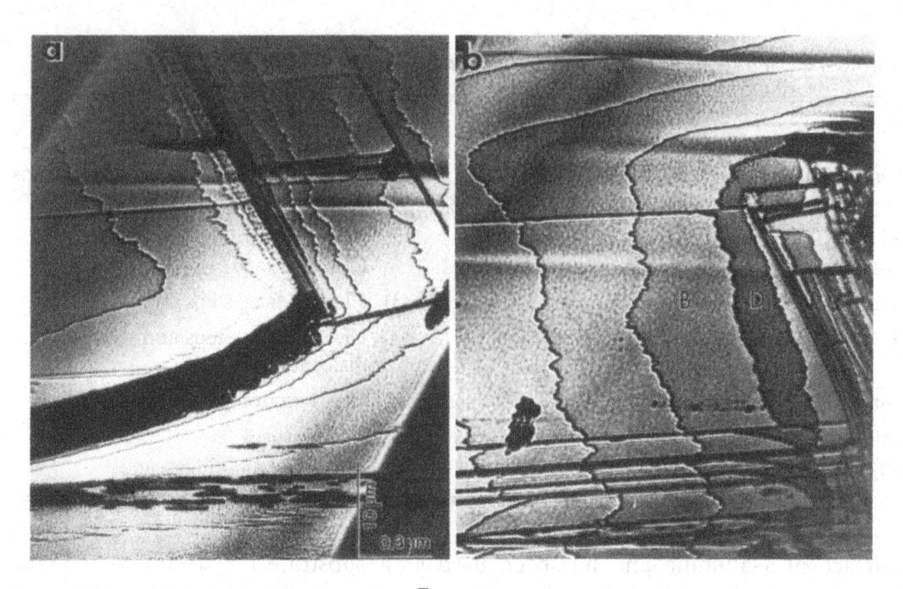

Figure 8.3 *REM images of the cleaved (01$\bar{1}$2) surface of α-alumina showing the distribution of surface steps, vacancy-type (v) and island-type (i) terraces. B and D indicate the surface terraces with different atomic terminations.*

make the dense arrangement of steps into a facet. This expected result has been shown in the REM images of annealed alumina surfaces (Ndubuisi *et al.*, 1992a and b) and TEM images of annealed alumina thin foils (Susnitzky *et al.*, 1986).

Lanthanum aluminate ($LaAlO_3$) is one of the optimum substrates for epitaxial growth of $YBa_2Cu_3O_{7-x}$ thin films. The {100} surfaces of annealed $LaAlO_3$ have been studied using REM (Figure 8.5). Numerous [010] and [001] 'saw-teeth' steps have been observed, and they are believed to be the lowest surface energy steps. The step heights have been determined to be two or three unit cells according to the mirror-imaging technique (see Section 5.7). For most cases, the surface step structures are undisturbed by the presence of the dark surface contaminants, which were found to be Si–La–O. However, it has occasionally been found that the morphology of surface steps can be drastically affected by the presence of the Si–La–O particles, as shown in Figure 8.5. Three features are seen. First, the particles are almost aligned along [001], the direction in which the beam strikes the surface. The surface structure is nearly symmetric on both sides of the particle line. Second, the particles are distributed on a 'long island', which is relatively flat and higher than the neighboring regions. Third, down steps are seen on both sides of the particles. At distances far away from the particles, the surface is covered by the [010] steps. Symbols 1–6 indicate some larger Si–La–O particles seen on the surface. Each particle is surrounded by down steps, and the step configuration is approximately symmetric on both sides of the particle. Each particle sits on top of the step terraces

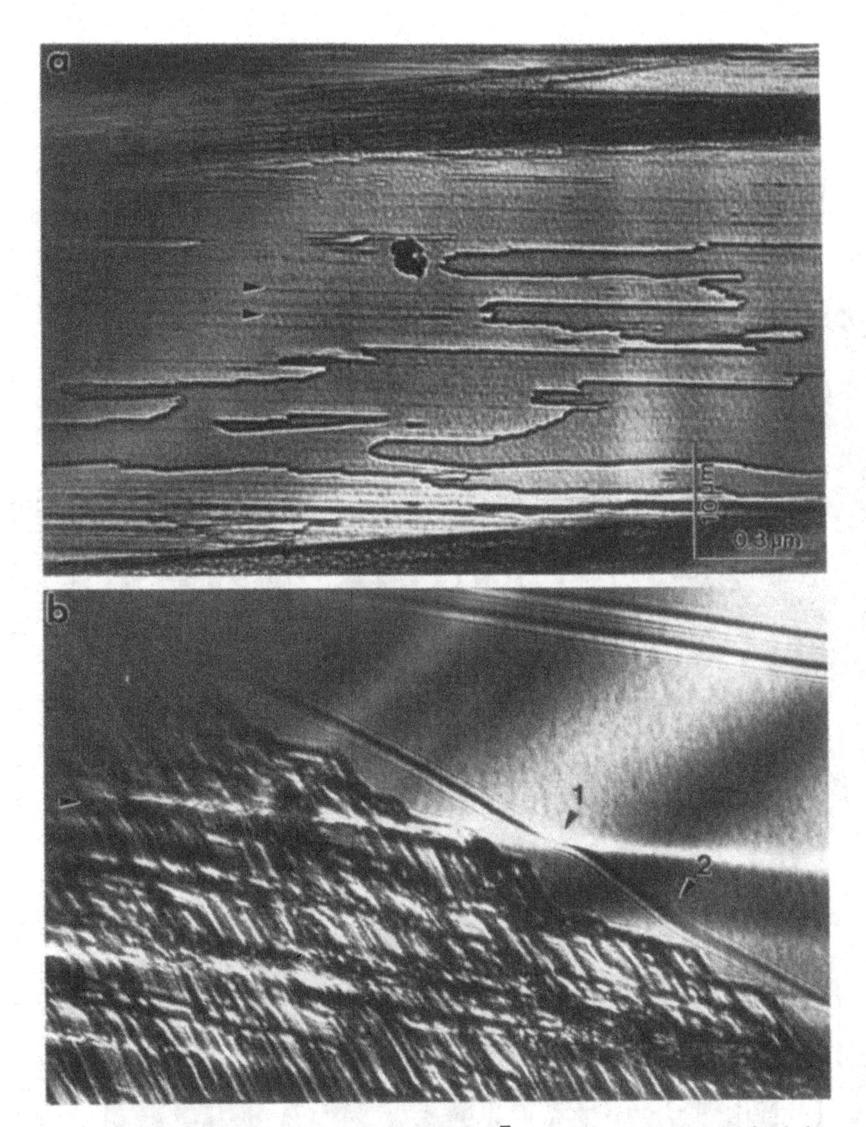

Figure 8.4 *REM images of a cleaved α-alumina (01$\bar{1}$2) surface showing irregular ledges and vacancy-type and island-type terraces. Arrowheads 1 and 2 indicate the bending of a step across a region of high surface strain.*

(or islands). Particles 1 and 3 are clearly surrounded by down steps, which are actually of 'circular' shape if the foreshortening effect is considered. The steps are apparently faceted along [010] and [001]. There are smaller sized particles distributed between particles 1 and 2, 2 and 3, 3 and 4, and 4 and 5. The symmetric steps distributed around particle 1 are almost unaffected by the nearby particles due to the large size of the particle and its long distance away from other particles. This observation has clearly demonstrated the effects of surface reaction products on the

Figure 8.5 *A REM image of the $LaAlO_3(100)$ surface showing the effect of the Si–La–O particles on the formation of surface steps. The image was recorded using the (10 0 0) specular reflected beam at 120 kV. This image is a match of two images recorded under different focusing in order to view the entire region clearly. The sharp line across particle 3 is the match line, the contrast variation across which is purely a photographic effect and has nothing to do with the structure of the surface. Particles 1–6 are the surface contaminants. Particle 7 is a fracture fragment standing on the surface.*

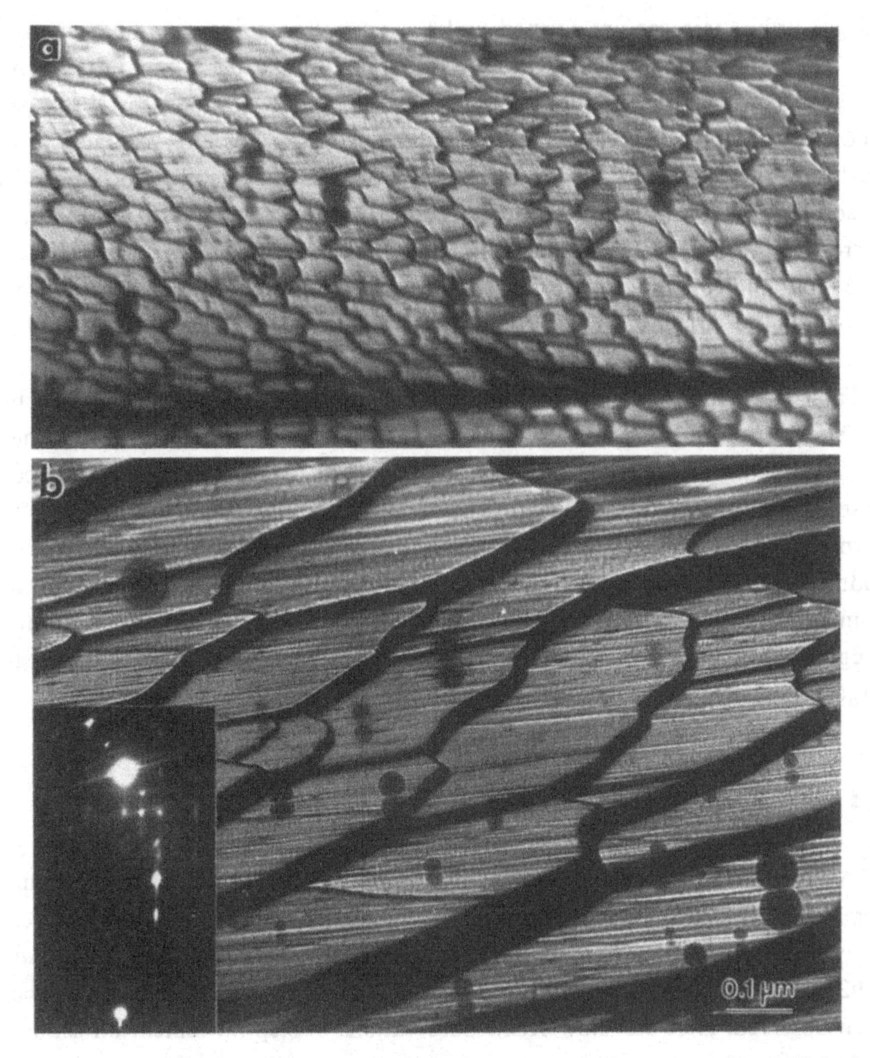

Figure 8.6 REM images of the LaAlO$_3$(110) surface annealed for (a) 10 h and (b) 20 h at 1500°C in air, showing the growth of large surface terraces.

growth of surface steps. It has been suggested by Wang and Shapiro (1995a) that the Si–La–O particles may create some nucleation sites, which tend to have the lowest energy, so that the atoms tend to migrate towards the particles, resulting in the formation of stair-type down ⟨100⟩ steps.

Structures of annealed ceramic surfaces can be dramatically different, depending on annealing temperature and duration of annealing. Figures 8.6(a) and (b) show two REM images recorded from the same LaAlO$_3$(110) specimen that had been annealed for 10 and 20 h, respectively, at 1500°C (Wang and Shapiro, 1995c). The (110) surface exhibits completely different structure to the (100) surface (Figure 8.5).

Large surface terraces are formed and the terrace area grows with increasing annealing time. Each terrace area, however, is not atomically flat, and there is no surface reconstruction based on the RHEED observation (see the inset). From the studies of the (100) surface, we have found that the ⟨100⟩ steps and {100} facets have the lowest surface energy. This result can be applied to interpret the morphological structure of the (110) surface. The fine structures observed on the (110) terraces may correspond to the formation of small-width (100) and (010) facets on the surface. The (100) and (010) facets are favored due to their low energy, and these facets remain on the surface even when the surface is annealed for an extensive period of time (Wang and Shapiro, 1995c).

Table 8.3 gives a summary of the ceramic surfaces that have been imaged by REM. Most of the annealed surfaces are flat with a few atom-high steps, but they consist of many facets of vicinal surfaces. Also, surface reconstructions have been observed on some of the annealed surfaces.

In some cases, the surface activation energy may be determined based on the width of surface terraces (Peng and Czernuszka, 1991a). However, it must be pointed out that the terrace width critically depends on the mis-cut and initial roughness of the surface. Thus, the activation energy measured may be subject to large errors.

8.4 *In situ* dynamic processes on ceramics surfaces

In situ observations of dynamic surface processes are important applications of REM and RHEED. This procedure allows direct observation of temperature-dependent surface structural changes. *In situ* experiments showing surface step movement have been carried out on Si(111) (Osakabe *et al.*, 1980; Murooka *et al.*, 1992) and Pt(11) (Ogawa *et al.*, 1987) surfaces, and direct surface sublimation has been observed. Here we show REM observations of step movement on ceramic surfaces.

Under the stimulation of heating at 1670 K, various surface processes can be generated on α-alumina $(01\bar{1}2)$ (Wang and Bentley, 1993a). REM images have shown that the dominant surface process is the expansion of surface vacancy-type terraces. An important surface process is desorption because of thermally induced vaporization. The prominent result is surface desorption layer by layer. It is likely that the desorption occurs mostly at the ledges of vacancy-type terraces, resulting in the expansion of the vacancy-type terraces. The nucleation of surface vacancy-type terraces appears to occur at surface regions with defects. Figure 8.7 shows a series of *in situ* REM images of an $\alpha\text{-}Al_2O_3(0\bar{1}12)$ surface at 1670 K. The growth of a vacancy-type terrace appears at surface regions with defects, as arrowed in Figure 8.7(a). The newly created vacancy-type terrace grew fairly quickly when its diameter

Table 8.3. *Ceramic surfaces studied by non-UHV REM. For ceramic surfaces, the annealing experiments were usually performed in air or in oxygen gas.*

Surface	References	Remarks
α-Al_2O_3(0001)	Kim and Hsu (1991)	Annealed face; facets, steps and screw dislocations
	Ndubuisi *et al.* (1992a)	Annealed face; terraces and high steps
α-Al_2O_3($11\bar{2}0$)	Hsu and Kim (1991)	Annealed face; facets and 4×1 reconstruction
	Hsu and Nutt (1986)	As-grown surface; 'rooftop' structure
	Hsu and Kim (1990)	Annealed surface; 'rooftop' structure and facets
	Ndubuisi *et al.* (1992b)	Annealed surface, steps and terraces
α-Al_2O_3($0\bar{1}11$)	Wang and Howie (1990)	Cleavage face; steps and screw dislocations
	Yao *et al.* (1989)	Beam radiation damage, surface atomic terminations
α-Al_2O_3($0\bar{1}12$)	Wang (1991)	Cleavage face; beam radiation damage, surface atomic terminations and beam-induced Al metal reduction from the domain terminated initially with oxygen
	Wang (1992b)	Cleavage face; surface steps, steps lower than 0.1 nm, dislocations, kinks and ledges
	Kim and Hsu (1992)	Annealed surface; surface morphology and beam damage effects
	Ndubuisi *et al.* (1992b)	Annealed surface; high steps and terraces
	Peng and Czernuszka (1991a)	Cleaved and annealed surfaces; steps and surface diffusion
	Wang and Bentley (1991d and 1993a)	*In situ* observation of surface step movement, surface desorption and diffusion at 1670 K
α-Al_2O_3(001)	Liu *et al.* (1992b)	Annealed surface; alumina-induced rutile (001) $5\sqrt{2} \times 1$ reconstruction
TiO_2(110)	Wang *et al.* (1989a)	Cleaved surface; beam radiation damage
TiO_2(100) and (001)	Wang L. *et al.* (1994)	Steps and surface roughening
MgO(100)	Wang (1988c)	Cleaved surface; steps and surface reactions
	Crozier *et al.* (1992)	Cleaved and annealed surfaces
	Uchida *et al.* (1984b)	Surface steps
	Crozier and Gajdardziska-Josifovska (1993)	Ca segregation

Table 8.3 (*cont.*)

Surface	References	Remarks
MgO(110)	Wang and Egerton (1989)	Cleavage surfaces
MgO(111)	Gajdardziska-Josifovska *et al.* (1991)	Annealed surface; $(\sqrt{3} \times \sqrt{3})R30°$ reconstruction
ZnO(0001) and $(1\bar{1}00)$	Peng and Czernuszka (1991b)	Cleaved faces; steps and dislocation movement
$LaAlO_3(100)$	Wang and Shapiro (1995a) Wang and Shapiro (1995b)	⟨100⟩ faceted steps, {100} twin boundary, step growth 5×5 reconstruction.
$LaAlO_3(110)$	Wang and Shapiro (1995c)	Steps
$BaTiO_3(100)$	Tsai and Cowley (1992)	Ferroelectric domain boundaries

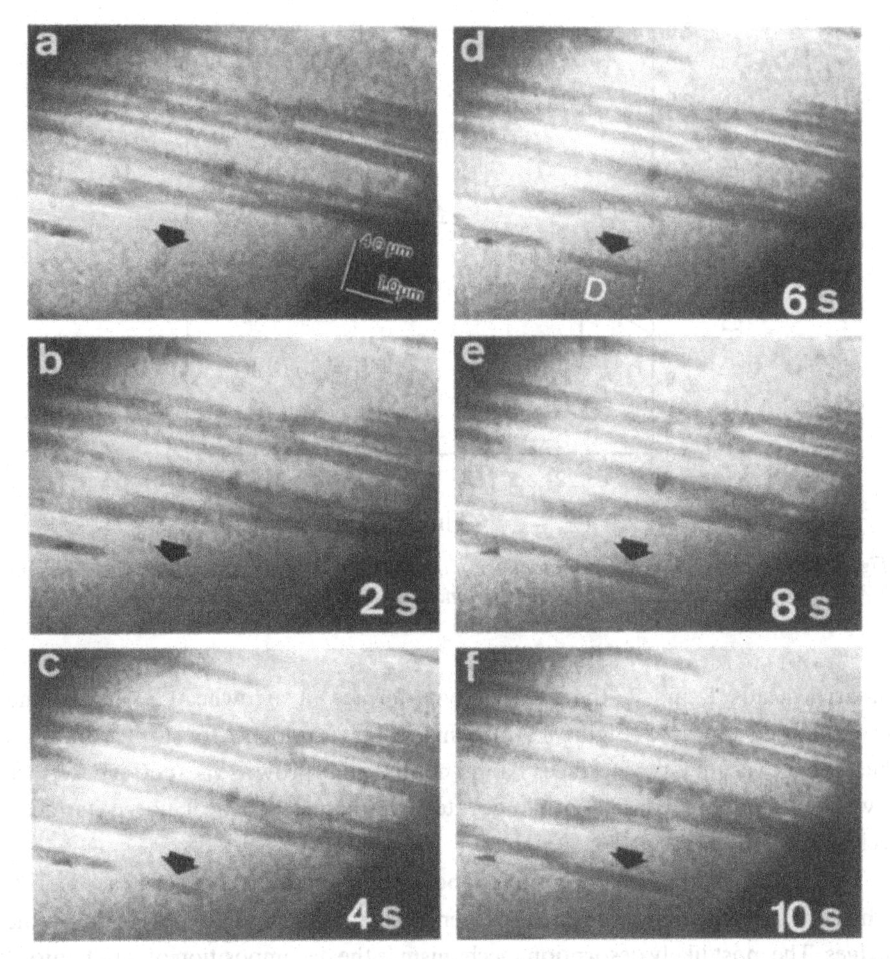

Figure 8.7 *A series of* in situ *REM images taken 2 s apart from the same area of a cleaved α-alumina (01$\bar{1}$2) surface at 1670 K, showing the creation and expansion of a vacancy-type terrace, indicated by the arrowhead (Wang and Bentley, 1991d).*

(ϕ_D) was less than about 2 μm. After about 2 s from the start of the growth (Figures 8.7(b)–(f) are at 2 s intervals), the growth rate of vacancy-type terrace diameter dD/dt was determined by reviewing the recording tape (Figure 8.8), and the results showed $d\phi_D/dt = 0.2\ \mu m\ s^{-1}$. The high initial growth rate at small ϕ_D may be due to easier desorption of the atoms located close to a defect because of their lower binding energy. After ϕ_D has become larger than about 1 μm, the influence of the original defects is negligible, and the desorption rate is determined primarily by the properties of the ledge around the vacancy terrace.

Simultaneously with desorption, surface diffusion can also be identified. Figure 8.9 shows a series of REM images taken at 10 s intervals from the same area of the (01$\bar{1}$2) surface at 1670 K. Accompanying the expansion of the vacancy-type terraces

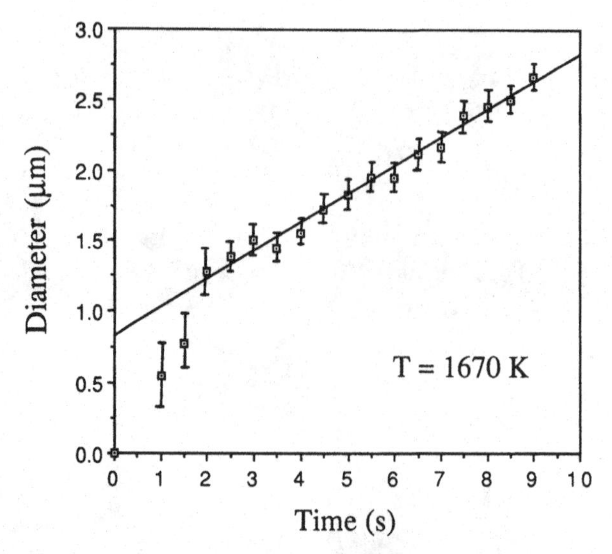

Figure 8.8 *A plot of the measured diameter of the newly created vacancy-type terrace in Figure 8.7 as a function of time, showing a constant increase of its diameter.*

at arrowheads 1 and 2, the vacancy-type terrace at arrowhead 3 shrank and eventually disappeared (Figure 8.9(c)). This process is evidence for surface diffusion because the vacancy-type terraces 1 and 2 continued to grow as vacancy-type terrace 3 was being annihilated. The shrinkage rate of the diameter of vacancy-type terrace 3 was about 0.14 $\mu m\ s^{-1}$.

The primary surface processes are vaporization and diffusion at surface ledges. Thus, the surface atoms are removed layer by layer as the result of desorption at the ledges. The most likely desorption mechanism is the decomposition of Al_2O_3 into O and AlO gas phases at high temperature (Wang and Bentley, 1993b; Brewer and Searcy, 1951). The direct sublimation process of alumina does not exist because Al_2O_3 molecules have never been detected. It is thus expected that the surface would undergo a decomposition process, resulting in vaporization of some other chemical molecules produced during the decomposition. There is evidence showing the existence at high temperatures of aluminum suboxides (Al_2O and AlO), which are produced by the reactions described below during Al_2O_3 decomposition.

Thermodynamics calculations suggest that gas phase AlO would be produced if Al_2O_3 were volatilized alone (Brewer and Searcy, 1951),

$$Al_2O_3 \Leftrightarrow 2AlO\uparrow + O\uparrow. \tag{8.1}$$

This reaction is likely to occur at $T > 1870$ K, whereas the desorption of surface atoms was observed at 1670 K. However, the critical temperature reported in the literature is applicable to 'bulk' reactions. For surface atoms at ledges, the average

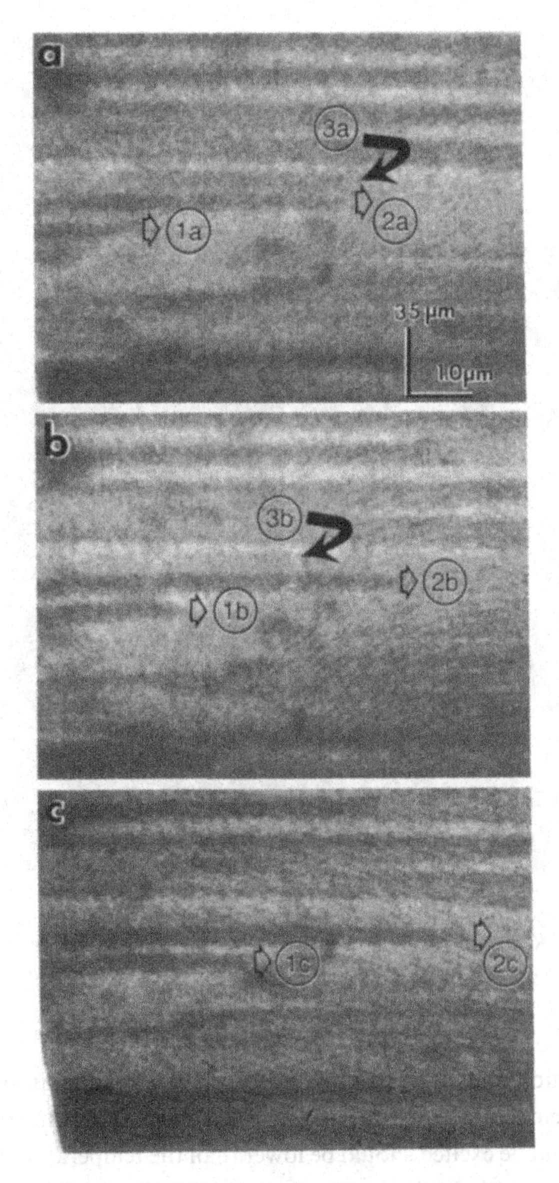

Figure 8.9 *A series of* in situ *REM images taken 10 s apart from the same area of a cleaved α-alumina* $(01\bar{1}2)$ *surface at 1670 K showing the annihilation of a vacancy-type terrace as the result of surface diffusion.*

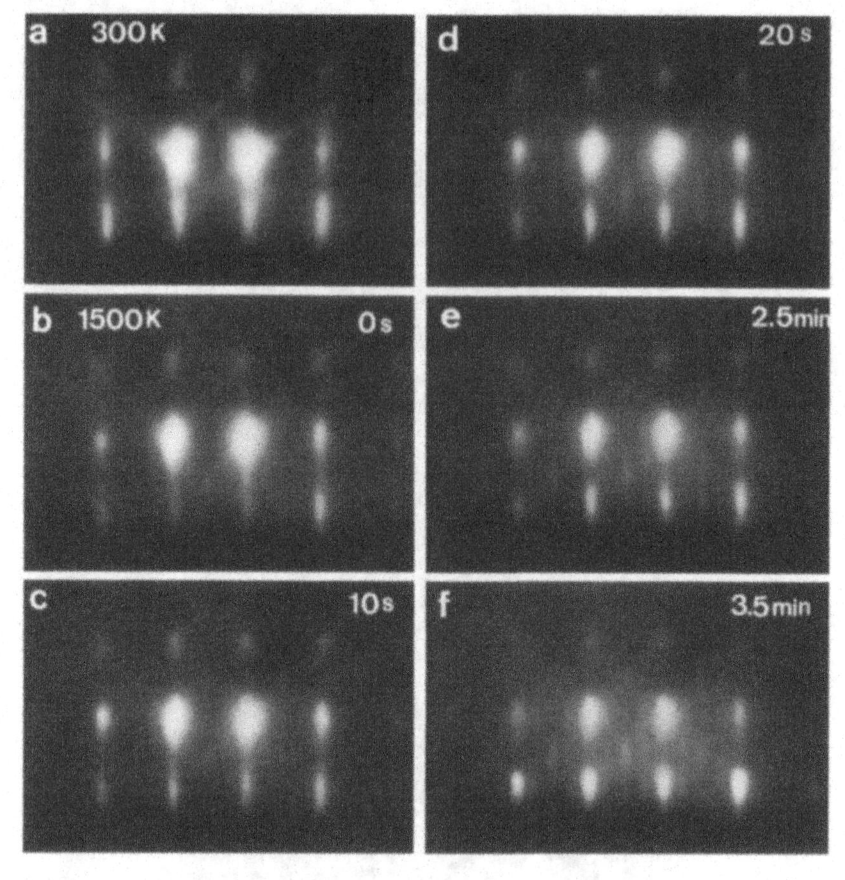

Figure 8.10 *A series of* in situ *RHEED patterns of a cleaved MgO(100) surface recorded with the crystal at 1500 K for different lengths of time. The growth of extra reflection rods, which characterize the phase transformation of the surface, is shown (Wang et al., 1992).*

number of atomic bonds is about two thirds of that for the atoms in the bulk. Therefore, the temperature at which the surface decomposition, sublimation and vaporization could be excited should be lower than the temperature for stimulating the same process in the bulk. Thus reaction (8.1) is considered to be possible in the *in situ* experiments.

In contrast to α-alumina, the MgO(100) surface does not show any significant desorption, but the corresponding crystallographic and surface composition structures undergo phase transformation at high temperatures (Wang *et al.*, 1992). A freshly cleaved MgO(100) bulk specimen was employed for *in situ* RHEED observations. No significant change in the RHEED pattern was observed when the temperature was lower than about 1500 K. However, extra 'rods' started to appear within 10 s after the specimen had reached 1500 K (Figure 8.10). With increasing

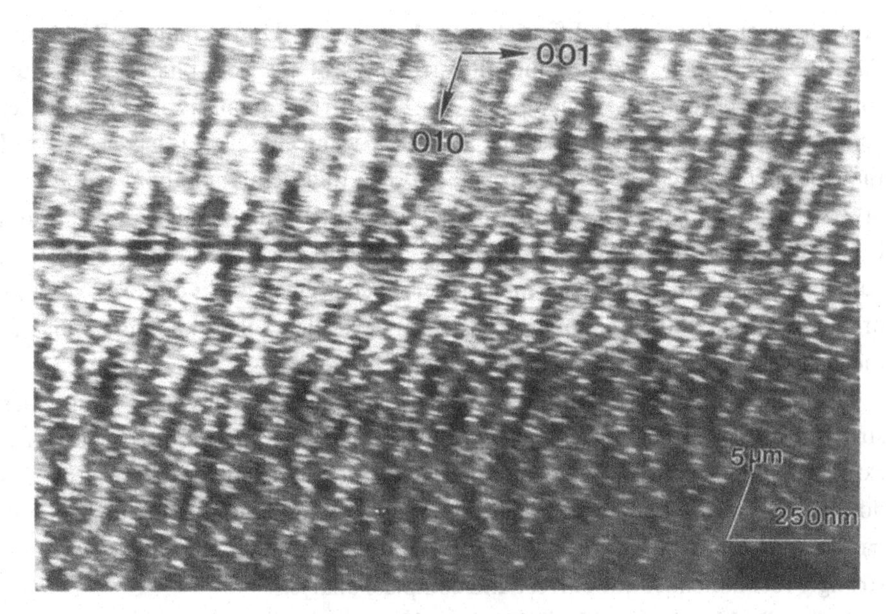

Figure 8.11 *A REM image of the MgO(100) surface after being heated to 1500 K for about 10 min, and then quenched to room temperature, showing the patch-type distribution of newly formed phases.*

time at 1500 K, these extra reflection rods became stronger and stronger, and gradually developed local maxima (Figure 8.9(f)). The separation between the newly formed rods is about (1/4){200}. Also, the intensity distribution along each rod is non-uniform, indicating the existence of a two-dimensional reflection pattern. The streaking of the beams is the result of the limited penetration into the surface of the electrons. An important fact shown by Figure 8.10 is that the intensity of the extra reflections is weaker than that of the MgO matrix reflections and that they are rod-shaped. This means that the thickness of the layer producing the reflection rods is no more than about 1–2 nm. The new structure formed on the MgO(100) surface was stable at room temperature, even after the specimen had been exposed to air at room temperature.

A REM image of the MgO surface after heating to 1500 K to form the new phase and then quenching to room temperature is shown in Figure 8.11. Since the image is formed by the (600) reflection beam, the bright contrast areas should correspond to the MgO matrix and the new phase if they generate equivalent reflections. More importantly, the new surface film may consist of small patches of diameter a few tens of nanometers. There are two possible explanations for the formation of the extra reflections in Figure 8.9. First, it is possible that new reflections represent a 4×4 surface reconstruction. However, the REELS analysis (as will be shown in Figure 10.2) did not support this assumption. Second, the new reflections are produced by a new phase or phases with different crystal structures. Unfortunately, REM and RHEED are unable to determine the structures of the new phases. One must sometimes use compensational techniques for surface analysis in TEM. After comprehensive studies using imaging and analytical techniques, it has been found that the super reflection rods are generated by a thin (less than about ten atomic layers) surface film, which is mainly composed of MgO_2 with about 5% calcium and chromium impurities (Wang *et al.*, 1992). The film covers the whole surface in patches a few tens of nanometers in diameter. Each MgO_2 patch is one of 12 variants oriented with its [110] direction and (001) plane parallel to the ⟨110⟩ direction and the {111} planes of the MgO matrix, respectively.

This study has also shown the important fact that the analysis of the surface structure using RHEED data alone may lead to incorrect information, because each extra diffraction beam (or rod) may come from different regions of the surface. Also double- and multiple-diffraction effects may create additional rods in RHEED patterns. A comprehensive use of imaging, diffraction, and analytical techniques is strongly recommended for correct and complete analysis.

Surface REM images of NaCl(100) have been obtained (Hsu, 1994), showing the presence of atom-high surface steps on the surface. Since the surface is very sensitive to beam damage, an electronic circuit was designed to block the beam when the recording plate was not placed at the viewing screen position. Image recording must be performed within 1–2 s.

8.5 Surface atomic termination and radiation damage

For cleaved α-alumina $(01\bar{1}1)$ and $(01\bar{1}2)$ surfaces, the surface termination can be a layer of either Al or O atoms, depending on where the bonds break. The surface termination effect could produce some contrast abnormality in the images (Yao *et al.*, 1989). Figure 8.12 shows a series of REM images of an α-alumina $(01\bar{1}1)$ surface after illumination by the electron beam with a dose rate of $(0.8\text{–}1.2) \times 10^{-3}$ A cm^{-2} for different lengths of time. The 'bright' (B) contrast domain has stronger reflected intensity and remains quite stable throughout. The contrast of the 'dark' (D) domain has changed from gray (Figure 8.12(a)) to dark (Figure 8.12(b)), then to bright (Figure 8.12(c)) and rough (Figure 8.12(d)). The B and D domains are separated by surface steps. The observed contrast can be interpreted as follows according to the Knotek–Feibelman process (Wang and Howie, 1990; Wang and Bentley, 1991c).

In studies of α-alumina solids in STEM (Berger *et al.*, 1987; Humphreys *et al.*, 1985), reduction to Al metal occurs under the electron beam. The reduction process can be attributed to the electron-beam-induced desorption of oxygen atoms after excitation of the internal Auger decay process (Knotek and Feibelman, 1978), or the

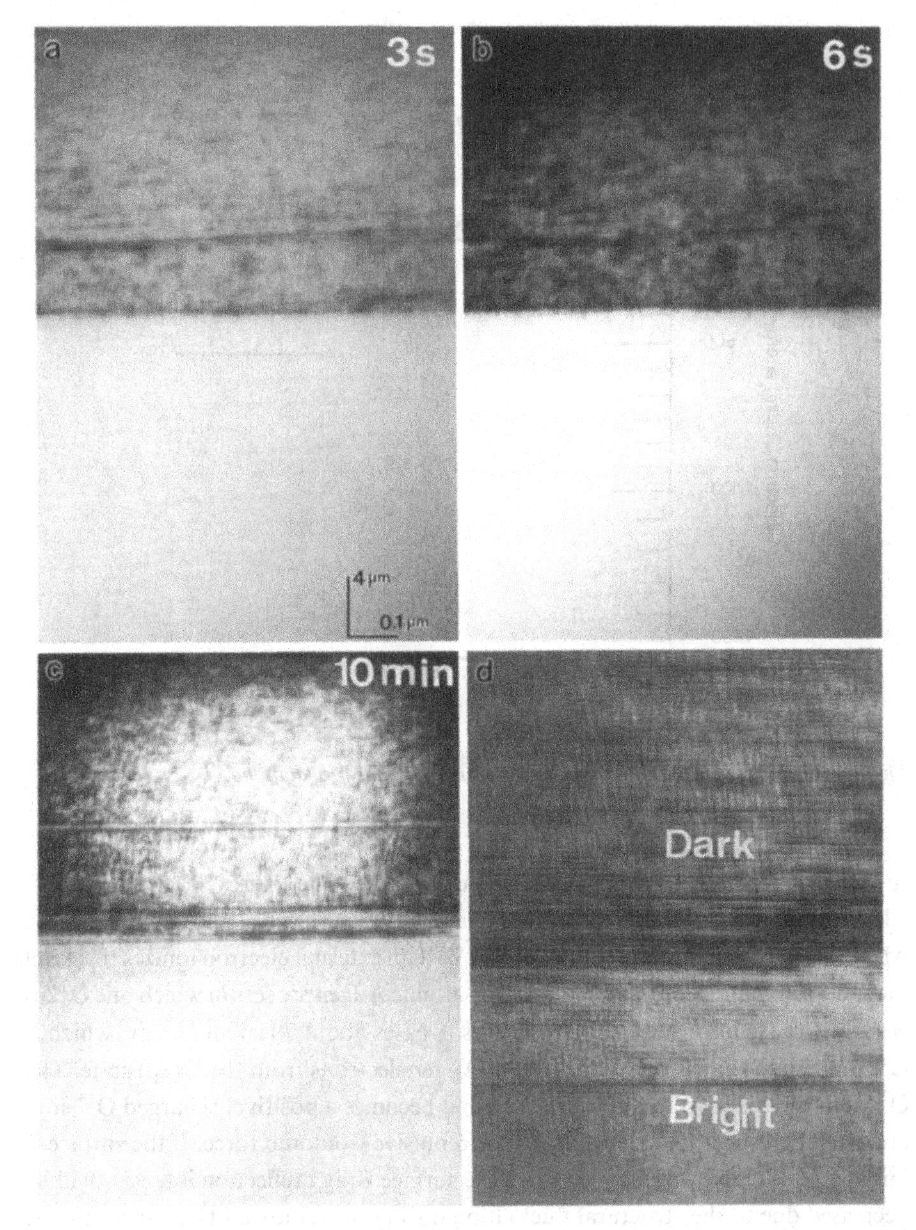

Figure 8.12 *Surface termination effects of α-alumina surfaces. (a)–(c) A series of REM images recorded from the α-alumina ($01\bar{1}1$) surface after being exposed to the electron beam for 3, 6 and 10 s, respectively, showing the radiation damage to the surface domains with different atomic terminations. (d) Is an image after the surface had been illuminated for more than 40 min.*

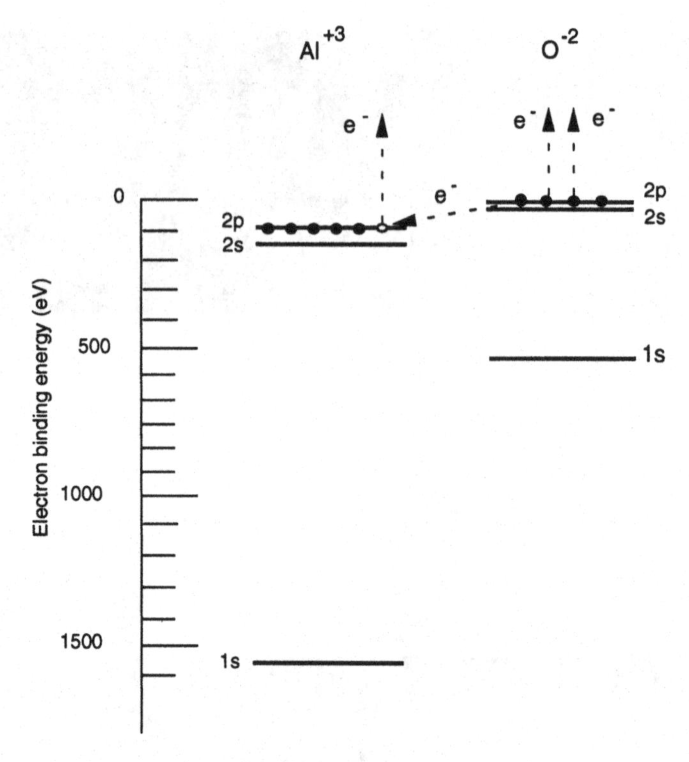

Figure 8.13 *The energy-level diagram of Al and O atoms in α-alumina.*

so-called Knotek–Feibelman process. According to the energy level diagram of Al_2O_3, as shown in Figure 8.13, the highest occupied level of the Al^{3+} ion is the Al(2p) level, with a binding energy of 73 eV. If an external electron ionizes this level then the dominant decay mode is an inter-atomic Auger process in which one O(2p) electron decays into the Al(2p) hole. This releases about 70 eV of energy, which is taken up by the emission of one or two Auger electrons from the O(2p) state. The O^{2-} ion thus loses up to three electrons and becomes a positively charged O^{1+} ion, and will be pushed off the surface by the repulsive Coulomb force. If the surface is initially terminated with oxygen, then the surface Bragg reflection intensity will be decreased due to the structural fluctuation during the removal of the oxygen ions, thereby showing darker contrast in REM images (i.e., the dark contrast regions in Figures 8.12(a) and (b). After the top layer of oxygen has been completely desorbed, exposing the Al layer beneath it, the surface desorption process is interrupted so that the surface reflection is dominated by the scattering of the exposed Al layer. This results in stronger Bragg reflection intensity and shows a bright contrast (the upper part of Figure 8.12(c)). By exposing the surface for a long time to an electron beam, surface knock-on damage may occur, resulting in rough contrast (Figure 8.12(d)).

Therefore, the dark contrast domains are terminated initially with an oxygen layer because the oxygen ions located at the surface are most easily desorbed.

For the domains initially terminated with aluminum layers, the resistance of the surface to electron beam damage can be analyzed as follows. In fact, it is believed that the fresh cleavage surface terminated initially with Al^{3+} ions is quickly oxidized in air before being put into the microscope. The adsorption rate may be much faster than the speed of formation of a reconstructed top surface; the Al^{3+} ions beneath the adsorbed oxygen then still preserve the two-dimensional periodic structure, resulting in stronger Bragg reflection. The adsorbed oxygen may form a thin amorphous layer with the Al^{3+} ions, which cannot be removed by electron beam radiation for the following reason. The adsorbed outermost oxygen atoms are probably not located next to the Al^{3+} ions of the original surface, to which the two Auger electrons of each oxygen atom would transfer. Thus the desorption process is unlikely to occur and this amorphous layer is effectively a protection layer. Therefore, the bright contrast domain is likely to be terminated initially with Al^{3+} and shows stable contrast. The existence of this adsorbed oxygen layer has been confirmed by REELS analysis, as will be shown in Figure 11.1. In REM images, the surface domains initially terminated with Al may not show the same contrast as the domains that are initially terminated with O even after the O layer has been removed by the electron beam through the internal Auger decay process. This is due to the presence of the adsorbed amorphous layer on the former but not on the latter under clean surface conditions, which is consistent with the observation shown in Figure 8.12(d).

The above interpretation provides a reasonable model for the stability and uniform contrast of the bright surface domains, but a remaining question is that of why the reflected intensities from the two surface domains are different. In order to provide an answer, the stability of the 'dark' domain after the Auger decay must be considered. The electrostatic repulsive force between the top Al^{3+} layer with the O^{1+} underneath opposes the force created between the O^{1+} ions and the Al^{3+} ions further inside the bulk. The top Al^{3+} layer is presumably sufficient to block or screen the underlying O atoms from being desorbed, but the top few layers are positively charged and may be very unstable. In this case, the Al^{3+} and the O^{1+} ions may tend to create positively charged surface clusters in order to decrease the surface energy, resulting in the formation of a new reconstructed surface, producing a low reflected intensity. After long exposures to the electron beam, the two surface domains appear with about the same brightness but the initial dark domains have more fine structure.

It is worth pointing out that the Knotek–Feibelman process is a common phenomenon in metal oxides. It has been observed for α-Al_2O_3, MgO and TiO_2. The reduction of metal from oxides is the basis of nano-probe lithography using an electron beam.

8.6 Reconstruction of ceramic surfaces

Reconstructions of ceramic surfaces are usually quite stable and appear to be unaffected by surface adsorption. It is possible to observe the reconstructed surface using conventional REM. Numerous reconstructions have been observed on surfaces of ionic oxides. A $(\sqrt{3} \times \sqrt{3})R30°$ reconstruction has been reported on the NiO(111) surface (Floquet and Dufour, 1983) and the annealed MgO(111) surface (Gajdardziska-Josifovska *et al.*, 1991). The annealed α-$Al_2O_3(1\bar{1}02)$ and $(1\bar{1}00)$ surfaces have been found to show reconstructions (Hsu and Kim, 1991). Reconstruction was also initiated when the $TiO_2(001)$ surface was contaminated by α-Al_2O_3 during annealing (Liu *et al.*, 1992).

As shown earlier in Figure 2.3, 5×5 structure was observed on an annealed $LaAlO_3(100)$ surface. REM can be directly applied to image the distribution of the reconstructed surface areas and their possible relation to surface steps. In the REM image shown in Figure 8.14(a), the reconstructed and unreconstructed surface areas are seen and show distinct contrast. Some of the reconstructed areas extend across the step without being interrupted. On some of the surface terraces, both reconstructed and unreconstructed surface areas co-exist and there is no step to isolate them. In Figure 8.14(b), many [001] and [010] steps are seen and the entire surface is almost covered by the reconstructed layer. The specimens used for recording Figures 8.14(a) and (b) were prepared under identical conditions.

Large plateau areas can be formed on the $LaAlO_3(100)$ surface. Figures 8.15(a) and (b) show REM images recorded from the left- and right-hand sides, respectively, of a plateau region. The top of the plateau is flat and displays 5×5 reconstruction. The surrounding of the plateau exhibits many steps, the distribution of which is similar to the distribution of 'tiles-on-a-roof'. Almost all the steps are along [001] and [010]. Most of the 'tiles' are covered by the 5×5 reconstruction. In Figure 8.15(a), the unreconstructed area is more pronounced at the edge of the plateau (see the bright contrast area). The REM results appear to show that the 5×5 reconstruction results from the unbalanced bonds of the surface atoms. The surface steps can enhance the surface reconstruction but are not essential for initiation of the reconstruction.

8.7 Imaging planar defects

Stacking faults, twins and slip planes are planar defects observed in REM. Lattice distortion and strain fields are usually associated with planar defects. The surface strain field of planar defects may be significantly different from that in the bulk because of the boundary relaxation of the defected region. We now use the twin

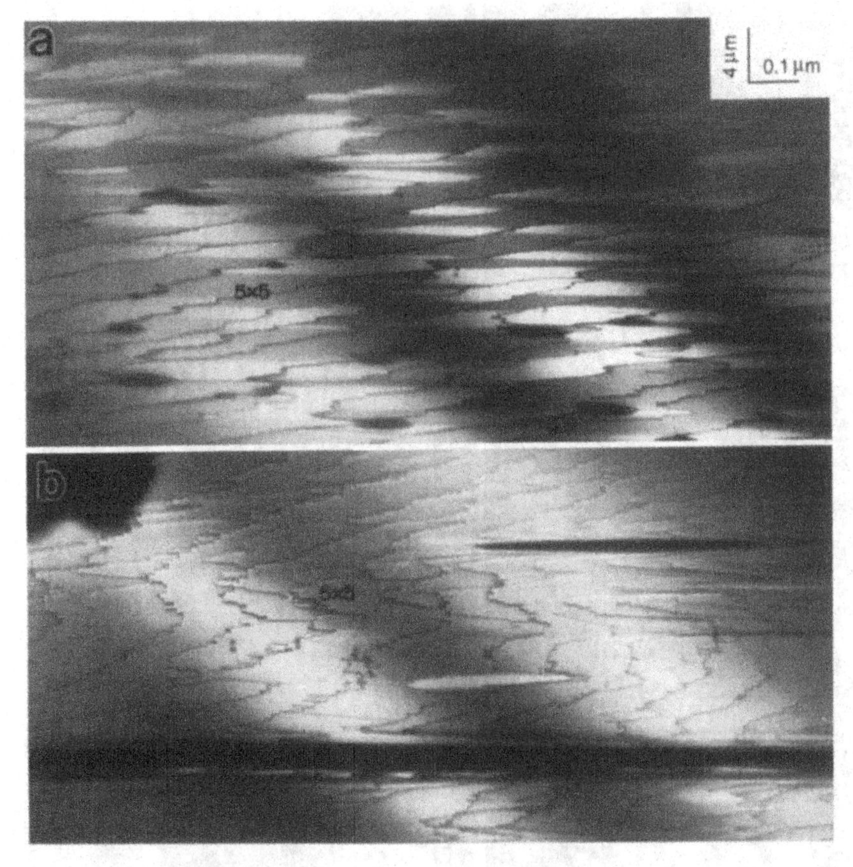

Figure 8.14 *REM images of the $LaAlO_3(100)$ surface showing the distribution of the reconstructed surface areas in the regions where there are many surface steps. The beam azimuth is [001].*

structure of $LaAlO_3$ to illustrate the features of planar defects. $LaAlO_3$ has {100} twin structure with twin angle about 0.12°. In REM, the strain field produced by a defect extending from the bulk introduces strong contrast in the image. Figure 8.16 shows a REM image of an annealed $LaAlO_3(100)$ surface. Both ends of the (001) twinned grains are seen, but the contrast is reversed. The contrast reversal can be interpreted based on the diffraction contrast mechanism of REM. If the left-hand twin-boundary region is set at the Bragg condition and shows bright contrast due to a strain modulation $\boldsymbol{R}(\boldsymbol{r})$, then the right-hand twin-boundary region must be out of the Bragg condition due to an opposite lattice modulation $-\boldsymbol{R}(\boldsymbol{r})$, resulting in dark contrast. The contrast of planar defects is dominated by diffraction contrast. Thus, contrast reversal is possible only when the diffracting condition is changed. The contrast has little dependence on the focus of the objective lens. This characteristic

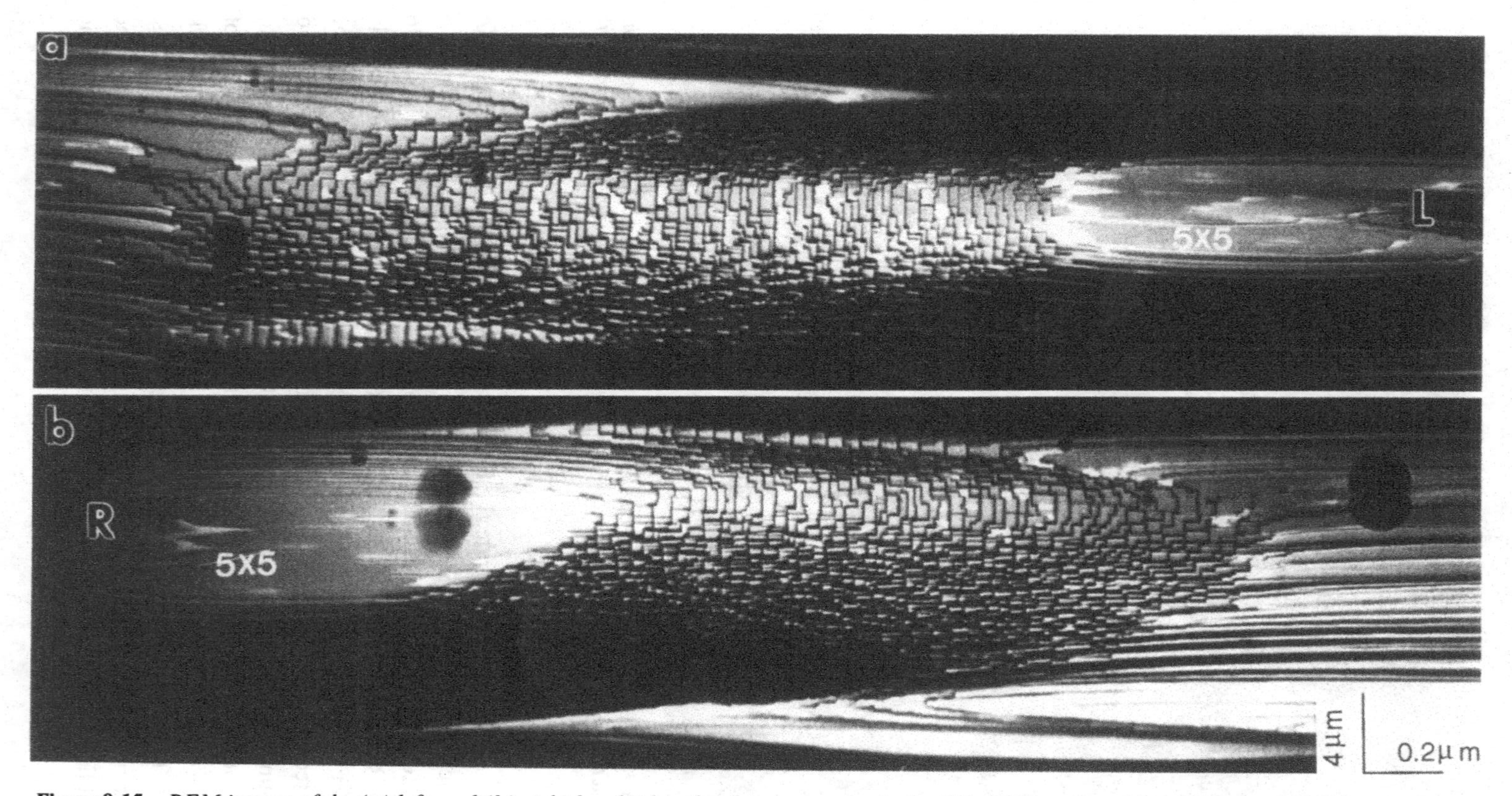

Figure 8.15 REM images of the (a) left- and (b) right-hand sides of a plateau area on the $LaAlO_3(100)$ surface. Many [001] and [010] steps are seen, and the surface exhibits the 'tiles on roof' structure. The beam azimuth is [001].

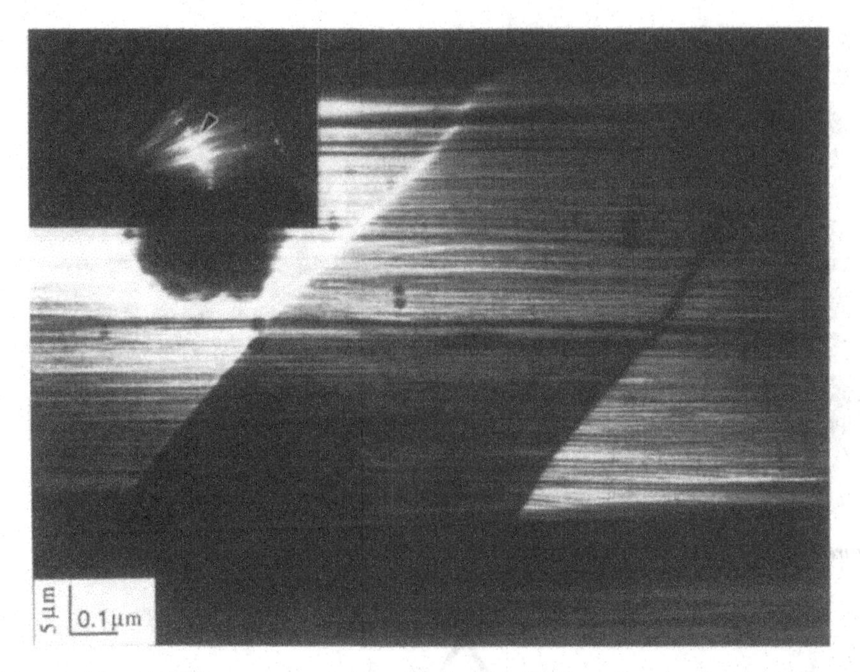

Figure 8.16 *A REM image recorded from the $LaAlO_3(100)$ surface showing the (010) twins. The beam azimuth is near [001]. The diffraction conditions for recording the image are shown in the inset.*

can be applied to distinguish planar defects from surface steps. Using this reflected electron technique, it is possible to image some details at the twin intersection (Hsu and Cowley, 1994) and stacking faults beneath the surface (Hsu and Cowley, 1985).

8.8 As-grown and polished surfaces

REM has also been applied to image as-grown surfaces and polished surfaces. Imaging of as-grown and polished natural diamond surfaces is an example. The synthetic diamonds to be studied are in the as-grown condition with grain sizes of 0.5–1 mm without chemical treatment or mechanical polishing. A REM image of diamond (111) is shown in Figure 8.17. The dark 'patches' on the surface are probably surface contaminants, which cannot be removed by ultrasonic washing in acetone. Besides these contaminants, unequally spaced contrast fringes, separated by approximately 3–6 nm and running parallel or almost parallel to the incident beam direction (from the top to the bottom of the micrograph), are observed. It appears that these fringes arise from surface topography rather than interference effects.

It was proposed that the diamond {111} surfaces were composed of small trigonal protrusions about 5 nm in size (Thornton and Wilks, 1976; Wilks and Wilks, 1972),

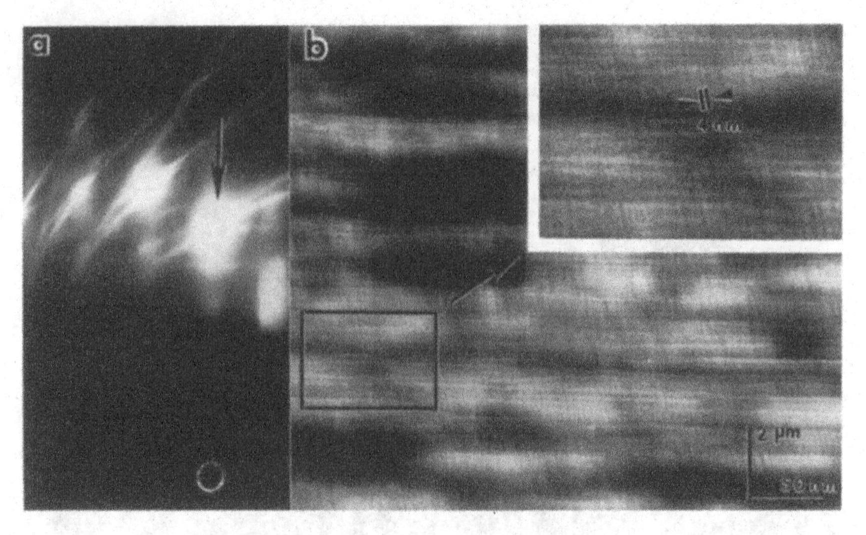

Figure 8.17 *A REM image of a synthetic diamond (111) surface.*

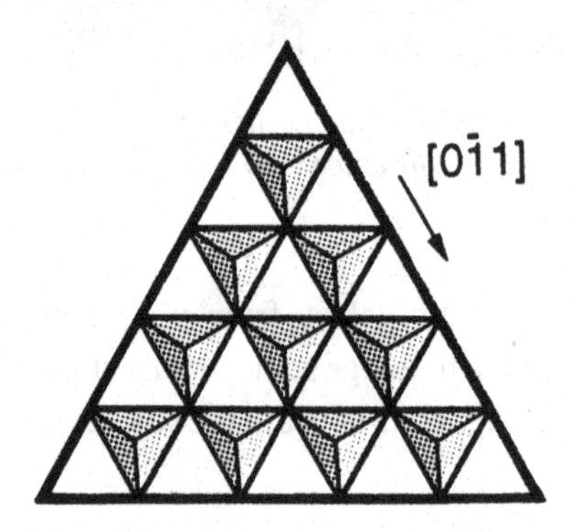

Figure 8.18 *A model showing expected nanometer-size trigonal protrusions on a diamond (111) surface.*

which are oriented, the corners pointing to the edges of the triangular outline of the octahedral faces (Figure 8.18). This model was proposed in order to interpret the orientationally dependent diamond-on-diamond friction coefficient measurements (Tabor, 1979). When observed along [0$\bar{1}$1], these features could appear in rows due to the foreshortening effect of REM, forming the observed topography fringes in Figure 8.17.

Single-crystal diamond thin films grown with a hot-filament-assisted chemical vapor deposition (CVD) system on single-crystal natural diamond {100} substrates have also been studied by REM. The as-grown film is a [100] single crystal. The surface microstructure appears complex in REM images (Figure 8.19). The features

Figure 8.19 *A REM image of as-grown CVD [100] single-crystal diamond film, showing nanometer-scale surface growth topographic features.*

can be summarized as: (1) regions of dark contrast (possibly contaminants); (2) lines of contrast running parallel to the beam direction [010] with separations of 6–11 nm; and (3) 'hill' and 'valley' topography, each hill showing piled-up steps. The observed fringes appear to be due to surface topographic contrast and may correspond to the existence of square-based pyramids on the (100) surface as expected (Thornton and Wilks, 1976; Wilks and Wilks, 1972; Wang *et al.*, 1993).

Various methods have been used to image the friction tracks left by a diamond needle after diamond-on-diamond friction measurements, but none of the methods has shown the sensitivity of REM for resolving the marks left by a single pass of friction measurement (Wang *et al.*, 1991). Shown in Figure 8.20 is a REM image of the friction tracks left by a single pass of the needle under an applied pressure of 13 GPa (below the critical pressure of about 21 GPa). The track width is about 4 μm, which corresponds to the real contact size of the needle on the surface. Those areas not touched, or very gently touched, by the needle can be easily identified as showing relatively strong reflection intensity. Several interesting points can be seen from Figure 8.20. First, the friction tracks are clearly visible and they do not run along

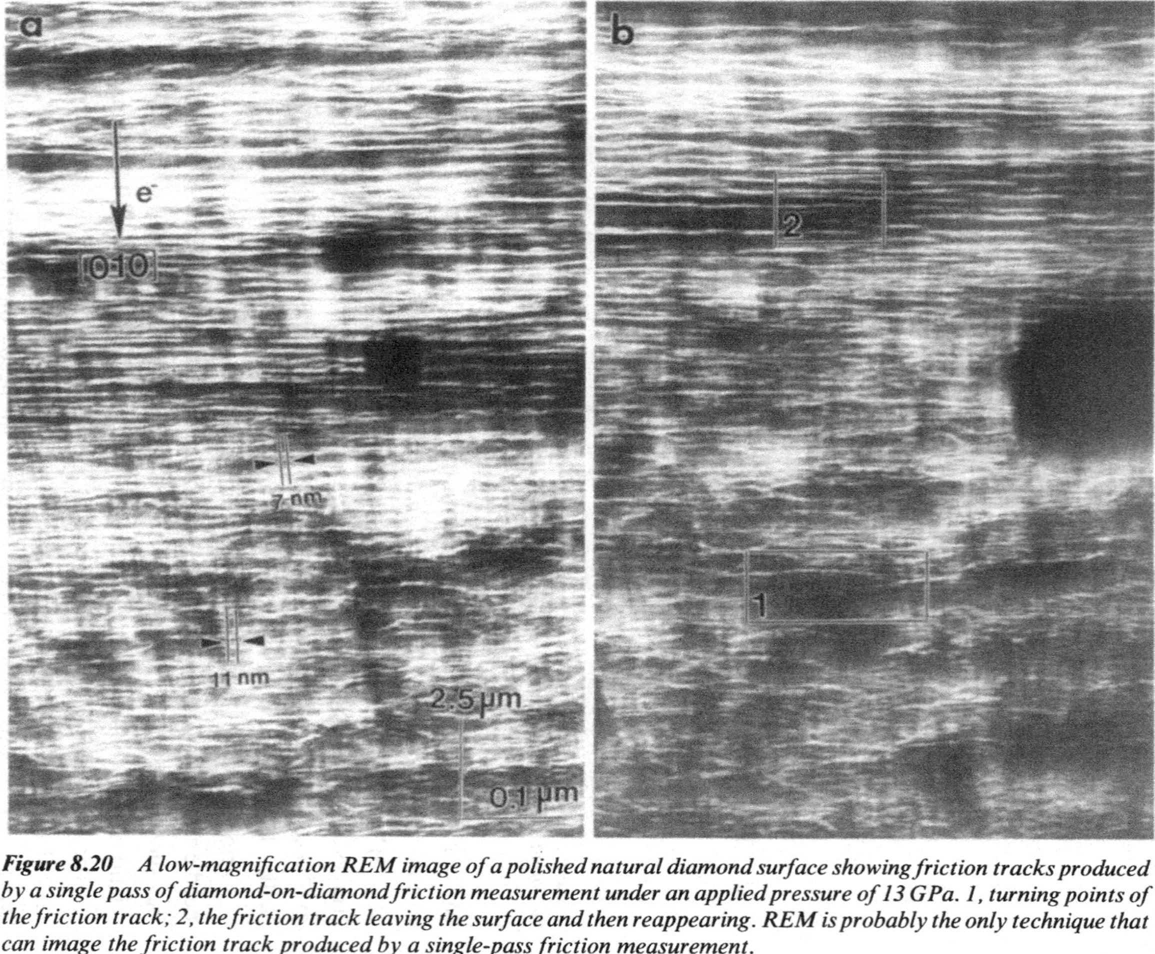

Figure 8.20 *A low-magnification REM image of a polished natural diamond surface showing friction tracks produced by a single pass of diamond-on-diamond friction measurement under an applied pressure of 13 GPa. 1, turning points of the friction track; 2, the friction track leaving the surface and then reappearing. REM is probably the only technique that can image the friction track produced by a single-pass friction measurement.*

Figure 8.21 *A REM image of a diamond surface after two friction passes under a higher applied pressure of 21 GPa, showing possible deformation on the diamond surface.*

straight lines. When the friction needle was temporarily blocked by a large polishing line, which ran perpendicular to the sliding direction of the needle, the driving force on the needle may not have been strong enough to fracture the diamond. The needle thus tended to slip along the polishing line for a short distance and then surpass it, resulting in a drift in the vertical direction as seen in the REM image. Second, the distribution of the image contrast inside of the friction track was not uniform. This was due to the non-uniform local contact of the needle cross-section with the diamond surface, even though the applied pressure was constant. Third, the polishing lines were continuous across the friction tracks and the large features of the surface had not yet been destroyed by a single pass measurement. Finally, the surface areas that had been contacted by the needle showed relatively non-uniform dark contrast, indicating the dependence of surface damage on the local contact pressure. REM is probably the only technique that is sensitive to the deformation on diamond surfaces.

Figure 8.21 shows a REM image of a diamond surface after two friction passes under a higher applied pressure of 21 GPa. As the needle runs along the surface, the polishing lines are interrupted and separated by a dark gap. These dark tracks can be formed when part of the needle is forced into the surface to create the grooves. As indicated in the figure, a friction track slips when crossing a large polishing line. The track that is left indicates the descending tendency of the diamond surface at that area. The mark left by the needle seems to indicate that some surface layers have been scratched off or stressed down. The continuous polishing lines are interrupted by grooves. Therefore, it is possible to have plastic deformation in diamond-on-diamond friction experiments, but no cracking is observed.

This example shows clearly how much useful information REM can provide for studying the contact of surfaces in friction experiments. In contrast to the theoretical model for measuring surface deformation, in which the diamond needle is assumed to be semi-spherical in shape with uniform contact with the surface, REM observation has clearly shown the non-uniform surface contact and that the local pressure could be much higher than the applied one. REM has also been applied to image the surfaces of synthetic diamonds, showing steps and dislocations (Kang, 1984).

This chapter has systematically reviewed the applications of REM in non-UHV TEMs for surface studies. Numerous examples have been shown to demonstrate the particular usefulness of the REM technique for revealing surface structure information in a conventional TEM. The materials that can be examined include metals, semiconductors and ceramic surfaces. The charging effect is not a factor that can influence the REM imaging of insulator surfaces. Atom-high surface steps and dislocations can be easily seen in REM images. Surface termination and reconstruction of ceramic surfaces can be directly imaged with REM.

PART C

Inelastic scattering and spectrometry of reflected electrons

CHAPTER 9

Phonon scattering in RHEED

Numerous inelastic scattering processes are involved in electron scattering. The mean-free-path length of inelastic scattering is about 50–300 nm for most materials, thus more than 50% of the electrons will be inelastically scattered if the specimen thickness is close to the mean-free-path length. Inelastic scattering not only affects the quality of REM images and RHEED patterns but also makes data quantification much more complex and inaccurate. In this chapter, we first outline the inelastic scattering processes in electron diffraction. Then phonon (or thermal diffuse) scattering will be discussed in detail. The other inelastic scattering processes will be described in Chapters 10 and 11.

9.1 Inelastic excitations in crystals

The interaction between an incident electron and the crystal atoms results in various elastic and inelastic scattering processes. The transition of crystal state is excited by the electron due to its energy and momentum transfers. Figure 9.1 indicates the main inelastic processes that may be excited in high-energy electron scattering. First, plasmon (or valence) excitation, which characterizes the transitions of electrons from the valence band to the conduction band, involves an energy loss in the range 1–50 eV and an angular spreading of less than 0.2 mrad for high-energy electrons. The decay of plasmons results in the emission of ultraviolet light. The cathodoluminescence (CL) technique is based on detection of the visible light emitted when an electron in a higher energy state (usually at an impurity) fills a hole in a lower state that has been created by the fast electron. Second, thermal diffuse scattering (TDS) or phonon scattering is the result of atomic vibrations in crystals. This process does not introduce any significant energy loss (<0.1 eV) but produces large momentum transfer, which can scatter the electrons out of the selecting angular range of the objective aperture. This is a localized inelastic scattering process. The electrons scattered to high angles are primarily due to TDS. Finally, atomic inner shell ionization is excited by the energy transfer of the incident electron, resulting in an electron being ejected from deep-core states (Figure 9.2(a)). This process introduces an energy loss in the range of a few tens to thousands of electron-volts with an angular spreading of the order $\vartheta_E = \Delta E/(2E_0)$, where ΔE is the electron energy loss and E_0 is the incident electron energy. Analogously to CL, the holes created at deep

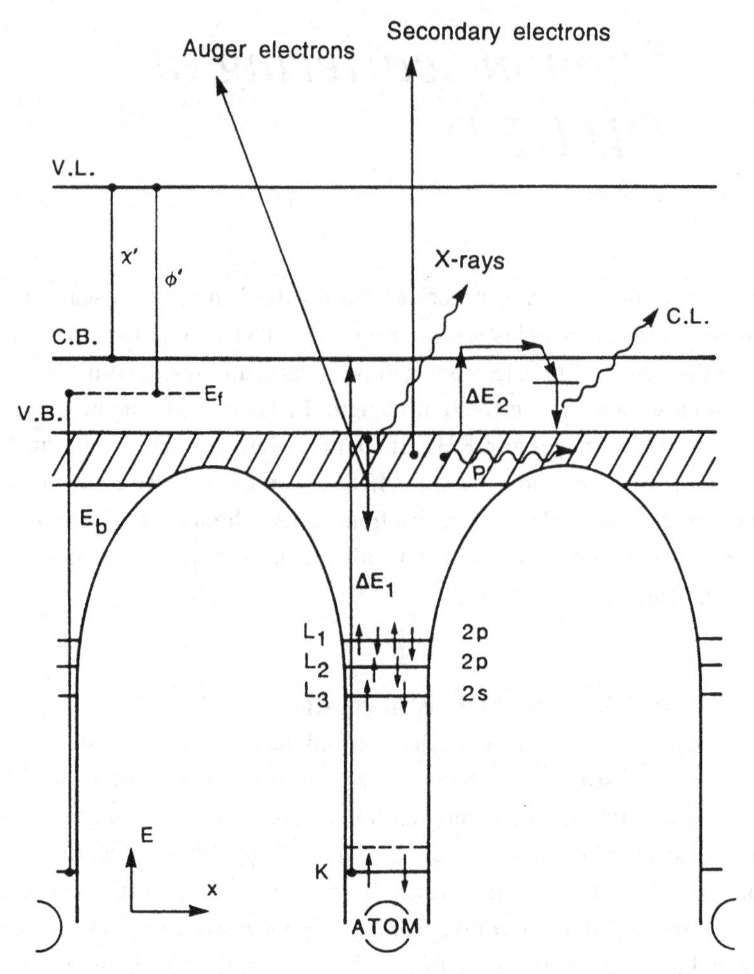

Figure 9.1 *A schematic one-electron energy level diagram plotted against the positions of atoms showing important excitations by an incident electron in a semiconductor material. Here, E_f is the Fermi level, E_b the binding energy, and C.B., V.B. and V.L. are the conduction-band minimum, valence-band maximum, and vacuum level, respectively. ΔE_1 is a K shell excitation; ΔE_2 is a single-electron excitation; C.L. is a cathodoluminescence photon; and P is a plasmon.*

core states tend to be filled by the core-shell electrons in high-energy levels, resulting in the emission of photons (or X-rays) (Figure 9.2(b)). The X-rays emitted are fingerprints of the elements, and thus can be used for chemical microanalysis. The holes, created by the ionization process, in deep-core states may be filled by the electrons in outer shells; the energy released in this process may ionize another outer shell electron (Figure 9.2(c)), resulting in the emission of Auger electrons. Accompanying these processes, secondary electrons can be emitted from the valence band, and they are useful to image surface morphology.

There are two continuous energy-loss processes in electron scattering. Contin-

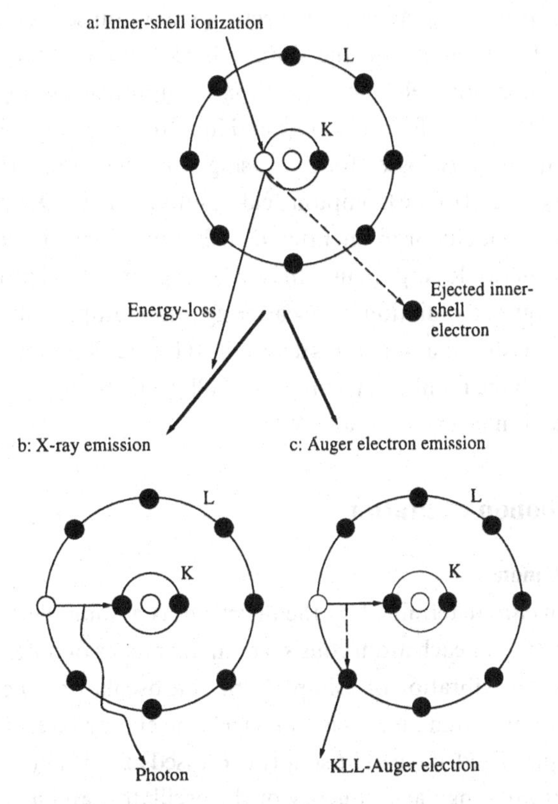

Figure 9.2 *Inelastic interactions of an incident electron with the atom-bound electrons, resulting in (a) inner-shell ionization, (b) X-ray emission and (c) Auger electron emission.*

uous energy-loss spectra can be generated by an electron that penetrates into the specimen and undergoes collisions with the atoms in it. The electromagnetic radiation produced is known as bremsstrahlung (German for 'braking radiation'). X-rays are usually generated. When penetrating the specimen, the force that acts on the electron is $\boldsymbol{F} = e\nabla V$, where V is the crystal potential. Thus, an incident electron will be accelerated and decelerated when approaching and leaving an atom due to electrostatic attraction and repulsion, respectively, resulting in electromagnetic radiation (Konopinski, 1981). Therefore, bremsstrahlung leads to continuous energy loss that increases with increasing scattering angle. The radiation energy is emitted in the form of X-rays. Thus bremsstrahlung is responsible for the background observed in energy dispersive X-ray spectroscopy (EDS).

The other process is energy transfer due to collision of the incident electron with a crystal electron. This process is similar to inelastic two-ball collision in classical physics, and is known as electron Compton scattering or electron–electron (e–e) scattering. Energy and momentum conservation are required in this process. Since large momentum transfer is involved in electron Compton scattering, this process may be responsible in part for the diffuse background observed in electron diffraction patterns of light elements. Electron Compton scattering leads to large losses (a broad peak from a few tens to several hundred electron-volts) with a peak that moves to high energy losses for increasing scattering angle (Williams *et al.*, 1984). The energy spread of the Compton peak results from the Doppler broadening due to the 'orbital' velocity or momentum distribution of the atomic electrons.

In an electron energy-loss spectrum, the signal background is primarily contributed by electromagnetic radiation, e–e scattering and multiple valence excitations. Which of these is the dominant factor is determined by the electronic structure of the material and the atomic number. A summary of all the inelastic scattering processes in electron diffraction is given in Table 9.1.

9.2 Phonon excitation

9.2.1 Phonons

Atomic vibrations are determined by the interactions among the atoms inside the crystal. The vibration of each atom consists of many modes of different frequencies and wavevectors. The vibration of a simple harmonic oscillator, for example, can be precisely solved in quantum mechanics, in which the energy levels of the oscillator are found to be quantized and can be simply expressed as $\varepsilon_n = \hbar\omega(n_s + \frac{1}{2})$, where ω is the classical vibration angular frequency of the oscillator, and n_s is an integer (or quantum number), which characterizes the n_sth excited state of the oscillator. n_s is actually the occupation number of phonons in the n_sth excited state. If the $n_s = 3$ state is excited, for example, this means the simultaneous creation of three phonons in this scattering process, which is usually referred to as multi-phonon scattering.

In crystals, the three-dimensional interactions of the atoms produce many phonon modes, each of which is characterized by a wavevector $\boldsymbol{q}$, polarization vector $\boldsymbol{e}$, frequency ω and dispersion surface j. Precisely speaking, a phonon is a vibration mode of the crystal lattice. The entire vibration of the lattice is a superposition of many of these modes (or phonons). Based on the harmonic oscillators and adiabatic approximations, in which all the atoms are assumed to interact with harmonic forces and the crystal electrons to move as though the ions were fixed in their instantaneous positions, one considers the phonon modes existing in a perfect crystal of an infinite number of unit cells. The equilibrium position vector of the nth unit cell relative to the origin is denoted by

$$\boldsymbol{R}_n \equiv n_1\boldsymbol{a} + n_2\boldsymbol{b} + n_3\boldsymbol{c}, \tag{9.1}$$

where n_1, n_2 and n_3 are any three integers. The instantaneous position of the αth atom in the nth unit cell is

Table 9.1. *A summary of inelastic scattering in electron diffraction.*

Inelastic process	Origin	Energy loss	Angular spreading	Character
Plasmon loss (or valence loss)	Valence band to conduction band	1–50 eV	<0.2 mrad	Strong peaks in EELS
TDS (phonon scattering)	Atomic vibrations	<0.1 eV	5–200 mrad	Diffuse background
Single-electron scattering	Atomic inner-shell ionization	A few tens to thousands of eV	$\vartheta = \Delta E/(2E_0)$	Absorption edges
Bremsstrahlung	Collision with atoms	Continuous	Large	Continuous background in EELS
Electron Compton scattering	Electron–electron collision	Continuous and high	Large	Very broad peaks

$$\boldsymbol{r}\binom{n}{\alpha} \equiv \boldsymbol{R}_n + \boldsymbol{r}(\alpha) + \boldsymbol{u}\binom{n}{\alpha}, \tag{9.2}$$

where $\boldsymbol{r}(\alpha)$ is the equilibrium position of the αth atom relative to the nth unit cell, and $\boldsymbol{u}\binom{n}{\alpha}$ is the time-dependent displacement vector of the atom. The lattice dynamics was initiated by Born (1942a and b) and has been introduced in many books (see Brüesch (1982) for example); here we outline some main results. The vibrational displacement of an atom is written as a sum of harmonic oscillators of momentum $\boldsymbol{q}$, frequency $\omega_j(\boldsymbol{q})$ and polarization vector $\boldsymbol{e}(\alpha|{}^{\boldsymbol{q}}_{j})$,

$$\boldsymbol{u}\binom{n}{\alpha} = \left(\frac{1}{2N_0 M_\alpha}\right) \sum_{\boldsymbol{q}} \sum_{j} \frac{1}{[\omega_j(\boldsymbol{q})]^{1/2}} \boldsymbol{e}(\alpha|{}^{\boldsymbol{q}}_{j}) \exp[2\pi i \boldsymbol{q} \cdot (\boldsymbol{R}_n + \boldsymbol{r}(\alpha))]$$
$$\times \{A({}^{\boldsymbol{q}}_{j}) \exp[-i\omega_j(\boldsymbol{q})t] + A^*({}^{-\boldsymbol{q}}_{j}) \exp[i\omega_j(\boldsymbol{q})t]\}, \tag{9.3}$$

or in normal coordinates,

$$\boldsymbol{u}\binom{n}{\alpha} = \left(\frac{\hbar}{2N_0 M_\alpha}\right) \sum_{\boldsymbol{q}} \sum_{j} \frac{1}{[\omega_j(\boldsymbol{q})]^{1/2}} \boldsymbol{e}(\alpha|{}^{\boldsymbol{q}}_{j}) \exp[2\pi i \boldsymbol{q} \cdot (\boldsymbol{R}_n + \boldsymbol{r}(\alpha))][a^+({}^{-\boldsymbol{q}}_{j}) + a({}^{\boldsymbol{q}}_{j})], \tag{9.4}$$

where N_0 is the total number of primitive unit cells in the crystal; M_α is the mass of the αth atom; j indicates phonon branches; $a^+({}^{-\boldsymbol{q}}_{j})$ and $a({}^{\boldsymbol{q}}_{j})$ are defined as the creation and annihilation operators of a phonon with wavevector $\boldsymbol{q}$ and frequency $\omega_j(\boldsymbol{q})$, respectively, and have the following properties:

$$a^+({}^{-\boldsymbol{q}}_{j})|n_s\rangle = (n_s + 1)^{1/2}|n_s + 1\rangle, \tag{9.5}$$

$$a({}^{\boldsymbol{q}}_{j})|n_s\rangle = \sqrt{n_s}|n_s - 1\rangle, \tag{9.6}$$

where n_s is the occupation number of phonons, with wavevector $\boldsymbol{q}$ and frequency ω_j, in the phonon state $|n_s\rangle$.

Phonon scattering is actually the creation and annihilation of phonons in crystals as a result of electron–crystal interactions. The momentum transferred from the incident electron (due to scattering) is taken by phonons. Each phonon is a mode of crystal vibration characterized by $(\omega_j, \boldsymbol{q}, \boldsymbol{e})$.

9.2.2 The effect of atomic vibrations on the crystal potential

Electron scattering is directly determined by the potential distribution in the crystal. The crystal potential is perturbed by the atomic vibrations that eventually create an incoherent or partially coherent intensity distribution in addition to the elastic Bragg scattering in the diffraction pattern. This is the electron–phonon interaction process in electron diffraction and is responsible for thermal diffuse scattering. The crystal potential is written as a time-independent and a time-dependent component; the former is determined by the equilibrium positions of the atoms and the latter depends on the instantaneous displacements of the atoms:

$$V(\boldsymbol{r},t)=V_0(\boldsymbol{r})+\Delta V(\boldsymbol{r},t). \tag{9.7}$$

The time-independent potential is a time average of the crystal potential, that is $V_0(\boldsymbol{r})=\langle V(\boldsymbol{r},t)\rangle$. The time-dependent crystal potential due to atomic vibrations is (Takagi, 1958a and b)

$$\Delta V(\boldsymbol{r},t)=\sum_n\sum_\alpha [V_\alpha(\boldsymbol{r}-\boldsymbol{R}_n-\boldsymbol{r}(\alpha)-\boldsymbol{u}(^n_\alpha))-V_{0\alpha}(\boldsymbol{r}-\boldsymbol{R}_n-\boldsymbol{r}(\alpha))], \tag{9.8}$$

where $V_{0\alpha}(\boldsymbol{r})$ is the time-averaged atomic potential and $V_\alpha(\boldsymbol{r})$ is the instantaneous atomic potential

$$V_\alpha(\boldsymbol{r})=\int \mathrm{d}\tau f^{\mathrm{e}}_\alpha(\tau)\exp(2\pi \mathrm{i}\tau\cdot\boldsymbol{r}), \tag{9.9}$$

$$V_0(\boldsymbol{r})=\langle V(\boldsymbol{r},t)\rangle=\sum_n\sum_\alpha V_{0\alpha}(\boldsymbol{r}-\boldsymbol{R}_n-\boldsymbol{r}(\alpha)), \tag{9.10}$$

$$V_{0\alpha}(\boldsymbol{r})=\int \mathrm{d}\tau (f^{\mathrm{e}}_\alpha(\tau)\langle\exp[2\pi\mathrm{i}\tau\cdot(\boldsymbol{r}-\boldsymbol{u}(^n_\alpha))]\rangle=\int \mathrm{d}\tau f^{\mathrm{e}}_\alpha(\tau)\exp(-W_\alpha)\exp(2\pi\mathrm{i}\tau\cdot\boldsymbol{r}), \tag{9.11}$$

where $W_\alpha=2\pi^2\langle|\tau\cdot\boldsymbol{u}(^n_\alpha)|^2\rangle$ is the Debye–Waller factor; a general relation of

$$\langle\exp x\rangle=\exp(-\langle x^2\rangle/2), \qquad \text{for } \langle x\rangle=0 \tag{9.12}$$

was used; and $\langle\rangle$ represents a time-average. The factor $\exp(-W_\alpha)$ is sensitive to the variation in specimen temperature. The Debye–Waller (DW) factor is always involved in elastic electron diffraction, because the crystal potential in elastic scattering is the time-averaged potential given by Eq. (9.10). Proper consideration of the DW factor is important in quantitative electron microscopy. The evaluation of the DW factor will be given in Section. 9.4.

9.2.3 Electron–phonon interactions

Electron–phonon interaction is an inelastic process in high-energy electron scattering. The perturbation of atomic vibrations on the crystal potential eventually affects the scattering behavior of the crystal. Phonons are created or annihilated in the process due to the energy (albeit very small) and momenta transfer of the incident electrons. Therefore, precisely speaking, phonon scattering is an inelastic excitation process, which involves transition between crystal vibration states.

The interaction Hamiltonian for creating a single phonon of momentum $\boldsymbol{q}$ and frequency $\omega_j(\boldsymbol{q})$ can be derived after expanding $\Delta V(\boldsymbol{r},t)$ in terms of atomic displacements (Whelan, 1965a and b). We have

$$H(\boldsymbol{r},\boldsymbol{q},\omega)=\langle n_{\mathrm{s}}(\boldsymbol{q},\omega)+1|[-e\Delta V(\boldsymbol{r})]|n_{\mathrm{s}}(\boldsymbol{q},\omega)\rangle$$
$$=e\langle n_{\mathrm{s}}(\boldsymbol{q},\omega)+1|\sum_n\sum_\alpha\{\boldsymbol{u}(^n_\alpha)\cdot\nabla V_\alpha(\boldsymbol{r}-\boldsymbol{R}_n-\boldsymbol{r}(\alpha))\}|n_{\mathrm{s}}(\boldsymbol{q},\omega)\rangle$$

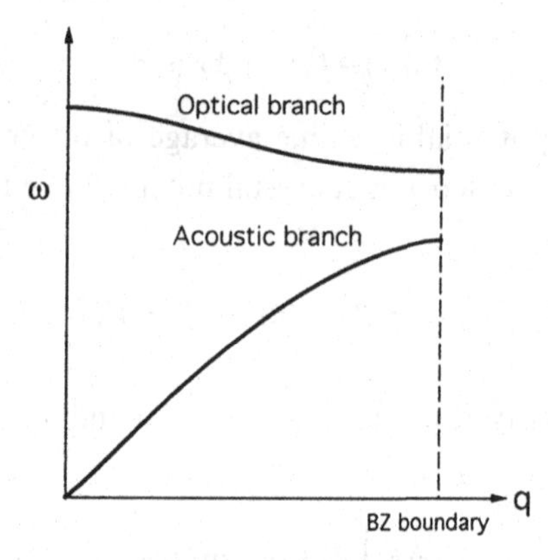

Figure 9.3 Optical and acoustical phonon branches of the dispersion relation for a one-dimensional diatomic linear lattice. BZ denotes the first Brillouin zone.

$$= e\sum_{\alpha}\sum_{n} A_{\alpha}(\omega_j(\boldsymbol{q}))\boldsymbol{e}(\alpha|_j^{\boldsymbol{q}})\cdot\nabla\int \mathrm{d}\boldsymbol{u} f_{\alpha}^{\mathrm{e}}(\boldsymbol{u})\exp[2\pi\mathrm{i}\boldsymbol{u}\cdot(\boldsymbol{r}-\boldsymbol{R}_n-\boldsymbol{r}(\alpha))]\exp[2\pi\mathrm{i}\boldsymbol{q}\cdot(\boldsymbol{R}_n+\boldsymbol{r}(\boldsymbol{a}))]$$

$$= 2\pi\mathrm{i}e\sum_{\alpha}\sum_{g} A_{\alpha}(\omega_j(\boldsymbol{q}))\boldsymbol{e}(\alpha|_j^{\boldsymbol{q}})\cdot(\boldsymbol{g}+\boldsymbol{q}) f_{\alpha}^{\mathrm{e}}(\boldsymbol{g}+\boldsymbol{q})\exp[2\pi\mathrm{i}(\boldsymbol{g}+\boldsymbol{q})\cdot\boldsymbol{r}]\exp[-2\pi\mathrm{i}\boldsymbol{g}\cdot\boldsymbol{r}(\alpha)], \tag{9.13}$$

where $\langle n_s \rangle$ is the average occupation number of the phonon state $|n_s(\boldsymbol{q},\omega)\rangle$,

$$\langle n_s(\boldsymbol{q},\omega)\rangle = \frac{1}{\exp[\hbar\omega_j/(k_B T)]-1}, \tag{9.14}$$

$$A_{\alpha}(\omega_j(\boldsymbol{q})) \equiv \left(\frac{\hbar(\langle n_s(\boldsymbol{q},\omega)\rangle+1)}{2\omega_j(\boldsymbol{q})M_{\alpha}N_0}\right)^{1/2} \tag{9.15}$$

is the atomic vibration amplitude in phonon mode $\omega_j(\boldsymbol{q})$. Equation (9.14) is the interaction Hamiltonian of phonon–electron interaction.

Phonon modes are classified as optical and acoustic branches. The optical branch, ω_+, for which ω remains constant when $\boldsymbol{q}$ approaches zero, may be approximated by the Einstein model, in which $\omega = \omega_0$ independent of $\boldsymbol{q}$. The acoustic branches, for which ω_- tends to zero when q approaches zero, may be approximated by the Debye model, in which $\omega_j = 2\pi v_j q$, where v_j is the phase velocity (Figure 9.3). The atomic displacement is mainly determined by the acoustic branches because A_{α}^2 is inversely proportional to ω_j and the acoustic frequency ω_j tends to zero for small wavevectors.

Phonon–electron interaction is the source of thermal diffuse scattering. Figure 9.4 shows a transmission electron diffraction pattern of a $BaTiO_3$ crystal. The crystal was rotated slightly off the [001] zone in order to reduce the zone-axis dynamical diffraction effect. Besides Bragg reflections, continuous, sharp $\langle 100 \rangle$ streaks are seen. These streaks are the result of phonon scattering. The line shape and directions

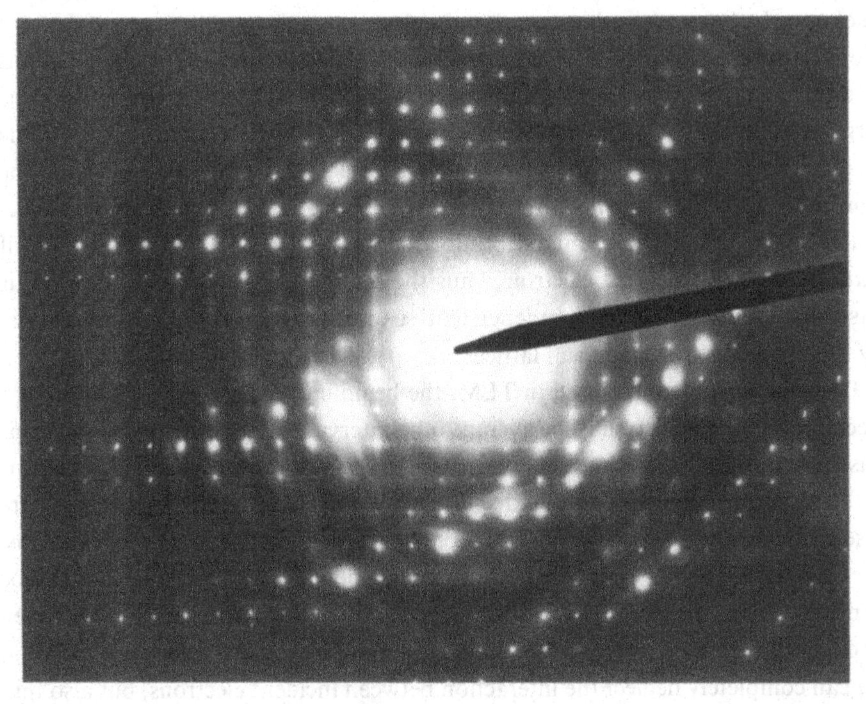

Figure 9.4 A transmission electron diffraction pattern of a $BaTiO_3$ crystal showing ⟨100⟩ TDS streaks near the [001] zone axis.

are determined by the phonon dispersion surface and polarization. Since crystal lattice vibration is not structurally periodic across the crystal, the diffusely scattered electrons are distributed at angles other than Bragg angles, introducing a 'background' in the diffraction pattern. The fine features in the background reflect the phase correlation between atom vibrations. A rigorous treatment of TDS in electron diffraction and imaging has been given by Wang (1992a and 1995).

9.3 The 'frozen' lattice model

We have described the creation and annihilation of phonons in electron diffraction. The mathematical treatment is rather lengthy, particularly when more than one phonon is involved. Since atomic vibration is a time-dependent process, its perturbation on the crystal potential is time-dependent. The first question here is that of whether we need to use the time-dependent Schrödinger equation to describe the diffraction behavior of TDS electrons. We must first examine the following factors before an answer is given.

In high-energy electron diffraction, the traveling velocity of the electron is larger than half the speed of light ($v = 0.53c$ for 100 keV electrons). For a thin crystal of thickness $d = 100$ nm, the duration that the electron interacts with the crystal atoms is on the order of $\Delta t \approx 6 \times 10^{-15}$ s. In general the vibration period T of crystal atoms is on the order of 10^{-13}–10^{-14} s (Sinha, 1973; Mitra and Massa, 1982). The large-amplitude, low-frequency acoustic waves, which produce most of the atomic displacements and are responsible for TDS, are many times slower. Thus the condition $\Delta t \ll T$ is satisfied. Therefore, the electron–crystal interaction time is much shorter than the atomic vibration period and the atoms are seen as if stationary by the incident electron. Thus the scattering of an incident electron is basically a quasi-elastic, 'time-independent' scattering process (without energy loss) by the distorted 'frozen' crystal lattices.

For an electron source used in TEM, the beam current is on the order of 10^{12} electrons per second, so that the average time interval between successive electrons passing through the specimen is about 100 atomic vibration periods. This delay is sufficiently large that the atomic displacements 'seen' by successive electrons are essentially uncorrelated. The physical process of accumulating millions of electrons, each of which has been scattered by an independent lattice configuration, can be considered as a Monte Carlo integration. In addition, the average distance between two successive incident electrons is on the order of 10^5 nm; this means not only that we can completely neglect the interaction between incident electrons, but also that the scattering of the electron is completely independent of that of the previous electron or the next incoming electron. In other words, single-electron-scattering theory holds.

Theoretical calculation has found that the life time of a phonon is on the order of 10^{-13} s (Björkman *et al.*, 1967; Woll and Kohn, 1962). The limited life time of phonons is due to anharmonic effects in atom vibrations. This is much larger than the interaction time Δt of the electron with the crystal. Thus there is no phonon decay during the scattering of an incident electron. The average time interval between two incident electrons is about 10^{-12} s, which is much longer than the life time of phonons. Therefore, the crystal can be considered as being in its ground state for each incident electron, and there is no need to consider the phonons that were created by the previous incident electron.

In practice, it usually takes more than 0.05 s to record a diffraction pattern or image. During this period of time, the crystal lattice has experienced millions of different vibration configurations. Thus the acquired electron diffraction pattern is an incoherent summation of independent scattering for millions of electrons. This is equivalent to a time-average over the intensity distribution produced by one electron scattering.

Therefore, an electron diffraction pattern or image is considered as the sum of the scattering intensities for many instantaneous pictures of the displaced atoms. In other words, TDS is actually a statistically averaged, quasi-elastic scattering of the electrons by crystals whose atoms are arranged in different configurations for every incident electron. The main task involved in the theory is to perform the time-averaging of the electron diffraction intensities for a vast number of different thermal vibration configurations. This calculation can be performed analytically before doing numerical calculations (see Section 9.6).

9.4 Calculation of the Debye–Waller factor

The Debye–Waller factor is an important quantity in electron scattering. This factor characterizes the temperature-dependence of the scattering data. We now calculate the DW factor using the expression for atom thermal displacement given by Eq. (9.4). Then, the result is simplified using the Debye model. At thermal equilibrium, no phonons are created or annihilated. Thus

$$\begin{aligned} W_\alpha &= 2\pi^2\langle[\tau\cdot \boldsymbol{u}(\begin{smallmatrix} n \\ \alpha \end{smallmatrix})]^2\rangle = 2\pi^2\langle n_s|[\tau\cdot \boldsymbol{u}(\begin{smallmatrix} n \\ \alpha \end{smallmatrix})]^2|n_s\rangle \\ &= \frac{\pi^2\hbar}{N_0 M_\alpha}\sum_{\boldsymbol{q}}\sum_j \frac{1}{\omega_j(\boldsymbol{q})}[\tau\cdot \boldsymbol{e}(\alpha|\begin{smallmatrix} \boldsymbol{q} \\ j \end{smallmatrix})]^2\langle n_s|[a^+(\begin{smallmatrix} -\boldsymbol{q} \\ j \end{smallmatrix})a(\begin{smallmatrix} \boldsymbol{q} \\ j \end{smallmatrix})]+[a(\begin{smallmatrix} -\boldsymbol{q} \\ j \end{smallmatrix})a^+(\begin{smallmatrix} \boldsymbol{q} \\ j \end{smallmatrix})]|n_s\rangle \\ &= \frac{2\pi^2\hbar}{N_0 M_\alpha}\sum_{\boldsymbol{q}}\sum_j \frac{1}{\omega_j(\boldsymbol{q})}[\tau\cdot \boldsymbol{e}(\alpha|\begin{smallmatrix} \boldsymbol{q} \\ j \end{smallmatrix})]^2\left(\langle n_s\rangle+\frac{1}{2}\right) \\ &= \frac{2\pi^2\hbar}{N_0 M_\alpha}\sum_{\boldsymbol{q}}\sum_j \frac{1}{\omega_j(\boldsymbol{q})}[\tau\cdot \boldsymbol{e}(\alpha|\begin{smallmatrix} \boldsymbol{q} \\ j \end{smallmatrix})]^2\left(\frac{1}{\exp[\hbar\omega_j(\boldsymbol{q})/(k_B T)]-1}+\frac{1}{2}\right). \end{aligned} \tag{9.16}$$

We now follow the method of Warren (1990) to simplify the calculation of (9.16). The following procedures are simply called the Warren approximation. We assume that all vibration waves can be considered as either pure longitudinal or pure transverse ones. The velocities of all longitudinal waves are replaced by an average longitudinal velocity, and the velocities of all transverse waves by an average transverse velocity. Each average velocity is considered to be a constant independent of the phonon wavevector $\boldsymbol{q}$.

To perform the summation in (9.16), the Brillouin zone is replaced by a sphere of radius q_m, whose volume is equal to that of the Brillouin zone $\frac{4}{3}\pi q_m^3 = V_{BZ} = 1/\Omega$. The density of points (or states) in the sphere is N_0/V_{BZ}, and the summation of $\boldsymbol{q}$ is replaced by an integration throughout the sphere. For each type of wave, longitudinal or transverse, the polarization vector $\boldsymbol{e}$ takes with equal probability all orientations relative to τ, and for all waves whose vector $\boldsymbol{q}$ terminates in the hollow sphere, we can use the average $\langle[\tau\cdot \boldsymbol{e}(\alpha|\begin{smallmatrix} \boldsymbol{q} \\ j \end{smallmatrix})]^2\rangle = \tau^2\langle\cos^2(\tau, \boldsymbol{e})\rangle = \tau^2/3$. With these approximations, (9.16) becomes

$$W_\alpha = \frac{2\pi^2\hbar\tau^2}{3N_0 M_\alpha}\sum_j \int_0^{q_m} dq\, 4\pi q^2 \frac{1}{\omega_j(q)}\left(\frac{1}{\exp[\hbar\omega_j(a)/(k_B T)]-1}+\frac{1}{2}\right)\frac{N_0}{V_{BZ}}. \tag{9.17}$$

In terms of a constant velocity v_j for each phonon branch, $\omega_j = 2\pi v_j q$. The integral variable q is substituted by ω, and (9.17) becomes

$$W_\alpha = \frac{2\pi^2\hbar\tau^2}{M_\alpha}\sum_j \frac{1}{\omega_{jm}^3}\int_0^{\omega_{jm}} \mathrm{d}\omega\,\omega\left(\frac{1}{\exp[\hbar\omega/(k_B T)]-1}+\frac{1}{2}\right), \tag{9.18}$$

where $\Omega_{jm} = 2\pi v_j q_m$. Let $x = \hbar\omega/(k_B T)$ and $x_j = \hbar\omega_{jm}/(k_B T)$, then (9.17) becomes

$$W_\alpha = \frac{2\pi^2 k_B T\tau^2}{M_\alpha}\sum_j \frac{1}{\omega_{jm}^2}\left(\frac{1}{x_j}\int_0^{x_j} \mathrm{d}x\,\frac{x}{\exp x - 1}+\frac{x_j}{4}\right). \tag{9.19}$$

The sum over j includes one longitudinal (ω_L) and two transverse (ω_T) waves. If we use an average Debye temperature defined as $T_D = \hbar\omega_D/k_B$, where the Debye frequency is determined from the longitudinal and transverse frequencies according to $3/\omega_D^2 = 1/\omega_L^2 + 2/\omega_T^2$, then (9.19) becomes

$$W_\alpha = \frac{6\pi^2\hbar^2\tau^2}{M_\alpha k_B T_D}\left[\left(\frac{T}{T_D}\right)^2 \int_0^{T_D/T} \mathrm{d}x\,\frac{x}{\exp x - 1}+\frac{1}{4}\right]. \tag{9.20}$$

The mean square vibration amplitude of the atom is

$$\overline{a_\alpha^2} = \frac{W_\alpha}{2\pi^2\tau^2} = \frac{3\hbar^2}{M_\alpha k_B T_D}\left[\left(\frac{T}{T_D}\right)^2 \int_0^{T_D/T} \mathrm{d}x\,\frac{x}{\exp x - 1}+\frac{1}{4}\right]. \tag{9.21}$$

Thus, the mean square vibration amplitude is inversely proportional to the atomic mass. The calculations of Eq. (9.20) and Eq. (9.21) can be performed if the Debye temperature is known.

9.5 Kinematical TDS in RHEED

Vibrations of crystal atoms may preserve certain phase relationships as characterized by the phonon dispersion surface and polarization vectors. Thus, the instantaneous displacements of atoms follow some patterns, determined by the crystal structure and interatomic forces, so that the diffuse scattering generated from different atom sites may retain some phase correlation. Thus, they are coherent or partially coherent. In this section, we outline the kinematical TDS theory with consideration given to the phase correlation between atom vibrations.

The total diffraction intensity is a sum of TDS and Bragg scattering, provided that there is no inelastic excitation. Thus, the TDS intensity distribution is the difference between the total scattering and the Bragg scattering. This statement is mathematically expressed as

$$\begin{aligned}
I_{TDS}(\tau) &= \langle|\sum_{\kappa} f_{\kappa}^{e}(\tau)\exp[2\pi i\tau\cdot(\boldsymbol{r}_{\kappa}+\boldsymbol{u}_{\kappa})]|^2\rangle - |\langle\sum_{\kappa} f_{\kappa}^{e}(\tau)\exp[2\pi i\tau\cdot(\boldsymbol{r}_{\kappa}+\boldsymbol{u}_{\kappa})]\rangle|^2 \\
&= \sum_{\kappa}\sum_{\kappa'} f_{\kappa}^{e} f_{\kappa'}^{e*}\exp[2\pi i\tau\cdot(\boldsymbol{r}_{\kappa}-\boldsymbol{r}_{\kappa'})]\{\langle\exp[2\pi i\tau\cdot(\boldsymbol{u}_{\kappa}-\boldsymbol{u}_{\kappa'})]\rangle - \langle\exp(2\pi i\tau\cdot\boldsymbol{u}_{\kappa})\rangle\langle\exp(2\pi i\tau\cdot\boldsymbol{u}_{\kappa'})\rangle\} \\
&= \sum_{\kappa}\sum_{\kappa'} f_{\kappa}^{e} f_{\kappa'}^{e*}\exp[2\pi i\tau\cdot(\boldsymbol{r}_{\kappa}-\boldsymbol{r}_{\kappa'})]\exp(-W_{\kappa}-W_{\kappa'})\{\exp[4\pi^2\langle(\tau\cdot\boldsymbol{u}_{\kappa})(\tau\cdot\boldsymbol{u}_{\kappa'})\rangle]-1\} \\
&= \sum_{\kappa}(f_{\kappa}^{e})^2[1-\exp(-2W_{\kappa})] \\
&\quad + \sum_{\kappa}\sum_{\kappa'\neq\kappa} f_{\kappa}^{e} f_{\kappa'}^{e*}\exp[2\pi i\tau\cdot(\boldsymbol{r}_{\kappa}-\boldsymbol{r}_{\kappa'})]\exp(-W_{\kappa}-W_{\kappa'})\{\exp[4\pi^2\langle(\tau\cdot\boldsymbol{u}_{\kappa})(\tau\cdot\boldsymbol{u}_{\kappa'})\rangle]-1\},
\end{aligned}$$

where $\boldsymbol{u}_{\kappa}$ is the time-dependent displacement of the κth atom. The first term is the incoherent scattering result and the second term is the coherent scattering. Only the first term remains if one uses the Einstein model, in which all the atoms are assumed to vibrate in random phases. A 'uniform' background occurs in the RHEED pattern if the Einstein model is used. The coherence between the thermal diffuse scattering generated from atom sites κ and κ' is determined by $\exp[4\pi^2\langle(\tau\cdot\boldsymbol{u}_{\kappa})(\tau\cdot\boldsymbol{u}_{\kappa'})\rangle]-1$.

Under the harmonic oscillators approximation, $\boldsymbol{u}_{\kappa}$ is given by Eq. (9.4), and we have

$$Y_{\kappa\kappa'} = 4\pi^2\langle(\tau\cdot\boldsymbol{u}_{\kappa})(\tau\cdot\boldsymbol{u}_{\kappa'})\rangle = \frac{4\pi^2\hbar}{N_0(M_{\kappa}M_{\kappa'})^{1/2}}\sum_{q}\sum_{i}\frac{\langle n_s\rangle+\frac{1}{2}}{\omega_i(\boldsymbol{q})}(\tau\cdot\boldsymbol{e}_{q,i})^2\cos(2\pi\boldsymbol{q}\cdot\Delta\boldsymbol{r}_{\kappa\kappa'}), \tag{9.22}$$

where $\Delta\boldsymbol{r}_{\kappa\kappa'} = \boldsymbol{r}_{0\kappa}-\boldsymbol{r}_{0\kappa'}$ is the interatomic vector. We now use the same approximations as introduced in the last section to calculate (9.22). The Brillouin zone is replaced by a sphere of radius q_m, and the orientation average $\langle[\tau\cdot\boldsymbol{e}(\alpha|_j^q)]^2\rangle = \tau^2/3$. Then (9.22) is approximated as

$$\begin{aligned}
Y_{\kappa\kappa'} &\approx \frac{4\pi^2\hbar\tau^2}{3N_0(M_{\kappa}M_{\kappa'})^{1/2}}\sum_j\int_0^{q_m} dq\, 2\pi q^2\int_0^{\pi} d\theta\sin\theta\,\frac{1}{\omega_j(q)}\left(\langle n_s\rangle+\frac{1}{2}\right)\cos(2\pi q\,\Delta r_{\kappa\kappa'}\cos\theta)\,\frac{N_0}{V_{BZ}} \\
&= \frac{\hbar\tau^2}{q_m^3 M_{\kappa}}\sum_j\frac{1}{v_j}\int_0^{q_m} dq\,\frac{\sin(2\pi q\Delta r_{\kappa\kappa'})}{\Delta r_{\kappa\kappa'}}\left(\frac{1}{\exp[2\pi\hbar v_j/(k_B T)]-1}+\frac{1}{2}\right).
\end{aligned} \tag{9.23}$$

The high-temperature limiting case of Eq. (9.23) is

$$Y_{\kappa\kappa'} \approx \frac{k_B T\tau^2}{q_m^2(M_{\kappa}M_{\kappa'})^{1/2}}\sum_j\frac{1}{v_j^2}\frac{\mathrm{Si}\,\Theta_{\kappa\kappa'}}{\Theta_{\kappa\kappa'}}, \tag{9.24}$$

where

$$\mathrm{Si}\,\Theta = \int_0^\Theta \mathrm{d}u \, \frac{\sin u}{u}, \tag{9.25}$$

and $\Theta_{\kappa\kappa'} = 2\pi q_\mathrm{m} |\Delta \boldsymbol{r}_{\kappa\kappa'}|$. On comparing this with the high-temperature limiting case of the DW factor,

$$W_\kappa \approx \frac{k_\mathrm{B} T \tau^2}{2 q_\mathrm{m}^2 M_\kappa} \sum_j \frac{1}{v_j^2}, \tag{9.26}$$

(9.24) is rewritten as

$$Y_{\kappa\kappa'} \approx 2 W_\kappa \left(\frac{M_\kappa}{M_{\kappa'}}\right)^{1/2} \frac{\mathrm{Si}\,\Theta_{\kappa\kappa'}}{\Theta_{\kappa\kappa'}} = 4\pi^2 \overline{a_\kappa^2} \left(\frac{M_\kappa}{M_{\kappa'}}\right)^{1/2} \tau^2 \frac{\mathrm{Si}\,\Theta_{\kappa\kappa'}}{\Theta_{\kappa\kappa'}}. \tag{9.27}$$

On substituting (9.27) into (9.21), the kinematical TDS intensity is

$$I_\mathrm{TDS} = \sum_\kappa (f_\kappa^\mathrm{e})^2 [1 - \exp(-2W_\kappa)] + \sum_\kappa \sum_{\kappa' \neq \kappa} f_\kappa^\mathrm{e} f_{\kappa'}^{\mathrm{e}*} \exp[2\pi \mathrm{i} \boldsymbol{\tau} \cdot (\boldsymbol{r}_\kappa - \boldsymbol{r}_{\kappa'})] \exp(-W_\kappa - W_{\kappa'}) \left\{ \exp\left[2W_\kappa \left(\frac{M_\kappa}{M_{\kappa'}}\right)^{1/2} \frac{\mathrm{Si}\,\Theta_{\kappa\kappa'}}{\Theta_{\kappa\kappa'}} \right] - 1 \right\}. \tag{9.28}$$

Therefore, the thermal diffuse scattering generated from atom sites κ and κ' is considered to be incoherent if the two atoms are separated far enough so that

$$4\pi^2 \overline{a_\kappa^2} \left(\frac{M_\kappa}{M_{\kappa'}}\right)^{1/2} \tau^2 \frac{\mathrm{Si}\,\Theta_{\kappa\kappa'}}{\Theta_{\kappa\kappa'}} \ll 1, \tag{9.29}$$

or

$$\frac{\mathrm{Si}\,\Theta_{\kappa\kappa'}}{\Theta_{\kappa\kappa'}} \ll \frac{(M_{\kappa'}/M_\kappa)^{1/2}}{4\pi^2 \overline{a_\alpha^2} \tau^2}. \tag{9.30}$$

The second term in (9.28) vanishes if condition (9.29) holds, that is the disappearance of coherence. The interatomic distance larger than which (9.29) holds is referred to as the coherence length, for distances smaller than which the TDS waves generated are considered to be coherent. The coherence volume is thus defined as a sphere whose diameter is the coherence length. It is apparent that the coherence length depends on the scattering angle ($\tau = 2\sin\vartheta/\lambda$).

Since $\mathrm{Si}\,\Theta_{\kappa\kappa'}/\Theta_{\kappa\kappa'}$ drops quickly with increasing interatomic distance, as shown in Figure 9.5, the coherence length for low-angle TDS is much smaller than that for high-angle TDS. The coherence length of light elements is larger than that of heavy

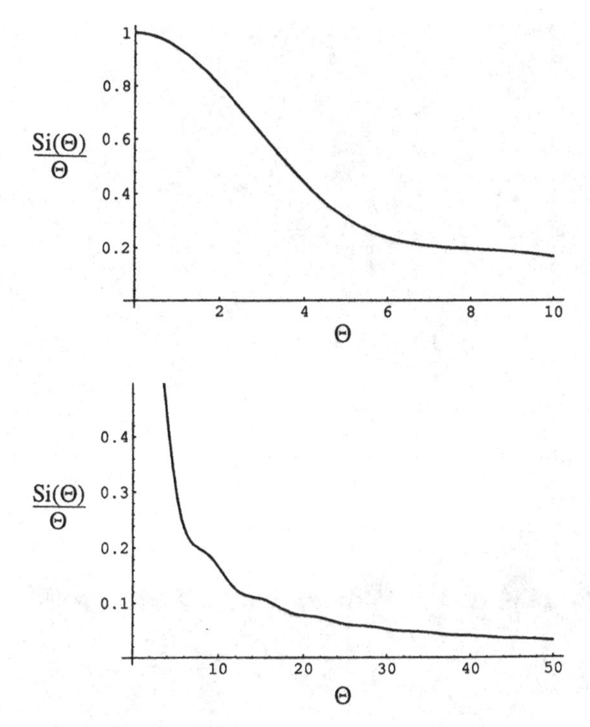

Figure 9.5 *Plots of the function* Si Θ/Θ.

elements because $\overline{a_\kappa^2}$ decreases with increasing atomic mass. For a general case, the coherent length is limited to three or four atoms.

9.6 Dynamical TDS in RHEED

As shown in Section 6.5, contributions made by electrons with energy losses larger than a few electron-volts can be removed from the RHEED pattern by use of an energy filter, but the TDS electrons remain. It is thus necessary to treat the diffraction of TDS electrons in RHEED. In this section, our ultimate goal is to introduce a formal dynamical theory, which can be applied to calculate the diffraction pattern of TDS electrons in RHEED. Sections 9.6.1 and 9.6.2 present some preliminary knowledge, which will be used in Section 9.6.3.

Before presenting the detailed theoretical results, we first investigate the significance of TDS in RHEED. Figure 9.6 shows two RHEED patterns recorded at 300 and 100 K, respectively. At $T=300$ K, the diffuse scattering is quite severe, particularly at high angles. When the temperature is reduced to 100 K, high-angle Kikuchi lines are clearly seen, and a circle near the first-order Laue zone with continuous intensity shows up. The formation of this ring-shaped intensity is closely related to one-dimensional electron channeling along atomic rows (Peng *et al.*, 1988;

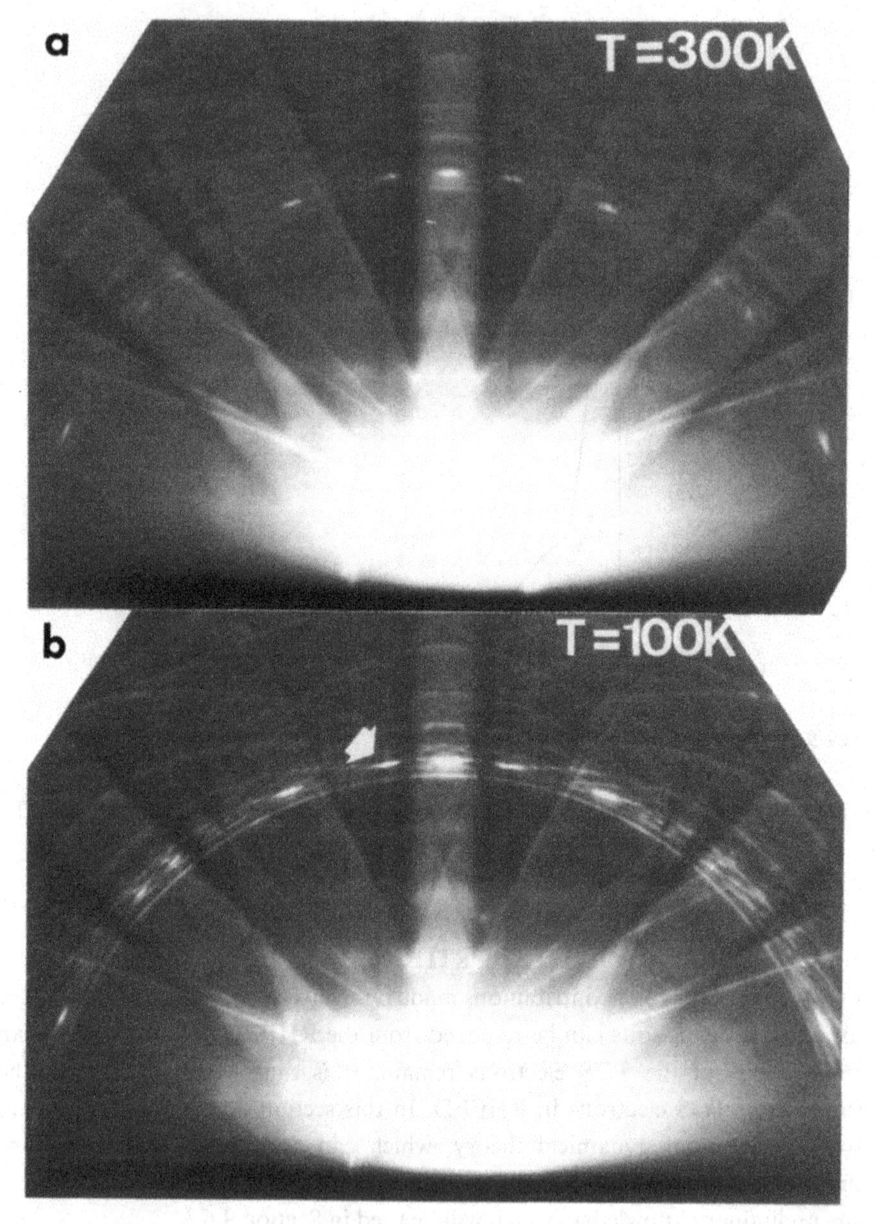

Figure 9.6 *RHEED patterns recorded at (a) 300 K and (b) 100 K from a GaAs(110) surface, showing the diffuse scattering at higher temperatures.*

Yao and Cowley, 1989). When channeling occurs along a single row of atoms, the electron 'sees' a one-dimensional potential channel. For simplicity, the potential of this channel is approximated as

$$V(\mathbf{r}) = \delta(x - x_i, y - y_i) \sum_j V(z - z_j), \tag{9.31}$$

where $z_j = c_0 j$ is the position of the jth atom in the row and c_0 is the interatomic distance along the z direction. The electron is assumed to travel along the z axis. Based on kinematical scattering theory, the diffraction of this channel is described by the Fourier transform of its potential. Since

$$|\mathrm{FT}[V(\mathbf{r})]|^2 = |\sum_j \mathrm{FT}[V(z - z_j)]|^2 = \sum_n [f^{(e)}(nc^*)]^2 \delta(u_z - nc^*), \tag{9.32}$$

the reflected intensity is distributed continuously in planes that are parallel to (u_x, u_y) and the interplanar distance is c^*. If the Ewald sphere cuts the planes in a direction nearly perpendicular to the planes, then the diffraction pattern shows some rings for FOLZ and HOLZs. This is just what was observed in Figure 9.6.

In kinematical electron scattering, the crystal scattering factor is modified by the Debye–Waller factor such that

$$V_g = \sum_\alpha f_\alpha^{(e)}(g) \exp(-W_\alpha).$$

For large g, V_g drops quickly with increasing W_α. On the other hand, the TDS intensity is proportional to $[f_\alpha^{(e)}(g)]^2[1 - \exp(-2W_\alpha)]$. Thus, a decrease in temperature will increase the elastic scattering factor V_g (i.e., the intensity of the continous ring) but greatly reduces the TDS intensity, resulting in a decrease in the diffuse background in RHEED. This simple discussion illustrates the importance of TDS in RHEED.

9.6.1 The reciprocity theorem

In electron imaging, reciprocity theory is commonly used to compare TEM with STEM. The reciprocity theorem is stated as follows (Figure 9.7). The amplitude of the disturbance at a point P due to radiation from a point Q, which has traversed any system involving elastic scattering processes only, is the same as the amplitude of the disturbance that would be observed at Q if the point source were placed at P.

The reciprocity theorem is the result of time-reversal in quantum mechanics, and it has been used to simplify some treatments related to electron diffraction. We now first prove the theorem, and then it is applied to calculate the Green function in Section 9.6.2. We choose the Born series method to prove the theory. If a point source is assumed to be located at point P in space (above the specimen as shown in Figure 9.7), then the wave function in space is determined by

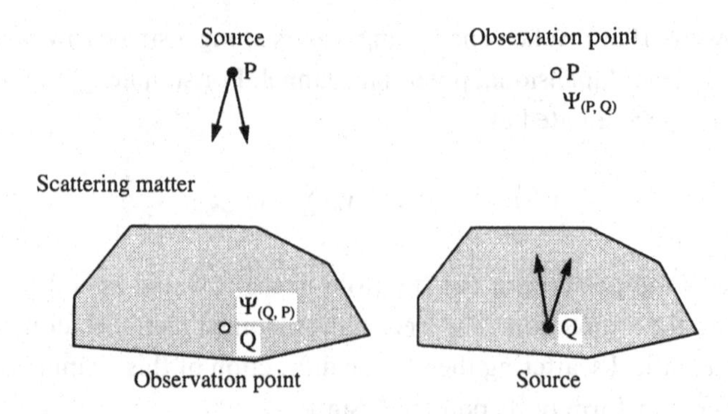

Figure 9.7 *An illustration of the reciprocity theorem. $\Psi(Q, P)$ is the wave 'observed' at Q if a point source is placed at P; $\Psi(P, Q)$ is the wave 'observed' at P if the point source is placed at Q. The reciprocity theorem indicates that $\Psi(Q, P) = \Psi(P, Q)$, i.e., the observation point and source can be exchanged without affecting the value of the 'observed' wave function.*

$$(\nabla^2 + 4\pi^2 K^2)\Psi_0(\boldsymbol{r}, \boldsymbol{r}_P) = -4\pi^2 U(\boldsymbol{r})\Psi_0(\boldsymbol{r}, \boldsymbol{r}_P) + \delta(\boldsymbol{r} - \boldsymbol{r}_P). \tag{9.33}$$

The solution of this equation can be written as

$$\Psi_0(\boldsymbol{r}, \boldsymbol{r}_P) = G_0(\boldsymbol{r}, \boldsymbol{r}_P) - 4\pi^2 \int \mathrm{d}\boldsymbol{r}'\, G_0(\boldsymbol{r}, \boldsymbol{r}') U(\boldsymbol{r}')\Psi_0(\boldsymbol{r}', \boldsymbol{r}_P), \tag{9.34}$$

where the Green function is

$$G_0(\boldsymbol{r}, \boldsymbol{r}') = -\frac{\exp(2\pi \mathrm{i} K_0 |\boldsymbol{r} - \boldsymbol{r}'|)}{4\pi |\boldsymbol{r} - \boldsymbol{r}'|}. \tag{9.35}$$

The solution of Eq. (9.34) can be obtained iteratively and is expressed in a Born series as

$$\begin{aligned}\Psi_0(\boldsymbol{r}_Q, \boldsymbol{r}_P) = {} & G_0(\boldsymbol{r}_Q, \boldsymbol{r}_P) - 4\pi^2 \int \mathrm{d}\boldsymbol{r}_1\, G_0(\boldsymbol{r}_Q, \boldsymbol{r}_1) U(\boldsymbol{r}_1) G_0(\boldsymbol{r}_1, \boldsymbol{r}_P) \\ & + (-4\pi^2)^2 \int \mathrm{d}\boldsymbol{r}_1 \int \mathrm{d}\boldsymbol{r}_2\, G_0(\boldsymbol{r}_Q, \boldsymbol{r}_1) G_0(\boldsymbol{r}_1, \boldsymbol{r}_2) U(\boldsymbol{r}_1) U(\boldsymbol{r}_2) G_0(\boldsymbol{r}_2, \boldsymbol{r}_P) \\ & + (-4\pi^2)^3 \int \mathrm{d}\boldsymbol{r}_1 \int \mathrm{d}\boldsymbol{r}_2 \int \mathrm{d}\boldsymbol{r}_3\, G_0(\boldsymbol{r}_Q, \boldsymbol{r}_1) G_0(\boldsymbol{r}_1, \boldsymbol{r}_2) G_0(\boldsymbol{r}_2, \boldsymbol{r}_3) U(\boldsymbol{r}_1) U(\boldsymbol{r}_2) U(\boldsymbol{r}_3) G_0(\boldsymbol{r}_3, \boldsymbol{r}_P) \\ & + \ldots \end{aligned} \tag{9.36}$$

Interchanging $\boldsymbol{r}_Q$ and $\boldsymbol{r}_P$ does not change the value of any order of the Born series, thus

$$\Psi_0(\boldsymbol{r}_Q, \boldsymbol{r}_P) = \Psi_0(\boldsymbol{r}_P, \boldsymbol{r}_Q). \tag{9.37}$$

Equation (9.37) is just the reciprocity theorem. Two points need to be emphasized here. First, no restriction was made on the form of the crystal potential, which can be a complex function if the anomalous absorption effect is included. Second, reciprocity holds for each order of the Born series. This must be so because the reciprocity theorem is the result of time-reversal in quantum mechanics. Therefore, the reciprocity theorem is exact for elastically scattered electrons regardless of the shape of the scattering object, the form of crystal potential, the scattering geometry, and the electron energy.

9.6.2 The Fourier transform of Green's function

The Green function to be used in high-energy electron diffraction is a solution of the following equation:

$$[\nabla^2 + 4\pi^2(U(\mathbf{r}) + K^2)]\hat{G}(\mathbf{r},\mathbf{r}_1) = \delta(\mathbf{r} - \mathbf{r}_P), \tag{9.38}$$

where the Dirac function $\delta(\mathbf{r} - \mathbf{r}_P)$ characterizes a point source at $\mathbf{r} = \mathbf{r}_P$. As will be shown in the next section, the two-dimensional Fourier transformation of Green's function is required for calculating the diffraction patterns of the inelastically scattered electrons,

$$\hat{G}(\mathbf{u}_b, z, \mathbf{r}_1) = \int d\mathbf{b} \exp(-2\pi i \mathbf{u}_b \cdot \mathbf{b})\, \hat{G}(\mathbf{r},\mathbf{r}_1), \tag{9.39}$$

In this section, the calculation in Eq. (9.39) is derived analytically.

Considering a case in which the observation point $\mathbf{r}$ is outside of the crystal and $\mathbf{r}_1$ is the point source of electrons inside the crystal, the reciprocity relation $\hat{G}(\mathbf{r},\mathbf{r}_1) = \hat{G}(\mathbf{r}_1,\mathbf{r})$ simply indicates that the same result would be obtained if the point source were moved to $\mathbf{r}$ outside of the crystal and the observion point $\mathbf{r}_1$ located inside the crystal. Therefore, the calculation of $\hat{G}(\mathbf{u}_b, z, \mathbf{r}_1)$ can be rewritten as

$$\hat{G}(\mathbf{u}_b, z, \mathbf{r}_1) = \int d\mathbf{b} \exp(-2\pi i \mathbf{u}_b \cdot \mathbf{b})\, \hat{G}(\mathbf{r}_1,\mathbf{r}), \tag{9.40}$$

where $\hat{G}(\mathbf{r}_1,\mathbf{r})$ is the wave function inside the crystal ($\mathbf{r}_1$) due to a point source outside of the crystal ($\mathbf{r}$), and can be expressed according to the scattering theory in quantum mechanics as

$$\hat{G}(\mathbf{r}_1,\mathbf{r}) = G_s(\mathbf{r}_1,\mathbf{r}) + \hat{G}_c(\mathbf{r}_1,\mathbf{r}), \tag{9.41a}$$

where $G_s(\mathbf{r}_1,\mathbf{r})$ is the wave emitted from the point source,

$$G_s(\mathbf{r}_1,\mathbf{r}) = \frac{m_0}{2\pi\hbar^2}\frac{\exp(2\pi i K|\mathbf{r} - \mathbf{r}_1|)}{|\mathbf{r} - \mathbf{r}_1|}, \tag{9.41b}$$

and $\hat{G}_c(\mathbf{r}_1,\mathbf{r})$ is the wave scattered by the crystal. For convenience in the following calculations, $G_s(\mathbf{r}_1,\mathbf{r})$ is converted into a Fourier integral form,

$$G_s(\mathbf{r}_1,\mathbf{r}) = -\frac{m_0}{2\pi^2\hbar^2}\int d\mathbf{u}\, \frac{\exp[2\pi i \mathbf{u}\cdot(\mathbf{r} - \mathbf{r}_1)]}{K^2 - u^2}. \tag{9.42}$$

The two-dimensional Fourier transform of $G_s(\boldsymbol{r}_1, \boldsymbol{r})$ can be performed analytically:

$$G_s(\boldsymbol{u}_b, z, \boldsymbol{r}_1) = \int \mathrm{d}\boldsymbol{b} \exp(-2\pi i \boldsymbol{u}_b \cdot \boldsymbol{b})\, G_s(\boldsymbol{r}_1, \boldsymbol{r})$$

$$= -\frac{i m_0}{\pi \hbar^2} \frac{\exp(-2\pi i \boldsymbol{u}_b \cdot \boldsymbol{b}_1)}{K|\cos\varphi_0|} \exp[2\pi i K \cos\varphi_0 (z - z_1)], \qquad (9.43)$$

where a substitution of $u_b = K_n \sin\varphi_0$ was made. We now return to the calculation of Eq. (9.40) using the relation given by (9.43),

$$\hat{G}_s(\boldsymbol{u}_b, z, \boldsymbol{r}_1) = -\frac{i m_0}{\pi \hbar^2} \frac{\exp(-2\pi i \boldsymbol{u}_b \cdot \boldsymbol{b}_1)}{K|\cos\varphi_0|} \exp[2\pi i K \cos\varphi_0 (z - z_1)]$$

$$+ \int \mathrm{d}\boldsymbol{b} \exp(-2\pi i \boldsymbol{u}_b \cdot \boldsymbol{b})\, \hat{G}_c(\boldsymbol{r}_1, \boldsymbol{r})$$

$$= a_z \exp(-2\pi i \boldsymbol{u} \cdot \boldsymbol{r}_1) + \int \mathrm{d}\boldsymbol{b} \exp(-2\pi i \boldsymbol{u}_b \cdot \boldsymbol{b})\, \hat{G}_c(\boldsymbol{r}_1, \boldsymbol{r}), \qquad (9.44)$$

where $u_z = K \cos\varphi_0$ and

$$a_z = -\frac{i m_0}{\pi \hbar^2} \frac{\exp(2\pi i \tau_z z)}{\tau_z}.$$

In Eq. (9.44), if the first term is taken as an incident plane wave of amplitude a_z and wavevector $-\boldsymbol{u}$, then the second term would be the wave scattered by the crystal due to the incident plane wave. Therefore, $\hat{G}_s(\boldsymbol{u}_b, z, \boldsymbol{r}_1)$ can be mathematically expressed (Dudarev *et al.*, 1993) as

$$\hat{G}(\boldsymbol{u}_b, z, \boldsymbol{r}_1) = a_z \Psi_0(-\boldsymbol{u}, \boldsymbol{r}_1), \qquad (9.45)$$

where $\Psi_0(-\boldsymbol{u}, \boldsymbol{r}_1)$ is the solution of the Schrödinger equation for an incident plane wave of wavevector $-\boldsymbol{u}$ (with $\tau_z = K\cos\varphi_0$). The negative sign of the wavevector means that the electron strikes the crystal along the $-z$ direction. The elastic scattering wave $\Psi_0(-\boldsymbol{u}, \boldsymbol{r}_1)$ can be obtained using the conventional dynamical theories introduced in Chapter 3. If the observation point is located at $z = \infty$, then the variable z can be dropped in Eq. (9.45); thus

$$G(\boldsymbol{u}_b, \boldsymbol{r}_1) = a_z \Psi_0(-\boldsymbol{u}, \boldsymbol{r}_1). \qquad (9.46)$$

This is an important result and will be used in the following discussion.

9.6.3 Green's function theory

We now consider the application of Green's function theory to calculating the diffraction pattern of the first-order TDS electrons. Elastic scattering of electrons from a frozen lattice configuration is treated as time-independent. The first theoretical objective in treating dynamical inelastic electron scattering is to statistically average over the quasi-elastic electron diffraction patterns produced by various crystal configurations before numerical calculations. Green's function theory is probably the method that is best suited for this purpose. For simplicity, we start from the equation of single-inelastic electron excitation taking into consideration the time-dependent (or configuration-dependent) thermal vibration perturbation on the crystal potential (ΔV),

$$\left(-\frac{\hbar^2}{2m_0}\nabla^2 - e\gamma V_0 - e\gamma\Delta V - E_0\right)\Psi_0 = 0. \tag{9.47}$$

Shifting the ΔV term to the right-hand side, Eq. (9.47) can be converted to an integral equation by use of the Green function

$$\Psi_0(\boldsymbol{r}) = \Psi_0(\boldsymbol{K}_0, \boldsymbol{r}) + \int \mathrm{d}\boldsymbol{r}_1\, \hat{G}(\boldsymbol{r}, \boldsymbol{r}_1)[e\gamma\Delta V(\boldsymbol{r}_1)\Psi_0(\boldsymbol{r}_1)], \tag{9.48}$$

where $\hat{G}$ is the Green function satisfying

$$\left(-\frac{\hbar^2}{2m_0}\nabla^2 - e\gamma V_0 - E_0\right)\hat{G}(\boldsymbol{r}, \boldsymbol{r}_1) = \delta(\boldsymbol{r} - \boldsymbol{r}_1), \tag{9.49}$$

and $\Psi_0(\boldsymbol{K}_0, \boldsymbol{r})$ is the elastic wave function of incident wavevector $\boldsymbol{K}_0$. On solving Eq. (9.48) iteratively, the elastically scattered electron wave is given by

$$\Psi_0(\boldsymbol{r}) = \Psi_0(\boldsymbol{K}_0, \boldsymbol{r}_1) + e\gamma \int \mathrm{d}\boldsymbol{r}_1\, \hat{G}(\boldsymbol{r}, \boldsymbol{r}_1)\Delta V(\boldsymbol{r}_1)\Psi_0(\boldsymbol{K}_0, \boldsymbol{r}_1) + \ldots, \tag{9.50}$$

where the first is the purely elastically scattered wave and the second term is the first-order TDS. Green's function theory has mathematically expressed the inelastic scattering in a simple way that can be adequately applied to treat TDS.

The physical meaning of Eq. (9.50) is clear. The second term, for example, is interpreted as follows. The elastic wave $[\Psi_0(\boldsymbol{K}_0, \boldsymbol{r}_1)]$ is diffusely scattered at point $\boldsymbol{r}_1$ due to thermal vibrations of crystal norms, $\Delta V(\boldsymbol{r}_1)$; then it is elastically re-scattered, $\hat{G}(\boldsymbol{r}, \boldsymbol{r}_1)$, from $\boldsymbol{r}_1$ to $\boldsymbol{r}$; the integration of $\boldsymbol{r}_1$ is to sum over the inelastic waves generated at different points inside the crystal.

Taking a Fourier transform of the second term in Eq. (9.50), the modulus squared of the result is the first-order TDS intensity scattered to $\boldsymbol{u}_b$ in reciprocal space:

$$I_{\mathrm{TDS}}(\boldsymbol{u}_b) = e^2\gamma^2 \int \mathrm{d}\boldsymbol{r}_1 \int \mathrm{d}\boldsymbol{r}_2\, \hat{G}(\boldsymbol{u}_b, \boldsymbol{r}_1)\hat{G}^*(\boldsymbol{u}_b, \boldsymbol{r}_2)\langle\Delta V(\boldsymbol{r}_1)\Delta V(\boldsymbol{r}_2)\rangle\Psi_0(\boldsymbol{K}_0, \boldsymbol{r}_1)\Psi_0^*(\boldsymbol{K}_0, \boldsymbol{r}_2). \tag{9.51}$$

The most important advantage of the Green function method is that the thermal average can be performed before numerical calculations (Dudarev *et al.*, 1991).

In general, the electron angular distribution is calculated by assuming that the observation point is far from the crystal (i.e., $z=\infty$), corresponding to the geometry of Fraunhofer diffraction. Thus the relation shown in Eq. (9.46) holds, and Eq. (9.51) becomes

$$I_{\mathrm{TDS}}(\boldsymbol{u}) = D_1 \int \mathrm{d}\boldsymbol{r}_1 \int \mathrm{d}\boldsymbol{r}_2\, \Psi_0(-\boldsymbol{K}_0-\boldsymbol{u}_b,\boldsymbol{r}_1)\Psi_0^*(-\boldsymbol{K}_0-\boldsymbol{u}_b,\boldsymbol{r}_2) \times \langle \Delta V(\boldsymbol{r}_1)\Delta V(\boldsymbol{r}_2)\rangle \Psi_0(\boldsymbol{K}_0,\boldsymbol{r}_1)\Psi_0^*(\boldsymbol{K}_0,\boldsymbol{r}_2), \tag{9.52}$$

with $D_1 = e^2\gamma^2 m_0[2\pi^2\hbar^2 E_n \cos^2\varphi_0]^{-1}$ is a constant factor. $\langle \Delta V(\boldsymbol{r}_1)\Delta V(\boldsymbol{r}_2)\rangle$ can be calculated by expressing $V(\boldsymbol{r}_1)$ in terms of the Fourier transform of the scattering factors,

$$\begin{aligned}\langle \Delta V(\boldsymbol{r}_1)\Delta V(\boldsymbol{r}_2)\rangle &= \langle [V(\boldsymbol{r}_1,t)-V_0(\boldsymbol{r}_1)][V(\boldsymbol{r}_2,t)-V_0(\mathbf{r}_2)]\rangle \\ &= \langle V(\boldsymbol{r}_1,t)V(\boldsymbol{r}_2,t)\rangle - V_0(\mathbf{r}_1)V_0(\boldsymbol{r}_2) \\ &= \int \mathrm{d}\boldsymbol{Q} \int \mathrm{d}\boldsymbol{Q}'\, S_{\mathrm{TDS}}(\boldsymbol{Q},\boldsymbol{Q}')\exp(2\pi \mathrm{i}\boldsymbol{Q}\cdot\boldsymbol{r}_1 - 2\pi\mathrm{i}\boldsymbol{Q}'\cdot\boldsymbol{r}_2),\end{aligned} \tag{9.53}$$

where

$$S_{\mathrm{TDS}}(\boldsymbol{Q},\boldsymbol{Q}') = \sum_\kappa \sum_{\kappa'} f_\kappa^{\mathrm{e}}(\boldsymbol{Q}) f_{\kappa'}^{\mathrm{e}}(-\boldsymbol{Q}')\exp(-2\pi\mathrm{i}\boldsymbol{Q}\cdot\boldsymbol{r}_\kappa + 2\pi\mathrm{i}\boldsymbol{Q}'\cdot\boldsymbol{r}_{\kappa'}) \times \exp[-W_\kappa(\boldsymbol{Q}) - W_{\kappa'}(\boldsymbol{Q}')]\{\exp[4\pi^2\langle(\boldsymbol{Q}\cdot\boldsymbol{u}_\kappa)(\boldsymbol{Q}'\cdot\boldsymbol{u}_{\kappa'})\rangle]-1\}. \tag{9.54}$$

The function $S_{\mathrm{TDS}}(\boldsymbol{Q},\boldsymbol{Q}')$ is the scattering factor of TDS. For a three-dimensional periodic crystal structure, the summation over κ can be separated into a summation over centers of unit cells $\boldsymbol{R}_n$ and a summation over atoms within a cell, $\boldsymbol{r}_\kappa = \boldsymbol{R}_n + \boldsymbol{r}_\alpha$. The quantity $\langle(\boldsymbol{Q}\cdot\boldsymbol{u}_\kappa)(\boldsymbol{Q}'\cdot\boldsymbol{u}_{\kappa'})\rangle$ is calculated as

$$\langle(\boldsymbol{Q}\cdot\boldsymbol{u}_\kappa)(\boldsymbol{Q}'\cdot\boldsymbol{u}_{\kappa'})\rangle = \Omega \int_{\mathrm{BZ}} \mathrm{d}\tau \sum_j \frac{\hbar(\langle n_s\rangle + \frac{1}{2})}{\omega_j(\tau)(M_\kappa M_{\kappa'})^{1/2}} [\boldsymbol{Q}\cdot\boldsymbol{e}(\alpha|_j^\tau)][\boldsymbol{Q}'\cdot\boldsymbol{e}(\alpha'|_j^{-\tau})]\exp[2\pi\mathrm{i}\tau\cdot(\boldsymbol{r}_\kappa - \boldsymbol{r}_{\kappa'})]. \tag{9.55}$$

Equation (9.55) can be further simplified using the Debye model and the Warren approximation (see Section 9.4). Since the orientation average

$$\langle[\boldsymbol{Q}\cdot\boldsymbol{e}(\alpha|_j^\tau)][\boldsymbol{Q}'\cdot\boldsymbol{e}(\alpha'|_j^{-\tau})]\rangle = \boldsymbol{Q}\cdot\boldsymbol{Q}'/3,$$

$$\langle(\boldsymbol{Q}\cdot\boldsymbol{u}_\kappa)(\boldsymbol{Q}'\cdot\boldsymbol{u}_{\kappa'})\rangle \approx \frac{\hbar\boldsymbol{Q}\cdot\boldsymbol{Q}'}{4\pi^2 q_{\mathrm{m}}^3(M_\kappa M_{\kappa'})^{1/2}} \sum_j \frac{1}{v_j} \int_0^{q_{\mathrm{m}}} \mathrm{d}q\, \frac{\sin(2\pi q\,\Delta r_{\kappa\kappa'})}{\Delta r_{\kappa\kappa'}} \left(\frac{1}{\exp[2\pi\hbar v_j/(k_{\mathrm{B}}T)]-1} + \frac{1}{2}\right). \tag{9.56}$$

At high temperatures, Eq. (9.56) is approximated as

$$\langle (\boldsymbol{Q}\cdot\boldsymbol{u}_\kappa)(\boldsymbol{Q}'\cdot\boldsymbol{u}_{\kappa'})\rangle \approx \frac{k_B T \boldsymbol{Q}\cdot\boldsymbol{Q}'}{4\pi^2 q_m^2 (M_\kappa M_{\kappa'})^{1/2}} \sum_j \frac{1}{v_j^2} \frac{\mathrm{Si}\,\Theta_{\kappa\kappa'}}{\Theta_{\kappa\kappa'}}$$

$$= \overline{a_\kappa^2} \left(\frac{M_\kappa}{M_{\kappa'}}\right)^{1/2} \boldsymbol{Q}\cdot\boldsymbol{Q}' \frac{\mathrm{Si}\,\Theta_{\kappa\kappa'}}{\Theta_{\kappa\kappa'}}. \qquad (9.57)$$

In RHEED, the fine details introduced by phonon dispersions are usually smeared out. Therefore, good agreement can be obtained between the calculated results based on the Einstein model and the experimentally observed results. Under the random vibrational phase approximation, Eq. (9.54) reduces to

$$S_{\mathrm{TDS}}(\boldsymbol{Q},\boldsymbol{Q}') = \sum_\kappa f_\kappa^e(\boldsymbol{Q}) f_\kappa^e(-\boldsymbol{Q}') \exp[-2\pi i(\boldsymbol{Q}-\boldsymbol{Q}')\cdot\boldsymbol{r}_\kappa]\{\exp[-W_\kappa(\boldsymbol{Q}-\boldsymbol{Q}')] - \exp[-W_\kappa(\boldsymbol{Q}) - W_\kappa(\boldsymbol{Q}')]\}. \qquad (9.58)$$

Equation (9.52) is a general result and it can, in principle, be applied to the scattering object of arbitrary shape as long as the solution of the elastic scattering wave can be found. To see the meaning of Eq. (9.58), we take $\boldsymbol{Q}=\boldsymbol{Q}'$, thus

$$S_{\mathrm{TDS}}(\boldsymbol{Q},\boldsymbol{Q}) = \sum_\kappa [f_\kappa^e(\boldsymbol{Q})]^2 \{1 - \exp[-2W_\kappa(\boldsymbol{Q})]\}. \qquad (9.59)$$

This is the familiar form of kinematical incoherent thermal diffuse scattering from crystal atoms using the Einstein model. Equation (9.59) can be easily understood as follows. The total kinematic scattering intensity of an atom is $[f_\kappa^e(\boldsymbol{Q})]^2$. The elastic scattering factor has to be multiplied by the Debye–Waller factor in order to include the temperature factor, that is $[f_\kappa^e(\boldsymbol{Q})]^2 \exp[-2W_\kappa(\boldsymbol{Q})]$. Subtracting the elastic scattering intensity from the the total scattering intensity, TDS intensity is obtained $[f_\kappa^e(\boldsymbol{Q})]^2\{1-\exp[-2W_\kappa(\boldsymbol{Q})]\}$.

Figure 9.8 shows the calculated elastic scattering intensity $(f^e)^2 \exp(-2W)$ and TDS intensity $(f^e)^2[1-\exp(-2W)]$ of silicon as a function of scattering vector s. The elastic scattering is dominant if the scattering vector $s = \sin\theta/\lambda$ is smaller than 1, and the TDS becomes dominant if $s > 1.25$. For crystalline materials, almost all of the elastic scattering intensity is accumulated at the Bragg beams, as determined by Bragg's law. The positions of the $\{l00\}$ systematic reflections, for example, are indicated by vertical solid lines. The excitations of the Bragg beams are determined by the diffracting conditions. The thermal diffusely scattered electrons are diffusely distributed between Bragg peaks.

We now transform Eq. (9.52) into a form that is best suited for numerical calculation. We use two representations of the elastic scattering wave. In the Bloch

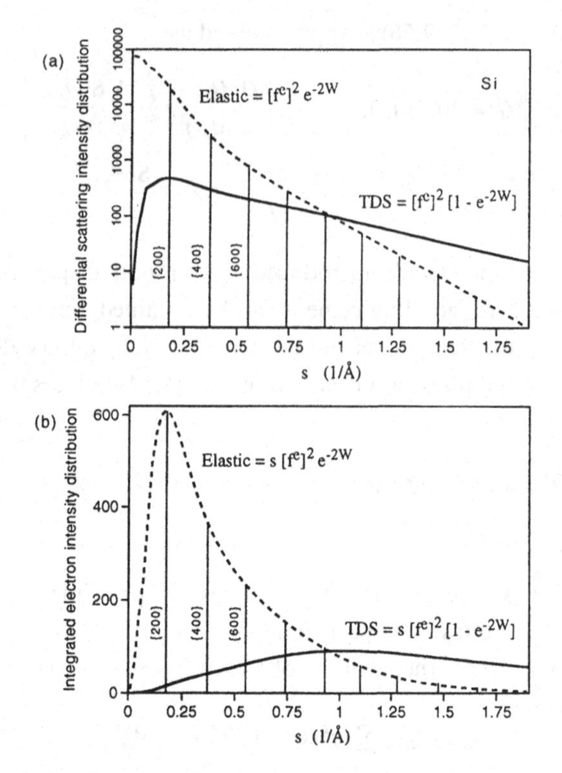

Figure 9.8 The calculated intensity distribution function for the electrons scattered elastically and thermal diffusely by a single silicon atom. (a) The angular distribution function and (b) the radial distribution function. The positions of the {100} row reflections are indicated by vertical solid lines in order to show the angle larger than which TDS is dominant. The mean vibration amplitude of Si is taken as $a_{Si} = 0.007$ nm.

wave representation, the elastic scattering wave of incident wavevector $\boldsymbol{K}$ is given by (see Eq. (3.8))

$$\Psi_0(\boldsymbol{K},\boldsymbol{r}) = \sum_i \sum_g \alpha_i(\boldsymbol{K}) C_g^{(i)}(\boldsymbol{K}) \exp[2\pi i(\boldsymbol{K}+\boldsymbol{g})\cdot\boldsymbol{r} + 2\pi i v_i z]. \tag{9.60}$$

On substituting (9.53) and (9.60) into (9.52), and performing the integrations, the TDS intensity at $\boldsymbol{u}$ in reciprocal space is

$$I_{TDS}(\boldsymbol{u}) = \frac{D_1}{4} \sum_i \sum_j \sum_{i'} \sum_{j'} \sum_g \sum_h \sum_{g'} \sum_{h'}$$
$$\alpha_i(-\boldsymbol{K})\alpha_j^*(-\boldsymbol{K})\alpha_{i'}(\boldsymbol{K}_0)\alpha_{j'}(\boldsymbol{K}_0) C_g^{(i)}(-\boldsymbol{K}) C_h^{(j)}(-\boldsymbol{K}) C_{g'}^{(i')}(\boldsymbol{K}_0) C_{h'}^{(j')}(\boldsymbol{K}_0) \times S_{TDS}(\boldsymbol{Q}_b, Q_z, \boldsymbol{Q}'_b, Q'_z), \tag{9.61a}$$

where

$$\boldsymbol{Q}_b = \boldsymbol{u}_b - \boldsymbol{g}_b - \boldsymbol{g}'_b, \tag{9.61b}$$

$$Q'_b = -u_b + h_b + h'_b, \tag{9.61c}$$

$$Q_z = -u_z - g_z - g'_z - v_i - v_{i'}, \tag{9.61d}$$

$$Q'_z = -u_z - h_z - h'_z - v_j - v_{j'}, \tag{9.61e}$$

and $K = K_0 + u_b$. The α coefficients are determined by the boundary conditions (see Eq. (3.19)). The sums of i, j, i' and j' are over all the Bloch waves, and the sums of g, h, g' and h' are over all the reciprocal-lattice vectors. The Bloch wave coefficients $C_g^{(i)}$ can be calculated using the standard matrix diagonalization method (Spence and Zuo, 1992).

In the parallel-to-surface multislice theory, the elastic scattering wave function is given by (see Eq. (3.23))

$$\Psi_0(K, r) = \sum_g \Psi_g(z) \exp[2\pi i(K_t + g) \cdot b], \tag{9.62}$$

and the TDS intensity scattered to u in reciprocal space is

$$I_{TDS}(u) = D_1 \sum_g \sum_h \sum_{g'} \sum_{h'} \int_0^\infty dz_1 \int_0^\infty dz_2\, \Psi_g(z_1)\Psi_h^*(z_2)\Psi_{g'}(z_1)\Psi_{h'}(z_2)$$
$$\times \int dQ_z \int dQ'_z\, S_{TDS}(Q_b, Q_z, Q'_b, q'_z) \exp[2\pi i(Q_z z_1 - Q'_z z_1)]. \tag{9.63}$$

Calculations according to either Eq. (9.61) or Eq. (9.63) have included the contribution made by HOLZ reflections.

9.6.4 A modified parallel-to-surface multislice theory

The perturbation theory presented in Section 3.5.1 can be modified to calculate the diffraction pattern of TDS electrons (Korte and Meyer-Ehmsen, 1993a and b). The theory is based on the first-order perturbation method. The Bragg reflections are considered to come only from the periodic potential V_0. The dynamical scattering is assumed to arise from V_0, and the scattering from ΔV is treated kinematically. We now consider TDS intensity at a non-Bragg position $u = g + q$. For each q, the reflection beams to be included in the calculations are (1) those of the periodic crystal reciprocal lattice array $\{g\}$ and (2) the 'diffuse' set $\{u = g + q\}$. For the N g beam case, the total beams to be included in the calculations are $2N$, as shown previously in Figure 3.12. The waves corresponding to the sharp reflections (g and g') are coupled by $V_{g-g'}$. The diffuse beams (u and u') interact via $V_{u-u'}$. The interaction between the g beam and the diffuse beam u is ΔV_{g-u}. The numerical calculation follows the same route as that given in (3.39)–(3.51). The entire calculation must be repeated for each q.

For TDS, the perturbation potential ΔV is time-dependent,

$$\Delta V(t) \approx \sum_{\kappa} \boldsymbol{u}_{\kappa}(t) \cdot \nabla V_{\kappa}(\boldsymbol{r}-\boldsymbol{r}_{\kappa}), \tag{9.64}$$

where $\boldsymbol{u}_{\kappa}(t)$ is the time-dependent displacement of the κth atom. Thus the calculated intensity has to be averaged over time. Under the Einstein model, the TDS intensity is the incoherent superposition of the scattered intensity of all atoms. Therefore, ΔV is approximated as a time-independent function

$$\langle \Delta V(t) \rangle \approx \sum_{\kappa} a_{\kappa} \frac{\partial V_{\kappa}(\boldsymbol{r}-\boldsymbol{r}_{\kappa})}{\partial r}. \tag{9.65}$$

The TDS intensity at $\boldsymbol{u}=\boldsymbol{g}+\boldsymbol{q}$ can be found by a single calculation. This has been shown by Korte and Meyer-Ehmsen (1993a), and the results show good agreement with experimental observations. In RHEED, the Einstein model is usually an excellent approximation. The perturbation theory introduced here can also be applied to calculate the diffraction of disordered surfaces.

Figure 9.9(a) shows the angular distribution of the calculated RHEED pattern of Pt(110) due to TDS (Korte and Meyer-Ehmsen, 1993b). The beam azimuth is $[1\bar{1}0]$, $\theta=95$ mrad and the beam energy is 19 keV. The atoms are assumed to vibrate independently along [001] and [110], the diffuse scattering intensity is superimposed incoherently, and the calculation is performed only for the first three atomic layers. Figure 9.9(b) is the experimentally observed diffuse scattering RHEED pattern under the conditions assumed in the calculation of Figure 9.9(a). The intensity of the sharp reflections has been artificially reduced for displaying the TDS intensity. An excellent agreement is obtained between Figures 9.9(a) and (b), although a very primitive model for the vibration has been used (no phase correlations, only three layers). Except for two bright regions at low exit angles in the calculated pattern, the modulations of the diffuse intensity are very well reproduced. This example demonstrates that the basic dynamical beam couplings for the diffuse scattering are caused by the periodic part of the whole scattering potential. This beam coupling is obviously mainly responsible for the observed modulations of the broad scattering distribution whereas the correlations between the thermal vibrations seem to play a minor role.

Thermal diffuse (or phonon) scattering is one of the inelastic scattering processes involved in electron diffraction. This process introduces negligible energy losses but large momentum transfers, leading to high-angle scattering. Thus, the TDS electrons remain in the RHEED pattern, even when an energy filter is used. In this chapter, we examined the nature of TDS and electron–phonon scattering. Kinematical and dynamical diffraction theories for calculating the RHEED patterns of TDS electrons were proposed based on the frozen lattice model. The theories have been

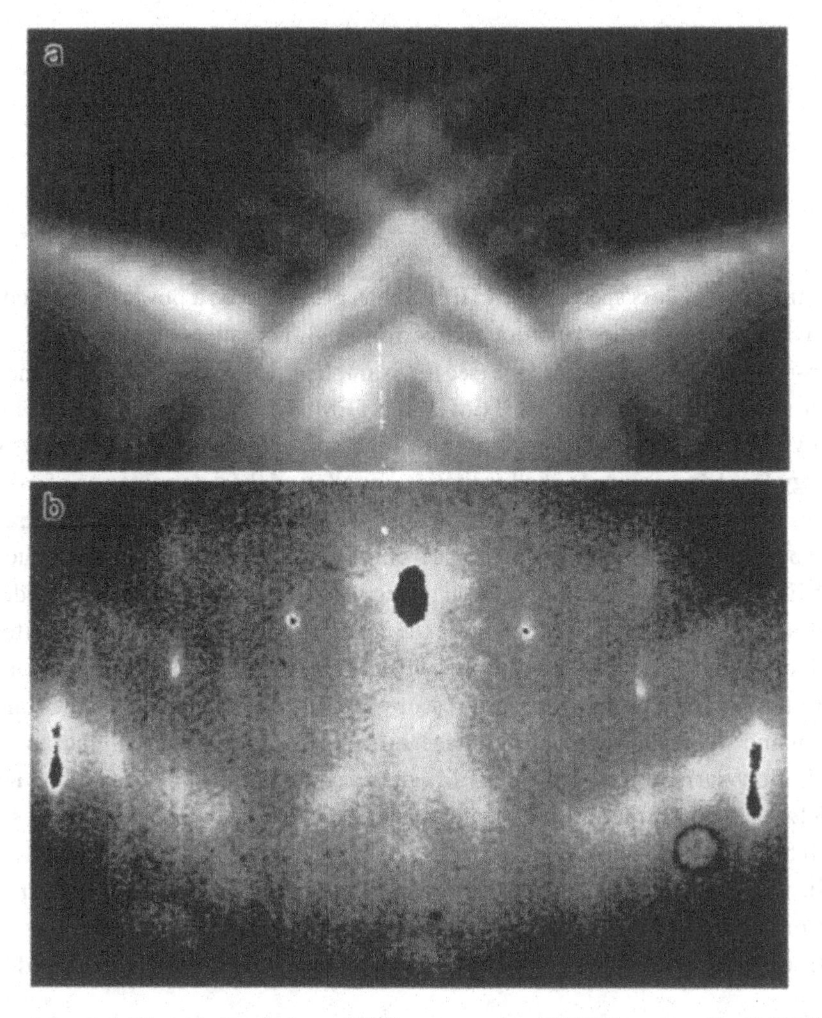

Figure 9.9 *(a) The calculated thermal diffuse scattering RHEED pattern for Pt(110) at 600 °C specimen temperature. Independent vibrations of the atoms in the top three layers (the Einstein model) have been assumed. The azimuth is [1$\bar{1}$0]; the beam energy is 19 keV; the incident angle is 95 mrad; the output angular range is 150 × 90 mrad; and the pixel resolution is 1 × 1 mrad. The lower margin corresponds to the polar exit angle 11 mrad; and the center of the displayed pattern corresponds to the exit azimuth [1$\bar{1}$0]. (b) The corresponding experimental diffuse scattering distribution. The intensity of the sharp reflection on the Laue circle is artificially reduced (courtesy of Drs U. Korte and G. Meyer-Ehmsen, 1993b).*

given in the form best suited for numerical calculations. A formula for calculating the Debye–Waller factor is given based on the Debye model and the Warren approximation. A coherence length was introduced to describe the coherence in TDS.

CHAPTER 10

Valence excitation in RHEED

Electron energy-loss spectroscopy (EELS) has proven to be a powerful method for studying the electronic structure and performing microanalyses of materials in a transmission electron microscope (see for example, Egerton (1986)). In conjunction with imaging of thin films by TEM and STEM, EELS has permitted chemical analysis of small specimen regions with high spatial resolution. The analysis of energy-loss edges for inner-shell excitation has allowed the determination of valence states of atoms from the energy-loss near-edge structure (ELNES) and determination of the local environments of atoms from the extended energy-loss fine structure (EXELFS). The use of EELS in the glancing incidence, surface-reflection mode for bulk samples is an attractive topic since the penetration of the electron beam into the surface, in general, is just a few atomic layers. This is the technique of high-energy reflection electron energy-loss spectroscopy (REELS). The analysis of the composition and structure of thin surface layers can form an important adjunct to high-resolution surface imaging by REM, or in combination with scanning REM (SREM), microdiffraction and secondary electron (SE) imaging, which are possible with STEM instruments.

In this chapter and Chapter 11 we describe high-energy REELS experiments performed in a TEM or STEM with high spatial resolution. The basic theory of valence excitation will be given and its applications to RHEED are described.

10.1 EELS spectra of bulk crystal surfaces

When an electron passes through a thin metal foil, the most noticeable energy loss is due to the plasmon oscillations in the sea of conduction electrons. For an ideal case in which the electrons can move 'freely' in the sea, the system can be treated as an electron gas. This case is best represented by aluminum. The outer-shell electrons can be considered as free electrons. Those negatively charged particles are mixed together with nuclei of positive charges, forming a solid state plasmon 'gas'. The resonance frequency of this plasmon is directly related to the density of electrons in the solid. This simple plasmon model may also be adopted to describe the valence excitation of semiconductor materials, such as Si. For non-conducting materials, this plasmon model does not hold and the excitation is referred to as valence band excitation. Therefore, the plasmon excitation is the valence excitation in conducting

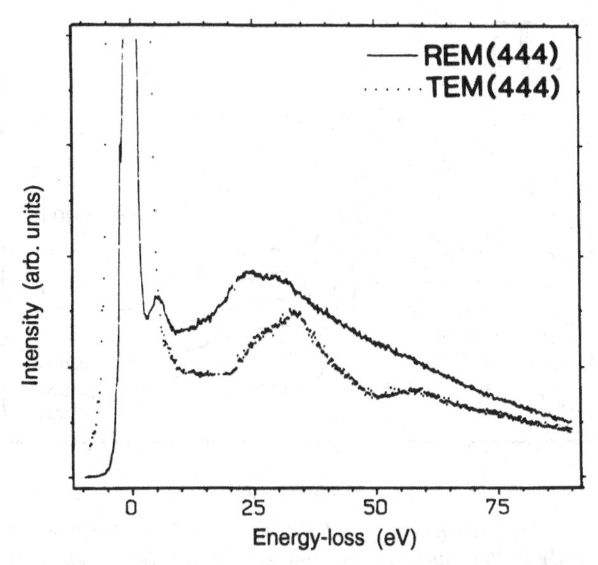

Figure 10.1 *A comparison of EELS spectra acquired in RHEED geometry from bulk Pt(111) (solid line) and in the transmission case from a thin Pt foil (dotted line), showing the sensitivity of REELS to surface electronic excitations. The beam energy is 120 kV.*

materials, in which the energy-loss spectrum is a sharp peak. The collective excitation of the electrons in the valence states will produce numerous low-energy excited states, resulting in energy loss from the electrons.

Valence excitation is particularly important in RHEED, because more than 50% of the reflected electrons have experienced multiple valence losses. The EELS spectrum can be acquired in RHEED geometry by permitting passage of the specularly reflected beam through the EELS spectrometer. The EELS spectrum displays the intensity distribution of the reflected electrons as a function of this energy loss. Figure 10.1 compares a REELS spectrum recorded in RHEED geometry from the Pt(111) surface with the one recorded from a thin Pt foil in transmission (which is referred to as transmission EELS or TEELS). The former is particularly sensitive to the electronic structure of surfaces, and the latter is the result of crystal volume excitation. Thus, the spectra are not identical. For example, a peak observed at 5 eV in REELS is absent in the TEELS spectrum acquired from the thin Pt foil. Also, the major TEELS peak at about 32 eV is shifted to lower energy in REELS. Figure 10.1 shows clearly that REELS can provide important structural information about crystal surfaces.

The valence excitation REELS spectrum is sensitive to the dielectric response of the solid and can be used to detect the presence of different phases, because structural modification of one or two atomic layers at the surface generally does not affect the spectrum shape (Wang and Howie, 1990). Figure 10.2 shows REELS

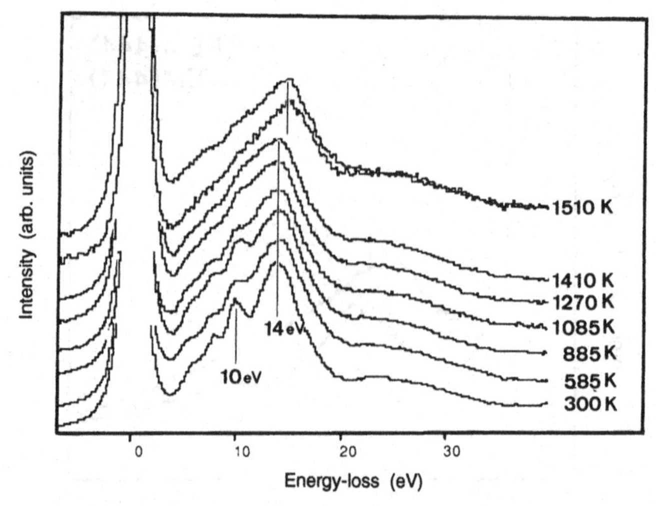

Figure 10.2 *A series of* in situ *valence-loss REELS spectra recorded with the (400) specularly reflected beam as a MgO(100) surface was being heated to different temperatures (see Fig. 8.10). The variation of the surface dielectric response properties can be clearly seen through the changes in the valence spectra. The topmost two spectra were acquired 3 min apart at 1510 K.*

valence spectra acquired during *in situ* experiments on MgO(100) at different temperatures, as shown earlier in Figure 8.10. At 300 K, the spectrum shows the typical excitation of a MgO surface. With an increase in temperature, the 10 eV energy-loss peak gradually disappeared, but the peak located at 14 eV remained at the same energy loss. This situation was preserved until the temperature reached 1500 K, at which a 0.8 eV shift in energy towards higher energy loss was observed for the 14 eV peak. This sharp variation must indicate a change in surface dielectric properties. Therefore, the formation of extra reflections that was shown in Figure 8.10 corresponds to the growth of a thin layer of new material on the surface. The two spectra at 1510 K taken about 3 min apart show almost identical shapes, indicating that the newly formed phase is stable.

From this example, we can see the sensitivity of valence-loss spectra to the formation of new phases at the surface. Therefore, quantitative analysis of REELS spectra can provide useful structural information. In the following sections, we first outline the theory of valence-loss REELS, then give the applications of the theory.

10.2 The dielectric response theory of valence excitations

Many sophisticated quantum mechanical many-body theories have been developed in order to consider the interactions of electrons in metals (Bohm, 1953; Nozières and Pine, 1958; Ferrell, 1956 and 1957). However, the theories become very complex if the solid is a non-metallic material so that the electron gas model is no longer valid. The valence band of a crystalline material is a collection of many electron energy states, and is usually characterized by a density of states (DOS) function. The valence electron wave functions are directly associated with chemical bonding. The energy loss of an electron results from the interband transition of the valence electrons to the conducting band. It is difficult, in general, to treat the excitation of each individual state of the valence band. Thus the statistical properties of the valence electrons are usually characterized by a dielectric function $\varepsilon(\omega, \boldsymbol{q})$, which depends on the frequency ω of the plasmon oscillation and the wavevector $\boldsymbol{q}$ of the disturbance (Ritchie, 1957). We will first derive the dielectric response theory. The equivalence of the theory with quantum theory will be proven in Section 10.7.

There are a few review articles on the dielectric excitation of surfaces: Kliewer and Fuchs (1974); Echenique *et al.* (1990); Schattschneider and Jouffrey (1995) and Wang (1996).

Valence-excitation spectra, acquired either in transmission or reflection geometry, can be quantitatively simulated with the use of either the dielectric response theory (Howie, 1983; Howie and Milne, 1984; Walls and Howie, 1989) or the quantum theory (Ferrell and Echenique, 1985). In classical theory, an incident electron is treated as a particle that moves along a certain trajectory. Its energy loss equals the work done by the induced charges distributed in the bulk and at the interface to slow down the electron motion. The fast incident electron is represented as a point charge with a well-defined path, as long as the changes of electron energy and momentum are small. For an infinitely large isotropic medium (Figure 10.3), the induced field generated by the moving electron is then determined by Poisson's equation,

$$\varepsilon(\omega, \boldsymbol{q}) \nabla^2 V_e(\boldsymbol{r}, t) = \frac{e}{\varepsilon_0} \delta(x)\delta(y)\delta(z - vt), \tag{10.1}$$

where the electron is assumed to travel along the z axis.

On transforming Eq. (10.1) into Fourier space and frequency space, we have

$$V_e(\boldsymbol{q}, \omega) = -\frac{e}{4\pi^2 \varepsilon_0 \varepsilon(\omega, \boldsymbol{q})} \frac{\delta(2\pi q_z v - \omega)}{q^2}, \tag{10.2a}$$

where

$$V_e(\boldsymbol{q}, \omega) = \frac{1}{2\pi} \int d\omega \int d\boldsymbol{r} \exp(-2\pi i \boldsymbol{q} \cdot \boldsymbol{r} + i\omega t)\, V_e(\boldsymbol{r}, t). \tag{10.2b}$$

The energy loss per unit path in the medium is given by

$$-\frac{dE}{dz} = -e \left. \frac{dV_e(\boldsymbol{r}, t)}{dz} \right|_{x=0, y=0, z=vt}$$

$$= \frac{e^2}{4\pi^2 \varepsilon_0 v} \int dq_x \int dq_y \int_0^\infty d\omega \frac{\omega}{q_x^2 + q_y^2 + [\omega/(2\pi v)]^2} \mathrm{Im}\left(-\frac{1}{\varepsilon(\omega, \boldsymbol{q})}\right). \tag{10.3}$$

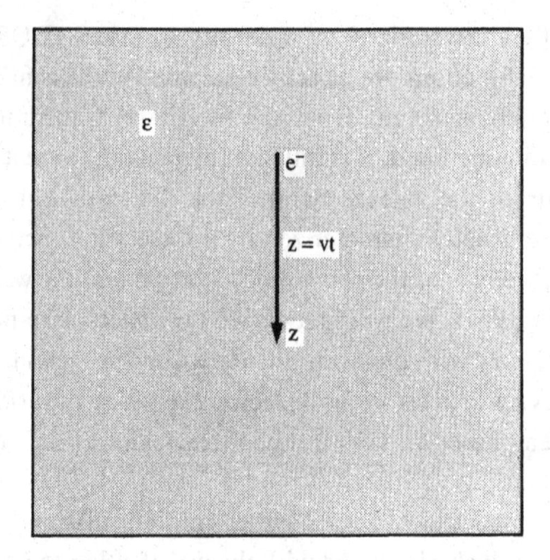

Figure 10.3 *The model of valence excitation by a fast electron in a homogeneous medium.*

Thus, the excitation probability of the valence states is

$$\frac{d^2P_v}{dz\,d\omega} = \frac{e^2}{4\pi^2\varepsilon_0\hbar v^2}\int_{-\infty}^{\infty} dq_x \int_{-\infty}^{\infty} dq_y \frac{1}{q_x^2+q_y^2+[\omega/(2\pi v)]^2}\,\text{Im}\left(-\frac{1}{\varepsilon(\omega,\boldsymbol{q})}\right), \tag{10.4}$$

where the integrations of q_x and q_y are to be summed over the electrons scattered to different angles. The angular distribution of valence-loss electrons is a Lorentzian function, $1/(\vartheta^2+\vartheta_E^2)$, with $\vartheta_E = \Delta E/(2E)$ as the characteristic angle of inelastic scattering. For a homogeneous medium, ε is independent of wavevector. Hence,

$$\frac{d^2P_v}{dz\,d\omega} = \frac{e^2}{4\pi^2\varepsilon_0\hbar v^2}\,\text{Im}\left(-\frac{1}{\varepsilon}\right)\ln\left[1+\left(\frac{2\pi q_c v}{\omega}\right)^2\right]$$
$$\approx \frac{e^2}{2\pi^2\varepsilon_0\hbar v^2}\,\text{Im}\left(-\frac{1}{\varepsilon}\right)\ln\left(\frac{2\pi q_c v}{\omega}\right), \tag{10.5}$$

where q_c is the cut-off value of the wavevector, and $\text{Im}\,[-1/\varepsilon(\omega,\boldsymbol{q})]$ is the energy-loss function. It is apparent that valence-excitation is a delocalized scattering process.

The mean free path Λ of valence-excitation is related to the excitation probability by

$$\frac{1}{\Lambda} = \int_0^{\infty} d\omega\,\frac{d^2P_v}{dz\,d\omega}. \tag{10.6}$$

For metals at 100 kV the mean free path is in the range 50–100 nm. For oxides, such as MgO, Λ can be as large as 250 nm. In RHEED, the mean traveling distance of the

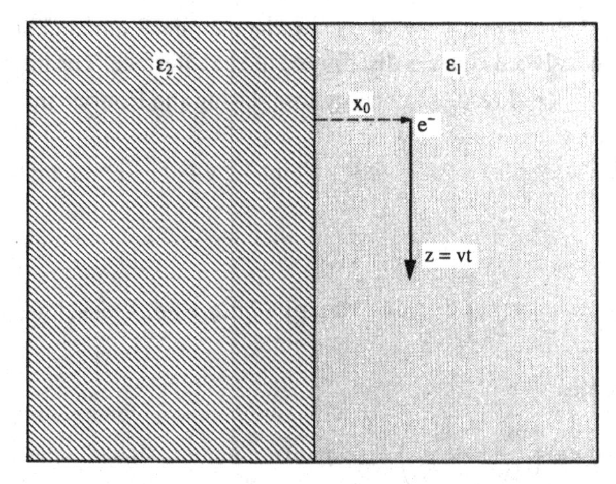

Figure 10.4 *The model of the dielectric response theory of interface excitation for two homogeneous media. The incident electron is traveling along the z axis at a distance x_0 from the interface.*

electrons along the surface is about 100 nm. Thus, valence-excitation is the most important inelastic scattering process in RHEED.

10.3 Interface and surface excitations

10.3.1 Classical energy-loss theory

The dielectric response theory introduced in the last section can also be applied to calculate the excitation of surfaces (or interfaces). In a more general case, an interface is formed by two media with dielectric constants ε_1 and ε_2 (Figure 10.4). The surface excitation can be considered as a one-electron excitation corresponding to the point probe case in STEM. The incident electron is assumed to travel at a distance x_0 in medium ε_1 parallel to the interface. Thus the field generated in space is the solution of

$$\varepsilon(\omega, \boldsymbol{q}) \nabla^2 V_e(\boldsymbol{r}, t) = \frac{e}{\varepsilon_0} \delta(x - x_0) \delta(y) \delta(z - vt). \tag{10.7}$$

The interface is assumed to be infinitely long in order to simplify the mathematical operation. We now consider the solution of Eq. (10.7) under the non-relativistic approximation (Howie, 1983). A more sophisticated treatment, taking into consideration the relativistic effect, will be given in Section 10.6.

The solution of Eq. (10.7) may be separated into two components,

$$V_e(\boldsymbol{r}, t) = V^{(s)}(\boldsymbol{r}, t) + V^{(id)}(\boldsymbol{r}, t), \tag{10.8}$$

where $V^{(s)}(\boldsymbol{r}, t)$ is the field generated by the electron itself and $V^{(id)}(\boldsymbol{r}, t)$ is the field generated by the induced charges distributed at the interface. The interaction of the electrons with $V^{(id)}(\boldsymbol{r}, t)$ results in energy loss. Using the Fourier transform of V_e,

$$V_e(\boldsymbol{r}, t) = \int d\omega \int dq_y \int dq_z \exp[2\pi i(q_y y + q_z z) - i\omega t]\, V_e(x, \boldsymbol{q}_t, \omega), \tag{10.9a}$$

or

$$V_e(x, \boldsymbol{q}_t, \omega) = \frac{1}{2\pi} \int dt \int dy \int dz \exp[-2\pi i(q_y y + q_z z) + i\omega t]\, V_e(\boldsymbol{r}, t), \tag{10.9b}$$

Eq. (10.7) becomes

$$\varepsilon(\omega, \boldsymbol{q}) \left(\frac{d^2}{dx^2} - 4\pi^2 q_t^2 \right) V_e(x, \boldsymbol{q}_t, \omega) = \frac{e}{\varepsilon_0} \delta(x - x_0) \delta(\omega - 2\pi q_z v), \tag{10.10}$$

where $\boldsymbol{q}_t = (q_y, q_z)$ and $q_t = (q_y^2 + q_z^2)^{1/2}$. The solution of Eq. (10.10) is

$$V_e(x, \boldsymbol{q}_t, \omega) = a(\boldsymbol{q}_t, \omega) \exp(2\pi q_t x) \qquad \text{for } x < 0, \tag{10.11a}$$

$$V_e(x, \boldsymbol{q}_t, \omega) = b(\boldsymbol{q}_t, \omega) \exp(-2\pi q_t x) + \frac{e}{4\pi\varepsilon_0\varepsilon_1 q_t} \exp(-2\pi q_t |x - x_0|) \delta(\omega - 2\pi q_z v) \qquad \text{for } x > 0. \tag{10.11b}$$

Matching the boundary condition at $x = 0$, the $a(\boldsymbol{q}_t, \omega)$ and $b(\boldsymbol{q}_t, \omega)$ coefficients are determined as

$$a(\boldsymbol{q}_t, \omega) = \frac{2\varepsilon_1}{\varepsilon_1 + \varepsilon_2} \frac{e}{4\pi\varepsilon_0\varepsilon_1 q_t} \exp(-2\pi q_t |x_0|) \delta(\omega - 2\pi q_z v), \tag{10.12a}$$

$$b(\boldsymbol{q}_t, \omega) = \frac{\varepsilon_1 - \varepsilon_2}{\varepsilon_1(\varepsilon_1 + \varepsilon_2)} \frac{e}{4\pi\varepsilon_0\varepsilon_1 q_t} \exp(-2\pi q_t |x_0|) \delta(\omega - 2\pi q_z v). \tag{10.12b}$$

The potential energy of the incident electron in its induced field is

$$\begin{aligned} E_p &= -eV^{(id)}(\boldsymbol{r}, t) \\ &= -\frac{e^2}{8\pi^2\varepsilon_0 v} \int_{-\infty}^{\infty} d\omega \int_{-\infty}^{\infty} dq_y \frac{\exp\{-2\pi\{q_y^2 + [\omega/(2\pi v)]^2\}^{1/2} |x + x_0|\}}{\{q_y^2 + [\omega/(2\pi v)]^2\}^{1/2}} \frac{\varepsilon_1 - \varepsilon_2}{\varepsilon_1(\varepsilon_1 + \varepsilon_2)} \\ &\quad \times \exp\left[-2\pi i q_y y + i\left(\frac{\omega}{v} z - \omega t \right) \right]. \end{aligned} \tag{10.13a}$$

On taking $x = x_0$, $y = 0$ and $z = vt$, the localized potential energy is

$$E_p = -\frac{e^2}{2\pi^2\varepsilon_0 v} \int_0^{\infty} d\omega \int_0^{\infty} dq_y \frac{\exp\{-4\pi\{q_y^2 + [\omega/(2\pi v)]^2\}^{1/2} |x_0|\}}{\{q_y^2 + [\omega/(2\pi v)]^2\}^{1/2}} \operatorname{Re}\left(\frac{\varepsilon_1 - \varepsilon_2}{\varepsilon_1(\varepsilon_1 + \varepsilon_2)} \right). \tag{10.13b}$$

E_p is a function of the electron impact distance x_0 and is at its maximum at the interface.

The energy loss of the electron results from the 'slowing down' effect of its induced charges, so that the rate at which the electron energy is reduced per unit distance is related to the differential excitation probability by

$$-\frac{dE}{dz}=-\frac{dE_p}{dz}\bigg|_{x=0,y=0,z=vt}=\int_0^{\infty} d\omega\,\hbar\omega\,\frac{d^2P_s}{d\omega\,dz}. \tag{10.14}$$

On substituting Eq. (10.13) into Eq. (10.14), we obtain

$$\frac{d^2P_s}{d\omega\,dz}=\frac{e^2}{4\pi^2\varepsilon_0\hbar v^2}\int_{-\infty}^{\infty} dq_y\,\frac{\exp\{-4\pi\{q_y^2+[\omega/(2\pi v)]^2\}^{1/2}|x_0|\}}{\{q_y^2+[\omega/(2\pi v)]^2\}^{1/2}}\,\mathrm{Im}\left(-\frac{\varepsilon_1-\varepsilon_2}{\varepsilon_1(\varepsilon_1+\varepsilon_2)}\right), \tag{10.15}$$

where a relation of $\varepsilon(-\omega,\boldsymbol{q})=\varepsilon^*(\omega,\boldsymbol{q})$ was used. If the dielectric function $\varepsilon(\omega,\boldsymbol{q})$ is assumed to be independent of wavevector $\boldsymbol{q}$, then the excitation probability of the interface is

$$\frac{d^2P_s}{d\omega\,dz}=\frac{e^2}{2\pi^2\varepsilon_0\hbar v^2}\,\mathrm{Im}\left(-\frac{\varepsilon_1-\varepsilon_2}{\varepsilon_1(\varepsilon_1+\varepsilon_2)}\right)K_0\left(\frac{2\omega|x_0|}{v}\right), \tag{10.16a}$$

where the modified Bessel function is

$$K_0\left(\frac{2\omega x_0}{v}\right)=\int_0^{\infty} du\,\frac{\exp\{-2[u^2+(\omega/v)^2]^{1/2}|x_0|\}}{[u^2+(\omega/v)^2]^{1/2}}. \tag{10.16b}$$

The localization of interface excitation is determined by the spatial variation of function $K_0(2\omega x_0/v)$. It has to be pointed out that Eq. (10.15) must be used if one needs to consider the dispersion relation of $\varepsilon(\omega,\boldsymbol{q})$.

The total excitation probability of the valence band is a sum of the volume and surface plasmons, from Eq. (10.5) and Eq. (10.16a), that is we have

$$\frac{d^2P}{dz\,d\omega}=\frac{e^2}{2\pi^2\varepsilon_0\hbar v^2}\left\{\mathrm{Im}\left(-\frac{1}{\varepsilon_1}\right)\left[\ln\left(\frac{2\pi q_c v}{\omega}\right)-K_0\left(\frac{2\omega x_0}{v}\right)\right]+\mathrm{Im}\left(-\frac{2}{\varepsilon_1+\varepsilon_2}\right)K_0\left(\frac{2\omega x_0}{v}\right)\right\}, \tag{10.17}$$

where the first term is the excitation probability of the volume plasmon as a function of the incident electron impact distance from the interface, and the second term is the excitation probability of the interface plasmon.

To understand the physical meaning of Eq. (10.17), a simple case of surface excitation is considered. If the electron is moving in vacuum, i.e., $\varepsilon_1 = 1$ and $\varepsilon_2 = \varepsilon$, then Eq. (10.17) reduces to

$$\frac{d^2P(x_0)}{dz\,d\omega} = \frac{e^2}{2\pi^2\varepsilon_0\hbar v^2}\,\mathrm{Im}\left(-\frac{2}{\varepsilon+1}\right)K_0\left(\frac{2\omega x_0}{v}\right). \tag{10.18}$$

Since $\mathrm{Im}\,[-2/(\varepsilon+1)]$ is the surface energy-loss function (Raether, 1980), no volume plasmon would be excited in this case. Valence losses are delocalized scattering processes, which occur even when the electrons are a few nanometers away from the surface in vacuum. If the electron penetrates into the crystal, i.e., $\varepsilon_2 = 1$ and $\varepsilon_1 = \varepsilon$, then Eq. (10.17) reduces to

$$\frac{d^2P(x_0)}{dz\,d\omega} = \frac{e^2}{2\pi^2\varepsilon_0\hbar v^2}\left\{\mathrm{Im}\left(-\frac{1}{\varepsilon}\right)\left[\ln\left(\frac{2\pi q_c v}{\omega}\right) - K_0\left(\frac{2\omega x_0}{v}\right)\right] + \mathrm{Im}\left(-\frac{2}{\varepsilon+1}\right)K_0\left(\frac{2\omega x_0}{v}\right)\right\}. \tag{10.19}$$

In Eq. (10.19), the $\mathrm{Im}(-1/\varepsilon)$ term is the excitation of the volume plasmon, which depends on the distance of the electron from the surface, and the second term is the excitation of the surface plasmon. The excitation probability of a volume plasmon drops to zero at $x_0 = 0$. If the scattering trajectory of the electron is known, then the excitation probability of the volume plasmon depends on the penetration depth into the crystal of the fast electron, and the excitation probability of the surface plasmon depends on both the penetration depth and the mean distance that electrons travel along the surface. On the other hand, these characteristics can be applied to measure these quantities (see Sections 10.9.2 and 10.9.3).

From the mathematical expression, the Bessel function $K_0(2\omega x_0/v)$ tends to infinity at $x_0 = 0$. This non-physical result came from two sources. One source is the assumption of $\varepsilon(\omega)$ being independent of q, i.e., no dispersion. The other source is the unlimited momentum transfer. For valence-excitation, there is a cut-off q_c for q, so that the upper limit of q integration in Eq. (10.16b) should be replaced by q_c.

Figure 10.5 shows the calculated differential excitation probabilities per unit distance of the electrons across a GaAs(110) surface. The dashed and solid lines represent the excitation probabilities of the volume and surface plasmons inside the crystal, respectively. The dotted line is the total excitation probability of the (surface and volume) plasmons. The excitation probabilities of surface and volume plasmons sensitively depend on the impact distance of the electron from the surface, but the total excitation probability is almost constant (for $x < 0$). Excitation of the volume plasmon occurs only when the electrons penetrate deep into the crystal. The excitation of the surface plasmon happens in a region about 1.5 nm from the surface. Only the surface plasmon is excited if the electron is outside the crystal. The electron–surface interaction is a long-range and non-localized one.

Figures 10.6(a) and (b) show the calculated valence-loss spectra (dashed lines)

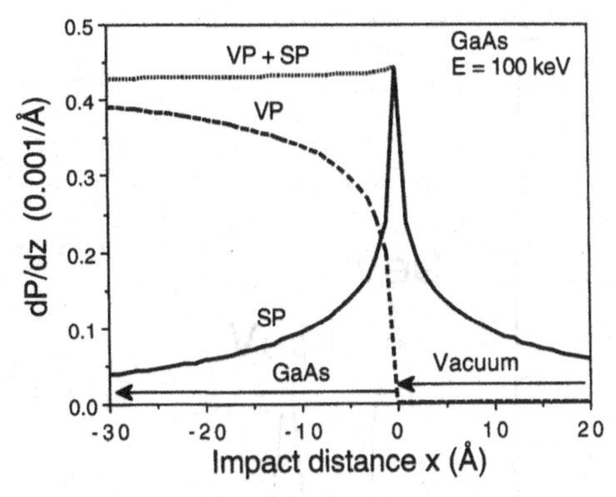

Figure 10.5 *Calculated excitation probability per unit distance across a GaAs surface as a function of the electron impact parameter x from the surface, with $x < 0$ inside the GaAs crystal and $x > 0$ in vacuum. The electron energy is 100 keV.*

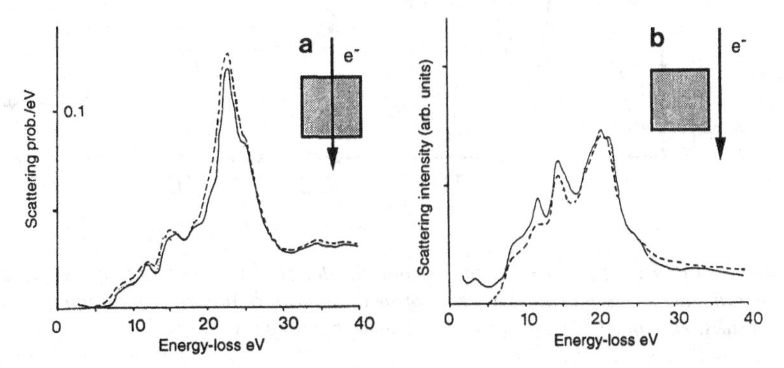

Figure 10.6 *(a) Volume and (b) surface excitation spectra acquired at 100 kV from a MgO cube of thickness about 120 nm. The dashed lines are the simulated spectra based on the dielectric response theory (Walls and Howie, 1989).*

and the experimentally observed spectra (solid lines) of a MgO cube. The volume excitation is dominant if the electron probe directly penetrates the cube (Figure 10.6(a)), the surface excitation is dominant if the electron probe is moved away from the cube (Figure 10.6(b)). The calculated spectra not only show the major features of the spectra but also exhibit the same absolute scattering probability in reference to the observed data.

10.3.2 Localization effects in surface excitation

For simplicity we consider a metal particle, which can be described by the free-electron model. It is well known that charges are induced at the metal particle surface when an incident electron approaches it. The charged wave can be excited to oscillate at particular frequencies. If the metal particle is sufficiently large and its

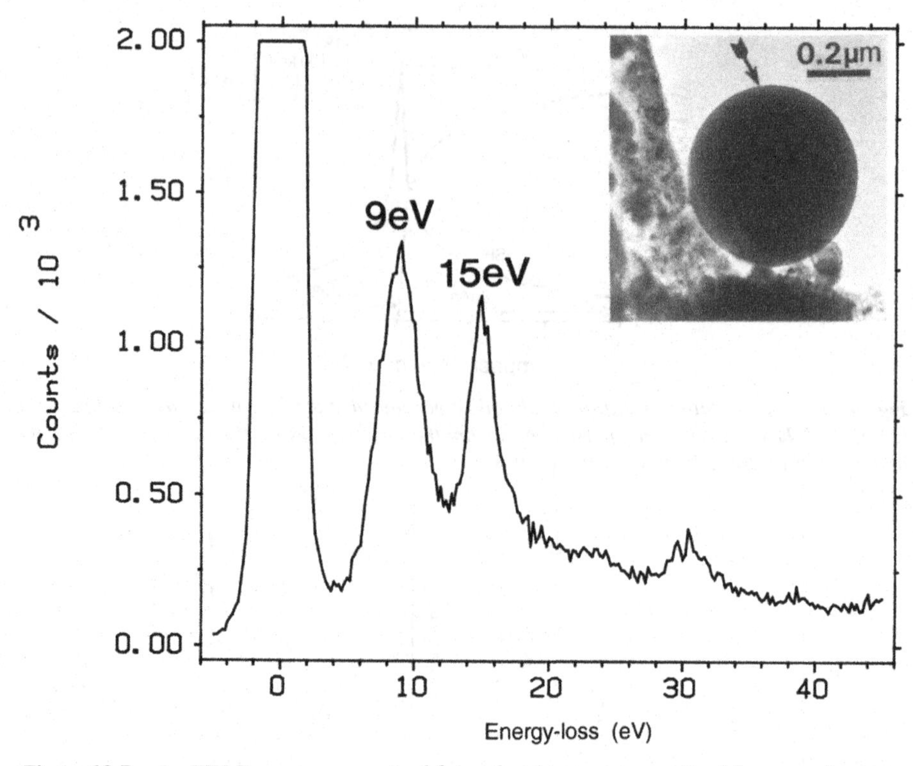

Figure 10.7 *An EELS spectrum acquired from the electrons transmitted from an aluminum sphere near the surface, as indicated by an arrowhead, showing both surface plasmon and volume plasmon peaks, at 9 and 15 eV, respectively. The beam energy is 120 kV.*

curvature is small, then the resonance frequency can be derived from the surface loss function

$$\mathrm{Im}\left(-\frac{2}{\varepsilon+1}\right)=\frac{2\varepsilon_i}{(\varepsilon_r+1)^2+\varepsilon_i^2}, \tag{10.20}$$

where $\varepsilon=\varepsilon_r+i\varepsilon_i$. For metals, $\varepsilon_r \gg \varepsilon_i$, thus, the resonance occurs if $\varepsilon_r+1=0$. Based on the free-electron model

$$\varepsilon_r=1-\frac{\omega_p^2}{\omega^2}, \tag{10.21}$$

where ω_p is the resonance frequency of the volume plasmon, the surface resonance frequency is $\omega_s=\omega_p/\sqrt{2}$. Figure 10.7 shows an EELS spectrum acquired from the electrons that transmit an aluminum particle at the surface, as indicated by an arrowhead. Besides the volume plasmon appearing at 15 eV, the surface plasmon located at 9 eV is seen. Since the particle is large, its resonance frequency is close to

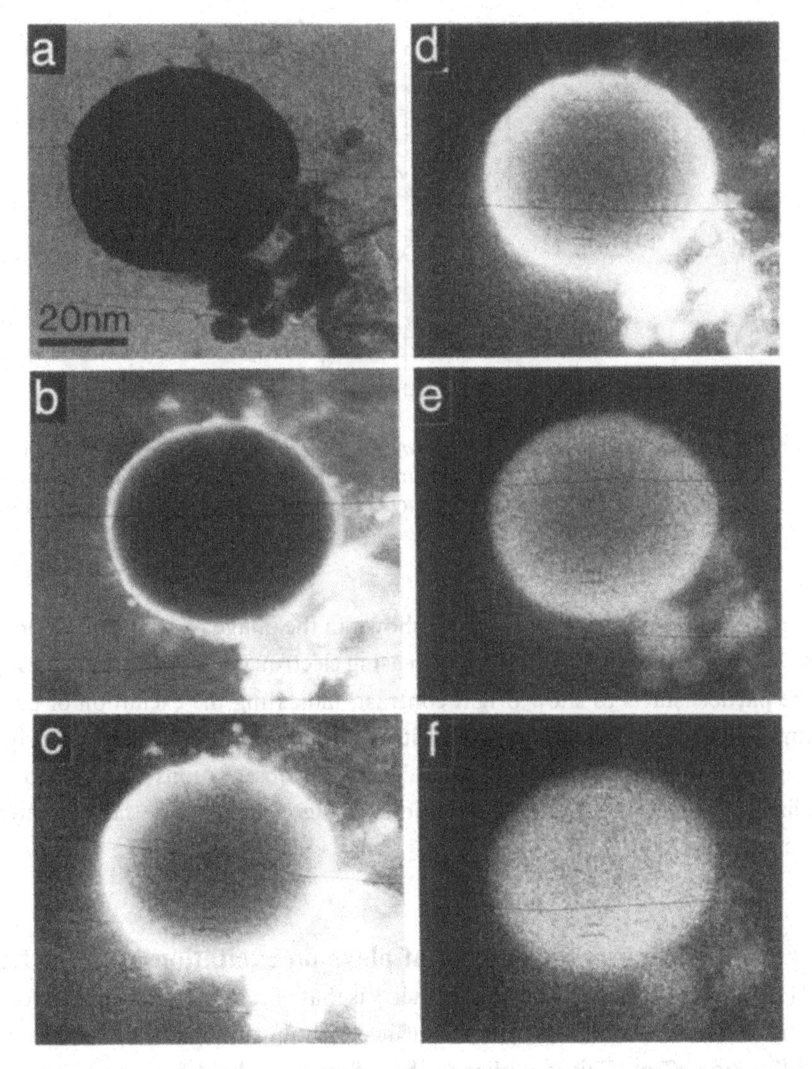

Figure 10.8 *Energy-filtered STEM images of an Al sphere using electrons with energy losses (a) 0, (b) 6, (c) 8, (d) 10, (e) 12 and (f) 15 eV, showing the localizations of different energy-loss processes. The beam energy is 100 kV.*

that of a planar metal surface. For small particles, the resonance frequencies are determined by $\omega_s = \omega_p[l/(2l+1)]^{1/2}$, with $l = 1, 2 \ldots$.

From Eq. (10.19), the excitation probablity of the surface plasmon drops quickly when the incident electron is far from the surface. This localization effect is best seen in the energy-filtered images formed by electrons with different energy losses, as shown in Figure 10.8. The width of the energy-selecting window was 1 eV. The bright-field image (Figure 10.8(a)) shows one large Al particle and a few small Al

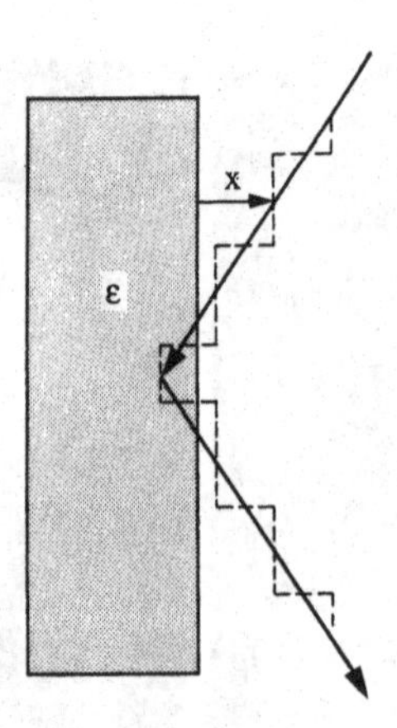

Figure 10.9 A stair-type electron traveling trajectory for simulating EELS spectra acquired in RHEED geometry. The dashed line indicates the approximated electron trajectory for theoretical calculations. The electron is considered to travel parallel to the surface within each stair step.

particles. The particle surface is clearly shown in the image formed by 6 eV energy-loss electrons (Figure 10.8(b)). With increasing electron energy loss, the central part of the particle starts to show bright contrast, indicating the excitation of particle volume. The volume excitation reaches its maximum in the image formed using 15 eV energy-loss electrons (Figure 10.8(f)). These observations clearly show the localization effect in plasmon excitation, in agreement with the theoretically expected results.

10.4 The average number of plasmon excitations in RHEED

In RHEED, the simplest scattering geometry is that of the mirror reflection model in which the electron is reflected from the surface without any penetration; thus there is no diffraction effect. This is similar to the reflection of light from a planar mirror surface. Equations (10.18) and (10.19) can be applied to simulate EELS spectra acquired in RHEED geometry by modifying the scattering trajectory of the electron as a staircase (Figure 10.9); the distance of the electrons from the surface is represented by a mean value for each stair step (Howie *et al.*, 1985), which is approximately valid if the stair step is chosen to be small. The path length of the electron within each stair segment is $\Delta x/\sin\theta$. This has been demonstrated as an effective experimental method for measuring electron penetration depth into the surface (Bleloch *et al.*, 1989a; Howie *et al.*, 1985; Wang, 1988d).

According to Eq. (10.18), only a surface plasmon is excited in the REELS spectrum when the electron is traveling in vacuum. This is because of delocalized excitation of surface plasmons due to long-range Coulomb interactions with surface induced charges when the electron approaches and leaves the surface, the probability of which is calculated as

$$\bar{m}_0 = 2\int_0^\infty \frac{dx_0}{\sin\theta}\frac{d^2P(x_0)}{dz\,d\omega} = \frac{e^2}{4\pi\varepsilon_0\hbar v\sin\theta}\int_0^\infty d\omega\,\frac{1}{\omega}\,\mathrm{Im}\left(\frac{-2}{\varepsilon+1}\right). \tag{10.22}$$

In practice, the finite penetration of the electron into the surface will introduce additional surface and volume excitations when the electron is propagating inside the crystal. This part is analogous to the valence excitation in transmission geometry. Therefore, the average number of plasmons to be excited is

$$\bar{m} = \bar{m}_0 + L_s/\Lambda, \tag{10.23}$$

where L_s is the average distance that the electron travels along the surface. $\bar{m}$ and $\bar{m}_0$ can be calculated using the FORTRAN program supplied in Appendix E.2.

It is important to note that $\bar{m}_0$ is inversely proportional to the electron incidence angle (for $\theta \ll 1$). This means that low-angle incidence may not always be the best choice for REELS, though at low angles surface sensitivity may be improved. The incidence angle also cannot be made too large because small angles are required in order to decrease the electron penetration depth. The excitation of localized atomic inner-shell ionization edges, however, occurs only when the electrons penetrate into the crystal. Therefore, the $\bar{m}_0$ term represents an extra excitation of surface losses in REELS compared with that in TEELS. The value of $\bar{m}_0$ is approximately 0.3–0.8. This effect decreases the signal-to-background ratio dramatically in REELS, and increases the acquisition time required to produce a good core-loss spectrum in the REM geometry. In the Bragg reflection case, the product $v\theta_g$ is approximately independent of the electron incidence energy, i.e. the average number of surface plasmon excitations for a non-penetrating mirror reflection model is independent of the electron energy.

10.5 Excitation of a sandwich layer

For the case of two media (of dielectric functions ε and ε'') with a sandwich layer (dielectric function ε') of thickness $2a$ between them, as shown in Figure 10.10, the excitation probability of the valence loss is given by Eq. (10.24) if the electron is located at a distance x ($x > a$) from the center of the sandwich layer in medium ε'' and moving parallel to the interface (Howie and Milne, 1985):

$$\frac{d^2P(x)}{d\omega\,dz} = \frac{e^2}{2\pi^2\varepsilon_0\hbar v^2}\int_0^{2\pi q_c} dq_y\,\frac{1}{q}\left[\mathrm{Im}\left(-\frac{1}{\varepsilon''}\right) + F(\varepsilon,\varepsilon',\varepsilon'',q,a)\exp(-2q|x-a|)\right], \tag{10.24a}$$

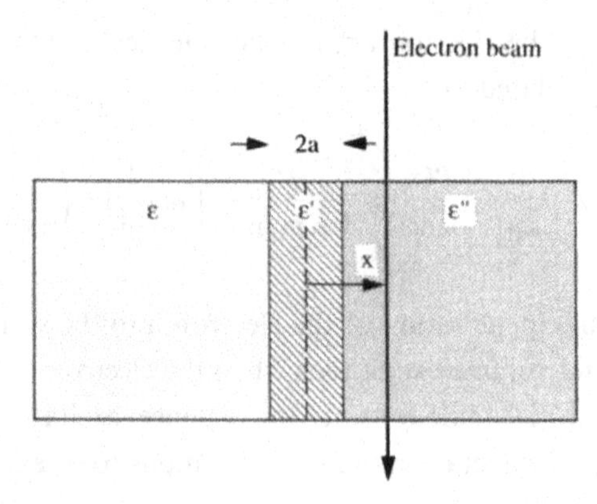

Figure 10.10 *Excitation of a 'sandwich' layer by an electron beam traveling parallel to the interface.*

where $q^2 = q_y^2 + \omega^2/v^2$ (see Howie and Milne (1985) for details), and

$$F = \mathrm{Im}\left(\frac{(\varepsilon'+\varepsilon)(\varepsilon'-\varepsilon'')\exp(2qa) - (\varepsilon'-\varepsilon)(\varepsilon'+\varepsilon'')\exp(-2qa)}{\varepsilon''[(\varepsilon'+\varepsilon)(\varepsilon'-\varepsilon'')\exp(2qa) - (\varepsilon'-\varepsilon)(\varepsilon'-\varepsilon'')\exp(-2qa)]}\right). \tag{10.24b}$$

These equations are useful for simulating the REELS spectra acquired from a surface if there is a thin layer of another phase.

Strictly speaking, Eq. (10.24) can only be applied to simulate the REELS spectra acquired using a very fine electron probe. For an electron probe of finite size, it needs to be modified by convolution with the shape function of the electron probe (Ritchie and Howie, 1988). It is also important to note that the above theory has been developed for high-energy electrons; for electrons of incident energy less than about 2 keV, the above theoretical model breaks down (Tougaard and Kraaer, 1991).

In the above discussion, we only considered the excitation of the surface (interface) that is parallel to the incident beam. For thin films in the transmission case, excitation of the top and bottom surfaces can also contribute to the valence spectra. For a thin slab of thickness d_0 and dielectric function ε with the incident beam perpendicular to the slab surface, the valence-loss excitation probability is (Ritchie, 1957)

$$\frac{\mathrm{d}P(x)}{\mathrm{d}\omega} = \frac{e^2}{2\pi^2\varepsilon_0\hbar v^2}\int_0^{2\pi q_c} \mathrm{d}u \frac{u^2}{[u^2+(\omega/v)^2]^2}$$
$$\times \mathrm{Im}\left(-\frac{1-\varepsilon}{\varepsilon}\,\frac{2(\varepsilon-1)\cos(\omega d_0/v) + (\varepsilon-1)\exp(-ud_0) + (1-\varepsilon^2)\exp(ud_0)}{(\varepsilon-1)^2\exp(-ud_0) - (\varepsilon+1)^2\exp(ud_0)}\right)$$
$$+ \frac{e^2 d_0}{2\pi^2\varepsilon_0\hbar v^2}\int_0^{2\pi q_c} \mathrm{d}u \frac{u}{u^2+(\omega/v)^2}\,\mathrm{Im}\left(-\frac{1}{\varepsilon}\right), \tag{10.25}$$

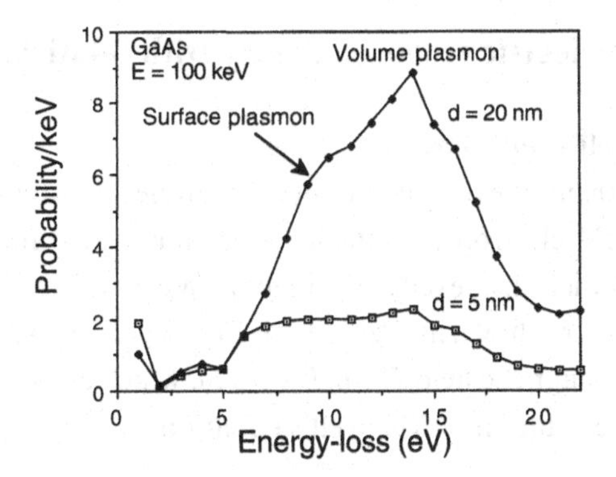

Figure 10.11 *Calculated transmission electron single-loss spectra of thin GaAs crystals of thicknesses 5 and 20 nm. The beam energy is 100 keV.*

where the first term is the surface excitation and the second term is the volume excitation. If the specimen is sufficiently thick so that the coupling between the top and bottom surfaces is weak and if the specimen thickness d_0 is sufficiently large, then Eq. (10.25) is approximated as

$$\frac{dP(x)}{d\omega} = \frac{e^2}{2\pi^2 \varepsilon_0 \hbar v^2} \int_0^{2\pi q_c} dq_y \frac{q^2}{[q^2 + (\omega/v)^2]^2} \operatorname{Im}\left(\frac{1}{\varepsilon} - \frac{4}{\varepsilon + 1}\right)$$

$$+ \frac{e^2 d_0}{2\pi^2 \varepsilon_0 \hbar v^2} \int_0^{2\pi q_c} du \frac{u}{u^2 + (\omega/v)^2} \operatorname{Im}\left(-\frac{1}{\varepsilon}\right)$$

$$\approx \frac{e^2}{4\pi^2 \varepsilon_0 \hbar v^2} \left[\arctan\left(\frac{2\pi q_c v}{\omega}\right) \operatorname{Im}\left(\frac{1}{\varepsilon} - \frac{4}{\varepsilon + 1}\right) + \ln\left[1 + \left(\frac{2\pi q_c v}{\omega}\right)^2\right] \operatorname{Im}\left(-\frac{1}{\varepsilon}\right)\right]. \tag{10.26}$$

A FORTRAN program that uses Eq. (10.25) for calculating the single-loss valence spectra in TEM is supplied in Appendix E.1. Figure 10.11 shows the calculated spectra for GaAs specimens of different thicknesses. When the specimen is thin, the coupling between the top and bottom surfaces is strong and the spectrum is dominated by the surface plasmon (SP) peak. When the specimen is thicker, the volume plasmon (VP) excitation is dominant. Since the volume plasmon shows a broad, strong peak, it is rather difficult to identify the surface plasmon peak. The increase in intensity at the energy-loss range of the surface plasmon results from the broad shoulder of the volume plasmon peak. It is apparent that more electrons suffer energy loss when the specimen is thicker.

10.6 The dielectric response theory with relativistic correction

10.6.1 Maxwell's equations

For high-energy electrons, the traveling velocity of an electron approaches the speed of light. For 100 keV electrons, $v=0.53c$. It is thus necessary to consider the retardation effect in dielectric excitation theory. We now consider the interface excitation in the geometry shown in Figure 10.4. The electric field $\boldsymbol{E}$ (or displacement vector $\boldsymbol{D}=\varepsilon\varepsilon_0\boldsymbol{E}$) and magnetic field $\boldsymbol{H}$ (or $\boldsymbol{B}=\mu_0\boldsymbol{H}$), which are excited in the space, are determined by the solution of Maxwell's equations

$$\nabla\times\boldsymbol{E}(\boldsymbol{r},t)=-\frac{\partial\boldsymbol{B}(\boldsymbol{r},t)}{\partial t}, \tag{10.27a}$$

$$\nabla\times\boldsymbol{H}(\boldsymbol{r},t)=\boldsymbol{J}(\boldsymbol{r},t)+\frac{\partial\boldsymbol{D}(\boldsymbol{r},t)}{\partial t}, \tag{10.27b}$$

$$\nabla\cdot\boldsymbol{D}(\boldsymbol{r},t)=\rho(\boldsymbol{r},t), \tag{10.27c}$$

$$\nabla\cdot\boldsymbol{B}(\boldsymbol{r},t)=0, \tag{10.27d}$$

where $\boldsymbol{J}$ is the current density of free charge and ρ is the free charge density function. We now introduce a transformation that converts the time-dependent quantity to a frequency-dependent quantity,

$$\boldsymbol{A}(\boldsymbol{r},\omega)=\int_{-\infty}^{\infty}\mathrm{d}t\exp(\mathrm{i}\omega t)\boldsymbol{A}(\boldsymbol{r},t), \tag{10.28a}$$

or

$$\boldsymbol{A}(\boldsymbol{r},t)=\frac{1}{2\pi}\int_{-\infty}^{\infty}\mathrm{d}\omega\exp(-\mathrm{i}\omega t)\boldsymbol{A}(\boldsymbol{r},\omega). \tag{10.28b}$$

Maxwell's equations are transformed into

$$\nabla\times\boldsymbol{E}(\boldsymbol{r},\omega)=\mathrm{i}\omega\boldsymbol{B}(\boldsymbol{r},\omega), \tag{10.29a}$$

$$\nabla\times\boldsymbol{H}(\boldsymbol{r},\omega)=\boldsymbol{J}(\boldsymbol{r},\omega)-\mathrm{i}\omega\boldsymbol{D}(\boldsymbol{r},\omega), \tag{10.29b}$$

$$\nabla\cdot\boldsymbol{D}(\boldsymbol{r},\omega)=\rho(\boldsymbol{r},\omega), \tag{10.29c}$$

$$\nabla\cdot\boldsymbol{B}(\boldsymbol{r},\omega)=0. \tag{10.29d}$$

We now introduce the Hertz vector $\boldsymbol{\Pi}$ (Stratton, 1941), using which $\boldsymbol{E}$ and $\boldsymbol{H}$ are expressed as

$$E(r,\omega)=\nabla(\nabla\cdot\Pi(r,\omega))+\frac{\varepsilon\omega^2}{c^2}\Pi(r,\omega), \tag{10.30a}$$

$$H(r,\omega)=-\mathrm{i}\omega\varepsilon\varepsilon_0\nabla\times\Pi(r,\omega). \tag{10.30b}$$

On substituting (10.30a) and (10.30b) into (10.29a) and (10.29b), we obtain the equation that determines the solution of Π

$$\left(\nabla^2+\frac{\varepsilon\omega^2}{c^2}\right)\Pi(r,\omega)=\frac{1}{\mathrm{i}\omega\varepsilon\varepsilon_0}J(r,\omega). \tag{10.31}$$

Before presenting the formal solution of Eq. (10.31) for the scattering geometry shown in Figure 10.4, we first prove that Eq. (10.29c) and Eq. (10.29d) are automatically satisfied with the E and H given in (10.30a) and (10.30b). From (10.29c)

$$\rho(r,\omega)=\nabla\cdot D(r,\omega)=\varepsilon\varepsilon_0\nabla\cdot E(r,\omega)=\varepsilon\varepsilon_0\nabla\cdot\left(\nabla^2+\frac{\varepsilon\omega^2}{c^2}\right)\Pi(r,\omega)=\frac{\nabla\cdot J(r,\omega)}{\mathrm{i}\omega}, \tag{10.32a}$$

or equivalently

$$\nabla\cdot J(r,t)+\frac{\partial\rho(r,t)}{\partial t}=0. \tag{10.32b}$$

This is just the law of charge conservation, which holds under any circumstance.

10.6.2 Valence excitation near an interface

We now apply Eq. (10.31) to calculate the excitation spectrum of an interface. The current density for a moving electron is expressed as

$$J(r,t)=-ev\hat{z}\delta(x-x_0)\delta(y)\delta(z-vt). \tag{10.33}$$

For the interface excitation as described in Figure 10.4, since the energy loss of the incident electron is independent of the origin of the y axis, we can introduce another transform:

$$\Pi(r,\omega)=\int du_y\int du_z\exp[2\pi\mathrm{i}(u_yy+u_zz)]\,\Pi(x,u_y,u_z,\omega). \tag{10.34}$$

On substituting (10.34) into (10.31), one has

$$\left[\frac{d^2}{dx^2}-\left(4\pi^2u^2-\frac{\varepsilon\omega^2}{c^2}\right)\right]\Pi(x,u_y,u_z,\omega)=\frac{1}{\mathrm{i}\omega\varepsilon\varepsilon_0}J(x,u_y,u_z,\omega), \tag{10.35a}$$

where

$$J(x,u_y,u_z,\omega)=-2\pi ev\hat{z}\delta(x-x_0)\delta(2\pi vu_z-\omega), \tag{10.35b}$$

and $u^2 = u_y^2 + u_z^2$. From symmetry, one may assume that Π has components only in x and z axes directions, $\Pi = (\Pi_x, 0, \Pi_z)$. The equation that one needs to solve for the case in which the beam is in medium ε_1 is

$$\left(\frac{d^2}{dx^2} - \chi_2^2\right) \Pi_z \ (x, u_y, u_z, \omega) = 0, \tag{10.36a}$$

$$\left(\frac{d^2}{dx^2} - \chi_1^2\right) \Pi_z^+ (x, u_y, u_z, \omega) = -\frac{2\pi ev}{i\omega\varepsilon_1\varepsilon_0} \delta(x - x_0)\delta(2\pi v u_z - \omega), \tag{10.36b}$$

$$\left(\frac{d^2}{dx^2} - \chi_2^2\right) \Pi_x^- (x, u_y, u_z, \omega) = 0, \tag{10.36c}$$

$$\left(\frac{d^2}{dx^2} - \chi_1^2\right) \Pi_x^+ (x, u_y, u_z, \omega) = 0, \tag{10.36d}$$

where Π^- and Π^+ represent the Hertz vectors for $x < 0$ and $x > 0$, respectively,

$$\chi_1^2 = 4\pi^2(u_y^2 + u_z^2) - \frac{\varepsilon_1\omega^2}{c^2}, \tag{10.37a}$$

$$\chi_2^2 = 4\pi^2(u_y^2 + u_z^2) - \frac{\varepsilon_2\omega^2}{c^2}. \tag{10.37b}$$

The solutions of (10.36a)–(10.36d) are written as

$$\Pi_z^- (x, u_y, u_z, \omega) = C \exp(\chi_2 x), \tag{10.38a}$$

$$\Pi_z^+ (x, u_y, u_z, \omega) = \frac{\pi ev}{i\omega\varepsilon_1\varepsilon_0\chi_1} \delta(2\pi v u_z - \omega) \exp(-\chi_1|x - x_0|) + A \exp(-\chi_1 x), \tag{10.38b}$$

$$\Pi_x^- (x, u_y, u_z, \omega) = D \exp(\chi_2 x), \tag{10.38c}$$

$$\Pi_x^+ (x, u_y, u_z, \omega) = F \exp(-\chi_1 x). \tag{10.38d}$$

Both χ_1 and χ_2 are required to have a positive real part in all subsequent manipulations, in order that the expressions in (10.38a)–(10.38d) converge at large x.

The four constants appearing in the solutions must be evaluated from the continuity of the tangential components of $\boldsymbol{E}$ and $\boldsymbol{H}$ parallel to the interface at the boundary $x = 0$. According to (10.30a) and (10.30b) the corresponding boundary conditions for the Hertz vectors at $x = 0$ are

$$\varepsilon_1 \Pi_z^+ = \varepsilon_2 \Pi_z^-, \tag{10.39a}$$

$$2\pi i u_z \Pi_z^+ + \frac{d\Pi_z^+}{dx} = 2\pi i u_z \Pi_z^- + \frac{d\Pi_x^-}{dx}, \tag{10.39b}$$

$$\varepsilon_1 \Pi_x^+ = \varepsilon_2 \Pi_x^-, \tag{10.39c}$$

$$\varepsilon_1 \frac{d\Pi_z^+}{dx} = \varepsilon_2 \frac{d\Pi_z^-}{dx}. \tag{10.39d}$$

For our purpose only the coefficients A and F are required:

$$A = \frac{\chi_1 - \chi_2}{\chi_1 + \chi_2} \zeta, \tag{10.40a}$$

where
$$F = 4\pi i u_z \chi_1 \zeta \frac{\varepsilon_2 - \varepsilon_1}{(\chi_1 + \chi_2)(\chi_2 \varepsilon_1 + \chi_1 \varepsilon_2)}, \tag{10.40b}$$

$$\zeta = \frac{\pi e v}{i\omega \varepsilon_1 \varepsilon_0 \chi_1} \delta(2\pi v u_z - \omega) \exp(-\chi_1 x_0). \tag{10.40c}$$

We now calculate the energy loss of the electron. The retarding force at the electron in the $-z$ direction is equal to its energy loss per unit path length. This is

$$\frac{d\Delta E}{dz} = -(-e)E_z(\boldsymbol{r},t)\bigg|_{\boldsymbol{r}=\boldsymbol{r}(t)} = \frac{e}{2\pi}\int_{-\infty}^{\infty} d\omega \exp(-i\omega t)\, E_z(\boldsymbol{r},\omega)\bigg|_{\boldsymbol{r}=\boldsymbol{r}(t)}$$

$$= \frac{e}{2\pi}\int_{-\infty}^{\infty} d\omega \exp(-i\omega t)\left(\frac{\partial}{\partial z}(\nabla\cdot\boldsymbol{\Pi}^+(\boldsymbol{r},\omega)) + \frac{\varepsilon_1\omega^2}{c^2}\Pi_z^+(\boldsymbol{r},\omega)\right)\bigg|_{\boldsymbol{r}=\boldsymbol{r}(t)}$$

$$= \frac{e}{2\pi}\int_{-\infty}^{\infty} d\omega \exp(-i\omega t)\int du_y \int du_z \exp(2\pi i u_z v t)$$

$$\times\left[2\pi i u_z\left(2\pi i u_z \Pi_z^+(x,u_y,u_z,\omega) + \frac{d}{dx}\Pi_x^+(x,u_y,u_z,\omega)\right) + \frac{\varepsilon_1\omega^2}{c_2}\Pi_z^+(x,u_y,u_z,\omega)\right]\Bigg|_{x=x_0}. \tag{10.41}$$

Also, the energy loss per unit path length is related to the differential excitation probability of valence losses by

$$\frac{d\Delta E}{dz} = \int_0^{\infty} d\omega\, \hbar\omega \frac{d^2P}{d\omega\, dz}. \tag{10.42}$$

Substituting the solutions given by (10.38a)–(10.38d) into (10.41), using the relation $\varepsilon(-\omega,\boldsymbol{u}) = \varepsilon^*(\omega,\boldsymbol{u})$, one finally has

$$\frac{d^2P}{d\omega\, dz} = \frac{e^2}{\pi\hbar v^2 \varepsilon_0}\int_0^{\infty} du_y \operatorname{Im}\left[\left(\frac{2(\varepsilon_2 - \varepsilon_1)\chi_1}{\varepsilon_1(\chi_1 + \chi_2)(\chi_2\varepsilon_1 + \chi_1\varepsilon_2)}\right.\right.$$

$$-\frac{[1-(v/c)^2\varepsilon_1](\chi_1-\chi_2)}{\varepsilon_1\chi_1(\chi_1+\chi_2)}\Bigg)\exp(-2\chi_1|x_0|)-\frac{1-(v/c)^2\varepsilon_1}{\varepsilon_1\chi_1}\Bigg], \qquad (10.43a)$$

where

$$\chi_1^2=4\pi^2u_y^2+\left(\frac{\omega}{v}\right)^2-\frac{\varepsilon_1\omega^2}{c^2}, \qquad (10.43b)$$

$$\chi_2^2=4\pi^2u_y^2+\left(\frac{\omega}{v}\right)^2-\frac{\varepsilon_2\omega^2}{c^2}. \qquad (10.43c)$$

The applications of Eq. (10.43a) for calculating single-loss REELS spectra and the integrated excitation probability P in RHEED geometry are given in Appendix E.2, where the FORTRAN source code is provided. A FORTRAN program that uses Eq. (10.43a) for calculating the single-loss valence spectra of surfaces in profile imaging case in TEM is given in Appendix E.3. The FORTRAN source code for applying Eq. (10.43) to calculate EELS spectra and excitation probability across the interface of two media in TEM is given in Appendix E.4.

To examine the consistency of Eq. (10.43a) with Eq. (10.15), one may check the limit $c\to\infty$, corresponding to a non-relativistic beam. Then $\chi_1^2=\chi_2^2=4\pi^2u_y^2+(\omega/v)^2$, and the classical excitation probability becomes

$$\frac{d^2P}{d\omega\, dz}=\frac{e^2}{2\pi^2\varepsilon_0\hbar v^2}\int_0^\infty du_y\, \mathrm{Im}\left\{\frac{\varepsilon_2-\varepsilon_1}{\varepsilon_1(\varepsilon_1+\varepsilon_2)}\exp\left\{-2\left[4\pi^2u_y^2+\left(\frac{\omega}{v}\right)^2\right]^{1/2}|x_0|\right\}-\frac{1}{\varepsilon_1}\right\}$$
$$\times\left[u_y^2+\left(\frac{\omega}{2\pi v}\right)^2\right]^{-1/2}. \qquad (10.44a)$$

If ε is independent of q,

$$\frac{d^2P}{d\omega\, dz}\approx\frac{e^2}{2\pi^2\varepsilon_0\hbar v^2}\left\{\mathrm{Im}\left(-\frac{1}{\varepsilon_1}\right)\left[\ln\left(\frac{2\pi q_c v}{\omega}\right)-K_0\left(\frac{2\omega x_0}{v}\right)\right]+\mathrm{Im}\left(-\frac{2}{\varepsilon_1+\varepsilon_2}\right)K_0\left(\frac{2\omega x_0}{v}\right)\right\}. \qquad (10.44b)$$

This is exactly the form of Eq. (10.17).

In comparison with the relativistic theory, the non-relativistic theory has ignored the retardation effect and Čerenkov radiation, which occurs when $\varepsilon(v/c)^2\geq 1$. When passing by a dielectric medium, a charged particle suffers both surface plasmon excitation and Čerenkov radiation. Figure 10.12 shows the ratio between the non-retarded and the retarded excitation probabilities for a 100 keV electron beam interacting externally with a MgO cube at different impact parameters with respect to its surface. This calculation can be performed using the FORTRAN program provided in Appendix E.3. It is seen that the retarded probability is always larger

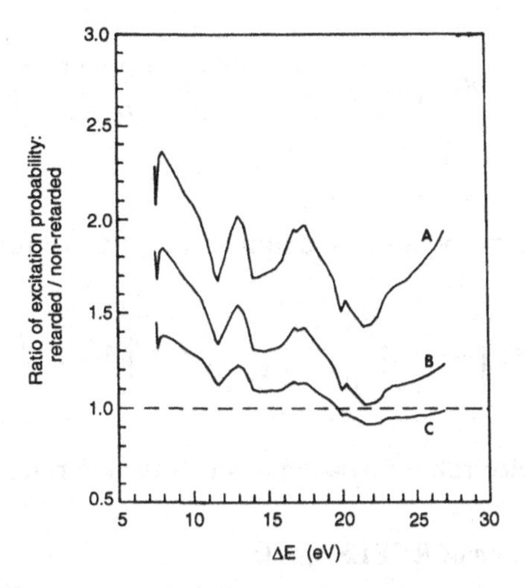

Figure 10.12 The ratio between the excitation probability for a relativistic beam (using Eq. (10.43)) and a non-relativistic beam (using Eq. (10.45)), traveling externally to a MgO flat surface. The distances from the surface to the beam are: A, $x_0=10$ nm; B, $x_0=4$ nm; and C, $x_0=0.5$ nm. The beam energy is 100 keV (courtesy of Dr R. Garcia-Molina et al., 1985).

than the non-retarded prediction, except for beams close to the surface. This is because the classical expression neglects retardation effects, and these will be appreciable at large beam–surface distances. In all cases, the effect of varying the distance x_0 from the beam to the target is larger than that of a variation in beam velocity.

10.6.3 The transverse force on an incident electron

It is also interesting to calculate the transverse force F_x, perpendicular to the electron's traveling direction, experienced by the electron due to the induced charge at the interface,

$$F_x = (-e)E_x(\mathbf{r}, t)\Big|_{\mathbf{r}=\mathbf{r}(t)} = -\frac{e}{2\pi}\int_{-\infty}^{\infty} \mathrm{d}\omega \exp(-\mathrm{i}\omega t)\, E_x(\mathbf{r}, \omega)\Big|_{\mathbf{r}=\mathbf{r}(t)}$$

$$= -\frac{e}{2\pi}\int_{-\infty}^{\infty} \mathrm{d}\omega \exp(-\mathrm{i}\omega t)\left(\frac{\partial}{\partial x}(\nabla\cdot\Pi^{+}(\mathbf{r}, \omega)) + \frac{\varepsilon_1\omega^2}{c^2}\Pi_x^{+}(\mathbf{r}, \omega)\right)\Bigg|_{\mathbf{r}=\mathbf{r}(t)}$$

$$= -\frac{e}{2\pi}\int_{-\infty}^{\infty} \mathrm{d}\omega \exp(-\mathrm{i}\omega t)\int \mathrm{d}u_y \int \mathrm{d}u_z \exp(2\pi\mathrm{i}u_z vt)$$

$$\times\left(2\pi\mathrm{i}u_z\frac{\mathrm{d}}{\mathrm{d}x}\Pi_z^{+}(x, u_y, u_z, \omega) + \frac{\mathrm{d}^2}{\mathrm{d}x^2}\Pi_x^{+}(x, u_y, u_z, \omega) + \frac{\varepsilon_1\omega^2}{c^2}\Pi_z^{+}(x, u_y, u_z, \omega)\right)\Bigg|_{x=x_0}$$

$$= -\frac{e^2}{\pi v \varepsilon_0} \int_0^\infty d\omega \int_0^\infty du_y \operatorname{Re}\left[\left(\chi_2 - \chi_1 + \frac{2(\varepsilon_2 - \varepsilon_1)[4\pi^2 u_y^2 + (\omega/v)^2]}{\chi_2 \varepsilon_1 + \chi_1 \varepsilon_2}\right) \frac{\exp(-2\chi_1 |x_0|)}{\varepsilon_1(\chi_1 + \chi_2)}\right]. \tag{10.45a}$$

Take a limit $c \to \infty$, corresponding to a non-relativistic electron,

$$F_x = -\frac{e^2}{\pi v \varepsilon_0} \int_0^\infty d\omega \int_0^\infty du_y \operatorname{Re}\left(\frac{\varepsilon_2 - \varepsilon_1}{\varepsilon_1(\varepsilon_1 + \varepsilon_2)}\right) \exp\left\{-2\left[4\pi^2 u_y^2 + \left(\frac{\omega}{v}\right)^2\right]^{1/2} |x_0|\right\}. \tag{10.45b}$$

Thus, $F_x < 0$ and the electron is attracted towards the interface.

10.6.4 Calculation of REELS spectra

We now apply Eq. (10.43a) to calculate the REELS spectra. For a case of surface excitation, the excitation probability can be directly written as follows. In vacuum ($x_0 > 0$):

$$\frac{d^2 P_o}{d\omega\, dz} = \frac{e^2}{2\pi^2 \hbar v^2 \varepsilon_0} \int_0^{2\pi q} dq_y \operatorname{Im}\left(\frac{2(\varepsilon - 1)\chi_0}{(\chi_0 + \chi)(\chi + \chi_0 \varepsilon)} - \frac{[1 - (v/c)^2](\chi_0 - \chi)}{\chi_0(\chi_0 + \chi)}\right) \exp(-2\chi_0 x_0), \tag{10.46a}$$

where

$$\chi_0^2 = q_y^2 + \left(\frac{\omega}{v}\right)^2 - \frac{\omega^2}{c^2}, \tag{10.46b}$$

$$\chi^2 = q_y^2 + \left(\frac{\omega}{v}\right)^2 - \frac{\varepsilon\omega^2}{c^2}. \tag{10.46c}$$

In the crystal ($x_0 < 0$):

$$\frac{d^2 P_i}{d\omega\, dz} = \frac{e^2}{2\pi^2 \varepsilon_0 \hbar v^2} \int_0^{2\pi q_c} dq_y \operatorname{Im}\left[\left(\frac{2(1 - \varepsilon)\chi}{\varepsilon(\chi_0 + \chi)(\chi + \chi_0 \varepsilon)} - \frac{[1 - (v/c)^2 \varepsilon](\chi - \chi_0)}{\varepsilon\chi(\chi_0 + \chi)}\right) \exp(2\chi_0 x_0) - \frac{[1 - (v/c)^2 \varepsilon]}{\varepsilon\chi}\right]. \tag{10.46d}$$

For simplicity, the scattering path of a reflected electron which contributes to the REELS spectrum is schematically shown in Figure 10.13. The mean penetration depth into the surface and the mean traveling distance along the surface under resonance condition are assumed to be D_p and L_s, respectively. The single-loss REELS spectra are calculated by integrating Eqs. (10.46a) and (10.46d) following the traveling path given in Figure 10.13,

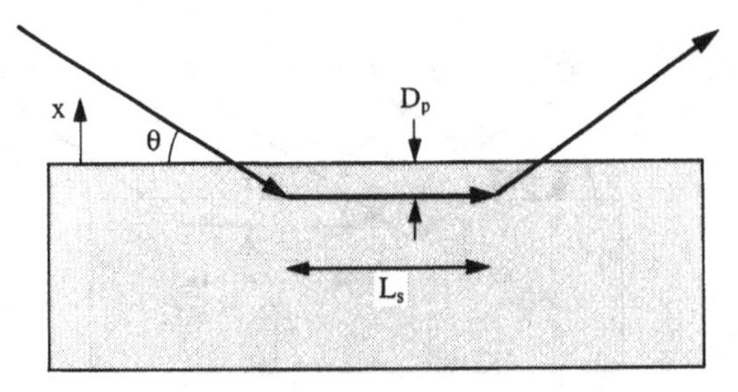

Figure 10.13 The resonance scattering path of the reflected electrons from a crystal surface. The effective penetration depth into the surface and the mean traveling distance along the surface are indicated by D_p and L_s, respectively.

$$\frac{dP}{d\omega}=2\int_0^{\infty}\frac{dx_0}{\sin\theta}\frac{d^2P_o}{d\omega\,dz}+2\int_0^{D_p}\frac{dx_0}{\sin\theta_g}\frac{d^2P_i}{d\omega\,dz}+L_s\left.\frac{d^2P_i}{d\omega\,dz}\right|_{x=-D_p}$$

$$=\frac{e^2}{2\pi^2\varepsilon_0\hbar v^2}\int_0^{2\pi q_c}dq_y\left\{\mathrm{Im}\left(\frac{2(\varepsilon-1)}{(\chi_0+\chi)(\chi+\chi_0\varepsilon)}-\frac{[1-(v/c)^2](\chi_0-\chi)}{\chi_0^2(\chi_0+\chi)}\right)\frac{1}{\sin\theta}\right.$$

$$+\mathrm{Im}\left[\left(\frac{2(1-\varepsilon)\chi}{\varepsilon(\chi_0+\chi)(\chi+\chi_0\varepsilon)}-\frac{[1-(v/c)^2\varepsilon](\chi-\chi_0)}{\varepsilon\chi(\chi_0+\chi)}\right)\left(\frac{1-\exp(-2\chi D_p)}{\chi\sin\theta_g}+L_s\exp(-2\chi D_p)\right)\right]$$

$$\left.-\mathrm{Im}\left(\frac{[1-(v/c)^2\varepsilon]}{\varepsilon\chi}\right)\left(\frac{2D_p}{\sin\theta_g}+L_s\right)\right\}. \quad (10.47a)$$

The integrated scattering probability

$$\bar{m}=P=\int_0^{\infty}d\omega\,\frac{dP}{d\omega}, \quad (10.47b)$$

where $\bar{m}$ is the average number of plasmon excitations defined in Eq. (10.23). The calculations of Eq. (10.47a) and Eq. (10.47b) are performed by the program provided in Appendix E.3.

Figure 10.14(a) shows three possible electron scattering trajectories. The mirror reflection path is the simplest case, in which the electron is reflected back without any penetration and propagation along the surface ($D_p=0$ and $L_s=0$). The non-resonance reflection is the path B, in which the electron penetrates into the surface without any resonance propagation along the surface ($L_s=0$). The case C is the resonance reflection, in which the electron is assumed to propagate along the surface for some distance before being reflected back. Path C is the most probable scattering geometry, as shown in Chapter 4. Figure 10.14(b) shows the calculated REELS spectra for these three cases. For path A, the spectrum is dominated by surface

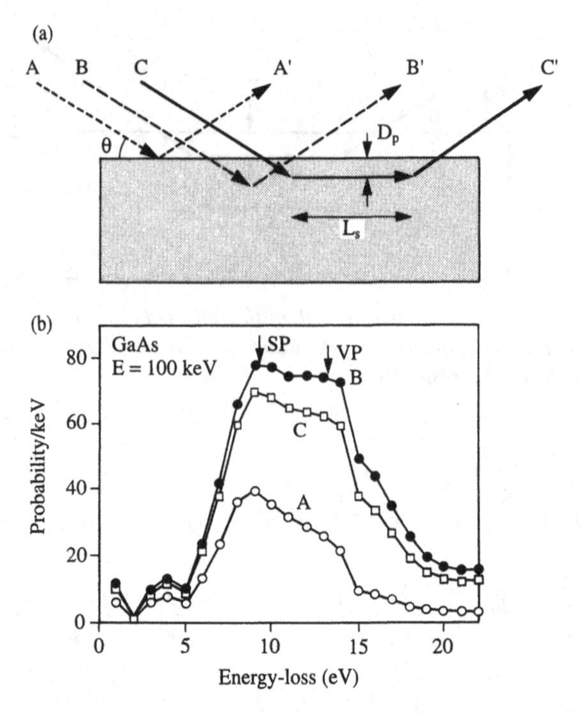

Figure 10.14 *(a) Electron scattering paths in RHEED and (b) the corresponding calculated single-loss REELS spectra for GaAs(110). Path A is the mirror reflection with* $D_p=0$ *and* $L_s=0$*; path B is the non-resonance reflection* $D_p=1.5$ *nm and* $L_s=0$*; and path C is the resonance reflection with* $D_p=0.5$ *nm and* $L_s=50$ *nm. The beam energy is 100 keV, the beam incident angle is 25 mrad.*

excitation. The volume excitation becomes significant as soon as the electron penetrates into the crystal (path B). The resonance reflection enhances not only the surface excitation but also the volume excitation (path C). The average numbers of plasmon excitations (total plasmon excitation probabilities) are 0.325, 0.845 and 0.716 for paths A, B and C, respectively.

10.7 The quantum theory of valence excitation

The classical dielectric response theory is an easy method for describing valence excitations in high-energy electron scattering. For metals, under the free-electron approximation, quantum mechanical charge-density fluctuation theory (Ashley *et al.*, 1974; Ashley and Ferrell, 1976) has shown equivalent results to those obtained using the dielectric response theory (Ritchie, 1981; Ritchie and Howie, 1988). In the excitations of small metal particles (Fujimoto and Komaki, 1968; Wang and Cowley, 1987a and b), both theories have shown identical results (Ferrell and

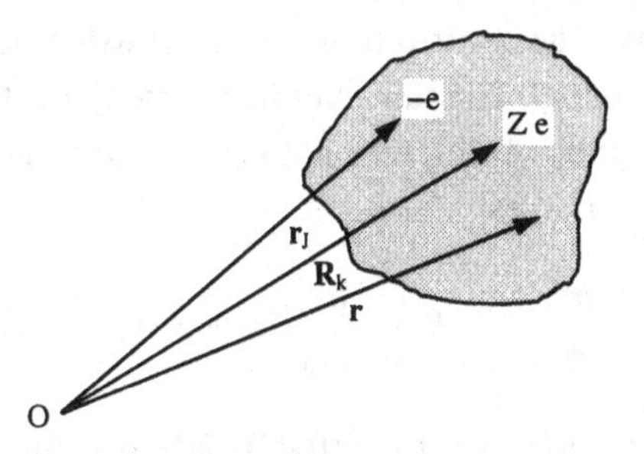

Figure 10.15 *Interaction of an external electron with the electrons and nuclei in the crystal.*

Echenique, 1985). In this section, we prove the equivalence of the dielectric response theory presented in Section 10.3 with the quantum theory.

10.7.1 The quantum mechanical basis of the classical theory

Figure 10.15 shows schematically a many-particle system of electrons undergoing perturbation by a fast electron with velocity $\boldsymbol{v}=v\hat{\boldsymbol{z}}$. The coordinates of the electrons are referred to some arbitrary origin at a point O. We omit spin indices for simplicity and only consider spin-independent interactions. Eigenstates and eigenenergies of the crystal electrons are represented by $|n\rangle$ and $\hbar\omega_n$, respectively. The $|n\rangle$ states are assumed orthonormal and complete, and the index n represents all observable characterizing quantum states of the system. We take a plane-wave basis set for the incident electron, $\exp(2\pi i\boldsymbol{K}\cdot\boldsymbol{r})$. The Hamiltonian H' characterizing interaction between the fast electron and the crystal electrons is

$$H'(\boldsymbol{r})=\sum_J\frac{e^2}{4\pi\varepsilon_0|\boldsymbol{r}-\boldsymbol{r}_J|}-\sum_k\frac{Z_ke^2}{4\pi\varepsilon_0|\boldsymbol{r}-\boldsymbol{R}_k|},\tag{10.48}$$

where $\boldsymbol{r}_J$ and $\boldsymbol{R}_k$ are the positions of the Jth crystal electron and the kth nucleus of charge Z_ke, respectively.

We apply standard first-order perturbation theory to obtain the cross-section for excitation of a transition of the many-electron system from its ground state $|0\rangle$ to a state $|n\rangle$, accompanied by the transition of the fast electron between momentum eigenstates characterized by wavevectors $\boldsymbol{K}$ (with $\boldsymbol{v}=h\boldsymbol{K}/m_0$) and $\boldsymbol{K}_\text{f}$. From the Golden rule we may write

$$\sigma_{n0}=\frac{2\pi}{\hbar^2 v}\sum_{K_\text{f}}\left|\langle n|\int \text{d}\boldsymbol{r}\exp(2\pi i\boldsymbol{k}\cdot\boldsymbol{r})\frac{1}{4\pi\varepsilon_0}\left(\sum_J\frac{e^2}{|\boldsymbol{r}-\boldsymbol{r}_J|}-\sum_k\frac{Z_ke^2}{|\boldsymbol{r}-\boldsymbol{R}_k|}\right)|0\rangle\right|^2\delta\left(\frac{E}{\hbar}-\frac{E_\text{f}}{\hbar}-\omega_{n0}\right),\tag{10.49}$$

where $\omega_{n0}=\omega_n-\omega_0$, $\boldsymbol{k}=\boldsymbol{K}-\boldsymbol{K}_\text{f}$, E and E_f stand for the energy of the incident electron before and after interacting with the crystal. The delta function stands for the conservation of energy. The contribution by nuclei will be dropped because of the orthonormal property of $|n\rangle$. The sum over K_f is to integrate the contributions made by the electron excitations of different momentum transfers but the same energy loss. We now change variables:

$$E-E_f=\frac{h^2}{2m_0}(K^2-K_f^2)=\frac{h^2}{2m_0}(2\boldsymbol{k}\cdot\boldsymbol{K}-k^2)=hk\cdot\boldsymbol{v}-\frac{h^2k^2}{2m_0}. \tag{10.50a}$$

Carrying out the integration over $\boldsymbol{r}$ in (10.49) and using the identity

$$\int d\boldsymbol{r}\,\frac{\exp(-2\pi i\boldsymbol{u}\cdot\boldsymbol{r})}{|\boldsymbol{r}-\boldsymbol{r}'|}=\frac{\exp(-2\pi i\boldsymbol{u}\cdot\boldsymbol{r}')}{\pi u^2}, \tag{10.50b}$$

we find

$$\sigma_{n0}=\frac{2\pi}{\hbar^2 v}\left(\frac{e^2}{4\pi\varepsilon_0}\right)^2\sum_{K_f}\frac{|\rho_{n0}(\boldsymbol{k})|^2}{\pi^2k^4}\,\delta\left(\frac{E}{\hbar}-\frac{E_f}{\hbar}-\omega_{n0}\right), \tag{10.51a}$$

where

$$\rho_{n0}(\boldsymbol{\tau})\equiv\langle n|\sum_j\exp(-2\pi i\boldsymbol{\tau}\cdot\boldsymbol{r}_j)|0\rangle \tag{10.51b}$$

is the matrix element of the density operator. For fast electrons, the sum over K_f can be replaced by an integral, thus

$$\sigma_{n0}=\frac{2\pi}{\pi\hbar^2 v}\left(\frac{e^2}{4\pi\varepsilon_0}\right)^2\int d\boldsymbol{k}\,\frac{|\rho_{n0}(\boldsymbol{k})|^2}{k^4}\,\delta\left(2\pi\boldsymbol{k}\cdot\boldsymbol{v}-\frac{2\pi^2\hbar k^2}{m_0}-\omega_{n0}\right). \tag{10.52}$$

To introduce an impact parameter conjugate for momentum transfer, we first neglect recoil of the incident electron; that is, we drop the term proportional to k^2 in the argument of the delta function (Ritchie, 1981). The larger v becomes, the less important will be the neglected term. Thus Eq. (10.52) is approximated as

$$\sigma_{n0}\approx\frac{2}{\pi\hbar^2 v}\left(\frac{e^2}{4\pi\varepsilon_0}\right)^2\int d\boldsymbol{k}\,\frac{\rho_{n0}(\boldsymbol{k})}{k^2}\int d\boldsymbol{u}\,\frac{\rho_{n0}^*(\boldsymbol{u})}{u^2}\,\delta(\boldsymbol{k}-\boldsymbol{u})\delta(2\pi\boldsymbol{k}\cdot\boldsymbol{v}-\omega_{n0}). \tag{10.53}$$

We now make use of the identity

$$\delta(\boldsymbol{k}_b-\boldsymbol{u}_b)=\int d\boldsymbol{b}\exp[2\pi i(\boldsymbol{k}_b-\boldsymbol{u}_b)\cdot\boldsymbol{b}] \tag{10.54}$$

in order to express Eq. (10.47) in terms of a spatial variable $\boldsymbol{b}$, where $\boldsymbol{b}=(b_x,b_y,0)$, $\boldsymbol{k}_b=(k_x,k_y,0)$ and $\boldsymbol{u}_b=(u_x,u_y,0)$. Equation (10.53) may be rewritten as

$$\sigma_{n0}=\frac{2}{\pi\hbar^2 v}\left(\frac{e^2}{4\pi\varepsilon_0}\right)^2\int d\boldsymbol{k}\,\frac{|\rho_{n0}(\boldsymbol{k})|^2}{k^4}\,\delta(2\pi\boldsymbol{k}\cdot\boldsymbol{v}-\omega_{n0})$$

$$=\frac{4}{\hbar^2}\left(\frac{e^2}{4\pi\varepsilon_0}\right)^2\int \mathrm{d}\boldsymbol{b}\left|\int \mathrm{d}\boldsymbol{k}\,\frac{\rho_{n0}(\boldsymbol{k})}{k^2}\exp[2\pi\mathrm{i}\boldsymbol{k}\cdot\boldsymbol{b}]\,\delta(2\pi k_z v-\omega_{n0})\right|^2. \quad (10.55)$$

Expressing the energy-conserving delta function in terms of an integral over time, i.e.,

$$\delta(2\pi k_z v-\omega_{n0})=\frac{1}{2\pi}\int \mathrm{d}t\exp(2\pi\mathrm{i}k_z vt-\mathrm{i}\omega_{n0}t), \quad (10.56)$$

and writing out the matrix element, we may express σ_{n0} as an integral over the impact parameter $\boldsymbol{b}$,

$$\sigma_{n0}=\int \mathrm{d}\boldsymbol{b}\,|a_{n0}(\boldsymbol{b})|^2, \quad (10.57\text{a})$$

where

$$a_{n0}(\boldsymbol{b})=\frac{1}{\mathrm{i}\pi\hbar}\frac{e^2}{4\pi\varepsilon_0}\int \mathrm{d}\boldsymbol{k}\,\frac{\rho_{n0}(\boldsymbol{k})}{k^2}\exp(2\pi\mathrm{i}\boldsymbol{k}\cdot\boldsymbol{b})\int \mathrm{d}t\exp(2\pi\mathrm{i}k_z vt-\mathrm{i}\omega_{n0}t)$$

$$=\frac{1}{\mathrm{i}\hbar}\int_{-\infty}^{\infty}\mathrm{d}t\,\langle n|\left(\sum_J\frac{e^2}{4\pi\varepsilon_0|\boldsymbol{b}+v\hat{\boldsymbol{z}}t-\boldsymbol{r}_J|}\right)|0\rangle\exp(-\mathrm{i}\omega_{n0}t)$$

$$=\frac{1}{\mathrm{i}\hbar}\int_{-\infty}^{\infty}\mathrm{d}t\,\langle n|H'|0\rangle\exp(-\mathrm{i}\omega_{n0}t). \quad (10.57\text{b})$$

Therefore, a_{n0} may be regarded as the probability amplitude that the crystal electron system will experience a transition under the influence of the Coulomb field of a classical point electron traveling with constant velocity v along a path specified by the impact parameter b beginning at $z=-\infty$ and ending at $z=\infty$. As is not surprising, Eq. (10.57b) agrees exactly with the result of a first-order quantum perturbation theoretical derivation of this probability amplitude, in which the interaction Hamiltonian H' is taken in the classical, prescribed, time-dependent form

$$H'=\sum_J\frac{e^2}{4\pi\varepsilon_0|\boldsymbol{b}+v\hat{\boldsymbol{z}}t-\boldsymbol{r}_J|}. \quad (10.58)$$

The result presented by Eq. (10.57) has been generalized by Ritchie and Howie (1988) for a case in which the incident electron beam is not a plane wave but rather a converged electron probe in STEM. They showed that, when all inelastically scattered electrons are collected, the measured probability of exciting a given transition may be computed theoretically as if the microprobe consisted of an incoherent superposition of classical trajectories distributed laterally to the beam direction according to the probe intensity function

10.7.2 The density operator and dielectric response theory

Before starting from Eq. (10.53) to derive the dielectric response theory, we first consider the density operator. In inelastic electron scattering, an important quantity is the so-called mixed dynamic form factor, which is related to the density operator by (Kohl and Rose, 1985; Kohl, 1983; Wang, 1995)

$$S(\boldsymbol{u},\boldsymbol{u}') \equiv \sum_{n\neq 0} \rho_{0n}(\boldsymbol{u})\rho_{n0}(-\boldsymbol{u}'). \tag{10.59}$$

The role played by $S(\boldsymbol{u},\boldsymbol{u}')$ in inelastic scattering is equivalent to that taken by the crystal potential V in elastic scattering. $S(\boldsymbol{u},\boldsymbol{u}')$ and $V(\boldsymbol{r})$ are the only two structurally related quantities that determine the behavior of electron scattering in crystals. The mixed dynamic form factor is directly related to the generalized dielectric function $\varepsilon_{\boldsymbol{uu}'}(\omega)$ (Kohl and Rose, 1985) by

$$S(\boldsymbol{u},\boldsymbol{u}') = \int_{-\infty}^{\infty} \mathrm{d}\omega \frac{2\pi \mathrm{i}\hbar\varepsilon_0}{e^2\{1-\exp[-\hbar\omega/(k_\mathrm{B}T)]\}}\left(\frac{\tau_2}{\varepsilon_{\boldsymbol{uu}'}(\omega)} - \frac{\tau'^2}{\varepsilon^*_{\boldsymbol{u}'\boldsymbol{u}}(\omega)}\right). \tag{10.60}$$

If one takes a special case with $\boldsymbol{u}=\boldsymbol{u}'$, Eq. (10.60) is simplified as

$$S(\boldsymbol{u},\boldsymbol{u}') = \sum_{n\neq 0} |\rho_{n0}(\boldsymbol{u})|^2 = \int_{-\infty}^{\infty} \mathrm{d}\omega \frac{2\pi \mathrm{i}\hbar\varepsilon_0}{e^2\{1-\exp[-\hbar\omega/(k_\mathrm{B}T)]\}} u^2 \mathrm{Im}\left(-\frac{1}{\varepsilon_{\boldsymbol{uu}}(\omega)}\right), \tag{10.61}$$

where $\mathrm{Im}[-1/\varepsilon_{\boldsymbol{uu}}(\omega)] = \mathrm{Im}[-1/\varepsilon(\omega,\boldsymbol{u})]$ is just the energy-loss function (Raether, 1980).

For valence excitation, the condition $|\hbar\omega| \gg k_\mathrm{B}T$ is always satisfied. Thus, the integration of ω from $-\infty$ to 0 vanishes in Eq. (10.61), since we have that $\{1-\exp(-\hbar\omega/k_\mathrm{B}T)\}^{-1} \approx 0$. Under this approximation, Eq. (10.61) becomes

$$S(\boldsymbol{u},\boldsymbol{u}) = \frac{4\pi\hbar\varepsilon_0}{e^2} \int_0^{\infty} \mathrm{d}\omega\, u^2 \mathrm{Im}\left(-\frac{1}{\varepsilon(\omega,\boldsymbol{u})}\right). \tag{10.62}$$

We now consider the total excitation probability of the crystal electrons. Summing over n in Eq. (10.53) for all valence states n $(n>0)$,

$$\sigma_\mathrm{t} = \sum_{n\neq 0} \sigma_{n0} = \frac{2}{\pi\hbar^2 v}\left(\frac{e^2}{4\pi\varepsilon_0}\right)^2 \sum_{n\neq 0} \int \mathrm{d}\boldsymbol{k} \frac{1}{k^4} \delta(2\pi k_z v - \omega_{n0})|\rho_{n0}(\boldsymbol{k})|^2. \tag{10.63}$$

This procedure is carried out in order to include the excitations of all the possible crystal valence states. The energy conservation law requires $\hbar\omega_{n0} = \hbar\omega$ (for an energy loss $\hbar\omega$). Substituting Eq. (10.59) into (10.63) and using (10.62), one finds

$$\frac{d\sigma_t}{d\omega} = \frac{e^2}{4\pi^2 \varepsilon_0 \hbar v^2} \int dk_x \int dk_y \frac{1}{k_x^2 + k_y^2 + (\omega/2\pi v)^2} \operatorname{Im}\left(-\frac{1}{\varepsilon(\omega, \boldsymbol{k})}\right). \qquad (10.64)$$

This is the exact result given by Eq. (10.4). Thus the equivalence of classical dielectric response theory and quantum transition theory is proven. Therefore, the classical theory can be accurately applied for simulating valence-loss spectra.

10.8 Determination of surface phases

In this section, we use an example, that of Al metal reduction from surface domains of α-alumina $(01\bar{1}2)$ initially terminated with oxygen, to illustrate the application of low-loss REELS in surface studies.

α-alumina (Al_2O_3) has a complex rhombohedral structure. The aluminum and oxygen atoms are arranged in a layered stacking sequence —Al—O—Al—O— parallel to $(01\bar{1}2)$. Thus, for cleaved α-alumina $(01\bar{1}2)$, the surface termination can be a layer of either aluminum or oxygen depending on where the bonds break. Radiation damage by the electron beam to the top layer of an α-alumina $(01\bar{1}2)$ surface can occur so rapidly that it is difficult to image the initial surface before the top oxygen layer is removed. It is possible to image the final reduced products. A cleaved α-alumina $(01\bar{1}2)$ surface initially showed uniform bright–flat contrast, but dark particles started to form on the surface immediately after it being illuminated for less than 30 s. More particles were formed in some particular areas although the entire surface was exposed to the electron beam. These particles grew very quickly and, upon being illuminated for about 10 min, accumulated to form large particles covering the upper surface, as shown in Figure 10.16. It is important to note that the radiation damage happens only at the dark surface domains (the upper part of Figure 10.16) but not at the bright–flat surface domains (the lower part of Figure 10.16). In Section 8.5, we have shown the radiation damage of surface domains terminated with different atomic layers.

The resistance to beam damage of the bright–flat domain can be seen through the REELS low-loss valence spectra, which are determined by the dielectric excitation property of the surface. A comparison of the REELS spectra acquired from a flat surface region after minimal exposure (solid line) and after being illuminated for an additional 8 min (dotted line) is shown in Figure 10.17. A broad peak is seen at about 22 eV with two weak shoulders on the low-loss side. These are characteristic excitations of an alumina crystal. The spectra retain the same shape and even exhibit no change in their relative magnitudes.

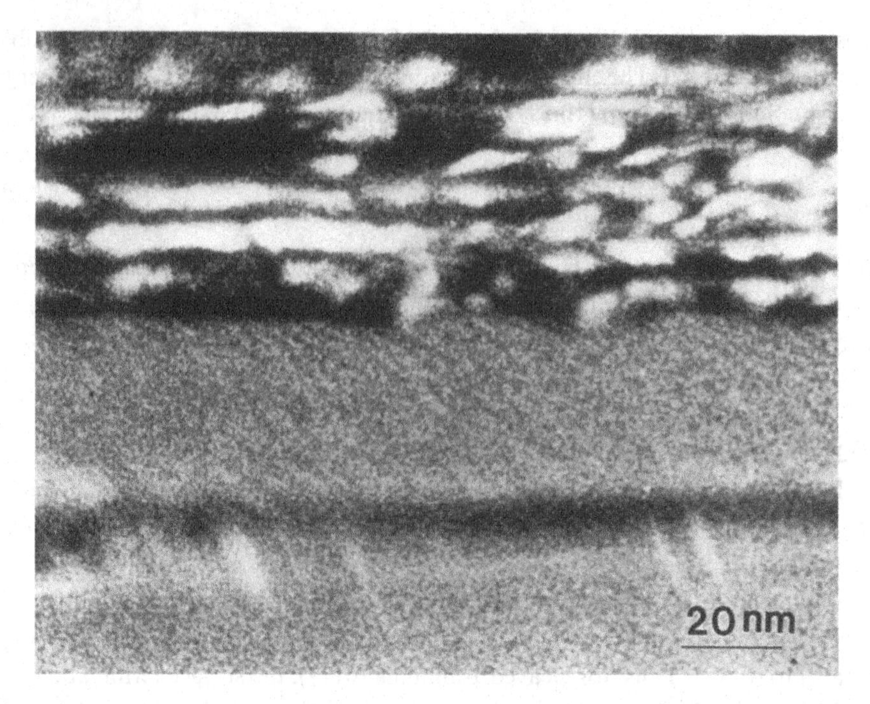

Figure 10.16 *A SREM image of an α-alumina $(01\bar{1}2)$ surface taken in a VG HB501 STEM, to show the formation of Al particles at the surface domains that were initially terminated with oxygen after having been illuminated for about 5 min. Al metal can be reduced from the dark-rough surface domains but not the bright–flat domains.*

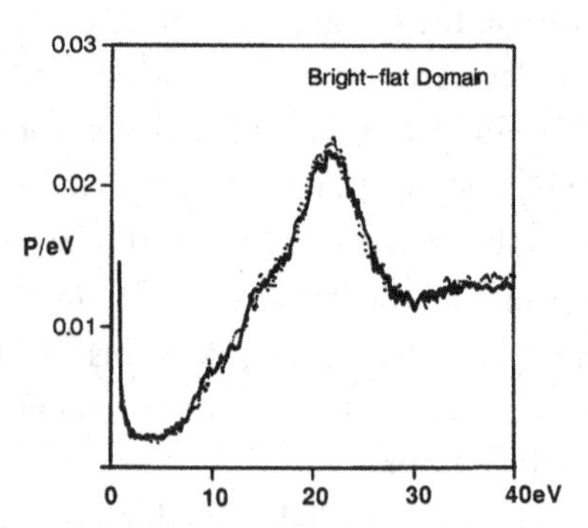

Figure 10.17 *REELS spectra acquired from the bright–flat surface domain of Figure 10.16 before and after illumination for 8 min. No evidence is found for the formation of Al metal particles on this domain.*

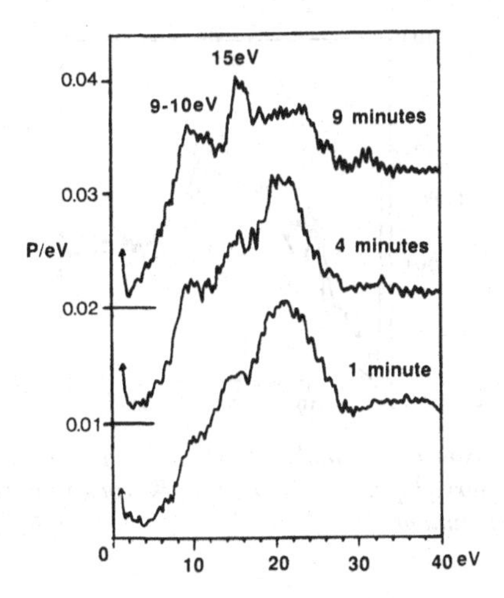

Figure 10.18 *A comparison of the REELS spectra acquired from the dark–rough surface domain of Figure 10.16 at different lengths of illumination time after their being first seen under the beam. The growth of the 9–10 eV (surface plasmon for Al particles) and 15 eV (volume plasmon of Al) peaks indicates the formation of Al metal particles at the rough surface domain.*

Figure 10.18 shows a comparison of the REELS spectra acquired from a single dark–rough surface domain after being illuminated for different lengths of time. When first seen under the beam, the spectrum shows about the same shape as those acquired from the bright–flat surface area (Figure 10.16). After the surface had been illuminated for about 4 min, the peak located at 9–10 eV clearly began to grow. This peak position can be related to the surface plasmon energy of small aluminum particles (see Section 10.3.2). The volume plasmon of Al located at 15 eV does not increase much in intensity at this stage. This is because the valence excitation is dominated by the surface effect if the particles are small. The alumina peak (22 eV) stays at about the same position. The 15 eV peak became more prominent after the surface had been illuminated for 9 min or more. This indicates the formation of larger aluminum particles. With the increase in particle size, the excitation probability of the volume plasmon increases. The excitation of the surface plasmon (9–10 eV) is also strengthened, because the total surface area of the particles is increased.

In the above case of Al metal formed by reduction of alumina, the sandwich layer is not a uniform dielectric layer but a mixture of small Al particles with vacuum. Since many small particles are covered by the illuminating beam, even in the small-probe SREM geometry, the dielectric property of this mixed layer can be statistically described by an effective dielectric function (Maxwell-Garnett, 1904 and 1906),

$$\varepsilon' = \varepsilon_0 \frac{(1+2\chi)\varepsilon_{\mathrm{Al}} + 2(1-\chi)\varepsilon_0}{(1-\chi)\varepsilon_{\mathrm{Al}} + (2+\chi)\varepsilon_0}, \tag{10.65}$$

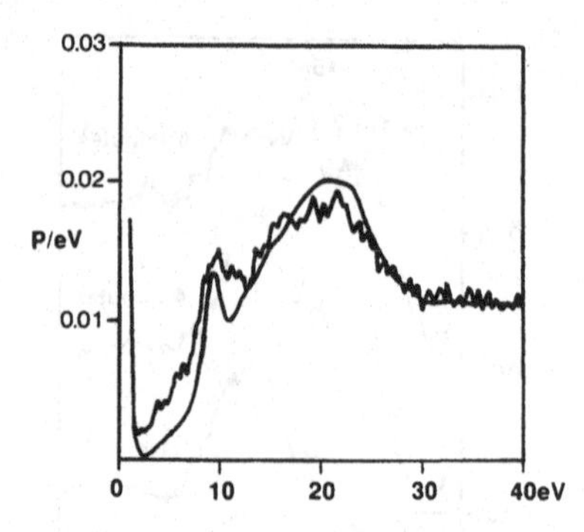

Figure 10.19 *A comparison of a simulated REELS spectrum according to the dielectric response theory and an observed spectrum from the dark–rough contrast domain in Figure 10.16 during the reduction of aluminum metal.*

where ε_{Al} is the dielectric function of the aluminum bulk, $\varepsilon_0 = 1$ and χ is the volume fraction of the Al particles in the sandwich layer. For the rough contrast domain, the structural fluctuations in the top few layers destroy the resonance propagation of the electrons along the surface in REM and the electron trajectory can be simplified to a direct 'mirror' reflection at a certain depth inside the bulk. In the computer simulation, the electron trajectory in the REM case is cut into may sections along the beam direction, so that the propagation of the electron in each section is considered to be parallel to the surface (see Figure 10.9).

Figure 10.19 shows a comparison of the calculated REELS spectrum with an observed spectrum from the rough contrast domain of Figure 10.16 during the reduction to Al metal. The effective penetration depth of the beam was chosen as $d = 1$ nm into alumina. The choice of d can sensitively affect the spectral match at higher energies as well as the peak at 22 eV. The thickness of the sandwich layer was taken as $2a = 3$ nm and the volume fraction of the Al particles was chosen to be $\chi = 0.07$. These parameters were chosen in order to match the height of the 10 eV surface plasmon peak and the 22 eV alumina plasmon. The choice of the thickness of the sandwich layer is correlated with the choice of the volume fraction χ, because the average thickness of Al metal is determined by $2a\chi$, which was estimated as 0.2 nm. This is approximately the thickness of the Al atoms in two or three Al (012) layers at the surface and is thus consistent with the Al particles being formed by the clustering of the top two or three Al atomic layers. The peak located at about 9–10 eV is present only when Al metal particles are a constituent of the sandwich layer, and is obviously the surface plasmon of the Al metal particles.

It must be pointed out that the effective medium theory breaks down if the particles are so big that the excitation property of each individual particle cannot be represented by the average excitation of a thin layer. Thus, the calculation cannot fit the spectrum displayed at the top of Figure 10.18, especially at the position of the 15 eV peak of the Al volume plasmon, because this peak comes from excitation in the Al bulk.

In order to understand the reduction process of aluminum from α-alumina surfaces under the electron beam, we consider the following mechanism. The reduction process was attributed to electron-beam-induced desorption of oxygen atoms, according to the Auger decay scheme (see Section 8.5). Under the electron beam, the oxygen atom tends to lose up to three electrons and becomes a positively charged O^{1+} ion, which may be pushed off the surface by the repulsive Coulomb force. In REM, the intensity of the Bragg reflections will decrease due to the structural fluctuation during the removal of the oxygen ions, thus giving rise to the dark contrast in REM images (see Figure 10.16). Thus, the dark domain is found to be terminated initially with oxygen. A complete removal of the top oxygen layers exposes fresh aluminum layers underneath, which may tend to reconstruct and form small Al particles in order to reduce the surface energy. The intense electron beam in STEM may cause the Al atoms to migrate so that small Al clusters nucleate, in turn exposing the oxygen layer underneath. The newly exposed oxygen layer will again be desorbed due to electrostatic repulsion after Auger decay, and the exposed Al layer atoms will then aggregate with the Al clusters to form particles. The repetition of this process will produce larger Al particles.

In contrast to the above-proposed process at surface domains initially terminated with oxygen, the resistance to electron beam damage of domains initially terminated with aluminum was attributed to the protection of the surface-adsorbed oxygen atoms. Since the adsorbed oxygen atoms are separated by one or two oxygen layers of the alumina from the top Al^{3+} ions, they will not undergo any Auger decay process because there is no adjacent Al (2p) energy level to which one of the O(2p) electrons can transit. Without the Auger decay process being stimulated, the oxygen atoms at the surface remain as neutral species or O^{2-} ions and will not be pushed off. They effectively act as a protective screen and prevent the oxygen atoms inside from being desorbed. The existence of this amorphous oxygen layer on the bright domain has been confirmed by REELS (Wang and Bentley, 1991a).

10.9 Multiple-scattering effects

In RHEED, the effective interaction distance between the electron and the surface is approximately the electron mean free path. Thus, multiple inelastic excitation becomes important. In this section, we first outline the theory that gives the energy and angular distribution laws of the inelastically scattered electrons. Then the theory is applied to measure some physical quantities that are of fundamental importance.

10.9.1 Poisson's distribution law

We first introduce the transport equation of Landau (1944) based on the principle of detailed balance. If $J(R, E_0-\Delta E)\,\mathrm{d}\Delta E$ is the fraction of the initially monoenergetic beam (of energy E_0) in the energy range $(E_0-\Delta E)\rightarrow(E_0-\Delta E-\mathrm{d}\Delta E)$ at position R in the specimen, and $w(R,\varepsilon)\,\mathrm{d}\varepsilon\,\mathrm{d}R$ (with $w(R,\varepsilon)=\mathrm{d}^2P/\mathrm{d}\varepsilon\,\mathrm{d}z$) is the probability that a given electron in the distribution $J(z, E_0-\Delta E)$ will lose energy in the range $\varepsilon\rightarrow\varepsilon+\mathrm{d}\varepsilon$ in a path length $\mathrm{d}R$, J obeys the equation (Landau, 1944)

$$\frac{\mathrm{d}J(R,E_0-\Delta E)}{\mathrm{d}R}=\int_0^\infty \mathrm{d}\varepsilon\, w(R,\varepsilon)[J(R,E_0-\Delta E+\varepsilon)-J(R,E_0-\Delta E)]. \tag{10.66}$$

We can introduce a generalized scattering parameter, which depends on the scattering trajectories of the electrons, as

$$m=\int_{\mathrm{Tr}}\mathrm{d}R\int_0^\infty \mathrm{d}\varepsilon\, w(R,\varepsilon), \tag{10.67}$$

where Tr characterizes the integration of R along the trajectory of the scattered electrons, and

$$w(R,\varepsilon)=\frac{\mathrm{d}^2P}{\mathrm{d}\varepsilon\,\mathrm{d}z}. \tag{10.68}$$

The first term on the right-hand side of (10.66) is the integrated contribution from the electrons with energy $(E_0-\Delta E+\varepsilon)$ after losing additional energy ε; the second term represents the decrease in electron intensity after propagating a distance $\mathrm{d}z$. In general, J is zero for negative energy loss and satisfies the normalization relation

$$\int_0^\infty \mathrm{d}\Delta E\, J(R,E_0-\Delta E)=1. \tag{10.69}$$

It is important to point out that $J(R, E_0-\Delta E)$ is the integrated electron intensity distribution of all the electrons scattered to different angles but with the same energy loss. The equation can be solved using the Laplace transform (LT)

$$J(R,p)=\mathrm{LT}(J)=\int_0^\infty \mathrm{d}\Delta E\exp(-p\Delta E)J(R,E_0-\Delta E), \tag{10.70}$$

and (10.66) becomes

$$\frac{\mathrm{d}J(R,p)}{\mathrm{d}R}=\mathrm{LT}(w(R,\varepsilon))J(R,p)-\frac{J(R,p)}{\Lambda}. \tag{10.71}$$

Equation (10.71) can be readily solved and the inverse Laplace transform (LT^{-1}) of the solution is

$$J(R, E_0-\varepsilon)=\exp(-m)\exp\left(-\frac{z}{\Lambda}\right)\mathrm{LT}^{-1}\left(\sum_{n=0}\frac{m^n}{n!}[\mathrm{LT}(f_{\mathrm{E}})]^n\right), \tag{10.72a}$$

with

$$f_{\mathrm{E}}(\varepsilon)=\frac{\int_T \mathrm{d}R\, w(R,\varepsilon)}{m} \tag{10.72b}$$

the single-loss function. Performing the inverse Laplace transform, Eq. (10.72a) becomes (Whelan, 1976)

$$J(R, E_0-\Delta E)=\exp(-m)\sum_{n=0}\frac{m^n}{n!}\{f_{\mathrm{E}}\}_n, \tag{10.73a}$$

with

$$\{f_{\mathrm{E}}\}_n=\{f_{\mathrm{E}}\otimes\ldots\otimes f_{\mathrm{E}}\}_n=\{f_{\mathrm{E}}\}_{n-1}\otimes f_{\mathrm{E}}. \tag{10.73b}$$

The nth power term corresponds to the nth-order multiple scattering. Therefore, the distribution of electrons as a function of energy loss obeys Poisson's distribution law. This relation is correct if all the electrons scattered to different angles are collected. Equation (10.73) has been applied to restore the single energy-loss function f_{E} from the experimentally measured electron energy-loss distribution function J (Johnson and Spence, 1974).

To illustrate the meaning of Eq. (10.73), we use the delta function approximation to represent the single-plasmon-loss function, i.e. $f_{\mathrm{E}}=f\delta(\varepsilon-\Delta E_{\mathrm{p}})$, where ΔE_{p} is the energy of plasmon loss. From Eq. (10.73a), the shape of the EELS spectrum is

$$\begin{aligned}J(d, E_0-\varepsilon)&=\exp(-m)\sum_{n=0}\frac{m^n}{n!}f^n\delta(\varepsilon-n\Delta E_{\mathrm{p}})\\ &=\exp(-m)\delta(\varepsilon)+\exp(-m)\frac{d}{\Lambda}f\delta(\varepsilon-\Delta E_{\mathrm{p}})+\exp(-m)\frac{m^2}{2!}f^2\delta(\varepsilon-2\Delta E_{\mathrm{p}})\\ &\quad+\ldots+\exp(-m)\frac{m^n}{n!}f^n\delta(\varepsilon-n\Delta E_{\mathrm{p}})+\ldots.\end{aligned} \tag{10.74}$$

Therefore, the relative intensity of the nth plasmon peak, located at $n\Delta E_{\mathrm{p}}$, with respect to the zero-loss peak is $(m^n/n!)f^n$.

The single-scattering function can be retrieved from the experimental data according to Eq. (10.73a). If we take a one-dimensional Fourier transform, Eq. (10.73a) becomes

$$\mathrm{FT}(J(R, E_0-\Delta E))=\exp(-m)\sum_{n=0}\frac{m^n}{n!}\{\mathrm{FT}(f_{\mathrm{E}})\}^n=\exp(-m)\exp[m\,\mathrm{FT}(f_{\mathrm{E}})]. \quad (10.75)$$

Thus the single-loss function is

$$\mathrm{FT}(f_{\mathrm{E}})=\frac{\ln[\mathrm{FT}(J(z, E_0-\Delta E))]}{m}+1, \quad (10.76a)$$

or

$$f_{\mathrm{E}}=\frac{\mathrm{FT}^{-1}\{\ln[\mathrm{FT}(J(z, E_0-\Delta E))]\}}{m}+\delta(\varepsilon). \quad (10.76b)$$

This is the Fourier–log method for retrieving the single-loss function from experimental data. However, it must be pointed out that the noise in the spectra could be enhanced due to the logarithm calculation involving a small number.

10.9.2 Measurement of electron penetration depth

In TEM, all the electrons can be assumed to have traveled for the same distance (approximately the specimen thickness) before leaving the crystal. To apply Eq. (10.73) in the geometry of RHEED, we need to known the scattering trajectory of the incident electron. For simplification, we can assume that the electron penetrates into the surface for a depth D_p, as shown earlier in Figures 5.14 and 10.14. Thus the scattering parameter m can be calculated as

$$w(R)=\int_0^\infty \mathrm{d}\varepsilon\, w(R, \varepsilon)=\int_0^\infty \mathrm{d}\omega\,\frac{\mathrm{d}^2P}{\mathrm{d}\omega\,\mathrm{d}z}. \quad (10.77)$$

From the dielectric response theory presented in Section 10.3 for non-relativistic scattering, we have

$$\frac{\mathrm{d}^2P}{\mathrm{d}\omega\,\mathrm{d}z}=\frac{e^2}{2\pi^2\hbar v^2\varepsilon_0}\int_0^{2\pi q_c}\mathrm{d}u_y\,L(u_y, x), \quad (10.78a)$$

with

$$L(u_y, x)=\frac{1-\exp(2qx)}{q}\,\mathrm{Im}\left(-\frac{1}{\varepsilon}\right)+\frac{\exp(2qx)}{q}\,\mathrm{Im}\left(-\frac{2}{\varepsilon+1}\right) \quad \text{for } x<0 \quad (10.78b)$$

(inside the crystal), and

$$L(x)=\frac{\exp(-2qx)}{q}\,\mathrm{Im}\left(-\frac{2}{\varepsilon+1}\right) \quad \text{for } x>0 \quad (10.78c)$$

(in vacuum), where $q=[u_y^2+(\omega/v)^2]^{1/2}$.

As shown earlier in Figure 10.5 for $x<0$, although the total excitation probability is almost constant in the crystal, the excitation probability of the volume plasmon rapidly decreases when the electron is near the surface, but the surface plasmon excitation increases dramatically. This effect will be used below to measure the mean penetration depth into the surface by the incident electrons.

For the scattering geometry shown in Figure 10.9, the differential path length $dR=dx/\sin\theta$, thus the scattering parameter is calculated from (10.67)

$$m=m_v+m_s, \tag{10.79}$$

where m_v is the scattering parameter contributed by volume plasmon excitation and m_s is that contributed by surface plasmon excitation

$$m_v=\frac{e^2}{2\pi^2\varepsilon_0\hbar v^2\sin\theta}\int_0^\infty d\omega\int_0^{2\pi q_c} du_y\,\frac{2qD_p-1+\exp(-2qD_p)}{q^2}\,\mathrm{Im}\left(-\frac{1}{\varepsilon}\right), \tag{10.80a}$$

$$m_s=\frac{e^2}{2\pi^2\varepsilon_0\hbar v^2\sin\theta}\int_0^\infty d\omega\int_0^{2\pi q_c} du_y\,\frac{1+(\sin\theta/\sin\theta_g)[1-\exp(-2qD_p)]}{q^2}\,\mathrm{Im}\left(-\frac{2}{\varepsilon+1}\right). \tag{10.80b}$$

The volume plasmon and surface plasmon excitations are separated based on the energy-loss function $\mathrm{Im}(-1/\varepsilon)$ for the bulk and $\mathrm{Im}[-2/(\varepsilon+1)]$ for the surface. The FORTRAN program provided in Appendix E.2 can perform the calculations of Eq. (10.80a) and Eq. (10.80b).

Precisely speaking, the Poisson law holds if the scattering trajectories are the same for all the collected electrons. This condition can be easily satisfied in TEM. In RHEED, however, not all the electrons would travel following the same trajectory. We now consider a case in which the entrance aperture of the EELS spectrometer is centered at the Bragg-reflected beam and the collection semi-angle is very small. In this case, all the collected electrons may be considered to have experienced the same scattering process (i.e., Bragg and resonance scattering). Thus the Poisson distribution law approximately holds, and the single-loss function can be retrieved using the Fourier–log method.

Figure 10.20 shows a comparison of the REELS spectra before (dashed line) and after (solid line) being treated using the Fourier–log deconvolution technique. The spectra in Figures 10.20(a)–(d) were acquired from GaAs(110) using the (440), (660), (880) and (10 10 0) specular reflections, respectively. The smoothness of the single-loss spectra indicates the success in applying the deconvolution technique. In the spectrum acquired using the (440) reflection ($\theta=15.6$ mrad for 120 keV), only the surface plasmon (SP) (11 eV) is seen, and there is no indication of the excitation

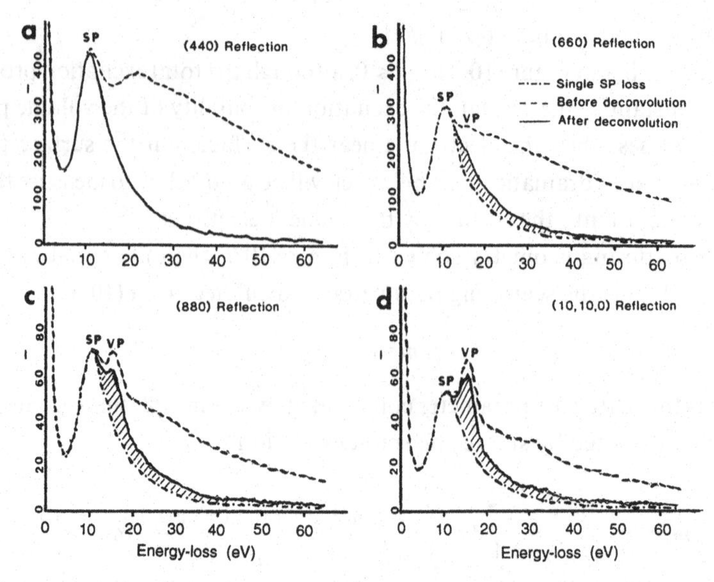

Figure 10.20 *A comparison of REELS spectra before (dashed line) and after (solid line) being treated by the Fourier–log deconvolution program for retrieving the single-loss function. In (a) to (d) are shown the spectra acquired from GaAs(110) at 120 keV under the specular reflections (440), (660), (880), (10 0 0), respectively. SP and VP indicate the peaks of surface and volume plasmons, respectively.*

of the volume plasmon (VP) (16 eV). This result is expected theoretically. The single-loss function obtained in this case can be taken as the standard spectrum for single excitation of a surface plasmon. In the spectrum recorded using (660) and higher order specularly reflected beams, the volume plasmon is excited due to deeper electron penetration into the surface. To find the contribution made by single-loss volume plasmon excitation, the scale of the single-loss spectrum in Figure 10.20(a) is magnified to match the photodiode counts at the position of the surface plasmon peak (11 eV) in Figures 10.20(b)–(d), and the results are indicated by dash–dot lines. The single-loss volume plasmon spectra can be obtained by subtracting the single-loss surface plasmon spectra (dash–dot lines) from the deconvoluted spectra (solid lines), and the results are indicated by the shadowed areas in Figure 10.20. The integrated intensities of the zero-loss peak (I_0) and single-loss volume plasmon spectrum (I_v) (the shadowed area) can be directly computed from the spectra. Thus, the measure $m'_v = I_v/I_0$ can be directly compared with the calculated m_v. The mean penetration depth D_p is determined if $m_v = m'_v$. For 120 kV electrons, the mean penetration depths into the GaAs(110) surface of the electrons under (660), (880) and (10 10 0) specular reflections were determined as 1.6, 2.0 and 3.0 nm, respectively (Wang, 1988d).

10.9.3 Measurement of electron mean traveling distance along a surface

In Chapter 4, an extensive study of electron resonance scattering from crystal surfaces was presented. Under surface resonance conditions, the electrons are believed to have traveled along the surface for some distance before being reflected back into vacuum. The traveling distances of electrons along the surface may be different, but it is possible to use a mean traveling distance to represent the average scattering behavior of the electron beam. We now intend to measure the mean distance that the electrons travel inside the crystal. From Eq. (10.73), the average number of plasmon excitations is

$$\bar{m} = \ln(I/I_0), \tag{10.81}$$

where I and I_0 are the integrated intensities for the total spectrum and the zero-loss peak, respectively. Using Eq. (10.23), we have

$$L_s \approx \frac{\ln(I/I_0) - \bar{m}_0}{\int_0^\infty d\omega \frac{d^2P}{d\omega\, dz}}, \tag{10.82}$$

where $d^2P/d\omega\, dz$ is the differential excitation probability of surface and volume plasmons near crystal surfaces. One of course may wonder about the variation of this quantity near the crystal surface. The calculated value of dP/dz, as a function of electron distance x from the surface, was given previously in Figure 10.5. Inside of the crystal ($x < 0$), dP/dz tends to a constant for large $|x|$, representing the excitation of volume plasmons, but when the electrons travel close to the surface, the dominant term in dP/dz arises from surface-plasmon excitation. As can be seen in Figure 10.5, the value of $dP/dz(= \mathrm{VP} + \mathrm{SP})$ at $x = 0.5$ nm is almost the same as the value inside the bulk (i.e., at large $|x|$). This means that no severe error could be introduced if dP/dz is represented by the excitation probability at $x = 0.5$ nm. This means that we could use the mean free path measured using a thin GaAs foil in TEM to quantify surface excitations. The value of the total inelastic mean free path $\Lambda_b (= 1/(dP/dz)$ at large $|x|$) for bulk scattering processes can be determined from the transmission energy-loss spectrum of a specimen of known thickness. In the case of a single GaAs foil (of thickness about 100 nm near the edge), thickness was determined by the convergent beam electron diffraction technique, from the measurement of the spacing of the interference fringes. The thickness d was found by this technique under the two-beam approximation to be

$$\frac{S_i^2}{n_i^2} + \frac{1}{n_i^2 \xi_g^2} = \frac{1}{d^2}, \tag{10.83}$$

where S_i is the deviation parameter for the (hkl) Bragg spot, ξ_g is the extinction distance for the (hkl) reflection and n_i is an integer, which has a value n_i for the first intensity minimum and increases by one for each successive minimum. A detailed description of this method had been given elsewhere (Kelly *et al.*, 1975). The bulk mean free path is then determined from

$$\Lambda_b = \frac{d_0}{\ln(I/I_0)}, \tag{10.84}$$

where d_0 is the specimen thickness. The result obtained was $\Lambda_b = 80 \pm 10$ nm for EELS entrance aperture $\beta_0 = 1.4$ mrad and $E_0 = 120$ keV.

The $\bar{m}_0$ term was calculated using Eq. (10.22) using the experimentally measured dielectric function for GaAs. For the (660) reflection, $\bar{m}_0 = 0.72$. The mean free path length is taken as 90 nm. Finally, the mean traveling distance of the electron along the crystal surface under (660) resonance reflection is 70–88 nm for 120 kV electrons. Similar experiments have been performed for MgO(100) under the (400) specular reflection condition. The mean traveling distance is 100–109 nm.

The measured mean traveling distance L_s has been applied to determine the absolute atomic concentration at crystal surfaces. An accuracy of better than 20% was obtained for Ga on the GaAs(110) surface and a 45% increase in oxygen concentration was determined at the MgO(110) surface (Wang and Egerton, 1989).

Valence-loss excitation, the most probable inelastic scattering process in electron diffraction, introduces an energy loss in the range 0–50 eV but causes only very small momentum transfers. In this chapter, we have introduced the dielectric response theories, with and without relativistic correction, for simulating the valence-loss spectra observed in TEM and REM. Details have been given to describe the surface and interface excitations under several scattering object geometries. The equivalence between classical dielectric response theory and quantum mechanical scattering theory has been proven. Following the multiple scattering theory, applications of valence-loss spectra for measuring some of the fundamental quantities in RHEED have been given.

CHAPTER 11

Atomic inner shell excitations in RHEED

11.1 Excitation of atomic inner shell electrons

For energy losses larger than about 50 eV, atomic inner shell excitations can be seen in EELS spectra. This is the process in which an atom-bound electron is excited from an inner shell state to a valence state accompanied by incident electron energy losses and momentum transfers. This is a localized inelastic scattering process, which occurs only when incident electrons are propagating in the crystal. The spectrum of the electron energy loss directly displays the excitation information of the atomic electrons based on the law of conversation of energy. Since the inner-shell energy levels are the unique features of the atom, the intensities of the ionization edges can be used effectively to analyze the chemistry of the specimen.

Inelastic scattering involves a transition of crystal state from the ground state $|0\rangle = a_0$ to an excited state $|n\rangle = a_n$ due to energy and momentum transfers. Using the first Born approximation, the differential cross-section for the transition is (Inokuti, 1971)

$$\frac{\mathrm{d}\sigma_{\mathrm{I}}}{\mathrm{d}\Theta} = \left(\frac{\gamma m_0}{\hbar^2}\right)^2 \frac{K_{\mathrm{f}}}{K_0} \left| \langle n| \int \mathrm{d}\boldsymbol{r} \exp(2\pi \mathrm{i}\boldsymbol{q}\cdot\boldsymbol{r}) \frac{1}{4\pi\varepsilon_0} \left(\sum_J \frac{e^2}{|\boldsymbol{r}-\boldsymbol{r}_J|} - \sum_k \frac{Z_k e^2}{|\boldsymbol{r}-\boldsymbol{R}_k|} \right) |0\rangle \right|^2, \tag{11.1}$$

where Θ is the solid angle, $\boldsymbol{K}_{\mathrm{f}}$ and $\boldsymbol{K}_0$ are the wavevectors of the incident electron before and after inelastic excitation, $h\boldsymbol{q} = h(\boldsymbol{K}_0 - \boldsymbol{K}_{\mathrm{f}})$ is the electron momentum transfer to the atom, $\boldsymbol{r}$ is the coordinate of the fast electron, $\boldsymbol{r}_J$ and $\boldsymbol{R}_k$ are the coordinates of crystal electrons and nuclei, respectively. Equation (11.1) can be converted to a form given by (11.2) following the same procedures as for (10.49) and (10.50):

$$\frac{\mathrm{d}\sigma_{\mathrm{I}}}{\mathrm{d}\Theta} = \left(\frac{\gamma m_0}{\hbar^2}\right)^2 \frac{K_{\mathrm{f}}}{K_0} \left(\frac{e^2}{4\pi\varepsilon_0}\right)^2 \frac{|\rho_{n0}(\boldsymbol{k})|^2}{\pi^2 k^4}, \tag{11.2a}$$

with

$$\rho_{n0}(\boldsymbol{\tau}) \equiv \langle n| \sum_J \exp(2\pi \mathrm{i}\boldsymbol{\tau}\cdot\boldsymbol{r}_J)|0\rangle. \tag{11.2b}$$

A generalized oscillator strength is introduced:

$$f_n(q) = \frac{\Delta E}{\mathfrak{R}} \frac{|\rho_{n0}(\boldsymbol{\tau})|^2}{(q a_{\mathrm{B}})^2}, \tag{11.3}$$

where ΔE is the electron energy loss, $\mathfrak{R} = (m_0 e^4/2)\,[1/(4\pi\varepsilon_0 \hbar)^2] = 13.6$ eV is the Rydberg energy and $a_B = 4\pi\varepsilon_0 \hbar^2/(m_0 e^2) = 0.0529$ nm is the Bohr radius. Equation (11.2a) is rewritten as

$$\frac{d\sigma_I}{d\Theta} = \frac{\gamma^2}{\pi^2}\frac{K_f}{K_0}\frac{\mathfrak{R}}{\Delta E}\frac{f_n(q)}{q^2}. \tag{11.4}$$

In many cases the energy-loss spectrum is a continuous rather than discrete function of the energy loss ΔE. It would be convenient to define a differential generalized oscillator strength, $f_n(q) = df(q)/d\Delta E$. The angular and energy-dependences of scattering are then specified by a double-differential cross-section

$$\frac{d^2\sigma_I}{d\Theta\, d\Delta E} = \frac{\gamma^2}{\pi^2}\frac{K_f}{K_0}\frac{\mathfrak{R}}{\Delta E}\frac{1}{q^2}\frac{df(q)}{d\Delta E}. \tag{11.5}$$

To obtain explicitly the angular distribution of inelastic scattering, the modulus of the scattering vector q is related to the scattering angle Θ. For small-angle scattering and high-energy electrons, it is a good approximation to assume that $K_f/K_0 = 1$, and

$$q^2 = K_0^2(\vartheta^2 + \vartheta_E^2), \tag{11.6}$$

where $\vartheta_E = \Delta E/(\gamma m_0 v^2) \approx \Delta E/(2E_0)$ is the characteristic angle of inelastic scattering, which approximately represents the full width at half maximum of the angular distribution. Finally,

$$\frac{d^2\sigma_I}{d\Theta\, d\Delta E} = \frac{\gamma^2}{\pi^2}\frac{\mathfrak{R}}{\Delta E K_0^2}\frac{1}{(\vartheta^2 + \vartheta_E^2)}\frac{df(q)}{d\Delta E}. \tag{11.7}$$

Since $df(q)/d\Delta E$ is almost independent of the electron scattering angle, the angular distribution of the inelastically scattered electron is described by the Lorentzian function, $1/(\vartheta^2 + \vartheta_E^2)$, provided that there is no crystal diffraction.

The cross-section normally used in quantitative EELS microanalysis is the integral form of (11.7),

$$\sigma_I(\beta, \Delta) = \int_{\Delta E}^{\Delta E + \Delta} d\Delta E \int_0^{2\pi} d\phi \int_0^{\beta} d\vartheta \sin\vartheta \frac{d^2\sigma_I}{d\Theta\, d\Delta E}, \tag{11.8}$$

where β is the collection semi-angle of the EELS spectrometer, and Δ is the energy width of the integration window.

11.2 Atomic inner shell excitation in reflection mode

In RHEED, atomic inner shell excitation occurs when the incident electron penetrates into the surface. Since the penetration depth into the bulk of the electrons

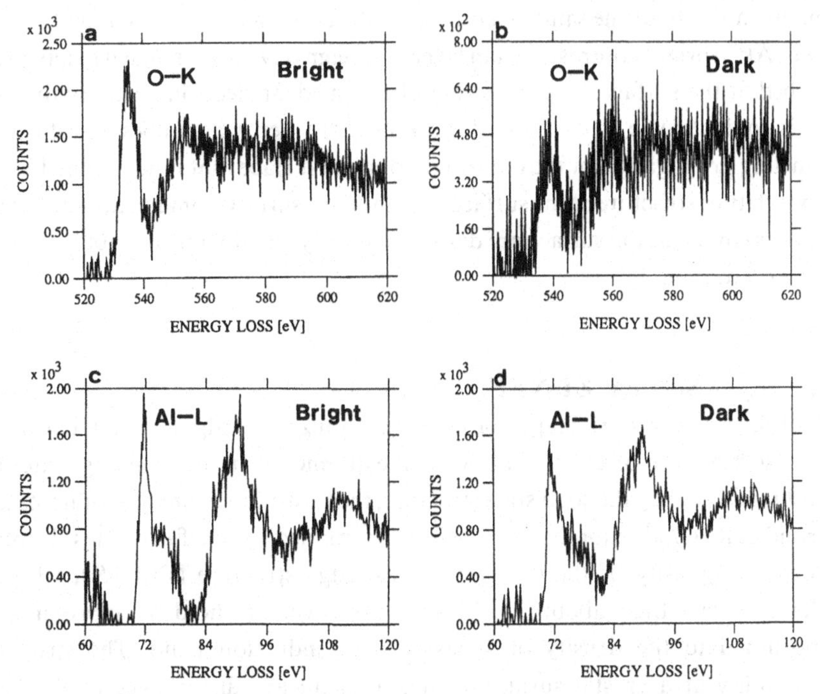

Figure 11.1 *REELS spectra acquired from (a) and (c) the 'bright' and (b) and (d) the 'dark' contrast domains of the α-alumina (0$\bar{1}$11) surface shown in Figure 8.12 during electron beam radiation damage. It is important to note that the scale of the O K edge decreases by a factor of more than three in (b) compared with (a), but the scales for the Al L edge are the same in (c) and (d), indicating that the surface regions of dark contrast are oxygen-rich (Wang and Bentley, 1991c).*

is limited to a few atomic layers, the inelastic signals in REELS are surface-sensitive. The EELS spectra of the reflected electrons can be acquired by selecting the specularly reflected beam using the entrance aperture of the EELS spectrometer. Shown in Figures 11.1(a) and (c), and (b) and (d) are the recorded REELS spectra from the bright and dark contrast domains of the α-alumina (0$\bar{1}$11) surface as shown earlier in Figure 8.12, respectively, under identical illumination and diffraction conditions so that the absolute counts of each spectrum can be compared with each other. Two important points can be noticed. First, the O K edge acquired from the bright domain has many more counts than the one from the dark domain. The spectrum (Figure 11.1(a)) appears with better shape and a stronger signal. It is therefore expected that there is a thin layer of oxygen adsorbed on the bright surface domain. The weaker excitation of the O K edge in the dark domain may indicate that the surface is terminated with Al^{3+} ions after the initial oxygen ions have been removed. This agrees with the discussions described in Section 8.5. Second, for the Al L edge, the integrated counts of the spectra acquired from the two different domains show about the same intensity. For the domain terminated initially with a layer of Al^{3+} (bright contrast), a thin layer of foreign oxygen is assumed to have been adsorbed at the surface. The scattering of the incident electron beam by this thin layer is equivalent to that of a uniform potential layer, which does not affect the channeling propagation of the electrons along the surface that is determined by the periodic lattice structure of the surface. For the dark surface domain, the few oxygen atoms remaining at the surface are desorbed quickly and do not contribute to the O K edge excitation.

11.3 Surface ELNES

In EELS, the near-edge shapes of core-shell ionization edges are related to the electronic structure of the crystal. Since the atomic inner-shell binding states are almost unaffected by the solid state structure, the transition from the atomic state to the solid state conduction band is mainly determined by the final states. In other words, the intensity of the energy-loss near-edge structure (ELNES), which is confined to less than about 20 eV from the edge threshold, is approximately proportional to the density of states in the conduction band. The structural discontinuity at a crystal surface creates a significant disturbance in the band structure. This difference may be obtained by comparing the shape of ELNES acquired in transmission mode from a thin crystal foil with that from a bulk crystal surface in REM (Wang and Cowley, 1988a; Wang *et al.*, 1989c). In REM, the effective penetration depth of a high-energy electron beam is usually limited to no more than a few atomic layers under surface resonance conditions (Wang *et al.*, 1989a). The information obtained in REELS is believed to come largely from the top few atomic layers. Using this method, the presence of an MgO(100) surface state has been observed (Wang and Cowley, 1988b, see Figure 11.4 later).

Structural changes at a crystal surface can also give different ELNES. Results from a $TiO_2(110)$ surface provide such an example. When it is first imaged in REM, the $TiO_2(110)$ surface appears atomically flat in broad regions between occasional surface steps, as shown in the region marked 'without reduction' in Figure 11.2. After being illuminated for about 15 min at 120 keV, faint rough-looking contrast appears as in the region labeled 'with reduction' in Figure 11.2. This beam-induced radiation damage process can also be detected through changes in the ELNES of the Ti L edge. Figures 11.3(a) and (b) show the Ti L edges from the areas without and with beam damage (reduction), respectively. The most important change is in the ratio of the L_2 to L_3 'white line' intensities. The white lines refer to the two peaks observed in the L shell ionization edge caused by the energy splitting due to spin–orbit coupling. This change may directly relate to a change in the surface density of states (DOS) (Wang *et al.*, 1989b).

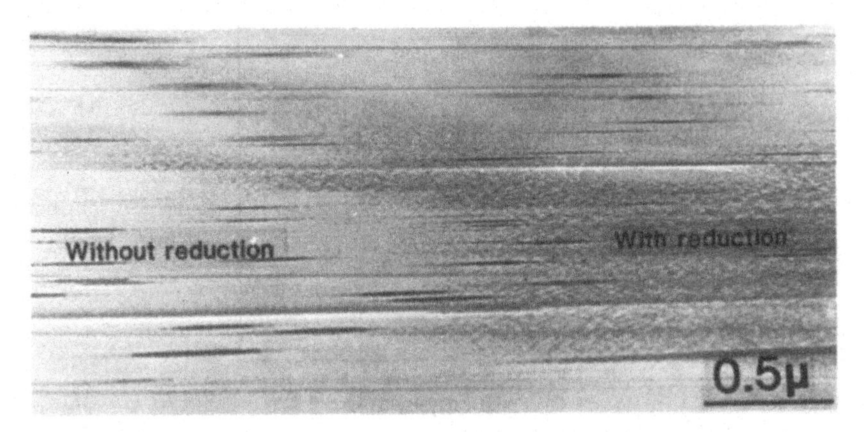

Figure 11.2 A REM image of $TiO_2(110)$ showing surface regions with and without metal reduction.

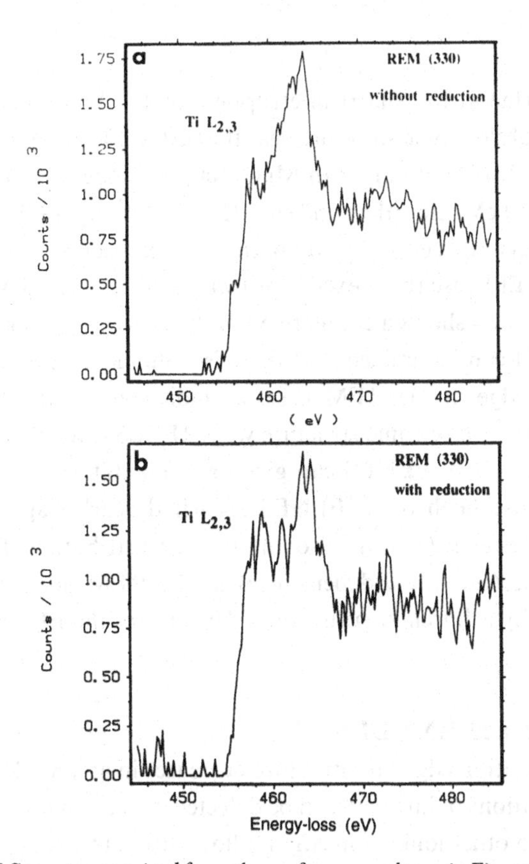

Figure 11.3 EELS spectra acquired from the surface area shown in Figure 11.2, (a) before and (b) after being damaged by the beam (Wang, Liu and Cowley, 1989b). The beam energy is 120 kV.

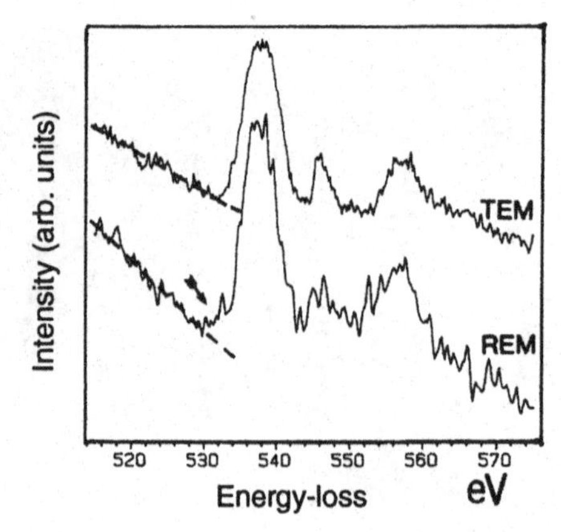

Figure 11.4 *A comparison of EELS spectra recorded in TEM and REM cases using a thin MgO foil and bulk MgO, respectively, showing the excitation of surface state in REELS (Wang and Cowley, 1988b). The spectra were acquired at 120 keV using a serial-detection EELS spectrometer.*

For crystal surfaces, the sharp interruption of the bulk crystal significantly perturbs the crystal electronic structure near the surface. Theoretical calculation for the electronic band structure of clean MgO(100) has suggested the existence of a surface state about 1 eV below the band edge (Lee and Wong, 1978). An attempt has been made to observe this effect by comparing EELS spectra recorded from a thin MgO foil in the TEM case (bulk excitation) with that of cleaved MgO(100) in the REM case. Figure 11.4 shows a comparison of the O K edge observed in TEM and REM geometry. It can be noticed that there is a shoulder peak located 1–1.5 eV before the O K edge in the REM case, as indicated by an arrowhead. Thus observation of surface electronic structure with REELS is feasible.

In practice, analysis of ELNES can give only qualitative or semi-quantitative information, because the theory of ELNES is still in the early stages of development (see for example Rez *et al.* (1991)). Also, the angular distribution of the inelastically scattered electrons (i.e., the $\boldsymbol{q}$-dependence) in the REM geometry may not be described simply by a Lorentzian function. This makes the ELNES analysis even more difficult.

11.4 Surface EXELFS

Although the ionization edge fine structure decreases in amplitude with increasing energy loss, oscillations of intensity can be detected over a range of several hundred electron-volts if no other ionization edges follow within this region. This extended energy-loss fine structure (EXELFS) is referred to as the scattering of the ejected core electron due to its interaction with nearest-neighbor atoms in the solid. If the energy of the ejected core electron is more than 50 eV, it behaves like a 'free' electron

Figure 11.5 *A schematic diagram showing the wave interference that gives rise to EXELFS. The core electron ejected from the center atom is approximately a spherical wave* $\exp(2\pi iqr)/r$*, which will be backscattered by the neighbor atoms.*

and can be backscattered by the neighboring atoms. Since the wavevector of the ejected electron $\boldsymbol{q}$ depends on the energy loss of the fast electron, the interference of the backscattered wave and the outgoing wave is alternately constructive and destructive, giving maxima and minima in the energy-loss spectrum (Stern, 1974; Sayers *et al.*, 1971), as shown in Figure 11.5. The periodicity and amplitude of the oscillation can be applied to determine the radial distribution function of crystal atoms (for a review see Egerton, (1986)). This has been shown to be a powerful technique for studying disordered and amorphous materials using EXELFS in X-ray absorption spectroscopy.

The atomic radial distribution functions around surface atoms are clearly grossly different from those of atoms in the bulk of the material. The rearrangement of surface atoms will result in major changes in the scattering of the ejected electrons during the ionization event, resulting in different interference features in EXELFS. These differences can be seen, for example, by comparing the EXELFS acquired from a thin MgO slab (TEM) with that from the (100) surface of a bulk MgO crystal (REM). Since most of the electrons have been multiply scattered by plasmons, it is necessary to retrieve the single-loss function using the Fourier–log deconvolution method (see Section 10.9.1). The REELS single-loss oxygen K edge from this surface after deconvolution data processing and background subtraction is shown in Figure 11.6(a) (the solid line). The REELS spectrum was recorded using the (400) specularly reflected beam; in TEM, the spectrum was also recorded using the (400) diffracted beam. Thus, the mean momentum transfer and scattering angles of the electrons are preserved. This may be important because differences in momentum transfer could produce distinct features in ELNES (Wang and Rez, 1987). Differences in the EXELFS compared with a spectrum from a thin MgO slab in the equivalent scattering geometry (the dashed line) are apparent. In the pre-edge part (530–535 eV), the edge in the REM case is shifted towards lower energies by about

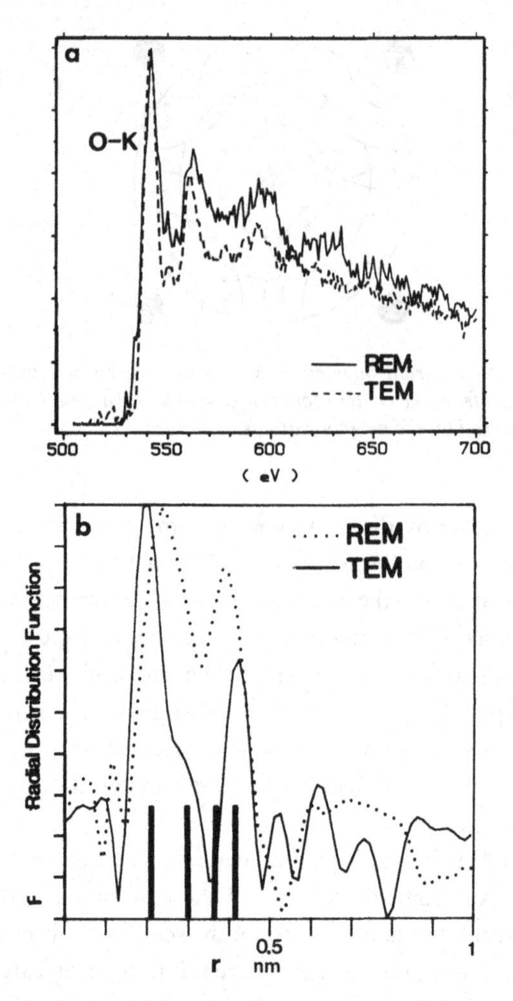

Figure 11.6 Surface EXELFS in REELS. (a) A comparison of the O K edges acquired from a thin MgO slab (TEM) and a MgO (100) bulk surface (REM) after removing the multiple scattering effect. (b) The atomic radial distribution function derived based on (a). The solid black bars on the horizontal axis represent the positions in order of increasing distance of atomic shells of Mg, O, Mg and O in the bulk. The spectra were acquired at 120 keV using a serial-detection EELS spectrometer.

1.5 eV. The fine structure near the edge has been changed and the relative heights of the edge peaks have been modified.

The data shown in Figure 11.6(a) were used for EXELFS analysis to determine the change in the atomic radial distribution function at the surface. The solid and dotted curves in Figure 11.6(b) represent the atomic radial distribution function around oxygen atoms inside the MgO bulk and at the MgO(100) surface, respectively. The solid black bars on the horizontal axis represent the positions in order of increasing distance of atomic shells of Mg, O, Mg and O in the bulk. In the TEM

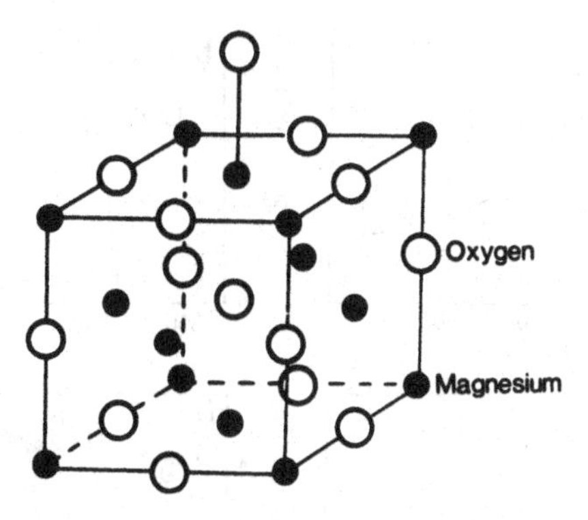

Figure 11.7 *A bonding model of the adsorbed oxygen on MgO(100).*

case, the O K edge data show the correct positions for the first, second and fourth shells but do not show the third shell. On comparing the two main peaks, represented by the dotted curve, for the surface case (the dotted curve) with that for the bulk case (the solid curve), the peaks originally located at about 0.21 and 0.42 nm are shifted to 0.25 and 0.39 nm, respectively. Also, the relative intensities of these two peaks have been changed in the REM case.

The changes in peak positions and relative heights in the data can be interpreted based on a surface adsorption model in which an oxygen atom is assumed to be located above a Mg atom, as shown in Figure 11.7 (Wang and Cowley, 1988b). This model is based on considerations of the conservation of the MgO structure and of the balance of the bonding forces. For this adsorbed oxygen atom, its first shell is one Mg atom, the second shell is four oxygen atoms, the third shell is four Mg atoms and the fourth shell is one oxygen atom. The relative increases in coordination of its second and third shells of atoms shifts the first peak to the right and the second peak to the left with an accompanying intensity increase. The effect of these additions from second and third nearest shells of the surface adsorbed O atom, at 0.30 and 0.38 nm respectively, is that the peak at 0.21 nm is broadened and shifted outwards and the peak at 0.42 nm is also broadened and shifted inwards. The resultant peak distributions should thus be modified to approach the form of the dotted line shape in Figure 11.6(b) if it is assumed that the surface atoms have a dominant effect in the REELS studied.

11.5 Surface chemical microanalysis

Compositional microanalysis is one of the most important applications of EELS. Figure 11.8 shows REELS spectra of O K and Mg K edges acquired from a MgO(100) surface before and after subtraction of background, with spectrometer collection semi-angle $\beta = 1.2$ mrad. Some near-edge structure is evident, but features

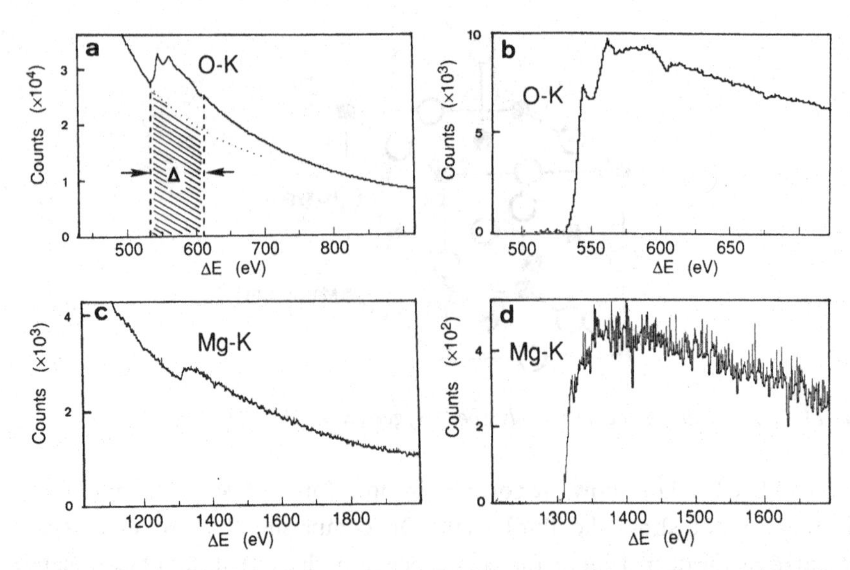

Figure 11.8 *REELS spectra of O K and Mg K edges acquired from a MgO(100) surface (a) and (c) before and (b) and (d) after the subtraction of background. The (600) specular reflection spot was selected and $\beta = 1.2$ mrad. The beam azimuth $B \approx [001]$ and the beam energy is 300 keV.*

in the extended energy-loss region are smeared out by multiple valence excitations. However, to a first-order approximation, surface compositional analysis is not affected. There are, in fact, many differences between the spectra acquired from a thin crystal in TEM and those from a bulk crystal surface in REM. Some details have to be examined before one applies the conventional theory developed for EELS in TEM to quantify REELS data.

In transmission EELS (TEELS), the excitation intensity of a core shell ionization edge is determined by (see for example Egerton (1986))

$$I_A(\beta, \Delta) = I(\beta, \Delta)\sigma_I(A, \beta, \Delta)n_A d_0, \tag{11.9}$$

where $I_A(\beta, \Delta)$ and $I(\beta, \Delta)$ are the integrated intensities for an energy window Δ of the core shell ionization edge and the low-energy-loss part, respectively; $\sigma_I(A, \beta, \Delta)$ is the integrated partial ionization cross-section of the element A; n_A is the atomic concentration of the element A; and d_0 is the specimen thickness. The composition ratio of two elements is given by

$$\frac{n_A}{n_B} = \frac{I_A(\beta, \Delta)}{I_B(\beta, \Delta)} \frac{\sigma_I(B, \beta, \Delta)}{\sigma_I(A, \beta, \Delta)}. \tag{11.10}$$

To apply Eq. (11.9) or Eq. (11.10) for surface microanalysis with the use of the atomic inner shell ionization edges, several points need to be addressed.

A TEELS microanalysis is usually performed with the collection aperture centered around the (000) spot. Equation (11.9) was derived on the assumption that contributions to the intensity of the inelastically scattered electrons from other Bragg spots are very small, and that dynamical scattering effects can be neglected. These assumptions can be reasonably satisfied for a thin sample with no strongly excited low-index reflections. However, in REELS analysis, the spectra are acquired from the diffracted spots. Intensities of neighboring spots are comparable to that of the specular reflection spot from which the spectra are taken.
B In TEELS, d_0 in Eq. (11.9) is approximately equal to the specimen thickness. In the REELS case, there is no simple way to define the electron propagation distance inside a crystal. A quantity L_s, which represents the mean distance that electrons travel along the surface, is introduced to replace the specimen thickness d_0 (Wang *et al.*, 1989a). The value of L_s can be obtained from valence excitation spectra together with numerical calculations (Wang and Egerton, 1989), see Section 10.9.3. In practice, L_s depends strongly on the diffraction (or channeling) conditions.
C The partial ionization cross-sections σ_1 for light elements are usually calculated by the SIGMAK and SIGMAL programs (Egerton, 1986), which are based on the assumption that the inelastically scattered electrons are distributed around the (000) spot in the form of a Lorentzian function and the atomic electron wave function is approximated by the hydrogen-like model. This assumption may be approximately valid in RHEED for some cases if β is small. However, in general, it is believed that the angular distribution around a reflected spot in RHEED is not spherically symmetric; it is thus amended to measure the ionization cross-section experimentally (Wang and Bentley, 1991b).
D Channeling effects should normally be avoided in EELS microanalysis and this is easily achieved in transmission EELS simply by tilting the specimen to avoid channeling conditions. Reasonable signal-to-background ratios can still be obtained. However, in REM the situation is totally different, because the channeling effect or surface resonance effect must be employed to improve REM image contrast and to produce visible REELS core-shell ionization signals (Wang, 1989d). Thus the channeling effect is unavoidably involved in REELS microanalysis and may yield apparently incorrect compositions if the different types of atoms are aligned in separate rows or planes when viewed along the direction of the beam. This situation happens, for example, for GaAs(110) with beam azimuth [001] (Wang *et al.*, 1987), α-alumina $(0\bar{1}11)$ (Wang and Bentley, 1991a) and MgO(100) with beam azimuth [001] (Wang and Bentley, 1991b).
E Dynamic calculations of RHEED are helpful in understanding the electron scattering processes at crystal surfaces. It has been found that most of the electrons will be trapped by the top few atomic layers near the surface for a distance of a few tens of nanometers before being reflected back into vacuum; the probability of direct

reflection without any penetration is small (Wang *et al.*, 1989a). It is believed that most of the information obtained in REELS comes from the outermost few layers near the surface; hence the surface sensitivity of REELS.

F Commonly in REELS, relatively poor signal-to-background ratios are observed compared with those in TEELS. This is partly because of delocalized excitation of surface plasmons due to long-range Coulomb interactions of electrons with surface induced charges when the electron approaches and leaves the surface. The low signal-to-background ratio can significantly decrease the visibility of ionization or core edges located above 1 keV, such as Mg K. Often only one clear ionization edge can be seen by REELS, such as O K from MgO(100). It is then desirable to determine the absolute atomic concentration near the surface with a single edge. The key step for this analysis is to determine the mean distance that the electrons travel inside the crystal (see Section 10.9.3).

Therefore, we must examine all the points illustrated above to quantify REELS data. Under strong diffraction conditions, the intensity of an atomic core-shell edge acquired in diffraction mode (for RHEED) can be written as

$$I_A(\beta, \Delta) = \int_0^{L_s} dz \int_\Sigma dx\, dy\, i(\boldsymbol{r}) n_A(\boldsymbol{r}) \sigma_I(A, \beta, \Delta), \qquad (11.11)$$

where $i(\boldsymbol{r})$ is the local channeling current density and Σ indicates a surface integration over the beam illumination area S. In the approximation of perfectly localized excitation, one has

$$n_A(\boldsymbol{r}) = \sum_{j_A} \delta(\boldsymbol{r} - \boldsymbol{R}_{j_A}), \qquad (11.12)$$

where $\boldsymbol{R}_{j_A}$ is the position of the jth A atom in the crystal. We define an effective ionization cross-section (σ_{eff}) as

$$\sigma_{eff}(A, \beta, \Delta) \equiv \sigma_I(A, \beta, \Delta) \frac{\sum_{j_A} i(\boldsymbol{R}_{j_A})}{i_0 N_A} K_A \equiv \sigma_I(A, \beta, \Delta) \frac{\overline{i_A}}{i_0} K_A, \qquad (11.13)$$

where i_0 is the average current density; N_A is the average atomic concentration of atom A; $\overline{i_A}$ is the average channeling current density at the A atomic sites; and K_A is introduced to take into account the deviation of the final inelastic electron angular distribution, $f(\vartheta)$, from the Lorentzian function, $L(\vartheta)$, due to dynamical diffraction effects. By assuming that the generalized oscillator strength is almost independent of the scattering angle ϑ,

$$K_A = \frac{\int_0^\beta d\vartheta \sin\vartheta f(\vartheta, \vartheta_E(A))}{\int_0^\beta d\vartheta \sin\vartheta L(\vartheta, \vartheta_E(A))}, \tag{11.14}$$

where L is the angular distribution of the electrons after being inelastically scattered by a single atom,

$$L(\vartheta) = \frac{1}{\vartheta^2 + \vartheta_E^2}. \tag{11.15}$$

The compositional ratio for two elements is thus

$$\frac{n_A}{n_B} = \frac{I_A(\beta, \Delta)}{I_B(\beta, \Delta)} \frac{\sigma_{eff}(B, \beta, \Delta)}{\sigma_{eff}(A, \beta, \Delta)} = \frac{I_A(\beta, \Delta)}{I_B(\beta, \Delta)} \frac{\sigma_I(B, \beta, \Delta)}{\sigma_I(A, \beta, \Delta)} \frac{\overline{i_B} K_B}{\overline{i_A} K_A}. \tag{11.16}$$

It must be pointed out that the newly defined σ_{eff} is determined not only by the properties of each single atom (σ_I) but also by both the dynamical elastic and the inelastic electron scattering (K_A) and the detailed channeling processes ($\overline{i_A}$) of the electrons. In REELS, for a general case, $\overline{i_A} \neq \overline{i_B}$ and $K_A \neq K_B$, because the elastic Bragg reflections and core-shell ionizations may not be independent events. This complicates the analysis of REELS data. In other words, calculated ionization cross-sections based on properties of an isolated atom are insufficient for REELS microanalysis; dynamical diffraction effects (K_A) and channeling discrimination effects ($\overline{i_A}$) generally have to be included. At this time it is not practical to calculate the latter two effects quantitatively, but it may be possible to measure their combined effects experimentally (see Section 11.8).

The relative concentration of two elements can be determined from the inner shell spectra if the effective ionization cross-section ratio can be determined experimentally:

$$\frac{\sigma_{eff}(B, \beta, \Delta)}{\sigma_{eff}(A, \beta, \Delta)} \equiv \frac{\sigma_I(B, \beta, \Delta)}{\sigma_I(A, \beta, \Delta)} \frac{\overline{i_B}}{\overline{i_A}} \frac{K_B}{K_A}. \tag{11.17}$$

Equation (11.17) was derived by assuming that the intensities of other diffracted spots are much smaller than that of the specularly reflected spot. If this is not the case, a small correction has to be made (see Section 11.6).

Before one presents the experimental results, it is necessary to discuss the meaning of surface chemical compositions if there are some adsorbates. If one assumes there is a monolayer of oxygen being adsorbed on the MgO(100) surface, the ratio n_O/n_{Mg}

would depend on the conditions under which the measurements were made, because the ratio of the inelastic signals provided by the incident electrons relies on the electron penetration depth into the surface. If the electrons penetrate up to only the first mixing O–Mg(100) layer, for example, then the measured surface composition ratio would be $n_O/n_{Mg}=2$. If the electrons penetrate into the surface up to two mixing O–Mg layers, then the surface composition determined would be $n_O/n_{Mg}=1.5$. Therefore, the surface composition may be a function of the electron penetration depth if there are some surface adsorbates. This is different from the composition measured in transmission EELS, which is a precise number. It is very important to remember this point when using REELS results.

11.6 The effect of strong Bragg beams

In RHEED, EELS spectra are usually acquired under strong resonance diffraction conditions, so that one or more Bragg beams are excited besides the specularly reflected beam that is to be used for acquiring the spectra. The intensities of other diffracted beams may be comparable to that of the specular beam. Therefore, the contributions of inelastic scattering from neighboring Bragg spots are significant and need to be taken into account. If the angular distribution of the inelastically scattered electrons around each Bragg beam is described by the Lorentzian function $1/(\vartheta^2+\vartheta_E^2)$, for an EELS entrance aperture with collection semi-angle β, then Eq. (11.9) should be generalized (Egerton, 1989) to

$$I_A(\beta,\Delta)=[I_0(\beta,\Delta)+\sum_{\boldsymbol{h}} I_{\boldsymbol{h}}(\beta,\Delta)\omega(\phi_{\boldsymbol{h}},\beta,\vartheta_E)]\sigma_I(A,\beta,\Delta)n_A d_0. \tag{11.18}$$

In this equation $\omega(\phi_{\boldsymbol{h}},\beta,\vartheta_E)$ is the ratio between the intensities collected by the EELS entrance aperture when it is placed at angle distances $\phi=\phi_{\boldsymbol{h}}$ and $\phi=0$ from the specular spot from which the spectra are taken,

$$\omega(\phi,\beta,\vartheta_E)=\frac{\int_0^{\beta}\mathrm{d}\vartheta\int_0^{2\pi}\mathrm{d}\alpha\,\dfrac{\vartheta}{(\vartheta^2+\phi^2-2\vartheta\phi\cos\alpha)+\vartheta_E^2}}{\int_0^{\beta}\mathrm{d}\vartheta\int_0^{2\pi}\mathrm{d}\alpha\,\dfrac{\vartheta}{\vartheta^2+\vartheta_E^2}}, \tag{11.19}$$

where $\phi(=\phi_{\boldsymbol{h}})$ is used to specify the angle between the Bragg spot $\boldsymbol{h}$ and the spot from which the spectra are taken, $I_{\boldsymbol{h}}$ is the integrated intensity for an energy window Δ of the EELS spectrum acquired from the $\boldsymbol{h}$ Bragg spot under equivalent conditions, α is the angle between angle distances (in reciprocal space) ϑ and ϕ, and the angular distribution of the inelastically scattered electrons is approximated by the Lorentzian function. Integrating α first, after a variable substitution of $y=\theta^2$, (11.19) yields

$$\omega(\phi, \beta, \vartheta_E) = \frac{\ln\{\{[(\beta^2+\phi^2+\vartheta_E^2)^2 - 4\phi^2\beta^2]^{1/2} + \beta^2 - \phi^2 + \vartheta_E^2\}/2\vartheta_E^2\}}{\ln(1+\beta^2/\vartheta_E^2)}. \tag{11.20}$$

It has been shown that an accuracy of better than 10% can be obtained for compositional microanalysis of BN with spectra acquired from the (200) diffracted spot using Eq. (11.18) (Wang, 1989d). The ω correction is significant in REELS microanalysis, especially when the edge threshold is located at higher energies.

As pointed out above, REELS surface microanalysis is mainly limited by the accuracy of the effective ionization cross-sections that are related to the angular distribution function, $f(\vartheta)$, of the inelastically scattered electrons inside the reflected spot and the dynamical scattering process of electrons from the surface. The $f(\vartheta)$ function is the result of dynamical multiple elastic and inelastic electron scattering under the specified diffraction conditions and cannot be described by a Lorentzian function, as in the transmission case under the kinematical approximation. To illustrate this point, one assumes that the electrons are channeling within the top two atomic layers under the Bragg-resonance reflection condition, as shown earlier in Figure 4.8; the transmitted and reflected coefficients of the wave through and from each layer are assumed to be constants R_t and R_r. The angular distribution of each localized inelastic scattering event can be approximately characterized by

$$L(\vartheta) = \frac{1}{\vartheta_x^2 + (\vartheta_z - \theta_g)^2 + \vartheta_E^2}, \tag{11.21}$$

where ϑ_x and ϑ_z are the scattering angles parallel and normal to the surface, respectively, and θ_g is the Bragg angle at which the electrons are incident. Thus the inelastic events occurring in the first layer when the electrons are being Bragg-reflected towards the vacuum will give an angular distribution $L(\vartheta)$ around the specularly reflected spot. Electrons inelastically scattered in the second layer will be (eventually) reflected to vacuum only if they satisfy the Bragg-reflection conditions. These electrons will be distributed in a narrow band with approximately the same width as the Kikuchi lines ($\Delta\vartheta_K$) in a form of $L(\vartheta)$. Electrons penetrating through the second layer into the crystal will be considered as being absorbed. By considering the multiple elastic Bragg scattering and single-incoherent inelastic scattering of the electrons within these two layers and assuming that the inelastic scattering is localized, the final angular distribution of the inelastically scattered electrons within the reflected spot, if $\vartheta_E \gg \Delta\vartheta_K$, may be approximately written as

$$f(\vartheta) = \frac{R_t^2 R_r}{1 - R_r^2}\left(\frac{1}{\vartheta_x^2 + (\vartheta_z - \theta_g)^2 + \vartheta_E^2} + \frac{\Delta\vartheta_K\,\delta(\vartheta_z - \theta_g)}{\vartheta_x^2 + (\vartheta_z - \theta_g)^2 + \vartheta_E^2}\right). \tag{11.22}$$

Integrating $f(\vartheta)$ for a collection aperture of radius β centered on $\vartheta_z = \theta_g$ and $\vartheta_x = 0$, one obtains

$$F(\beta) = \frac{R_t^2 R_r}{1 - R_r^2} \ln\left(1 + \frac{\beta^2}{\vartheta_E^2}\right) + \frac{R_t^2 R_r}{1 - R_r^2} \frac{2\Delta\vartheta_K}{\vartheta_E} \arctan\left(\frac{\beta}{\vartheta_E}\right), \tag{11.23}$$

where the first term gives the isolated-atom Lorentzian distribution and the second term gives the deviation caused by diffraction effects. Thus the K_A factor defined in Eq. (11.14) is

$$K_A = 1 + \frac{2\Delta\vartheta_K \arctan(\beta/\vartheta_E)}{\vartheta_E \ln(1 + \beta^2/\vartheta_E^2)}. \tag{11.24}$$

It is impossible, for a general case, to give an analytical expression for $f(\vartheta)$, because $f(\vartheta)$ is determined partially by dynamical scattering effects and partially by inelastic multiple plasmon excitations and thermal diffuse scattering. At the present time, it is thus necessary to measure σ_{eff} experimentally.

11.7 Resonance and channeling effects

In TEM, the scattering of an incident electron beam can produce strong channeling effects along the atomic columns or planes if certain diffraction conditions are satisfied. Differences in potential can generate different channeling currents at different atomic sites, producing different inelastic scattering signals, which are not proportional to the atomic concentration in the solid. This is the channeling effect that is used in identifying the sites of impurity atoms by atomic location by the channeling-enhanced microanalysis (ALCHEMI) technique (Spence, 1993). Channeling effects should normally be avoided in EELS microanalysis and this is easily achieved in transmission EELS simply by tilting the specimen to avoid channeling conditions. Reasonable signal-to-background ratios can still be obtained. In RHEED, however, the situation is different because the channeling effect or surface resonance effect must be employed to improve the REM image contrast and to produce visible REELS core-shell ionization signals. Thus the channeling effect is unavoidably involved in REELS microanalysis and may yield apparently incorrect compositions if the different types of atoms are aligned in separate rows or planes when viewed along the direction of the beam.

REELS spectra are sensitive to the channeling states of electrons at crystal surfaces (Liu and Cowley, 1989) and the analysis of multiple inelastic electron scattering behavior can serve as a sufficient condition for the confirmation of surface resonance (Wang *et al.*, 1989c). Figure 11.9 shows a comparison of the REELS spectra acquired from a synthetic diamond (111) surface in on- and off-resonance conditions, achieved by a slight deflection of less than 2 mrad of the incident beam.

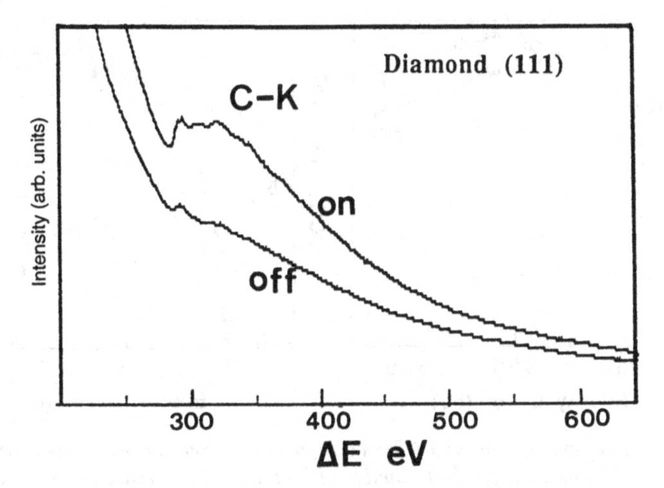

Figure 11.9 *A comparison of the C K edges acquired from a synthetic diamond (111) surface under the on- and off-resonance conditions. The diamond surface is in the as-grown condition. The spectra were acquired at 300 kV using a parallel-detection EELS system. The electron beam was deflected by an angle less than 2 mrad in the direction parallel to the surface in order to achieve the on- and off-resonance conditions.*

The whole diffraction pattern is not significantly affected but the signal-to-background ratio can be changed by a factor of up to three. Also, it is only under the resonance conditions that optimized REM image contrast can be obtained. The increase in inner shell ionization intensity under the resonance condition is due to the increased traveling distance of the electron inside the crystal.

Channeling effects are important if different elements are arranged in separate rows or planes along the beam direction. It is possible to incorrectly determine surface compositions if the channeling currents along different rows or planes of atoms are different, resulting in differences in inelastic signals. This occurs for a MgO(100) surface if the incident beam azimuth is approximately parallel to [011]. Figure 11.10 shows a comparison of the O K and Mg K edges acquired from the MgO(100) surface when the beam azimuth is nearly parallel to [001] or [011]. Since the oxygen and magnesium atoms are both present alternately in each row when the beam azimuth is close to [001], surface channeling effects should not affect the results of surface microanalysis. However, by rotating the same surface 45° around the surface normal so that the beam azimuth is [011], the intensity ratio of O K to Mg K is decreased by a factor of about two compared with that for beam azimuth [001]. This is a consequence of the stronger channeling current along the Mg rows parallel to [011].

In RHEED, the channeling effect is usually unavoidable in order to increase the signal intensity. Caution has to be exercised if one intends to draw some quantitative

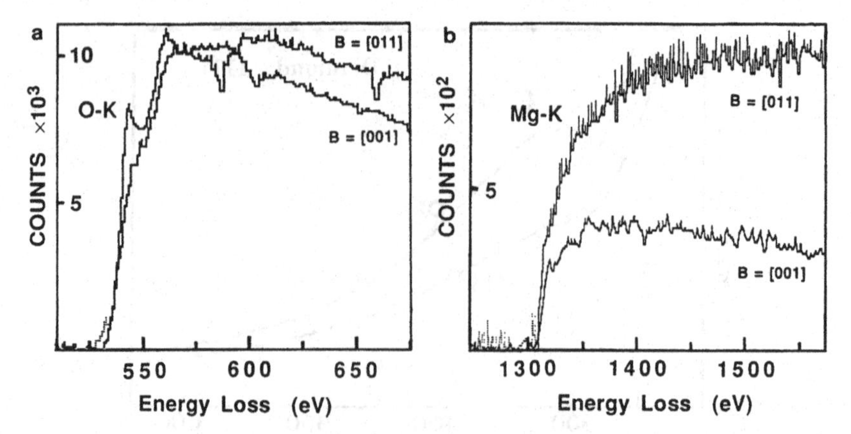

Figure 11.10 A comparison of REELS spectra acquired from the same MgO(100) surface when the beam azimuths are along [001] and [011], showing the increase in Mg K edge intensity due to the stronger channeling current along the Mg atomic rows along [011]. The spectrum background has been subtracted (Wang and Bentley, 1993b).

conclusion. It is expected that the channeling effects in quantifying REELS data can be eliminated by using the σ_{eff} ratios measured in the TEELS experiments by the method to be described in the next section. However, the channeling effect has been used to estimate the thickness of a surface adsorbed oxygen layer on α-alumina $(01\bar{1}1)$ (Wang and Bentley, 1991a) (see Section 11.10).

11.8 Effective ionization cross-sections

We have illustrated in Section 11.5 that the effective ionization cross-section to be used for surface microanalysis has to be measured experimentally. It is necessary to seek an appropriate method that can be applied to measure σ_{eff} accurately. As shown by several authors (Lehmpfuhl and Dowell, 1986; Yao and Cowley, 1989), the characteristic features, such as Kikuchi line distribution and resonance parabolas, in a RHEED pattern can also be observed in a transmission high-energy electron diffraction (THEED) pattern of the same material under identical diffraction conditions. Thus, it should be possible to define the angular distribution of the inelastically scattered electrons within a reflected beam from measurements made in the transmission geometry on a thin foil of the same crystal with the incidence beam azimuth tilted to conditions equivalent to those in the RHEED case. For a specimen with thickness equal to the mean distance that electrons travel along the surface in the RHEED case, the effects of dynamical scattering and channeling (or resonance) in the RHEED geometry should be equivalently generated in the THEED geometry and thus be automatically included in TEELS measurements of σ_{eff} ratios for the equivalent diffraction beam. Experiments of this type were performed on a

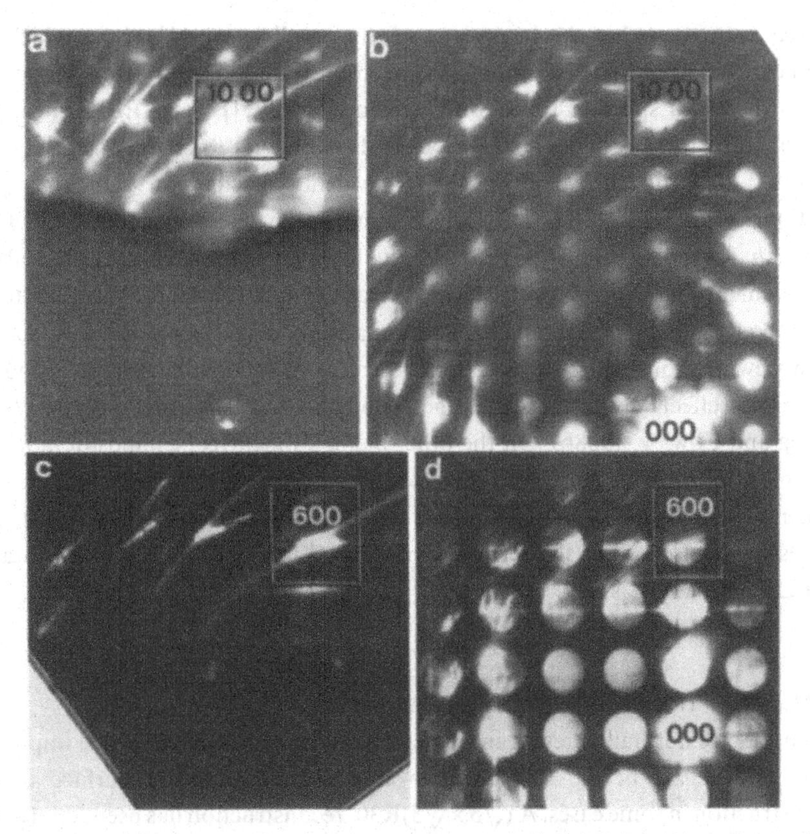

Figure 11.11 *A comparison of (a) the RHEED pattern of the MgO(100) surface with the beam azimuth close to [011] under the (10 0 0) specular reflection condition and (b) THEED pattern from a thin MgO foil. In (c) and (d) are shown RHEED and THEED patterns of MgO(100) with the beam azimuth close to [001] under the (600) specular reflection condition.*

MgO(100) surface, as shown in Figures 11.10(a) and (b). The electron distribution around the (10 0 0) beam is qualitatively similar in both RHEED and THEED.

For the MgO(100) surface and under the (600) specular reflection case along $B=[001]$ (Figure 11.11(c)), the REELS measured intensity ratio for core losses is $I(\mathrm{O})/I(\mathrm{Mg})=18.0\pm0.4$ for semi-collection angle $\beta=1.2$ mrad, $E=300$ kV and $\Delta=100$ eV. For $B=[001]$, the channeling effect would not discriminate between the O and Mg atoms belonging to the bulk but would discriminate the O adsorbates on the surface. The σ_{eff} ratio measured in the equivalent transmission geometry (Figure 11.11(d)) gives $\sigma_{\mathrm{eff}}(\mathrm{O})/\sigma_{\mathrm{eff}}(\mathrm{Mg})=9.0\pm0.6$, provided that the bulk MgO composition is $n_{\mathrm{O}}/n_{\mathrm{Mg}}=1.0$. It is very important to point out that the measured value of $\sigma_{\mathrm{eff}}(\mathrm{O})/\sigma_{\mathrm{eff}}(\mathrm{Mg})$ has little dependence on specimen thickness. The surface composition determined by Eq. (11.16) is $n_{\mathrm{O}}/n_{\mathrm{Mg}}=2.0\pm0.1$. Excess oxygen atoms are adsorbed on the surface.

To confirm the above result, the specimen was then rotated 45° toward the surface normal so that the beam was nearly parallel to [011], along which the channeling effect would discriminate between the excitations of O and Mg atoms. Under the (400) specular reflection case along $B=[011]$, the REELS measured intensity ratio for core losses is $I(\mathrm{O})/I(\mathrm{Mg})=17.4\pm0.4$ for $\beta=1.2$ mrad and $\Delta=100$ eV. The σ_{eff} ratio measured in the equivalent transmission geometry gives $\sigma_{\mathrm{eff}}(\mathrm{O})/\sigma_{\mathrm{eff}}(\mathrm{Mg})=12.2\pm0.6$. The surface composition determined by Eq. (11.16) is thus $n_{\mathrm{O}}/n_{\mathrm{Mg}}=1.43\pm0.1$. This result is qualitatively consistent with the above measurements but with relatively less oxygen adsorption. This is probably related to the ordered or disordered arrangement of the adsorbed oxygen atoms on the surface when viewing along different directions, because an ordered monolayer of adsorbates would effectively decrease the penetration into the surface of the electron beam (Wang and Cowley, 1988b). On the other hand, an incorrect composition, $n_{\mathrm{O}}/n_{\mathrm{Mg}}\approx0.76$, would be obtained if the σ_{eff} ratio, $\sigma_{\mathrm{eff}}(\mathrm{O})/\sigma_{\mathrm{eff}}(\mathrm{Mg})=22.9\pm1.1$, were measured from the (000) transmitted beam under weak diffraction conditions. This proves the usefulness of the term σ_{eff} introduced in Eq. (11.17) for quantitative REELS microanalysis.

11.9 Impurity segregation at surfaces

Impurities are normally present in single-crystalline oxides. Some of the impurities tend to segregate at the surface after annealing, resulting in surface atomic reconstruction in some cases. A $(\sqrt{3}\times\sqrt{3})\mathrm{R}30^\circ$ reconstruction has been reported on the NiO(111) surface and is attributed to the formation of a Ni–O–Si compound, where Si has diffused from the bulk (Floquet and Dufour, 1983). Experimental data have indicated segregation of calcium, chromium, aluminum, iron, titanium, silicon and lanthanum at MgO surfaces (Kingery *et al.*, 1979; Berthelet *et al.*, 1976). Of these impurities, Ca is most often observed. Calcium has been found to diffuse to the surface when the sample is heated in the range 900–1400 K. REELS can be applied to define the distribution of Ca impurities on the surface. The MgO(100) surface annealed in air exhibits atomic flatness with steps. REELS spectra are acquired using a small electron probe, which is made to converge on and off the step, and the results are shown in Figure 11.12. The O K edge signal from the flat terrace is about 70% stronger than that from the step region. This observation, however, cannot be simply interpreted as a reduction in the oxygen concentration in the step region. The drop in O K signal may be caused by the drop in intensity of the specularly reflected beam at the step, which was used to acquire the spectra. The Ca L edge signal obtained from the same two regions, however, shows a different behavior. The Ca L edge signal increases in spite of the net drop in reflectivity. The Ca : O ratio is about 2.6 at the step compared with 1.2 in the terrace region, indicating that the step is Ca-rich.

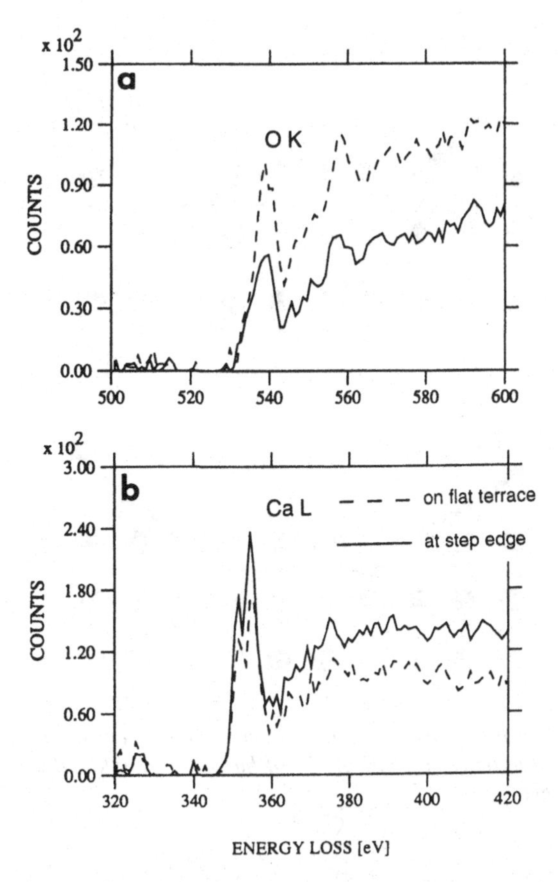

Figure 11.12 *REELS spectra acquired from surface step (solid line) and terrace (dashed line) regions of an annealed MgO(100) surface. The spectral background has been subtracted (courtesy of Drs P. A. Crozier and M. Gajdardziska-Josifovska, 1993).*

The tendency for Ca impurities to diffuse to the MgO surface may be explained by the theoretical calculations of Colbourn *et al.* (1983). They found the substitution energy of a Ca^{2+} ion at the planar MgO(100) surface to be 1.10 eV lower than that in the bulk. It is expected that the Ca^{2+} ions tend to segregate at the surface if the temperature is high enough to permit diffusion to the surface.

11.10 Oxygen adsorption on surfaces

Although the channeling effect is not beneficial to surface microanalysis, it has been used to estimate the thickness of a surface-adsorbed oxygen layer on α-alumina $(01\bar{1}1)$ (Wang and Bentley, 1991c). In this section, we will quantify the spectra shown in Figure 11.1 in order to determine the thickness of the adsorbed oxygen layer on the α-alumina $(01\bar{1}1)$ surface. Since the surface domains terminated with

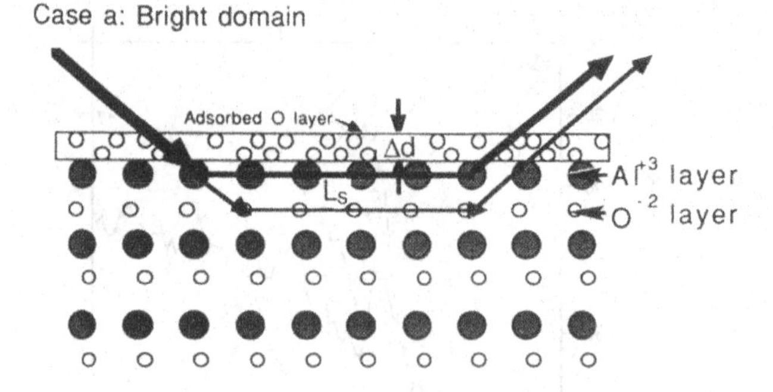

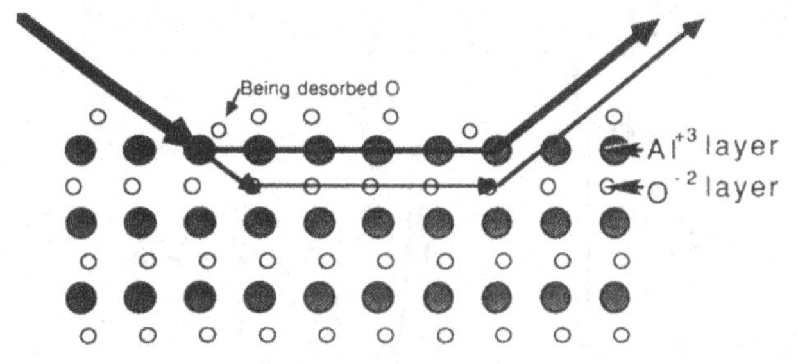

Figure 11.13 *Schematic models of electron resonance reflections from the α-alumina (01$\bar{1}$1) surfaces terminated initially (a) with Al^{3+} but having been oxidized and (b) with O^{2-} but having been damaged.*

different atomic layers undergo completely different radiation damage processes, analysis of the compositions of the two surface domains was undertaken and is described below. The channeling effect will be used to provide 'extra' information about the surface.

A schematic diagram is shown in Figure 11.13 to illustrate the scattering of high-energy electrons from a layered alumina crystal surface according to the dynamic calculation shown in Chapter 4. For the domain terminated initially with a layer of Al^{3+}, a thin oxygen layer of thickness Δd is assumed to be adsorbed at the surface. The presence of this thin layer, the scattering from which is equivalent to that from a uniform potential layer, does not affect the channeling propagation of the electrons along the surface, which is determined by the periodic structure. If the resonance conditions are satisfied, then the electrons that ultimately are reflected into the vacuum are postulated to propagate for a mean distance L_s parallel to the surface along the atomic rows within the top two layers. Then the channeling current (i_{Al}) along the Al^{3+} layer nearest to the surface may be different from that along the O^{2-} layer below (i_O). Thus the excited signals of the O K and Al L edges from the 'bright' domain are approximately

$$I_{\mathrm{OK}}(\Delta)=(i_{\mathrm{O}}+i_{\mathrm{Al}})\sigma_{\mathrm{I}}(\mathrm{O},\Delta)n'_{\mathrm{O}}\frac{2\Delta d}{\theta}+i_{\mathrm{O}}\sigma_{\mathrm{I}}(\mathrm{O},\Delta)n_{\mathrm{O}}L_{\mathrm{s}}, \tag{11.25}$$

$$I_{\mathrm{Al\,L}}(\Delta)=i_{\mathrm{Al}}\sigma_{\mathrm{I}}(\mathrm{Al},\Delta)n_{\mathrm{Al}}\,L_{\mathrm{s}}, \tag{11.26}$$

where n_{O}, n_{Al} and n'_{O} are the average numbers of atoms per unit volume (i.e., concentration) of oxygen in the O^{2-} layer, aluminum in the top Al^{3+} layer and oxygen in the adsorbed thin layer on the surface, respectively. On the right-hand side of Eq. (11.25), the first term indicates the excitation of the adsorbed oxygen layer and the second term is for the excitation of the oxygen in the bulk. It is important to point out that Eq. (11.25) and Eq. (11.26) are first-order approximations to the results of electron scattering and do not include the contributions of electron scattering from deeper layers. This model is therefore clearly only qualitative.

For the 'dark' surface domain, as shown in Figure 11.13(b), the few remaining oxygen atoms at the surface are being desorbed quickly and are assumed not to contribute to the O K edge excitation. In the electron microscope vacuum of about 10^{-8} Torr, the freshly exposed Al^{3+} layer may be covered by carbon rather than by oxygen, with no affect on the results. The measured ionization signals for the O K and Al L edges can thus be written approximately as

$$I'_{\mathrm{OK}}(\Delta)=i'_{\mathrm{O}}\sigma_{\mathrm{I}}(\mathrm{O},\Delta)n_{\mathrm{O}}\,L_{\mathrm{s}}, \tag{11.27}$$

$$I'_{\mathrm{Al\,L}}(\Delta)=i'_{\mathrm{Al}}\sigma_{\mathrm{I}}(\mathrm{Al},\Delta)n_{\mathrm{Al}}\,L_{\mathrm{s}}, \tag{11.28}$$

where a prime is introduced in some of the quantities to distinguish case (b) of Figure 11.13 from case (a). If one further assumes that the ratio of the channeling current $i_{\mathrm{O}}/i_{\mathrm{Al}}$ is not affected by the presence of an 'amorphous' oxygen layer at the surface, then

$$\frac{i_{\mathrm{O}}}{i_{\mathrm{Al}}}\approx\frac{i'_{\mathrm{O}}}{i'_{\mathrm{Al}}}, \tag{11.29}$$

On solving Eqs. (11.25)–(11.29), one has

$$N'_{\mathrm{O}}=n'_{\mathrm{O}}\Delta d=\frac{L_{\mathrm{s}}\theta n_{\mathrm{O}}}{2}\frac{\dfrac{I_{\mathrm{OK}}}{I_{\mathrm{Al\,L}}}-\dfrac{I'_{\mathrm{OK}}}{I'_{\mathrm{Al\,L}}}}{\dfrac{\sigma_{\mathrm{I}}(\mathrm{O},\Delta)(n_{\mathrm{O}}}{\sigma_{\mathrm{I}}(\mathrm{Al},\Delta)n_{\mathrm{Al}}}+\dfrac{I'_{\mathrm{OK}}}{I'_{\mathrm{Al\,L}}}}, \tag{11.30}$$

where N'_{O} is the density of the oxygen atoms adsorbed per unit surface area. For a general REM case, the mean distance that the electron travels along the surface is in the range 50–100 nm. If one takes n_{O} and n_{Al} as the average atomic concentrations in bulk α-alumina and $\theta=20$ mrad, then the integrated intensities from the spectra shown in Figure 11.1, and the calculated ionization cross-section from SIGMAK

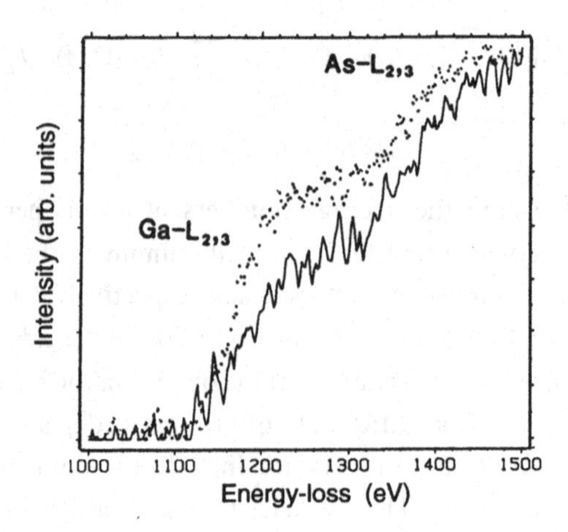

Figure 11.14 *Two types of REELS spectra acquired at 120 kV from GaAs(100) surface regions terminated with Ga (dotted curve) and As (solid curve) layers. The diffraction condition was kept the same whilst acquiring the spectra.*

and SIGMAL programs (Egerton, 1986), yield $N'_0 = 0.5$–1 Å^{-2}. Thus about a half to one oxygen atom per ångström unit squared is adsorbed on the brightly imaged domain of an α-alumina (01$\bar{1}$1) surface.

As shown in Chapter 4, resonance occurs within a few atomic layers next to the surface. Thus, the reflected intensity is most sensitive to the structure of the top few atomic layers. A cleaved GaAs(100) surface is terminated with either a Ga layer or As layer. Figure 11.14 shows the two typical REELS spectra of Ga $L_{2,3}$ and As $L_{2,3}$ ionization edges. The spectrum with the stronger Ga L edge excitation is often seen (the dotted curve), and the one with the stronger As L edge excitation is seen less often (the solid curve). These types of spectra represent the terminations of the surfaces with Ga and As atomic layers, respectively.

11.11 REELS in MBE

A REELS spectrometer has been directly installed on a MBE chamber to perform on-line surface chemical analysis during thin film growth, thus providing a sensitive method for monitoring monolayer growth on surfaces and a quantitative determination of surface alloy composition (Atwater and Ahn, 1991, Nikzad *et al.*, 1992). Figure 11.15 shows a series of REELS spectra of the Ge L edge recorded at 30 kV during MBE growth of Ge on Si. It is clear that a 0.15 nm thick Ge layer on the Si surface can be sensitively detected. Further analysis of the Ge L edge intensity as a function of beam deviation from the ⟨110⟩ zone axis has shown a decrease in Ge L signal when the beam is off the exact zone axis, suggesting that the mean channeling distance of the electrons along the surface is the maximum at the exact zone axis

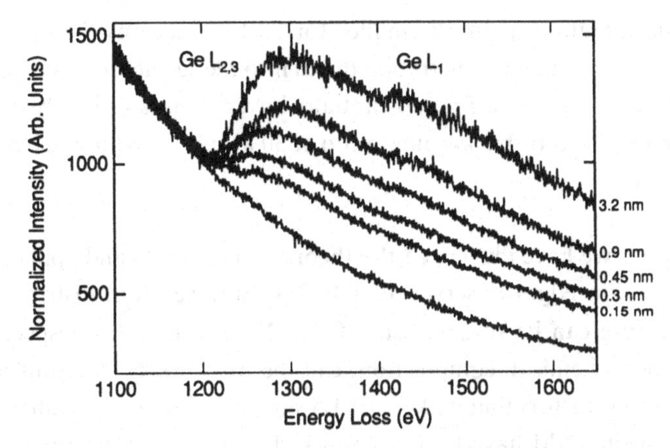

Figure 11.15 *REELS spectra acquired during* in situ *MBE growth of Ge on Si, illustrating the change in Ge $L_{2,3}$ edge intensity with the increase in thickness of the deposited Ge layer. The surface sensitivity of REELS is clearly shown (courtesy of Drs C. C. Ahn and H. A. Atwater).*

(Atwater *et al.*, 1991). EXELFS analysis is also possible in this case, and it has been applied to determine the radial distribution function of Sn on Si(100), where one or two monolayers of Sn were deposited on Si(100), which shows Sn(3 × 1)/Si(100) reconstruction.

Weak features in EELS spectra can be enhanced in the spectra using the double-differential technique. It has been found by Atwater *et al.* (1993) that the detection limit of carbon on Si surfaces is 1% of a monolayer.

It can be noticed in Figure 11.15 that the REELS acquired at 30 kV from the Ge/Si surface shows much better signal-to-background ratio than those acquired in REM using atomically flat surfaces (see Figure 11.1 for example). There are several differences for the two cases. First, the ionization cross-section for 30 kV electrons is significantly larger than that for 100 kV (or higher energy) electrons, but the effective penetration into the surface and mean traveling distance along the surface are smaller for 30 kV electrons. Second, the probability of surface plasmon excitation for 30 kV electrons is higher than that for 100 kV electrons, but the Bragg angle is larger for 30 kV electrons. Thus, the product of $v\theta_g$ is approximately independent of electron energy. Third, the specimen surface used to acquire REELS spectra at 100 kV is much smoother than that grown in the MBE chamber, so that there are no islands or protrusions at the surface. In MBE, however, islands (such as Ge/Si) are usually formed at the surface, and the reflected electrons will penetrate the deposited islands first before reaching the EELS spectrometer. Thus, the ionization signal of the Ge L edge is enhanced with respect to that of the Si edge. The decreased reflection intensity due to the presence of surface islands cannot affect the signal-to-noise ratio, because a large beam current is normally used for RHEED in the MBE chamber. Finally, the probability for continuous energy-loss processes, such as

Bremsstrahlung radiation, is much smaller for 30 kV electrons than for 100 kV electrons because of a smaller penetration depth into the surface. Consequently, the background intensity at 100 kV is substantially lower. This is probably the reason why the observed signal-to-background ratio at 30 kV is better than that at 100 kV or higher energies.

In this chapter, we have illustrated the theory, experiments and applications of core-shell ionization edges observed in REELS. Surface atom distribution and electronic structure can be studied using EXELFS and ELNES, respectively, in REELS. Surface chemical composition can be reasonably determined using REELS. Numerous factors that make REELS surface microanalysis different from thin foil analysis in TEM have been examined, leading to the introduction of an effective ionization cross-section, which can be measured experimentally using a technique that has been introduced. Diffraction and channeling effects in REELS and their applications in some particular cases have been demonstrated.

CHAPTER 12

Novel techniques associated with reflection electron imaging

There are two basic requirements for REM imaging of surfaces. The specimen is strongly preferred to be a single-crystalline material so that strong Bragg-reflected beams can be generated. The surface has to be flat enough to permit grazing angle imaging. The foreshortening effect along the beam direction, however, is a major disadvantage of REM, which limits the application of REM for imaging a relatively rough surface. It is thus desirable to enhance the potential of this technique by using it in conjunction with other surface imaging and analytical techniques. We first examine the interaction of an electron beam with the surface.

Various processes can be excited when a fine electron probe interacts with a specimen, as shown in Figure 12.1. The reflected primary electrons from the surface can be used to form the surface image. The analysis of electron energy loss can provide rich chemical information about crystal surfaces. Secondary electrons (SE) are generated wherever the electron interacts with the specimen. The energy of secondary electrons is less than about 50 eV. SE signals have been widely used in scanning electron microscopy (SEM) for obtaining surface topographic information because their effect escape depth is about 1 nm. Nanometer-resolution SE imaging has been performed in a dedicated STEM instrument, which generates an electron probe as small as 0.5 nm. Under UHV conditions, Auger electrons emitted from the surface can be analyzed to provide surface-sensitive chemical information. Scanning Auger microscopy allows the detection of small metallic particles as small as 0.5 nm (Liu *et al.*, 1992a, b). X-rays are also emitted from the surface and can be applied to examine the surface chemistry under certain incident conditions. Photons are emitted from the illuminated specimen area.

In this chapter, novel techniques using the various signals described above for surface studies are reviewed. Examples are shown to exhibit their applications to surface science.

12.1 Scanning reflection electron microscopy

12.1.1 Imaging surface steps

As shown earlier in Figure 5.5, SREM is equivalent to REM based on the reciprocity theorem. There are two methods by which to form SREM images. One is to use

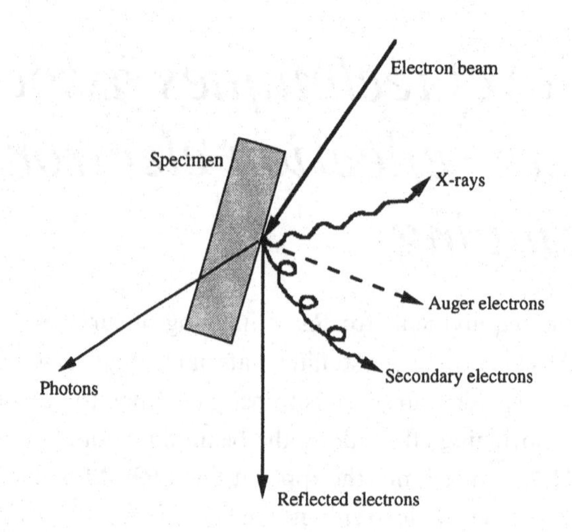

Figure 12.1 *Various processes excited by an electron beam at crystal surfaces.*

STEM, in which signals are collected when a small electron probe scans across the sample surface. Modern electron optics has made it possible to generate a high-brightness electron probe of diameter less than 0.5 nm, so that surface analysis with high spatial resolution is feasible (Milne, 1987; Ou and Cowley, 1987; Liu and Cowley, 1989). A detailed description of SREM has recently been given by Liu and Cowley (1993), who outline the various features of SREM performed in a dedicated STEM instrument. The other method is in conjunction with scanning RHEED in a MBE chamber in order to perform *in situ* observations of crystal growth. SREM images are formed by detecting the intensity of a reflected beam when the electron probe scans across the surface at a grazing angle. In this apparatus, the growth of the surface film is not interrupted during the recording of surface images. This has provided an important method to image *in situ* surface film growth of semiconductors on semiconductors and metals on semiconductors (Ichikawa and Doi, 1987; Isu *et al.*, 1990).

Figure 12.2 shows a SREM image of recrystallized copper vicinal surfaces composed of (110) (in bright contrast) and (111) (in gray contrast) facets. The surface ridge runs approximately along $[1\bar{1}1]$. It is clear that facets occur not only along $[1\bar{1}1]$ but also along $[\bar{1}12]$ as shown by the formation of rectangular-like (110) surface areas. Both (110) and (111) facets have many small steps. The fine-scale atomic level structure may be due to the oxidation of the copper surface, because the surface was prepared in air before examination by SREM.

SREM imaging can also be performed in MBE chambers during film growth. This is the most important application of the reflected electron imaging technique. Figure 12.3 shows SREM images of Si growth on a fully Ga-adsorbed surface at different substrate temperatures and constant deposition rate. It is apparent that the denuded

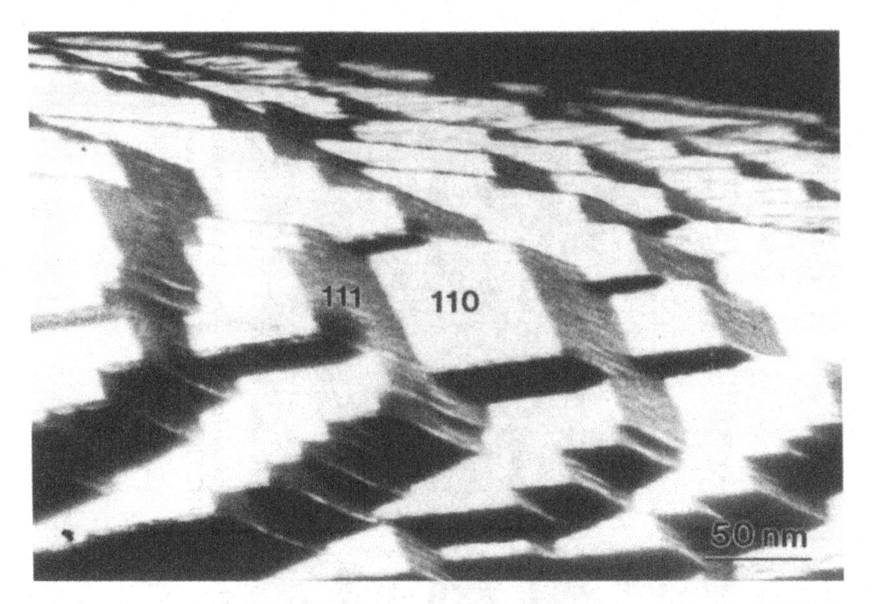

Figure 12.2 *A SREM image of Cu(110) and vicinal surfaces showing (110) and (111) facets. This image was taken with a 0.5 nm electron probe at 100 kV in a VG HB5 STEM (courtesy of Drs J. Liu and J. M. Cowley 1989).*

zone became larger with increasing substrate temperature. SREM provides a powerful technique for imaging *in situ* thin film growth at high temperatures.

The advantages of SREM over REM can be summarized as follows. First, the foreshortening factor can, in principle, be reduced by signal processing. However, this may not improve the content of the image. Second, dynamical focusing can be used to reduce the out-of-focus effect found in REM, so that the entire SREM image is taken into focus. Third, it is possible to image a small surface area with a high-brightness electron probe and the image contrast can be artificially improved using the electronic gain control. Microdiffraction from small surface particles and chemical analysis using energy-dispersive spectroscopy (EDS) and EELS are possible. Fourth, the larger space that is available around the specimen allows for the introduction of various surface-treatment and analytical techniques, such as Auger electron spectroscopy (AES) (Ichikawa *et al.*, 1985). Fifth, chromatic aberration does not affect SREM images but does affect REM images, because the condenser lenses in STEM are placed before the electrons interacting with the specimen (see below). Thus, the SREM image may have better resolution than REM. Finally, it is possible to collect large-angle TDS electrons in the RHEED geometry to form compositionally sensitive surface images (Liu and Cowley, 1991; Wang and Bentley, 1991a).

Strictly speaking, SREM has characteristic differences from REM because of differences in the optic systems. The contribution of plasmon-loss electrons in the

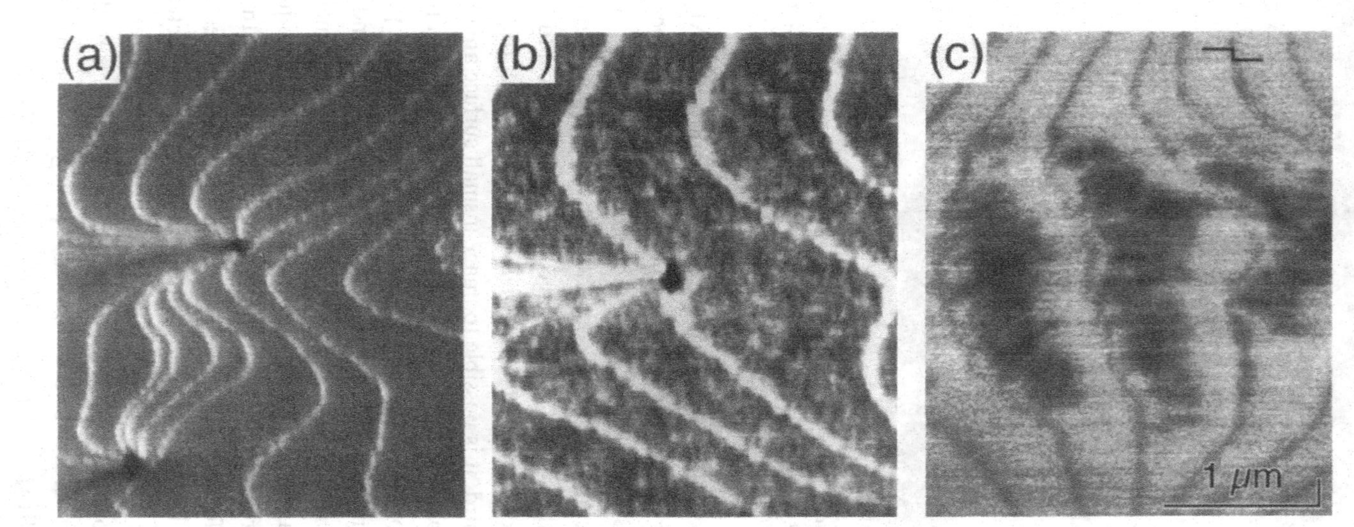

Figure 12.3 *SREM images recorded during the growth of Si on a Ga/Si(111) $\sqrt{3}\times\sqrt{3}$ surface, showing the temperature-dependence of the denuded zone at constant growth rate (0.12 bi-layer per minute). (a) 450°C, (b) 500°C and (c) 550°C. The beam energy is 20 keV (courtesy of Dr H. Nakahara et al., 1993).*

images of SREM is different from that in REM images. Theoretical calculations have found that the main effect of valence excitation (with energy losses 5–30 eV) is to introduce a shift in focus in the lens transfer function due to chromatic aberration (Wang and Bentley, 1991a). The shift in focus can be as large as a few tens of nanometers. The contrast of valence-loss electrons would be the same as that of the elastically scattered electrons in SREM but not in REM. This is because the contrast transfer effect of the condenser and objective lenses in the SREM (or STEM) and REM (or TEM) cases occurs before and after, respectively, the electrons interact with the crystal, so that the inelastic valence excitation within the specimen would have almost no effect on the contrast of SREM images but would on that of REM images. Image contrast can be artificially improved by using electronic gain controlling systems in SREM but not in REM, making it more difficult to enhance the contrast of weak features in REM images.

However, it is generally sufficient to use the same basis of diffraction and imaging theory for REM and SREM and then consider the modifications to the results that are needed to take account of the practical situation in each case (Cowley, 1987).

12.1.2 Imaging dislocations

In REM, images of surface emergent dislocations can be easily obtained because the electrons scattered at the exact Bragg angle (i.e., $\Delta\theta = \vartheta - \theta_g = 0$) and at angles slightly different from the Bragg angle ($\Delta\theta > 0$ and $\Delta\theta < 0$) are propagating along slightly different directions and both come from different regions of the surface, resulting in diffraction contrast. In SREM, however, the size of the imaging detector determines the visibility of dislocations. If the detector is a point detector, the SREM image is identical to the REM image. The dislocation may not show any contrast in the image if the detector size is significantly larger than the maximum deviation angle $\Delta\theta_{max}$ of the local lattice from the Bragg angle. The dislocations are expected to show up in the image if the detector size is smaller than $\Delta\theta_{max}$. In SREM, the RHEED pattern is a convergent beam diffraction pattern and the dislocation contrast can be interpreted using Figure 12.4. If the detector is placed at the exact Bragg angle (i.e., $\Delta\theta = 0$), the image of the screw dislocation would show symmetric dark contrast in the SREM image. If the detector is placed at an angle corresponding to $\Delta\theta > 0$, the screw dislocation would show bright–dark contrast. The dislocation contrast is reversed if the detector is placed at an angle corresponding to $\Delta\theta < 0$. These conditions have been tested experimentally (Liu and Cowley, 1993).

12.2 Secondary electron imaging of surfaces

Secondary electrons are emitted when high-energy electrons interact with a specimen. Incoherent SE images can be formed by collecting these electrons as a small

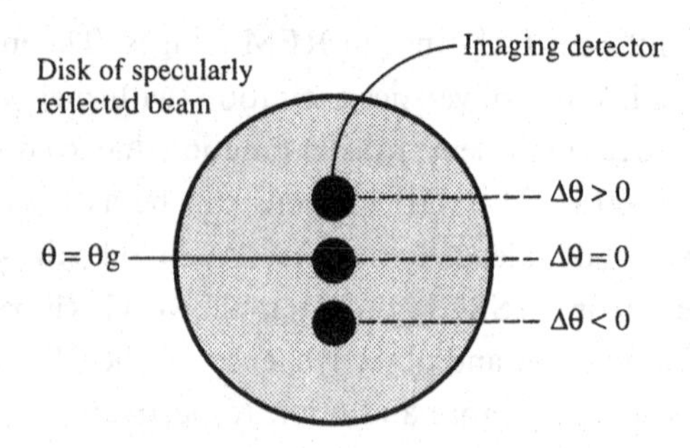

Figure 12.4 *The detector geometry for forming dislocation images in SREM.*

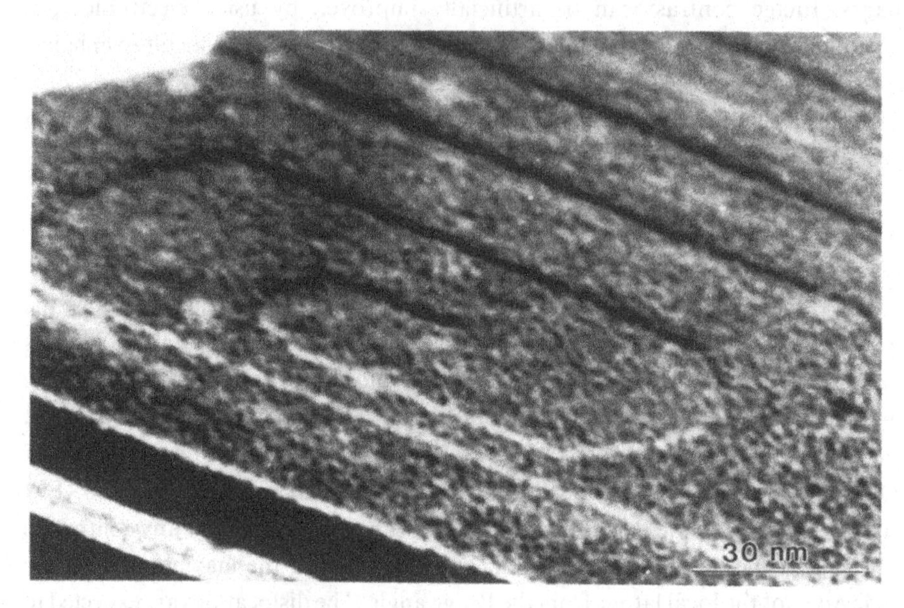

Figure 12.5 *A secondary electron image obtained at 100 kV in a UHV VG501 STEM of a MoO_3 crystal, showing a spiral step associated with growth around a screw dislocation (courtesy of Drs J. Liu and J. M. Cowley, 1991).*

electron probe scans across a specimen (Imeson *et al.*, 1985; Liu and Cowley, 1988; Bleloch *et al.*, 1989b). The SE images are sensitive to surface morphology and can thus be used to characterize surface structures. The SE images can resolve atom-high surface steps on crystal surfaces (Milne, 1989; Drucker *et al.*, 1991; Homma *et al.*, 1991b; Uchida *et al.*, 1992b). Figure 12.5 shows an SE image of a MoO_3 surface, exhibiting steps spiraling around a screw dislocation. With respect to the incident beam direction, the up and down steps show dark and bright contrast, respectively. This observation can be interpreted using Figure 12.6.

The emission rate of SE depends on the incident electron effective path length along which SE can escape within a distance shorter than the SE escape depth (Liu *et*

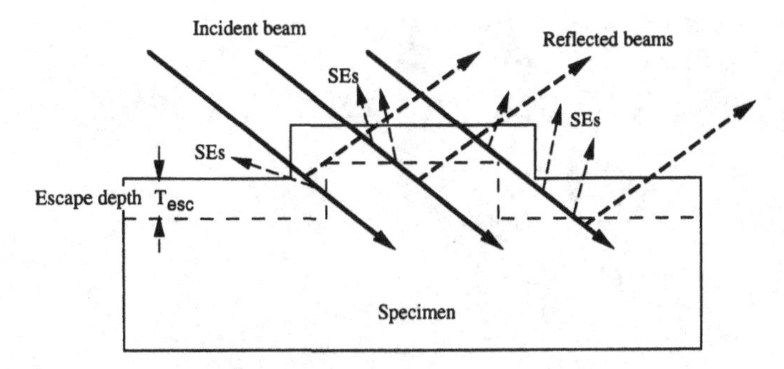

Figure 12.6 *The contrast mechanism for imaging one-atom-high surface steps using secondary electrons.*

al., 1990; Milne, 1989). When the incident beam strikes the surface, some electrons will penetrate into the surface. Secondary electrons can be generated by both transmitted and reflected electrons along their path length. The secondary electrons that will contribute to the image are those generated with an escape depth T_{esc} from the top surface. For a perfectly flat surface, the mean path length of the incident electron within the escape depth T_{esc} is $L_{path} = T_{esc}/\sin\theta$. This path length is called the effective path length along which the secondary electrons generated will contribute to the image. The electron effective path length for an up step is relatively short compared with that for a down step. Therefore, more secondary electrons are generated at a down step than at an up step, resulting in the contrast observed in Figure 12.5. It is natural that the step contrast decreases with an increase in electron incidence angle.

It must be pointed out that the mechanism shown in Figure 12.6 is based on the assumption that all the secondary electrons emitted within the escape depth will be entirely detected by the detector, and that there is no effect from the detector geometry and detector position. However, caution must be exercised when interpreting secondary electron images. Experiments by Homma *et al.* (1993a) have shown that the orientation of the step direction with respect to the SE detector plays a crucial role in forming the atomic step contrast. Shadows are created by the atomic steps facing away from the SE detector due to the low detection efficiency of SEs. It has recently been found that the secondary electron yield is sensitive to the atomic configuration of the surface; contrast is observed between the 2×1 and 1×2 reconstructed Si(001) surface regions (Homma *et al.*, 1993b). This remarkable result clearly demonstrates the close relationship between surface structure and SE emission, suggesting the great potential of SE imaging.

SE imaging can also be performed in UHV SEM. High-resolution SE imaging has been shown as a powerful technique for imaging atom adsorption on surfaces and surface reconstructions (Endo and Ino, 1993a and b; Endo, 1993). Figure 12.7 shows

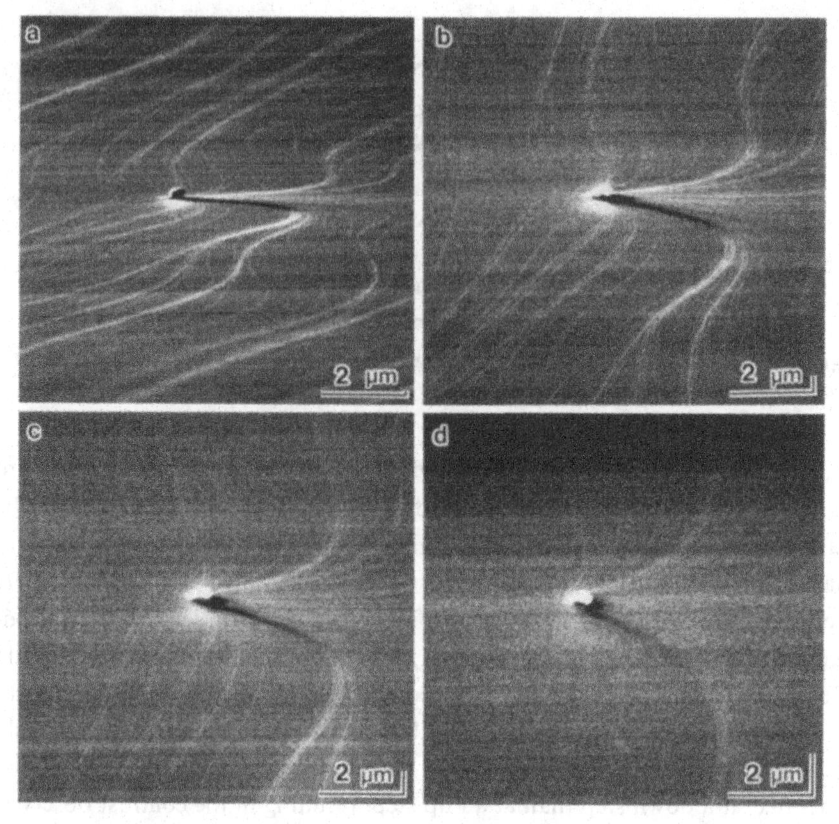

Figure 12.7 *SE images of Si(111) with different angles of primary electron incidence: (a) 2.6°, (b) 8°, (c) 15°, and (d) 27°. The electron beam is incident from the bottom, and steps are down from right to left. The SE detector was located on the left-hand side of the SE images. The beam energy is 25 keV (courtesy of Dr Y. Homma et al., 1991b).*

atomic-step images of Si(111) for incidence angles up to 27°. The steps are distributed around a SiC particle. Atomic steps are clearly seen up to 15°, and are faint but still discernible even at 27°. The foreshortening effect has been eliminated. It is anticipated that SE imaging will play an important role in studying surface dynamic processes.

Imaging with secondary electrons has several important advantages over SREM. First, the image contrast does not depend sensitively on diffracting conditions. It is thus not necessary to sustain the resonance condition as discussed in Chapter 4. In general, no Bragg-diffracting condition is required in order to perform SE imaging. Second, surfaces of non-crystalline materials can be successfully imaged. However, for SREM, only single-crystalline materials can be imaged. Third, it is possible to image surfaces at higher incident angles (about 25° or larger), so that the foreshortening effect is eliminated. Similar to REM, SREM is also limited by the foreshort-

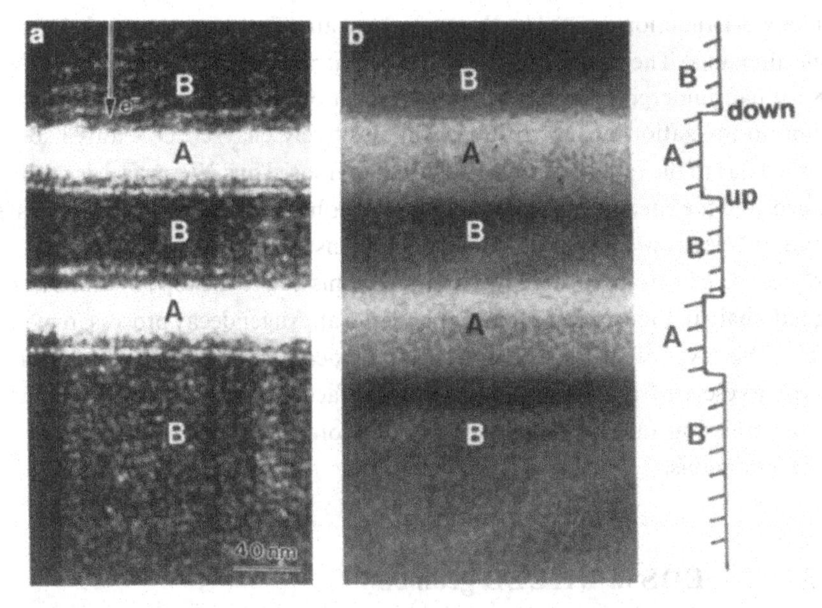

Figure 12.8 *A comparison of (a) a SREM image and (b) the corresponding SE image from the same surface area to show the difference of secondary electron emissions from different surface domains of α-alumina (01$\bar{1}$1) initially terminated with oxygen (dark contrast domain) and aluminum (bright contrast domain), respectively (Wang and Howie, 1990).*

ening effect. Finally, SE imaging can be applied to specimens that may have rough surfaces.

Secondary electrons, emitted from solids during the scattering of high-energy electrons, can be effectively used to show surface morphology because of the influence of surface geometry and composition on SE generation and escape probabilities. The secondary electron emission rate also depends on the work function of the surface, which may be affected by the structure of the surface layers. Figure 12.8 shows a comparison of the SREM image and the SE image recorded almost simultaneously from the same area of a cleaved α-alumina (01$\bar{1}$1) surface after its having been damaged by the beam to a certain extent (see Section 8.5 for the detailed damage mechanism). The bright and dark domains visible in the SREM image (Figure 12.8(a)) result from electron beam damage of the surfaces initially terminated with aluminum and oxygen, respectively. These contrast effects can be understood based on the internal Auger decay process (or electrostatic desorption process) for surface domains terminated with different atoms (see Section 8.5). A similar bright and dark contrast effect is observed in the SE image, with the bright domain in the SREM image having a larger secondary electron emission rate than the dark domain (Figure 12.8(b)).

The SE image contrast can be interpreted by considering the difference in the surface work function created by the presence of an amorphous oxygen layer on the bright domains. The domain showing dark contrast in the SREM image (Figure 12.8(a)) had undergone an electrostatic desorption due to internal Auger decay, leading to ionization of O^{2-} to O^{1+}. The positively charged O^{1+} ions had been desorbed due to the repulsive force from the Al^{3+} ions. Thus, the surface is positively charged, and the effective surface work function is higher, resulting in a lower rate of secondary electron emission. For the domains showing bright contrast, the adsorbed oxygen atoms located one or more atoms above the top Al^{3+} ions (see the bright domain in Figure 12.8(b)) do not undergo an Auger decay process, remaining as O^{2-}. This layer of negative charge effectively decreases the work required for an electron to escape from the bright (SREM) surface domain, so that the secondary electron emission rate is higher at the bright contrast domains than at the dark contrast domains.

12.3 EDS in RHEED geometry

Energy-dispersive X-ray spectroscopy (EDS) is a well-established technique for chemical microanalysis in TEM. In RHEED, we have attempted to acquire EDS spectra when the bulk crystal surface faces the EDS detector as in TEM. Figure 12.9 shows a comparison of EDS spectra recorded in RHEED geometry (solid line) from a cleaved GaAs(110) surface and the one obtained from a thin GaAs foil in TEM (dashed line). In the RHEED case, the take-off angle of the EDS detector is about 80° with respect to the crystal surface. It is apparent that the X-ray signals observed in RHEED geometry are much different from those seen in the TEM case. This discrepancy is not due to any difference in surface chemical composition but to absorption and fluorescence effects (Wang, 1988f). The resonance condition has little effect on the Ga K and As K lines. In RHEED, X-rays can be generated by the electrons that are transmitted into the bulk crystal and do not contribute to the RHEED intensity. Owing to the large escape depth of the X-rays, X-rays generated at deep depths can still reach the detector. These X-rays suffer strong absorption and fluorescence effects due to the excitation of lower energy X-rays. Therefore, EDS may not be sensitive to surface chemistry in conventional TEM. However, a technique to be described in Section 12.7 is highly surface-sensitive if the detector take-off angle is very small.

12.4 Electron holography of surfaces

Image resolution in TEM is limited by the effect of spherical aberration of the electron optic lenses. The principle of electron holography was introduced by Gabor (1949) in an attempt to exceed the point-to-point resolution of an electron

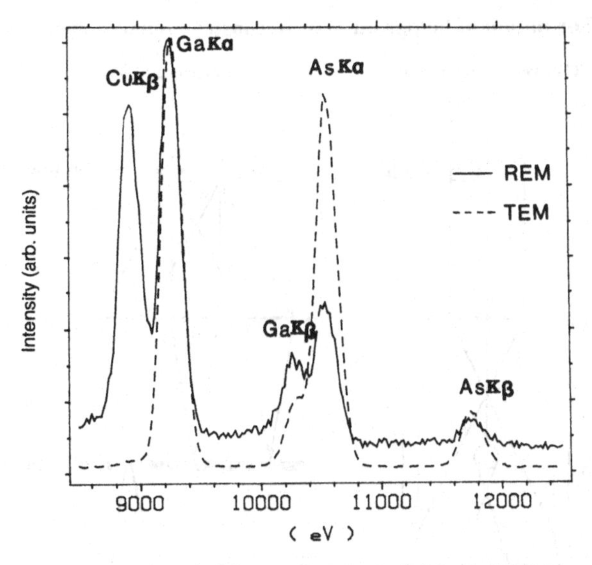

Figure 12.9 EDS spectra recorded from a thin GaAs foil in the TEM case (dashed line) and from a bulk crystal GaAs(110) surface in the REM case. The beam energy is 80 keV. The copper signal came from the specimen grid.

microscope. Holography is based on the interference and diffraction properties of waves, thereby producing a 'true' image of an object (including amplitude and phase) without any distortion of the lenses. The development of high-brightness, high-coherence electron sources has made it possible to obtain holograms using electron waves in TEM (Tonomura, 1987 and 1990; Lichte, 1986). A point-to-point resolution of better than 0.1 nm has been achieved in a 200 kV TEM (Völkl and Lichte, 1990). In this section, the basic principle of off-axis electron holography is outlined and its applications for surface imaging in REM are described.

12.4.1 Principles and theory

In a normal TEM image, phase information is lost in the image recording process. An important aspect of holography is to recover the phase distribution function. For this purpose, electron holography is designed as a two-step imaging technique. The first step is to form the interference hologram as shown in Figure 12.10(a). The other step is to reconstruct the hologram to get the real object image. Image-plane off-axis holograms are recorded by means of an electron biprism inserted between the back focal plane of the objective lens and the intermediate image plane. The specimen is positioned to cover half of the image plane, leaving the other half for the reference wave. The wave function of the specimen, after the objective lens, is assumed to be $\exp(i2\pi\boldsymbol{K}\cdot\boldsymbol{r})A(x,y)\exp[i\phi(x,y)]$, where A is the wave amplitude function, ϕ is the phase distribution and $\boldsymbol{K}$ is the wavevector of the incident wave. By

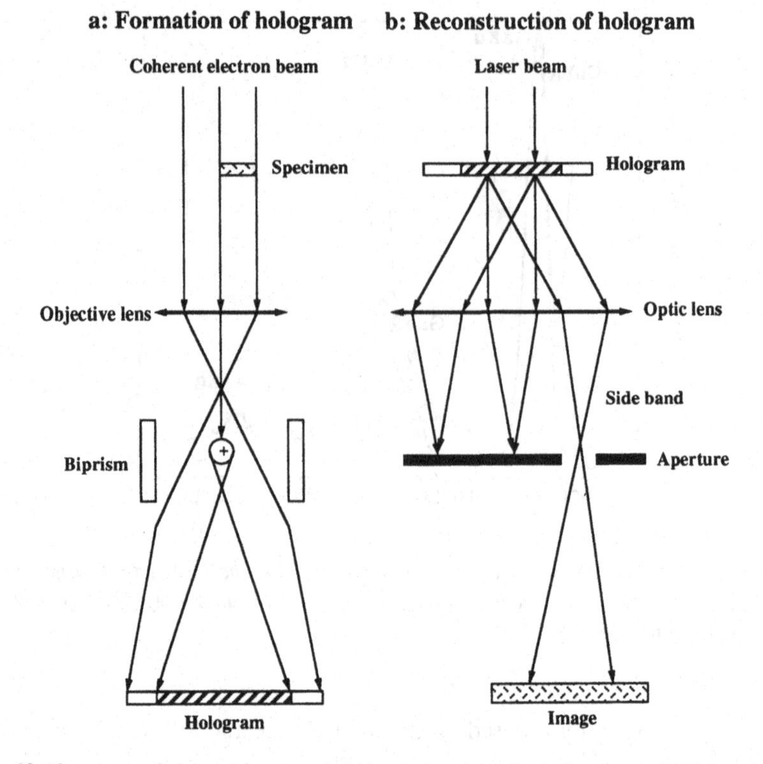

Figure 12.10 *A ray diagram showing (a) the formation of a hologram in TEM and (b) the reconstruction of the hologram using an optic system.*

applying a positive voltage to the filament of the biprism, the waves on both sides of the biprism are deflected towards each other, forming an interference pattern in the image plane, i.e., the hologram, the intensity distribution of which is

$$\begin{aligned} I_{\text{hol}}(x, y) &= |\exp(\mathrm{i}2\pi \boldsymbol{K}\cdot\boldsymbol{r} - \mathrm{i}\pi u_{\mathrm{T}} x) + \exp(\mathrm{i}2\pi \boldsymbol{K}\cdot\boldsymbol{r} + \mathrm{i}\pi u_{\mathrm{T}} x)\, A(x, y) \exp[\mathrm{i}\phi(x, y)]|^2 \\ &= 1 + A^2(x, y) + 2A(x, y) \cos[\phi(x, y) + 2\pi u_{\mathrm{T}} x], \end{aligned} \tag{12.1}$$

where u_{T} is the spatial carrier frequency and is determined by the wave overlapping angle γ as $u_{\mathrm{T}} = \gamma/\lambda$. The phase shift $\pi u_{\mathrm{T}} x$ corresponds to the phase shift of the electron wave when passing through an electrostatic potential (produced by the biprism), because of the Aharonov–Bohm effect (Aharonov and Bohm, 1959).

In the second step, as shown in Figure 12.10(b), image reconstruction is performed by taking a Fourier transform (FT) of the intensity distribution of the hologram. This procedure can be performed either optically or digitally. We start the numerical reconstruction of (12.1)

$$\begin{aligned} \mathrm{FT}[I_{\text{hol}}(x, y)] &= \delta(\boldsymbol{u}) + \mathrm{FT}[A^2(x, y)] + \mathrm{FT}[A(x, y) \exp(\mathrm{i}\phi(x, y))] \otimes \delta(u_x + u_{\mathrm{T}}) \\ &\quad + \mathrm{FT}[A(x, y) \exp(-\mathrm{i}\phi(x, y))] \otimes \delta(u_x - u_{\mathrm{T}}), \end{aligned} \tag{12.2}$$

where the first two terms are located near the center of the diffraction pattern; and the last two terms are centered at $u_x = -u_T$ and $u_x = u_T$, respectively, and are so-called side-bands (Figure 12.10(b)). In practice, this procedure can be performed using an optic lens, the spherical aberration of which is negligible. If the carrier spatial frequency u_T is sufficiently large, the side-bands do not overlap with the center band (i.e., the first two terms in Eq. (12.2)). If the aperture allows only one side-band, say $u_x = -u_T$, to pass through, then the inverse Fourier transform of this band would give the original wave function $A(x,y)\exp(i\phi(x,y))$. Finally, the electron wave function after transmitting the specimen can be recovered by removing the aberration effect of the objective lens

$$\Phi(x,y) = \mathrm{FT}^{-1}\{\mathrm{FT}[A(x,y)\exp(i\phi(x,y))]/T_{obj}(\boldsymbol{u})\}. \tag{12.3}$$

The resultant $\Phi(x,y)$ is the real image of the specimen without distortion by the lens and can be compared directly with the calculated electron wave function of the allowed beams by the objective aperture. The hologram reconstruction can also be performed with the use of a computer. The result of (12.3) is critically affected by the precision of C_s and Δf. For atomic resolution image reconstruction, the C_s has to be accurate to better than 1%.

12.4.2 Surface holography

Electron holography has also been applied for surface imaging in REM. The only difference is that the reference wave is replaced by the wave reflected from flat surface areas without steps and defects. The interference of this wave with the waves from steps or defects forms the holograms (Osakabe *et al.*, 1989). Figure 12.11(a) shows an as-recorded hologram of the Pt(111) surface. Interference fringes of separation about 1 nm are seen. The bending of the fringes across surface steps indicates the electron phase shift at the step. Figures 12.11(b) and (c) are the reconstructed amplitude and phase images of the surface. Obviously, the defocus has a remarkable influence on the amplitude of the image wave in the regions of the steps, which vanishes only in the focused region. The phase pattern, in contrast, is easily visible in this region. The phase shift can be directly measured from the phase image and the result shows good agreement with that calculated based on Eq. (6.2) (Banzhof and Herrmann, 1993). The phase modulation appears to be determined mainly by the geometrical shape of the surface whereas the amplitude may exhibit disturbances of the interference condition due to lattice distortions at the top of the step.

The surface holography is a quantitative technique for measuring the strain field associated with surface emergent dislocations (Osakabe *et al.*, 1993). Figure 12.12 shows a reconstructed reflection electron interferogram of a screw dislocation on a GaAs(110) surface in REM geometry. The vertical displacement around the

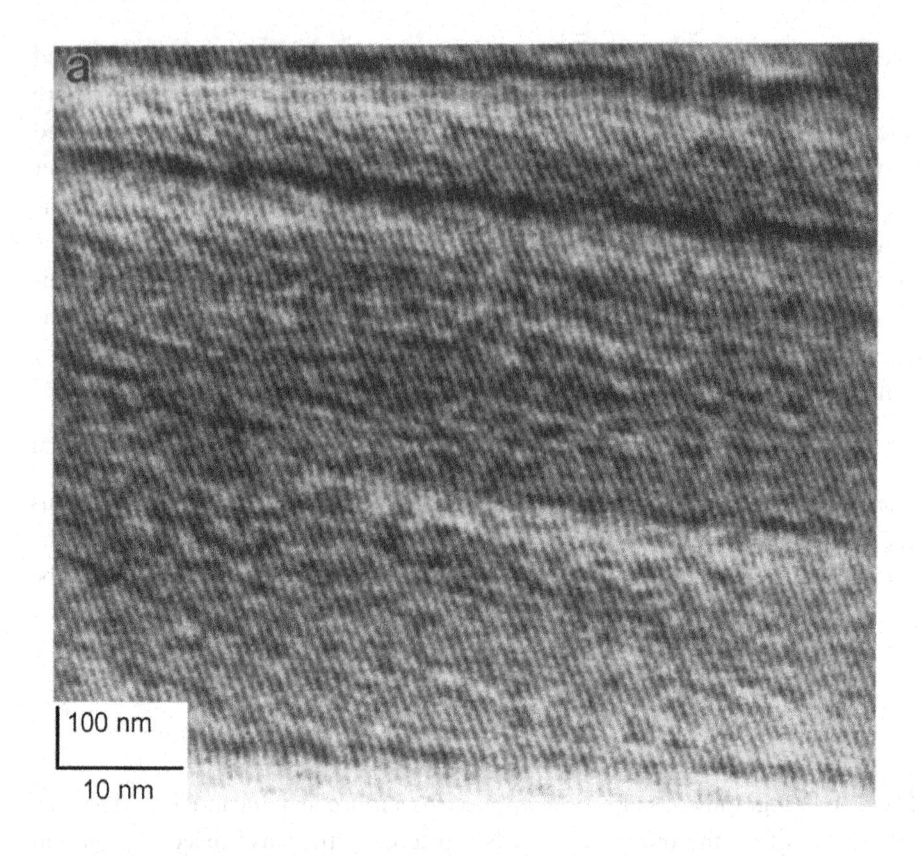

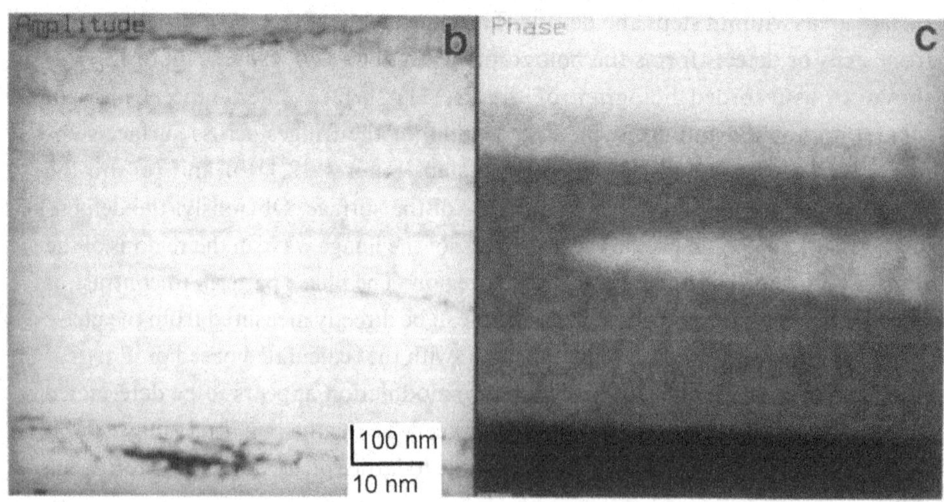

Figure 12.11 *(a) A reflection hologram of the Pt(111) surface recorded at 200 kV using the (666) specular reflection near [11$\bar{2}$] azimuth. In (b) and (c) are shown the digitally reconstructed amplitude and phase images, respectively (courtesy of Drs H. Banzhof and K. H. Herrmann, 1993).*

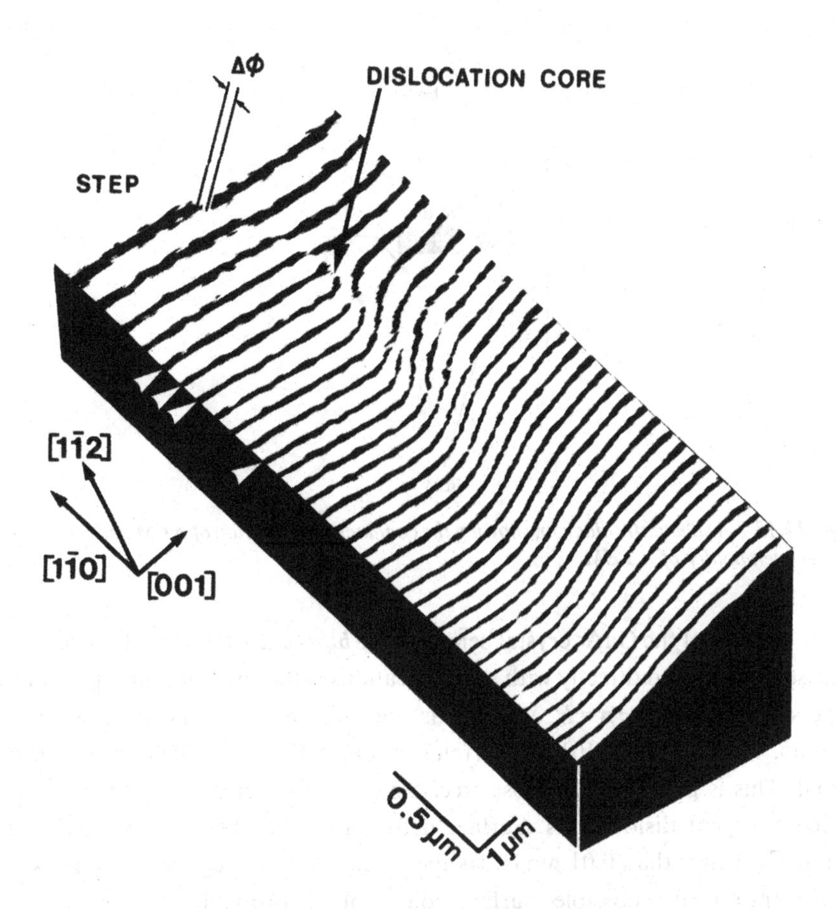

Figure 12.12 *A reconstructed reflection electron interferogram of a dislocation on a GaAs(110) surface in REM geometry, for determining the dislocation Burgers vector and step height. The foreshortening factor is 4 in the beam direction (courtesy of Dr N. Osakabe).*

dislocation can be directly observed as a shift of the interference fringes from the parallel fringes. One fringe shift corresponds to only 0.05 nm in height variation. The phase shift between the fringes on the upper terrace and on the lower terrace of the step is 0.4π, which is a fraction of the calculated value 7.6π for a step height of $H = 0.2$ nm ($= d_{220}$) according to Eq. (4.1) by assuming an inner potential of 13.2 eV. The total phase shift can be directly determined from the terminating interference fringes pointed out by the white arrows in Figure 12.12. These four fringes should be connected to the upper terrace at the surface step if observed with enough lateral resolution. Hence, the total phase shift is $8\pi - 0.4\pi = 7.6\pi$. It is thus possible to accurately determine the height of surface steps by reflection electron holography (Banzhof *et al.*, 1992; Osakabe, 1992).

The Burgers vector can also be determined from the observed hologram. Possible

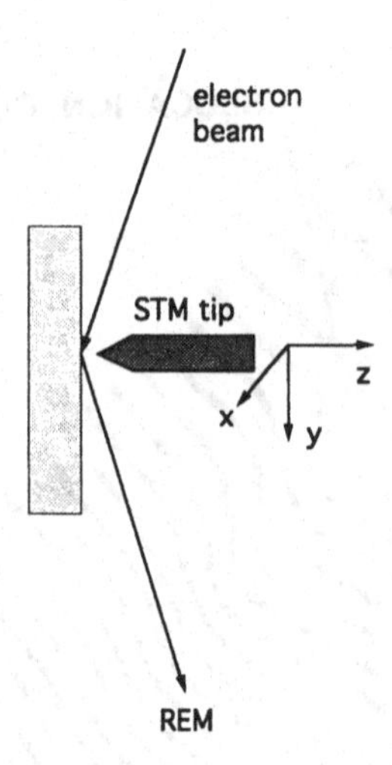

Figure 12.13 *A schematic diagram showing the conjunction operation of REM and STM inside the specimen holder of a TEM.*

Burgers vectors of the GaAs crystal belong to the $\boldsymbol{b}_B = (a/2)\langle 110\rangle$ family. The height of the step associated directly with the dislocations is the vertical component of the Burgers vector, i.e. $\boldsymbol{b}_B \cdot \boldsymbol{n} = H$, where $\boldsymbol{n}$ is the unit vector of the surface normal direction. Thus $\boldsymbol{b}_B = (a/2)[101]$, which intersects with the surface at 60° to the surface normal. This is probably the most precise method for determining the nature of surface emergent dislocations. Surface deformation has been determined to an accuracy of better than 0.01 nm (Osakabe *et al.*, 1989). A quantitative analysis of surface strain is also possible. Surface holography is probably the most accurate method for measuring surface step height and dislocation strain field.

12.5 REM with STM

Scanning tunneling microscopy (STM) is a powerful technique for direct imaging of surface structures at atomic resolution (Binnig *et al.*, 1983). The discovery of STM and atomic force microscopy (AFM) (Binnig *et al.*, 1986) has opened the era of scanning probe techniques for surface studies. In REM, large surface areas can be examined but its applications are limited by the foreshortening effect and the finite image resolution. An optimum technique would be to combine STM with REM. The idea of building a STM that operates inside a specimen holder for TEM (i.e. a STM REM holder), as illustrated schematically in Figure 12.13 was initiated by Spence (1988b). The use of glancing-angle-reflected electron imaging allows observations of the region directly beneath the tunneling tip to be made. REM and STM are operated simultaneously. In this case, the distribution of surface steps and dislocations can be seen in REM images, and local atomic-resolution surface images

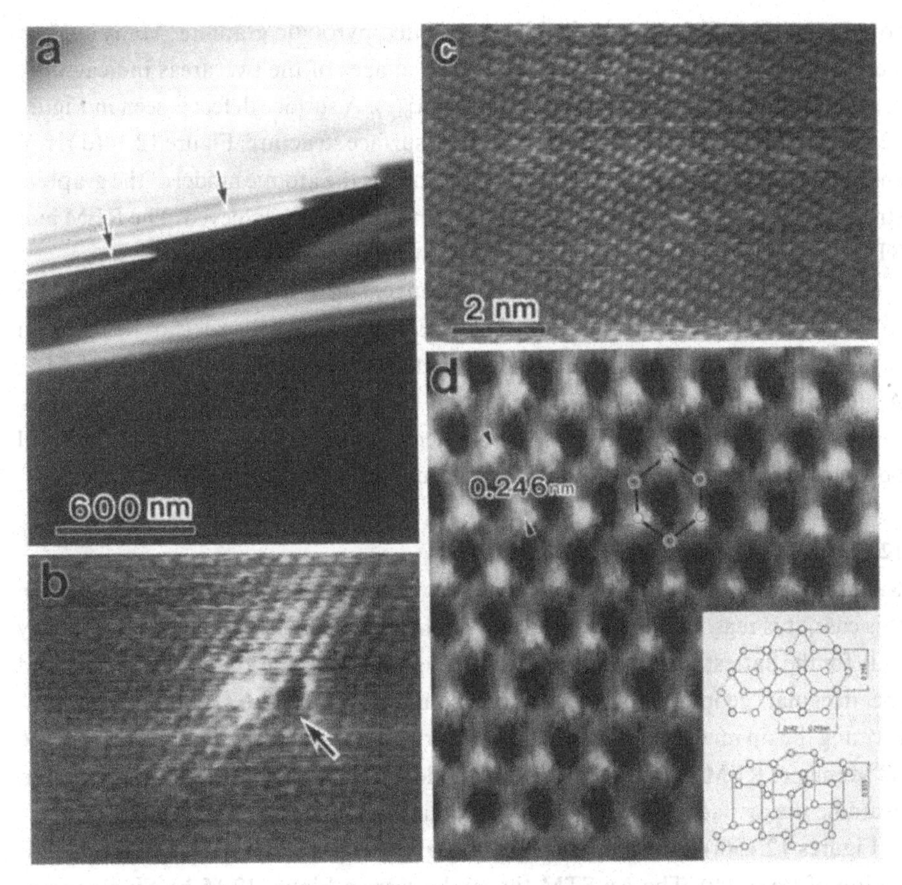

Figure 12.14 (a) A REM image and (b), (c) and (d) the STM images of graphite (0001), showing a defect area in (b) and perfect atomic arrangement in (c) and (d) (courtesy of Dr M. Iwatsuki et al., 1991).

can be obtained with STM. The electron beam contribution to the tunneling current is constant in time and so may be subtracted electronically. Using this technique, the interaction of a STM tip with a graphite (0001) surface has been studied (Spence *et al.*, 1990).

12.5.1 Atomic-resolution surface imaging

The REM–STM technique is aimed at obtaining atomic resolution images of crystal surfaces independently using REM and STM. This goal has been achieved by Lo and Spence (1993) and Iwatsuki *et al.* (1991). It has been shown by Lo and Spence (1993) that the specially built REM–STM holder is stable and allows atomic-resolution imaging of graphite (0001) surfaces. The experiments of Iwatsuki *et al.* (1991) were performed in an UHV TEM, in which they observed 7×7 Si(111) surfaces with atomic resolution. Figure 12.14 shows a REM image and the corresponding STM images of a highly oriented pyrolytic graphite. Many surface steps are seen in the REM image. The STM images of the two areas indicated by arrowheads are shown in Figures 12.14(b) and (c). A surface defect is seen in Figure 12.14(b), and Figure 12.14(c) exhibits perfect surface structure. Figure 12.14(d) is an enlargement of Figure 12.14(c), in which the inset is the atomic model of the graphite structure unit cell. Atomic resolution has been successfully achieved. The REM and STM images can be used complementarily to define surface structures.

STM can be used to determine atomic arrangement at the surface but it is unable to provide chemical information. The REM–STM technique can be used in conjunction with other microanalysis and diffraction techniques, such as REELS, Auger spectroscopy and RHEED, so that the surface chemistry and crystallography can be uniquely determined. It is anticipated that the REM–STM technique will become one of the important tools for surface structure determination.

12.5.2 Artifacts in STM imaging

Reliable surface structure images can be provided by STM if the surface is flat. In this case, it is reasonable to assume that tunneling occurs from a single tip asperity only for the flattest surface. For relatively rough surfaces, the finite size of the tip and the irregular geometry of the surface commonly introduce multiple contacts, forming the tip image. This makes tip and surface features difficult to separate in the STM image. REM with STM provides an ideal technique, which can be applied to examine the artifacts introduced in STM imaging (Lo and Spence, 1993).

Figures 12.15(a) and (b) are STM and REM images, respectively, of a different region of PbS(100). The Au STM tip can be seen in Figure 12.15(b). Single atom-high steps visible in the REM image are not resolved in the STM image due to large-amplitude noise. Figures 12.15(c) and (d) are the corresponding STM and REM images after voltage pulsing. Major differences are found between the REM and STM images. The most striking difference is in the shape of the protrusion. The REM image shows a tall protrusion (170 nm tall, 60 nm wide at the top), which is tilted to one side. Next to it is a small mound. It is most likely to be a region of PbS damaged by the voltage pulse. REM images of the protrusion show that its overall shape was preserved after STM imaging.

The STM image of the protrusion does not reproduce its tilt. From the STM image, the width at the base of the two mounds is about 400 nm, significantly greater than the actual width of 270 nm. This is the result of tip convolution (or tip shape). Most significant is the distortion of the objects' heights. The STM cross-section of the large mound yields a height of 90 nm. This is much less than the actual height of 170 nm. Thus, compression of soft structures by the STM tip may be a common occurrence. This can be a serious problem for STM imaging of biological specimens. The REM–STM technique is probably the only one to directly 'see' this problem. Therefore, care must be exercised when interpreting STM images.

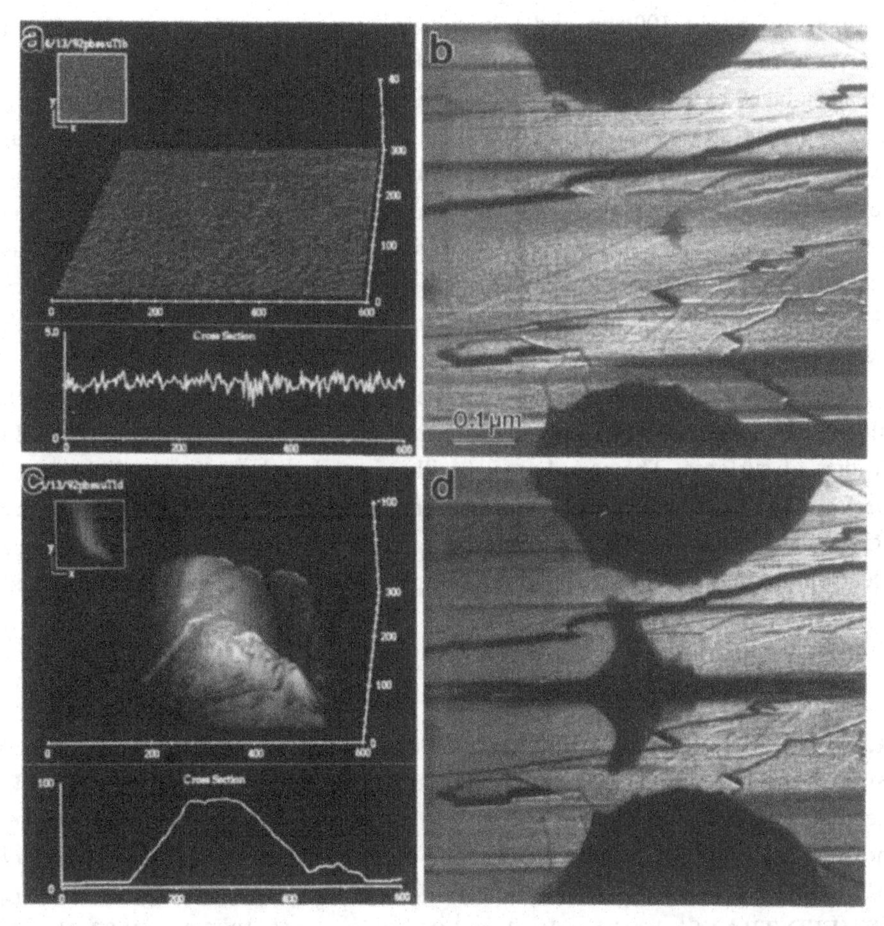

Figure 12.15 (a) A 600 nm × 300 nm STM image of a PbS(100) surface before application of a voltage pulse. (b) A 100 kV REM image corresponding to (a). The large, dark features at the top and bottom of this image are the Au STM tip and its mirror image, respectively. The scale mark indicates 100 nm. (c) A 600 nm × 300 nm STM image after application of a voltage pulse to the sample during tunneling. Two mounds formed by the voltage pulse can be seen. The large mound is 90 nm tall. The width across the base of both mounds is 400 nm. The scale along the axis is in nanometers. The actual height of the protrusion is 170 nm, not 90 nm as indicated in the STM image. The many step-like contrasts seen on the large mound in (c) are due to tip imaging. (d) A REM image of the surface after the tunneling experiment. The scale marker indicates 100 nm (courtesy of Drs W. Lo and J. C. H. Spence, 1993).

12.6 Time-resolved REM and REM with PEEM

Imaging of ultra-fast surface processes is a new application of REM. This technique employs a laser-pulsed thermal electron gun, which delivers high-brightness electron pulses with a duration of < 20 ns (Bostanjoglo and Heinricht, 1987), and allows imaging of laser-pulse-induced surface processes with a time resolution better than 20 ns (Heinricht, 1989). The laser pulse is focused onto the specimen surface to a diameter of about 100 μm and initiates the melting process on gold. After a selectable delay time, the electron pulse is shot onto the laser-irradiated area. This area is imaged by REM, and the image is intensified 10^5 times and then stored in a digital frame storage system. Ultra-short-time exposure images revealed that evaporation, melting, and flow processes on GaAs and Au–Si layers occur within several hundred nanoseconds. This technique allows recording of surface dynamic processes with time resolution and high spatial resolution. This technique has recently been applied to image the melting process of Si and GaAs surfaces (Heinricht and Bostanjoglo, 1992). They have found that the melt expands with radial velocities in the range 60–110 m s^{-1}.

A new UHV TEM has recently been designed by Kondo *et al.* (1991), which allows simultaneous REM and photon emission electron microscopy (PEEM) studies of surfaces. The sample can be illuminated with ultraviolet light when performing REM experiments. Surface crystallography can be determined by RHEED and REM, while PEEM observations give spatial changes of surface electronic structures. Surface bands and single steps on Si(111) have been observed by PEEM (Ohkawa *et al.*, 1994).

12.7 Total-reflection X-ray spectroscopy in RHEED

Surface chemistry can also be analyzed by acquiring X-ray spectra in the RHEED geometry (Sewell and Cohen, 1967). The surface sensitivity of this technique, however, depends critically on the selection of X-ray detector take-off angle (Ino *et al.*, 1980), and a technique, total-reflection-angle X-ray spectroscopy in RHEED (RHEED-TRAXS), has been developed (Hasegawa *et al.*, 1985; Ino, 1987; Yamanaka *et al.*, 1993; Ino and Yamanaka, 1993; Ino *et al.*, 1993). The physical principle of this technique can be described as follows.

We first consider the grazing angle refraction of X-rays at surfaces. In the case of X-rays, the refractive index is determined by the scattering factor $f_\alpha^x(0)$ for 0° forward scattering at a wavelength λ (Warren, 1990):

$$n_x = 1 - \lambda^2 e^2 \left(\sum_\alpha f_\alpha^x(0) \right) \Big/ 2\pi m_0 c^2 \Omega. \tag{12.4}$$

As the X-ray scattering factor at zero scattering angle is simply the atomic number Z, Eq. (12.4) can be written as

$$n_x = 1 - \delta n = 1 - 2.70 \times 10^{-9} \rho \lambda^2 \frac{\sum_\alpha Z_\alpha}{\sum_\alpha M_\alpha}, \tag{12.5}$$

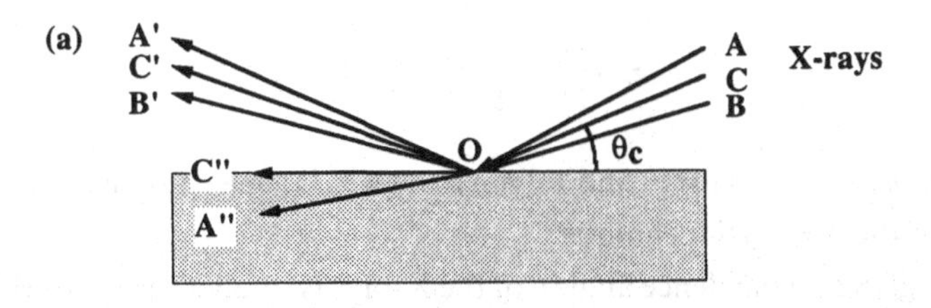

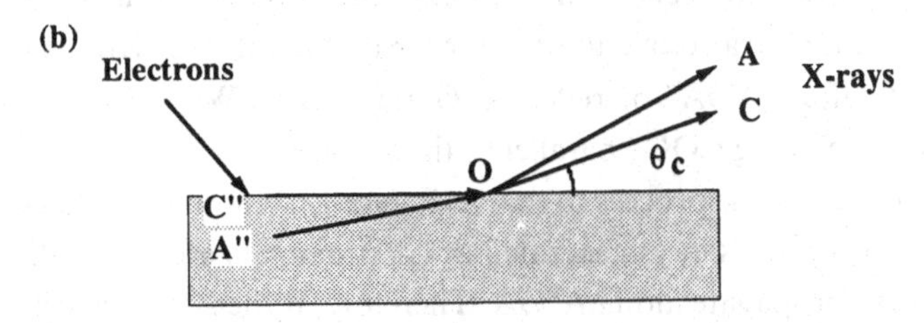

Figure 12.16 *(a) A ray diagram showing the total external reflection of X-rays at a solid surface for incidence angles less than θ_c. This is different from the total reflection of light, which occurs only inside a solid. In (b) is shown the reverse process to that in (a), in which X-rays are emitted inside the solid when it is excited by an electron beam. Surface-sensitive information can only be obtained when the X-ray detector is placed at a take-off angle equal to or smaller than θ_c.*

where M_α are the atomic masses, and ρ is the mass density in kg m^{-3}. Snell's law is written as

$$\cos\theta = (1 - \delta n)\cos\theta_\text{I}, \tag{12.6}$$

where θ_I is the internal refraction angle, and δn is on the order of 10^{-5}. Since the refraction index of X-rays is slightly less than unity (Zachariasen, 1945), it is possible to generate total external reflection when the X-rays strike a crystal surface, as illustrated in Figure 12.16(a). The critical angle at which total X-ray reflection occurs is determined by

$$\theta_c \approx (2\delta n)^{1/2} = 7.35 \times 10^{-5}\lambda\left(\rho\frac{\sum\limits_\alpha Z_\alpha}{\sum\limits_\alpha M_\alpha}\right)^{1/2}, \tag{12.7}$$

leading to critical angles of a few milliradians (and thus a few tenths of a degree) for typical X-ray wavelengths of about 0.15 nm. Using the small-angle approximation, Snell's law is rewritten as

$$\theta_\text{I} = (\theta^2 - \theta_c^2)^{1/2}. \tag{12.8}$$

Thus, for $\theta < \theta_c$, θ_I is imaginary and the component of the wavevector perpendicular to the surface ($K_z = K_0 n_x \sin\theta_\text{I}$) is also imaginary, which is characteristic of an evanescent wave perpendicular to the surface. The intensity of the X-ray wave field in the near-surface regions therefore decays with a characteristic length of

$$z_c = \frac{1}{2K_z} \approx \frac{\lambda}{2}(\theta^2 - \theta_c^2)^{-1/2}.$$

Typical values are 3.2 nm for Si and 1.2 nm for Au. z_c is a measurement of the surface sensitivity of the TRAXS technique.

When the glancing incidence angle θ of the X-rays is smaller than the critical angle θ_c, total X-ray reflection occurs. The incident X-rays reflect at the surface point O and propagate along trajectory BOB′. For $\theta > \theta_c$, the incident X-rays refract at O and propagate along AOA″ or reflect at O along OA′. When $\theta = \theta_c$, the incident X-rays propagate along COC″ parallel to the surface.

Now, when the reverse process to that in Figure 12.16(a) is considered, as shown in Figure 12.16(b), the X-rays generated at C″ propagate along C″OC, and those generated at A″ propagate along A″OA. Therefore, if the X-ray detector D with a fine slit is located at the position corresponding exactly to the total reflection critical angle θ_c, then only signals generated at C″ would arrive at the detector. The X-rays emitted inside the bulk could not arrive at the detector unless after propagating a long distance inside the crystal in order to reduce their glancing exit angles. However, the absorption of the crystal effectively reduces the contribution of this part of the X-rays. For $\theta < \theta_c$, only X-rays generated from the top surface layer are detected. This is, in principle, a surface-sensitive technique. However, in practice, the critical angle θ_c may be less than 1°, so that a slight variation in surface topography can destroy the surface sensitivity of this technique.

Figure 12.17 shows the existence of the critical take-off angle and its dependence on X-ray energy. The sample used is a clean Si(111) surface onto which one monolayer of Ag has been evaporated. With the increase in take-off angle θ_t, the intensity of the Si Kα line from the substrate increases slowly at first, then rapidly and subsequently levels off, making a shoulder-like form at about 1.0° of θ_t. The critical angle for Ag Lαβ is about 0.6°.

To see the change in relative value of Ag Lαβ against Si Lα, the values of the ratio of Ag Lαβ to Si Lα are plotted against the change in θ_t (see Figure 12.18). These curves correspond to the dependence on take-off angle of the detection sensitivity at the surface. For $\theta_t > 1.0°$, the detection sensitivity of one monolayer of Ag at the surface is not high. However for $\theta_t < 1.0°$, the sensitivity to Ag Lαβ is dramatically improved. Under optimum detection conditions, the intensity of the characteristic Ag Lαβ X-rays of only one monolayer of Ag atoms is comparable to that of the Si Lα from the substrate. This clearly demonstrates the surface sensitivity of RHEED-TRAXS under the condition illustrated above. This provides a powerful technique for determining the composition of the top surface monolayer and makes it possible to solve the structure of surface reconstruction due to adsorption or deposition.

As reported in RHEED patterns shown in Figures 2.13 and 2.14, the $(10,1) \times (\bar{3},4)$

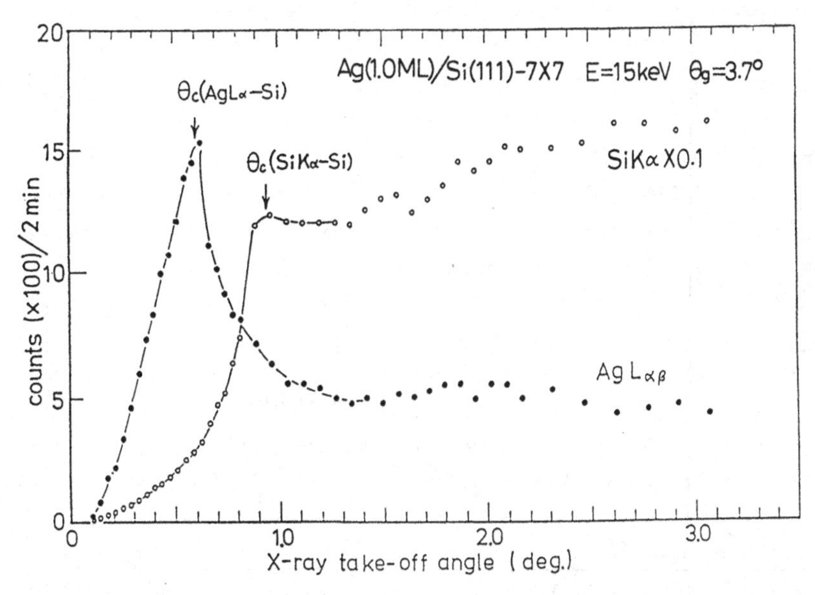

Figure 12.17 *The dependence of the absolute intensities (peak heights) of the characteristic X-rays on the X-ray take-off angle. The spectra were recorded from a Si(111)–Ag(1 ML) surface. Arrowheads indicate the critical take-off angles for total reflection of each characteristic X-ray signal (courtesy of Dr S. Ino, 1987).*

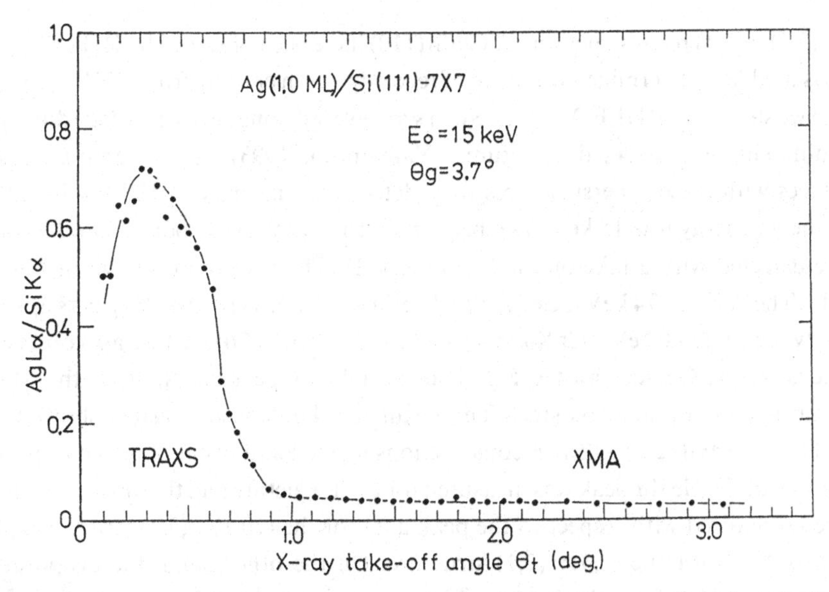

Figure 12.18 *Relative intensities of the characteristic X-ray (the Ag Lαβ line) from the deposit with respect to that (the Si Lα line) from the substrate versus the X-ray take-off angle θ_t in RHEED-TRAXS (courtesy of Dr S. Ino, 1987).*

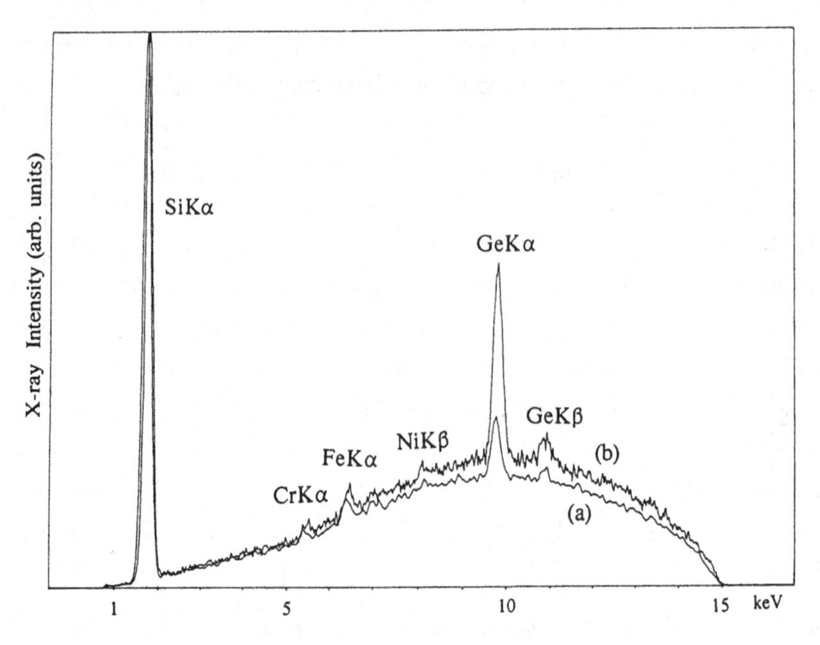

Figure 12.19 *RHEED-TRAXS spectra for the superstructure (a) (10,1) × ($\bar{3}$,4) and (b) 2,3) × ($\bar{2}$,1) of Ge/Si(110) observed in Figures 2.13 and 2.14, respectively (courtesy of Dr Y. Yamamoto, 1993).*

and (2,3) × ($\bar{2}$,1) reconstructions of Ge/Si(110) have been observed by RHEED, but one is unable to determine the nature of reconstruction entirely from RHEED data. Here we show that RHEED-TRAXS can sensitively distinguish the surface domains of different reconstructed structures (Yamamoto, 1993). X-rays emitted from surfaces with these superstructures were detected by means of RHEED-TRAXS. The beam energy was 15 keV. The beam incidence angle was about 2°, and X-rays were detected with a take-off angle of about 1°. The results are shown in Figure 12.19. The Si Kα (1.74 keV), Ge Kα (9.87 keV) and Ge Kβ (10.98 keV) peaks appear clearly. Fe Kα (6.40 keV), Cr Kα (5.41 keV) and Ni Kβ (8.26 keV) are also observed besides Si Kα, Ge Kα and Ge Kβ. The extra X-ray peaks came from the MBE chamber made of stainless steel. The result for the deposited layers shows that (10,1) × ($\bar{3}$,4) and (2,3) × ($\bar{2}$,1) reconstructions formed as a result of Ge adsorption. The area of the Ge Kα peak was measured for each structure and the obtained values were normalized with respect to the peak area for the (10,1) × ($\bar{3}$,4) structure. The ratio is 1:3.5 for the (2,3) × ($\bar{2}$,1) structure. On the other hand, the evaporated amount was 0.6 ML for the (10,1) × ($\bar{3}$,4) structure, and 1.8 ML for the (2,3) × ($\bar{2}$,1) structure, giving 0.6 : 1.8 = 1 : 3, in agreement with RHEED-TRAXS measurements. From the ratio of the signal peak to noise, the detection limit for Ge is estimated to be better than 0.05 ML.

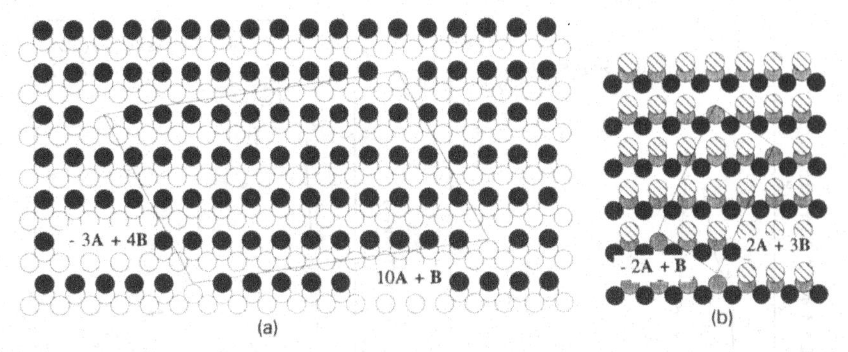

Figure 12.20 *Atom arrangements determined by RHEED-TRAXS for the (a) (10,1) × ($\bar{3}$,4) and (b) (2,3) × ($\bar{2}$,1) superstructures of the Ge/Si(110) system shown in Figures 2.13 and 2.14, respectively. The open circles represent Si atoms, the solid and gray circles Ge atoms, and the shaded circles Ge atoms on the Ge(110) plane (courtesy of Dr Y. Yamamoto, 1993).*

Based on RHEED (Figures 2.13 and 2.14) and RHEED-TRAXS observations, a structural model of (10,1) × ($\bar{3}$,4) Ge/Si(110) is shown in Figure 12.20(a), in which the open circles represent Si atoms forming the Si(110) plane, and the solid circles Ge atoms adsorbed on them up to an amount of 0.49 ML. At the corner of the unit mesh, Ge atoms do not exist. This structural model is consistent with the experimental result that the (10,1) × ($\bar{3}$,4) structure was observed in the range of 0.41–0.94 ML of Ge.

Likewise, the intensity of every reciprocal lattice rod was equal for the (2,3) × ($\bar{2}$,1) structure. Figure 12.20(b) shows the atomic arrangement for this structure; solid and gray circles represent Ge atoms on the Si(110) plane and the shaded ones Ge atoms on the Ge(110) surface. Again, Ge atoms do not appear at the corners of the unit mesh. The coverage is 1.35 ML. The fact that the (2,3) × ($\bar{2}$,1) structure was observed at coverages above 0.94 ML indicates that this model is reasonable.

12.8 Surface wave excitation Auger electron spectroscopy

As shown in Chapter 4, the reflected electrons are generated from the scattering of the top few surface layers under the resonance condition. The Auger electron and secondary electron yields of the surface are enhanced under the resonance condition owing to the confined parallel or nearly parallel propagation of the electron beam along the surface. A technique called surface wave excitation Auger electron spectroscopy (SWEAES) has been developed to collect the Auger electron emitted from the surface under different incident angles (Nakayama *et al.*, 1990). SWEAES can provide surface-sensitive atomic level information about surface valence electron states and surface composition of ultra-thin heterostructures (Nakayama *et al.*, 1993; Ueda *et al.*, 1991).

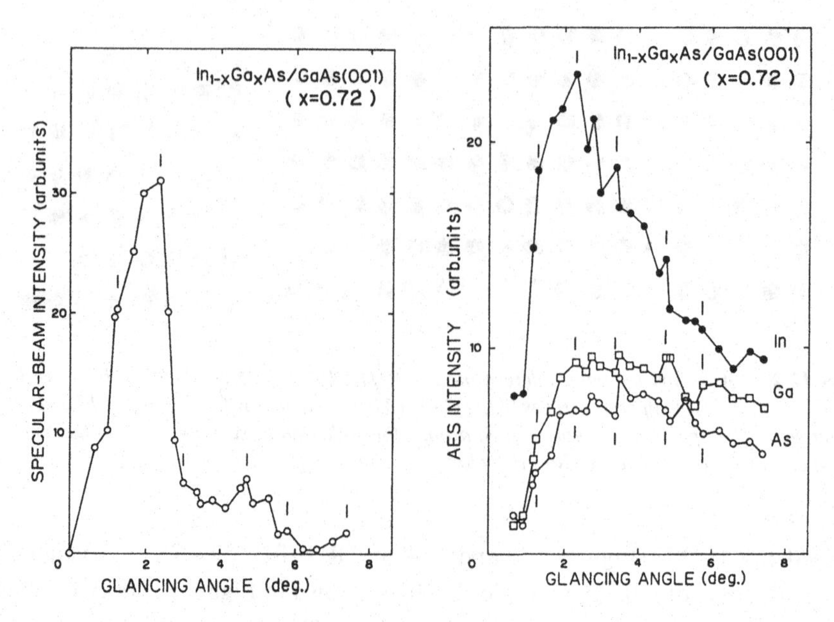

Figure 12.21 *(a) A plot of the specular beam intensity of the (00) rod as a function of the beam incident angle. (b) Plots of Auger intensity of In, Ga and As as a function of the incident beam angle (courtesy of Dr H. Nakayama).*

We now use the growth of InGaAs on GaAs(001) to illustrate the application of SWEAES. Figure 12.21(a) shows a plot of the specular beam intensity as a function of the beam incident angle for the $In_{1-x}Ga_xAs/GaAs(001)$ surface. The beam azimuth is [100]. Plots of the corresponding Auger signal intensities of In, Ga and As as a function of incidence angle are shown in Figure 12.21(b). These Auger intensities were measured as peak-to-peak heights of Auger spectra of In[3d,4d,4d] (400 eV), Ga[2p,3d,3d] (1070 eV) and As[2p,3d,3d] (1228 eV) principal Auger peaks. The primary beam energy was 9 keV. It is seen that the Auger intensity profiles of In and Ga fluctuate and have their maxima around the angle at which the RHEED specular beam intensity exhibits its maximum. In contrast, the As Auger signal exhibits a minimum.

The data shown in Figure 12.21(b) are analyzed in order to get some compositional information about the surface. Figure 12.22 shows the plots of Auger intensities in the form of the normalized intensities in the form As/(In + Ga) and In/(In + Ga), as a function of the incident angle. The calculations were performed with consideration of the difference in Auger sensitivity factors (equivalent to ionization cross-sections), which are 0.98, 0.14 and 0.11 for In, Ga and As, respectively. Thus, the ratios shown in Figure 12.22 are approximately proportional to the concentration ratios. The As/(In + Ga) ratio is minimum at the angle at which the specular beam intensity is maximum. This indicates that the surface resonance beams are

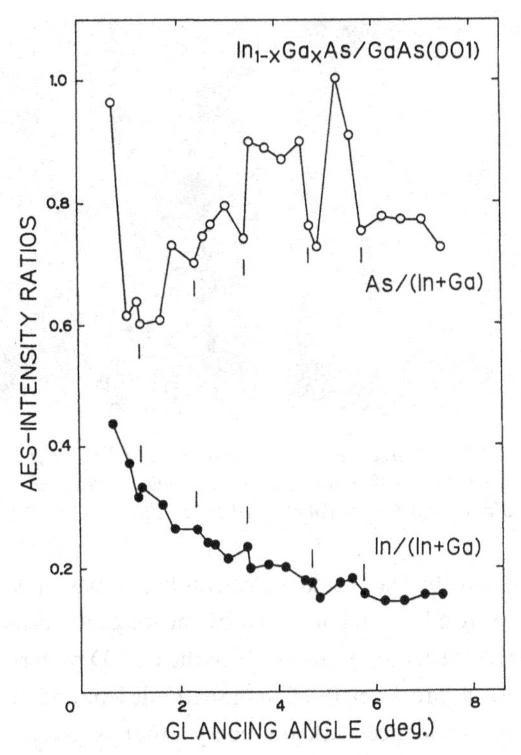

Figure 12.22 Plots of Auger intensity ratios In/(In+Ga) and As/(In+Ga), with the correction for the Auger sensitivity factors considered, so that the intensity ratios correspond to the concentration ratios of the probed region (courtesy of Dr H. Nakayama).

localized in the In/Ga plane rather than the As planes under diffraction conditions corresponding to the experimentally observed specular beam peak. It also shows that the electron wave localization is due to the presence of the well-ordered In/Ga outermost plane. This information is useful to identify the surface reconstruction using RHEED data.

12.9 LEED and LEEM

Low-energy electron diffraction (LEED) is a well-established technique for surface studies. Low-energy electrons, typically about 100 eV, hit the specimen almost at normal angle and are diffracted by the specimen before being scattered at more than 150° into vacuum. The backscattered electrons carry structural information about the surface. No magnetic lens is involved in LEED.

A similar idea has been developed to form images of crystal surfaces using low-energy electrons in LEED geometry, and it is known as low-energy electron microscopy (LEEM) (Bauer, 1994; Bauer *et al.*, 1989). The specimen is illuminated

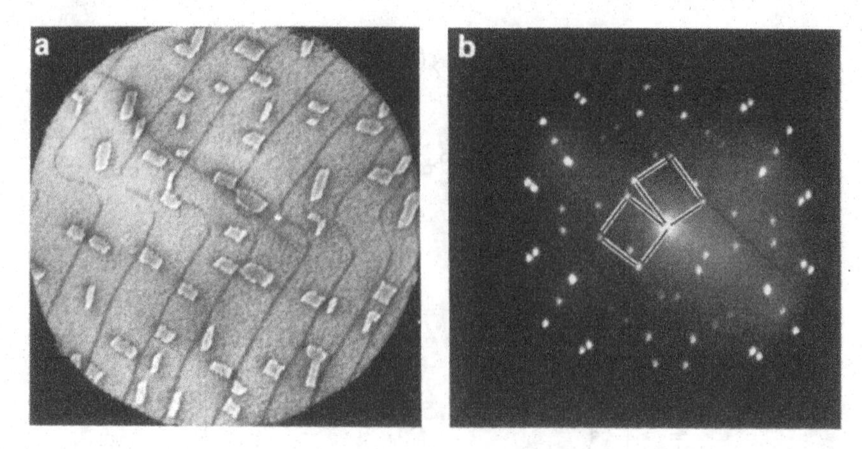

Figure 12.23 *(a) A LEEM image (E=9 eV) and (b) a LEED pattern (E=11 eV) of two-dimensional molybdenum carbide formed by cracking of residual gas at 1000 K on Mo(110). The image diameter is 6 μm (courtesy of Dr E. Bauer et al., 1989).*

through the cathode lens by the incident electron beam, which is produced by an electron gun, focused by a lens and deflected by the magnetic beam separator into the back focal plane of the cathode lens in which the LEED pattern of the specimen is recorded. The electrons are decelerated in the cathode lens and hit the specimen at (or nearly at) normal incidence with the energy that is given by the potential difference between cathode and specimen. The elastically backscattered electrons are re-accelerated to their original energy and focused by the optic system to form the LEEM image.

Both diffraction and imaging information can be simultaneously obtained from LEEM. This is a particularly powerful feature of this technique. Figure 12.23 shows a LEEM image of the Mo(110) surface and the corresponding LEED pattern. Formation of two-dimensional Mo carbide by cracking of residual gas and segregation from the bulk is seen. The crystallography of the particles formed on the surface can be obtained from the LEED pattern, and their real-space distribution can be seen in the LEEM image. Two types of crystal orientations have been identified from the LEED pattern (Figure 12.23(b)).

There are three important contrast mechanisms in LEEM. The dominant contrast is caused by diffraction within the specimen. Geometric phase contrast is due to different optical path lengths (in vacuum) of the electrons reflected from terraces adjoining a step. Quantum size contrast is caused by mutual interference of the electrons reflected from the parallel top and bottom faces of the surface layer. Constructive and destructive interferences are determined by the ratio of wavelength λ to step height or film thickness, which thus can be measured by varying the energy of the electrons. LEEM is a powerful technique for studying *in situ* real-time surface phase transformation and surface–gas reactions. LEEM has two important

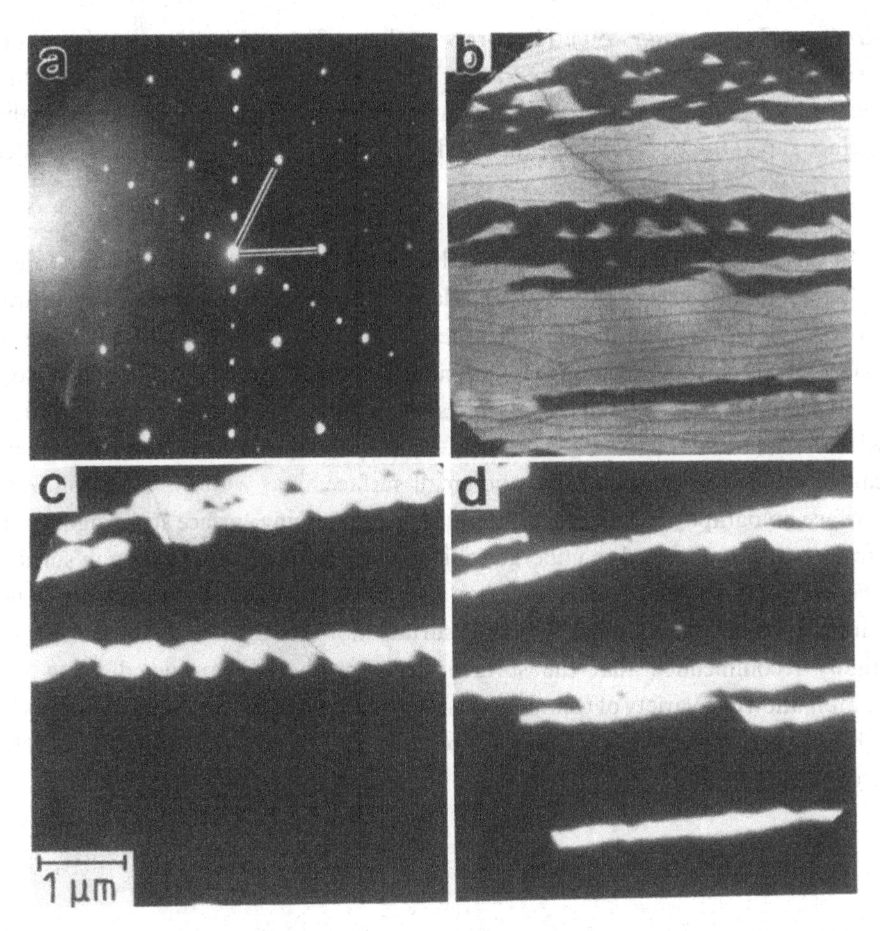

Figure 12.24 (a) A LEED pattern (30 eV), (b) bright-field (6 eV) and (c and d) dark-field (6 eV) LEEM images of Au deposition on Si(111). The dark-field images were recorded using the surface superlattice reflection beams of the two (5 × 1) orientations coexisting with the ($\sqrt{3} \times \sqrt{3}$)R30° structure seen in (a). In the bright-field image (a), the surface areas with ($\sqrt{3} \times \sqrt{3}$)R30° and 5 × 1 structures are in bright and dark contrast, respectively (courtesy of Dr E. Bauer).

advantages in comparison with REM. There is no foreshortening effect in LEEM, allowing distortion-free observation of surface structures. The specimens to be studied can be polycrystalline materials. However, the resolution of LEEM may be limited due to the low energy of the electrons used. A comprehensive review of this technique has been given by Bauer (1994).

Figure 12.24 shows some LEEM images of Au deposition on Si(111) (Swiech *et al.*, 1991). Two 5 × 1-oriented superstructures are seen in the LEED pattern (Figure 12.24). The distribution of surface steps and Au-covered surface areas is revealed by the bright-field image recorded using the (00) backscattered beam (Figure 12.24(b)). The area that shows bright contrast in Figure 12.24(b) exhibits the ($\sqrt{3} \times \sqrt{3}$)R30°

structure (Swiech *et al.*, 1991). The corresponding diffraction basis vectors from this structure are indicated in Figure 12.24(a). The distributions of the surface areas with the 5×1 structure are imaged using the dark-field imaging technique by selecting the surface superlattice reflection beams, and the results are shown in Figures 12.24(c) and (d).

In this chapter, we have demonstrated the techniques that have been developed either in conjunction with REM or related with reflected electrons for surface structure determination. Each of these techniques has its own unique advantages. The RHEED-TRAXS technique is particularly useful for determining the chemistry of surfaces, leading to a reliable atomic structure model of surface reconstruction with the use of RHEED data. The REM–STM technique is an exciting one, which can be applied to examine large and small surface areas with atomic resolution. Surface holography is a unique technique for determining surface step heights and strain, which is expected to undergo rapid development and application in surface science. Secondary electron imaging has demonstrated resolution better than 1 nm. This is a powerful technique for imaging surface morphology at high resolution. It is highly recommended that the surface structure should be studied using the combination of a variety of techniques. It is to be anticipated that the applications of RHEED and REM will be extensively broadened if these novel techniques can be used in conjunction.

Physical constants, electron wavelengths and wave numbers

The fundamental physical constants used in this book are:

$c = 2.99792458 \times 10^{8}$ m s^{-1}
$e = 1.6021892 \times 10^{-19}$ C
$h = 6.626196 \times 10^{-34}$ J s
$\hbar = h/2\pi = 1.054592 \times 10^{-34}$ J s
$k_B = 1.380622 \times 10^{-23}$ J K^{-1}
$m_0 = 9.109534 \times 10^{-31}$ kg
$\varepsilon_0 = 8.8541878 \times 10^{-12}$ C V^{-1} m^{-1}

The following table gives, as a function of accelerating voltage U_0, the relativistic electron wavelength λ, wave number $K = 1/\lambda$, relativistic factor $\gamma = m_e/m_0$, and velocity $\beta = v/c_0$.

E (kV)	λ (Å)	K (Å^{-1})	γ	β
0.001	12.26	0.0815	1.000002	0.0020
0.01	3.878	0.2579	1.000019	0.0063
0.1	1.226	0.8154	1.000196	0.0198
0.5	0.5483	1.824	1.000978	0.0625
1	0.3876	2.580	1.00196	0.0806
2	0.2740	3.650	1.00391	0.0882
3	0.2236	4.473	1.00587	0.1079
4	0.1935	5.167	1.00783	0.1244
5	0.1730	5.780	1.00978	0.1389
6	0.1579	6.335	1.01174	0.1519
7	0.1461	6.845	1.01370	0.1638
8	0.1366	7.322	1.01566	0.1749
9	0.1287	7.770	1.01761	0.1852
10	0.1220	8.194	1.01957	0.1950
20	0.0859	11.64	1.0391	0.2719
30	0.0698	14.33	1.0587	0.3284
40	0.0602	16.62	1.0783	0.3741
50	0.0536	18.67	1.0978	0.4127
60	0.0487	20.55	1.1174	0.4462
70	0.0448	22.30	1.1370	0.4759
80	0.0418	23.95	1.1566	0.5024
90	0.0392	25.52	1.1761	0.5264
100	0.0370	27.02	1.1957	0.5482
200	0.0251	39.87	1.3914	0.6953
300	0.0197	50.80	1.5871	0.7765
400	0.0164	60.83	1.7828	0.8279

Table (*cont.*)

E (kV)	λ (Å)	K (Å^{-1})	γ	β
500	0.0142	70.36	1.9785	0.8629
600	0.0126	79.57	2.1742	0.8879
700	0.0113	88.56	2.3698	0.9066
800	0.0103	97.38	2.5655	0.9209
900	0.0094	106.1	2.7912	0.9321
1000	0.0087	114.7	2.9569	0.9411

APPENDIX B

The crystal inner potential and electron scattering factor

In this appendix, we derive the relationship between the crystal inner potential and the electron scattering factor. Taking an inverse Fourier transform of Eq. (1.16), the crystal potential is written as

$$V(\boldsymbol{r}) = \sum_{g} V_g \exp(2\pi i \boldsymbol{g} \cdot \boldsymbol{r}). \tag{B.1}$$

Thus, the mean inner potential of the crystal is

$$\bar{V}_0 = \int \mathrm{d}\boldsymbol{r}\, V(\boldsymbol{r}) = V_{g=0}, \tag{B.2}$$

where $V_{g=0}$ is the crystal structure factor for $g=0$,

$$V_{g=0} = \frac{1}{\Omega} \sum_{\alpha} f_{\alpha}^{e}(0). \tag{B.3}$$

From Mott's formula (Eq. (1.15)), one has

$$f_{\alpha}^{e}(0) = \lim_{s \to 0} \frac{e}{16\pi^2 \varepsilon_0} \frac{[Z - f_{\alpha}^{x}(s)]}{s^2} = \frac{e}{16\pi^2 \varepsilon_0} \sum_{i=1}^{4} a_i^{(\alpha)} b_i^{(\alpha)}, \tag{B.4}$$

where $a_i^{(\alpha)}$ and $b_i^{(\alpha)}$ are the Doyle–Turner fitting parameters, which are listed in the table below. Substituting (B.4) into (B.3), the mean inner potential of the crystal is

$$\bar{V}_0 = \frac{e}{16\pi^2 \varepsilon_0 \Omega} \sum_{\alpha} \sum_{i=1}^{4} a_i^{(\alpha)} b_i^{(\alpha)}, \tag{B.5}$$

where superscript α is introduced in order to denote the atoms in the unit cell. The mean inner potential changes significantly when a neutral atom is ionized. Thus, the measurement of $\bar{V}_0$ can provide useful information about the bonding in crystals. The calculation of inner potential has been considered in detail by O'Keefee and Spence (1994), who have pointed out several factors that can easily introduce artifacts into this calculation.

The Doyle–Turner fitting parameters for calculating the X-ray scattering factors of neutral and ionized atoms (Doyle and Turner, 1968).

Z Element	a_1	b_1	a_2	b_2	a_3	b_3	a_4	b_4	c
1 H	0.48992	20.6593	0.26200	7.74039	0.196767	49.5519	0.049879	2.20159	0.0013
2 He	0.8734	9.1037	0.6309	3.3568	0.3112	22.9276	0.178	0.9821	0.0064
3 Li	1.1282	3.9546	0.7508	1.0524	0.6175	85.3905	0.4653	168.261	0.0377
3 Li^{+}	0.6968	4.6237	0.7885	0.9557	0.3414	0.6316	0.1563	10.0953	0.0157
4 Be	1.5919	43.6427	1.1278	1.8623	0.5391	103.483	0.7029	0.542	0.0385
4 Be^{2+}	6.2603	0.0027	0.8849	0.8313	0.7993	2.2758	0.1647	5.1146	−6.1092
5 B	2.0545	23.2185	1.3326	1.021	1.0979	60.3498	0.7068	0.1403	−0.1932
6 C	2.31	20.8439	1.02	10.2075	1.5886	0.5687	0.865	51.6512	0.2156
7 N	12.2126	0.0057	3.1322	9.8933	2.0125	28.9975	1.1663	0.5826	−11.529
8 O	3.0485	13.2771	2.2868	5.7011	1.5463	0.3239	0.867	32.9089	0.2508
9 F	3.5392	10.2825	2.6412	4.2944	1.517	0.2615	1.0243	26.1476	0.2776
10 Ne	3.9553	8.4042	3.1125	3.4262	1.4546	0.2306	1.1251	21.7184	0.3515
11 Na	4.7626	3.285	3.1736	8.8422	1.2674	0.3136	1.1128	129.424	0.676
11 Na^{+}	3.2565	2.6671	3.9362	6.1153	1.3998	0.2001	1.0032	14.0390	0.4040
12 Mg	5.4204	2.8275	2.1735	79.2611	1.2269	0.3808	2.3073	7.1937	0.8584
12 Mg^{2+}	3.4988	2.1676	3.8378	4.7542	1.3284	0.1850	0.8497	10.1411	0.4853
13 Al	6.4202	3.0387	1.9002	0.7426	1.5936	31.5472	1.9646	85.0886	1.1151
14 Si	6.2915	2.4386	3.0353	32.3337	1.9891	0.6785	1.541	81.6937	1.1407
15 P	6.4345	1.9067	4.1791	27.157	1.78	0.526	1.4908	68.1645	1.1149
16 S	6.9053	1.4679	5.2034	22.2151	1.4379	0.2536	1.5863	56.172	0.8669
17 Cl	11.4604	0.0104	7.1964	1.1662	6.2556	18.5194	1.6455	47.7784	−9.5574
17 Cl^{+}	18.2915	0.0066	7.2084	1.1717	6.5337	19.5424	2.3386	60.4486	−16.3776
18 Ar	7.4845	0.9072	6.7723	14.8407	0.6539	43.8983	1.6442	33.3929	1.4445
19 K	8.2186	12.7949	7.4398	0.7748	1.0519	213.187	0.8659	41.6841	1.4228
19 K^{+}	7.9578	12.6331	7.4917	0.7674	6.3590	−0.0020	1.1915	31.9128	−4.9978
20 Ca	8.6266	10.4421	7.3873	0.6599	1.5899	85.7484	1.0211	178.437	1.3751
20 Ca^{2+}	15.6348	−0.0074	7.9518	0.6089	8.4372	10.3116	0.3537	25.9905	−14.8751
21 Sc	9.189	9.0213	7.3679	0.5729	1.6409	136.108	1.468	51.3531	1.3329
22 Ti	9.7595	7.8508	7.3558	0.5	1.6991	35.6338	1.9021	116.105	1.2807

23 V	10.2971	6.8657	7.3511	0.4385	2.0703	26.8938	2.0571	102.478	1.2199
23 V^{2+}	10.1060	6.8818	7.3541	0.4409	2.2884	20.3004	0.0223	115.1221	1.2298
24 Cr	10.6406	6.1038	7.3537	0.392	3.324	20.2626	1.4922	98.7399	1.1832
25 Mn	11.2819	5.3409	7.3573	0.3432	3.0193	17.8674	2.2441	83.7543	1.0896
25 Mn^{2+}	10.8061	5.2796	7.3620	0.3435	3.5268	14.3430	0.2184	41.3235	1.0874
26 Fe	11.7695	4.7611	7.3573	0.3072	3.5222	15.3535	2.3045	76.8805	1.0369
26 Fe^{3+}	11.1764	4.6147	7.3863	0.3005	3.3948	11.6729	0.0724	38.5566	0.9707
26 Fe^{2+}	11.0424	4.6538	7.3740	0.3053	4.1346	12.0546	0.4399	31.2809	1.0097
27 Co	12.2841	4.2791	7.3409	0.2784	4.0034	13.5359	2.3488	71.1692	1.0118
27 Co^{2+}	11.2296	4.1231	7.3883	0.2726	4.7393	10.2443	0.7108	25.6466	0.9324
28 Ni	12.8376	3.8785	7.292	0.2565	4.4438	12.1763	2.38	66.3421	1.0341
28 Ni^{2+}	11.4166	3.6766	7.4005	0.2449	5.3442	8.8730	0.9773	22.1626	0.8614
29 Cu	13.338	3.5828	7.1676	0.247	5.6158	11.3966	1.6735	64.8126	1.191
29 Cu^{+}	11.9475	3.3669	7.3573	0.2274	6.2455	8.6625	1.5578	25.8487	0.8900
30 Zn	14.0743	3.2655	7.0318	0.2333	5.1652	10.3163	2.41	58.7097	1.3041
30 Zn^{2+}	11.9719	2.9946	7.3862	0.2031	6.4668	7.0826	1.3940	18.0995	0.7807
31 Ga	15.2354	3.0669	6.7006	0.2412	4.3591	10.7805	2.9623	61.4135	1.7189
32 Ge	16.0816	2.8509	6.3747	0.2516	3.7086	11.4468	3.683	54.7625	2.1313
33 As	16.6723	2.6345	6.0701	0.2647	3.4313	12.9479	4.2779	47.7972	2.531
34 Se	17.0006	2.4098	5.8196	0.2726	3.9731	15.2372	4.3543	43.8163	2.8409
35 Br	17.1789	2.1723	5.2358	16.5796	5.6377	0.2609	3.9851	41.4328	2.9557
35 Br^{-}	17.1718	2.2059	6.3335	19.3345	5.5754	0.2871	3.7272	58.1535	3.1776
36 Kr	17.3555	1.9384	6.7286	16.5623	5.5493	0.2261	3.5375	39.3972	2.825
37 Rb	17.1784	1.7888	9.6435	17.3151	5.1399	0.2748	1.5292	164.934	3.4873
37 Rb^{+}	17.5816	1.7139	7.6598	14.7957	5.8981	0.1603	2.7817	31.2087	2.0782
38 Sr	17.5663	1.5564	9.8184	14.0988	5.422	0.1664	2.6694	132.376	2.5064
38 Sr^{2+}	18.0874	1.4907	8.1373	12.6963	2.5654	24.5651	−34.1929	−0.0138	41.4025
39 Y	17.776	1.4029	10.2946	12.8006	5.72629	0.125599	3.26588	104.354	1.91213
40 Zr	17.8765	1.27618	10.948	11.916	5.41732	0.117622	3.65721	87.6627	2.06929
41 Nb	17.6142	1.18865	12.0144	11.766	4.04183	0.204785	3.53346	69.7957	3.75591
42 Mo	3.7025	0.2772	17.2356	1.0958	12.8876	11.004	3.7429	61.6584	4.3875
43 Tc	19.1301	0.864132	11.0948	8.14487	4.64901	21.5707	2.71263	86.8472	5.40428
44 Ru	19.2674	0.80852	12.9182	8.43467	4.86337	24.7997	1.56756	94.2928	5.37874

Table (*cont.*)

Z Element	a_1	b_1	a_2	b_2	a_3	b_3	a_4	b_4	c
45 Rh	19.2957	0.751536	14.3501	8.21758	4.73425	25.8749	1.28918	98.6062	5.328
46 Pd	19.3319	0.698655	15.5017	7.98929	5.29537	25.2052	0.605844	76.8986	5.26593
47 Ag	19.2808	0.6446	16.6885	7.4726	4.8045	24.6605	1.0463	99.8156	5.179
48 Cd	19.2214	0.5946	17.6444	6.9089	4.461	24.7008	1.6029	87.4825	5.0694
49 In	19.1624	0.5476	18.5596	6.3776	4.2948	25.8499	2.0396	92.8029	4.9391
50 Sn	19.1889	5.8303	19.1005	0.5031	4.4585	26.8909	2.4663	83.9571	4.7821
50 Sn^{2+}	19.1094	0.5036	19.0548	5.8378	4.5648	23.3752	0.4870	62.2061	4.7861
50 Sn^{4+}	18.9333	5.7640	19.7131	0.4655	3.4182	14.0049	0.0193	−0.7583	3.9182
51 Sb	19.6418	5.3034	19.0455	0.4607	5.0371	27.9074	2.6827	75.2825	4.5909
52 Te	19.9644	4.81742	19.0138	0.420885	6.14487	28.5284	2.5239	70.8403	4.352
53 I	20.1472	4.347	18.9949	0.3814	7.5138	27.766	2.2735	66.8776	4.0712
53 I^-	20.2332	4.3579	18.9970	0.3815	7.8069	29.5259	2.8868	84.9304	4.0714
54 Xe	20.2933	3.9282	19.0298	0.344	8.9767	26.4659	1.99	64.2658	3.7118
55 Cs	20.3892	3.569	19.1062	0.3107	10.662	24.3879	1.4953	213.904	3.3352
55 Cs^+	20.3524	3.5520	19.1278	0.3086	10.2821	23.7128	0.9615	59.4565	3.2791
56 Ba	20.3361	3.216	19.297	0.2756	10.888	20.2073	2.6959	167.202	2.7731
57 La	20.578	2.94817	19.599	0.244475	11.3727	18.7726	3.28719	133.124	2.14678
58 Ce	21.1671	2.81219	19.7695	0.226836	11.8513	17.6083	3.33049	127.113	1.86264
59 Pr	22.044	2.77393	19.6697	0.222087	12.3856	16.7669	2.82428	143.644	2.0583
60 Nd	22.6845	2.66248	19.6847	0.210628	12.774	15.885	2.85137	137.903	1.98486
61 Pm	23.3405	2.5627	19.6095	0.202088	13.1235	15.1009	2.87516	132.721	2.02876
62 Sm	24.0042	2.47274	19.4258	0.196451	13.4396	14.3996	2.89604	128.007	2.20963
63 Eu	24.6274	2.3879	19.0886	0.1942	13.7603	13.7546	2.9227	123.174	2.5745
64 Gd	25.0709	2.25341	19.0798	0.181951	13.8518	12.9331	3.54545	101.398	2.4196
65 Tb	25.8976	2.24256	18.2185	0.196143	14.3167	12.6648	2.95354	115.362	3.58324
66 Dy	26.507	2.1802	17.6383	0.202172	14.5596	12.1899	2.96577	111.874	4.29728
67 Ho	26.9049	2.07051	17.294	0.19794	14.5583	11.4407	3.63837	92.6566	4.56796
68 Er	27.6563	2.07356	16.4285	0.223545	14.9779	11.3604	2.98233	105.703	5.92046
69 Tm	28.1819	2.02859	15.8851	0.238849	15.1542	10.9975	2.98706	102.961	6.75621

70 Yb	28.6641	1.9889	15.4345	0.257119	15.3087	10.6647	2.98963	100.417	7.56672
71 Lu	28.9476	1.90182	15.2208	9.98519	15.100	0.261033	3.71601	84.3298	7.97628
72 Hf	29.144	1.83262	15.1726	9.5999	14.7586	0.275116	4.30013	72.029	8.58154
73 Ta	29.2024	1.77333	15.2293	9.37046	14.5135	0.295977	4.76492	63.3644	9.24354
74 W	29.0818	1.72029	15.43	9.2259	14.4327	0.321703	5.11982	57.056	9.8875
75 Re	28.7621	1.67191	15.7189	9.09227	14.5564	0.3505	5.44174	52.0861	10.472
76 Os	28.1894	1.62903	16.155	8.97948	14.9305	0.382661	5.67589	48.1647	11.0005
77 Ir	27.3049	1.59279	16.7296	8.86553	15.6115	0.417916	5.83377	45.001	11.4722
78 Pt	27.0059	1.51293	17.7639	8.81174	15.7131	0.424593	5.7837	38.6103	11.6883
79 Au	16.8819	0.4611	18.5913	8.6216	25.5582	1.4826	5.86	36.3956	12.0658
80 Hg	20.6809	0.545	19.0417	8.4484	21.6575	1.5729	5.9676	38.3246	12.6089
81 Tl	27.5446	0.65515	19.1584	8.70751	15.538	1.96347	5.52593	45.8149	13.1746
82 Pb	31.0617	0.6902	13.0637	2.3576	18.4420	8.618	5.9696	47.2579	13.4118
83 Bi	33.3689	0.704	12.951	2.9238	16.5877	8.7937	6.4692	48.0093	13.5782
84 Po	34.6726	0.700999	15.4733	3.55078	13.1138	9.55642	7.02588	47.0045	13.677
85 At	35.3163	0.68587	19.0211	3.97458	9.49887	11.3824	7.42518	45.4715	13.7108
86 Rn	35.5631	0.6631	21.2816	4.0691	8.0037	14.0422	7.4433	44.2473	13.6905
87 Fr	35.9299	0.646453	23.0547	4.17619	12.1439	23.1052	2.11253	150.645	13.7247
88 Ra	35.763	0.616341	22.9064	3.87135	12.4739	19.9887	3.21097	142.325	13.6211
89 Ac	35.6597	0.589092	23.1032	3.65155	12.5977	18.599	4.08655	117.02	13.5266
90 Th	35.5645	0.563359	23.4219	3.46204	12.7473	17.8309	4.80703	99.1722	13.4314
91 Pa	35.8847	0.547751	23.2948	3.41519	14.1891	16.9235	4.17287	105.251	13.4287
92 U	36.0228	0.5293	23.4128	3.3253	14.9491	16.0927	4.188	100.613	13.3966
93 Np	36.1874	0.511929	23.5964	3.25396	15.6402	15.3622	4.1855	97.4908	13.3573
94 Pu	36.5254	0.499384	23.8023	3.26371	16.7707	14.9455	3.47947	105.98	13.3812
95 Am	36.6706	0.483629	24.0992	3.20647	17.3415	14.3136	3.49331	102.273	13.3592
96 Cm	36.6488	0.645154	24.4096	3.08997	17.399	13.4346	4.21665	88.4834	13.2887
97 Bk	36.7881	0.451018	24.7736	3.04619	17.8919	12.8946	4.23284	86.003	13.2754
98 Cf	36.9185	0.437533	25.1995	3.00775	18.3317	12.4044	4.24391	83.7881	13.2674

APPENDIX C.1

Crystallographic structure systems

Listed below are the distance (d_g) between the adjacent atomic planes with Miller indices (hkl), the angle (Θ_p between planes ($h_1k_1l_1$) and ($h_2k_2l_2$), and the relationship between plane normal direction [uvw] and the plane indices, for the seven crystal systems as illustrated in Figure C.1. A more complete description of the crystallography has been given by Jackson (1991). A FORTRAN program for calculating these quantities is given in Appendix C.2.

1 *Cubic*

Three equal axes at right angles: $a=b=c$, $\alpha=\beta=\gamma=90°$:

$$\frac{1}{d_g^2}=\frac{h^2+k^2+l^2}{a^2},$$

$$\cos\Theta_p=\frac{h_1h_2+k_1k_2+l_1l_2}{(h_1^2+k_1^2+l_1^2)^{1/2}(h_2^2+k_2^2+l_2^2)^{1/2}}$$

$$\frac{u}{h}=\frac{v}{k}=\frac{w}{l}.$$

2 *Hexagonal*

Two axes at 120°, third axis at right angles to both: $a=b\neq c$, $\alpha=\beta=90°$, $\gamma=120°$:

$$\frac{1}{d_g^2}=\frac{4}{3}\left(\frac{h^2+hk+k^2}{a^2}\right)+\frac{l^2}{c^2},$$

$$\cos\Theta_p=\frac{4}{3a^2}\left(h_1h_2+k_1k_2+\frac{1}{2}(h_1k_2+h_2k_1)+\frac{3a^2}{4c^2}l_1l_2\right)d_{g1}d_{g2},$$

where d_{g1} and d_{g2} are the interplanar distances for ($h_1k_1l_1$) and ($h_2k_2l_2$), respectively,

$$\frac{u}{h}=\frac{v}{k}=\frac{w}{3a^2l/2c^2}.$$

An analysis of transmission electron diffraction from hexagonal systems has been reviewed by Edington (1976).

3 *Trigonal (or rhombohedral)*

Three equal axes, equally inclined: $a=b=c$, $\alpha=\beta=\gamma\neq 90°$:

$$\frac{1}{d_g^2}=\frac{(h^2+k^2+l^2)\sin^2\alpha+2(hk+kl+hl)(\cos^2\alpha-\cos\alpha)}{a^2(1-3\cos^2\alpha+2\cos^3\alpha)},$$

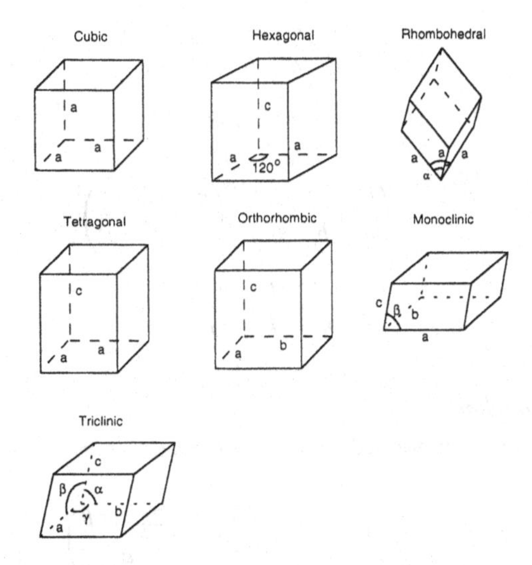

Figure C.1 *The seven crystallographic systems.*

$$\cos\Theta_p = \frac{d_{g1}d_{g2}}{3a^2}\,\xi[h_1h_2+k_1k_2+l_1l_2+(h_1k_2+h_1l_2+k_1h_2+l_1h_2+l_1k_2)\cos\alpha^*],$$

where

$$\xi = \frac{4+5\cos\alpha}{\cos\alpha-\cos(2\alpha)}, \qquad \cos\alpha^* = \frac{3-2\zeta^2}{3+4\zeta^2}, \qquad \zeta = \left(\frac{3}{2}\,\frac{1+2\cos\alpha}{1-\cos\alpha}\right)^{1/2},$$

$$\frac{u}{h+(h-k-l)\cos\alpha} = \frac{v}{k+(k-l-h)\cos\alpha} = \frac{w}{l+(l-h-k)\cos\alpha}.$$

4 *Tetragonal*

Three axes at right angles: $a=b\neq c$, $\alpha=\beta=\gamma=90°$:

$$\frac{1}{d_g^2} = \frac{h^2+k^2}{a^2}+\frac{l^2}{c^2},$$

$$\cos\Theta_p = \frac{\dfrac{h_1h_2+k_1k_2}{a^2}+\dfrac{l_1l_2}{c^2}}{\left(\dfrac{h_1^2+k_1^2}{a^2}+\dfrac{l_1^2}{c^2}\right)^{1/2}\left(\dfrac{h_2^2+k_2^2}{a^2}+\dfrac{l_2^2}{c^2}\right)^{1/2}},$$

$$\frac{u}{h}=\frac{v}{k}=\frac{w}{l}\,\frac{c^2}{a^2}.$$

5 *Orthorhombic*

Three orthogonal unequal axes: $a\neq b\neq c$, $\alpha=\beta=\gamma=90°$:

$$\frac{1}{d_g^2}=\frac{h^2}{a^2}+\frac{k^2}{b^2}+\frac{l^2}{c^2},$$

$$\cos\Theta_p=\frac{\dfrac{h_1h_2}{a^2}+\dfrac{k_1k_2}{b^2}+\dfrac{l_1l_2}{c^2}}{\left(\dfrac{h_1^2}{a^2}+\dfrac{k_1^2}{b^2}+\dfrac{l_1^2}{c^2}\right)^{1/2}\left(\dfrac{h_2^2}{a^2}+\dfrac{k_2^2}{b^2}+\dfrac{l_2^2}{c^2}\right)^{1/2}},$$

$$\frac{ua^2}{h}=\frac{vb^2}{k}=\frac{wc^2}{l}.$$

6 *Monoclinic*

Three unequal axes, one pair not orthogonal: $a\neq b\neq c$, $\alpha=\gamma=90°\neq\beta$.

$$\frac{1}{d_g^2}=\frac{1}{\sin^2\beta}\left(\frac{h^2}{a^2}+\frac{k^2\sin^2\beta}{b^2}+\frac{l^2}{c^2}-\frac{2hl\cos\beta}{ac}\right).$$

$$\cos\Theta_p=d_{g1}d_{g2}\,\frac{\dfrac{h_1h_2}{a^2}+\dfrac{k_1k_2\sin^2\beta}{b^2}+\dfrac{l_1l_2}{c^2}-\dfrac{(h_2l_1+h_1l_2)\cos\beta}{ac}}{\sin^2\beta},$$

$$\frac{u}{hb^2c^2-lab^2c\cos\beta}=\frac{v}{kc^2a^2\sin^2\beta}=\frac{w}{ha^2b^2-hab^2c\cos\beta}.$$

7 *Triclinic*

Three unequal axes, none at right angles: $a\neq b\neq c$, $\alpha\neq\beta\neq\gamma$. The equations given for this case are applicable to any crystal system:

$$\frac{1}{d_g^2}=h^2a^{*2}+k^2b^{*2}+l^2c^{*2}+2hka^*b^*\cos\gamma^*+2klb^*c^*\cos\alpha^*+2lhc^*a^*\cos\beta^*,$$

$$\cos\Theta_p=[(h_1h_2a^{*2}+k_1k_2b^{*2})+l_1l_2c^{*2}+(h_1k_2+h_2k_1)a^*b^*\cos\gamma^*$$
$$+(k_1l_2+k_2l_1)b^*c^*\cos\alpha^*+(h_1l_2+h_2l_1)c^*a^*\cos\beta^*]d_{g1}d_{g2},$$

where

$$a^*=\frac{bc\sin\alpha}{\Omega},\qquad b^*=\frac{ca\sin\beta}{\Omega},\qquad c^*=\frac{ab\sin\gamma}{\Omega},$$

$$\cos\alpha^*=\frac{\cos\beta\cos\gamma-\cos\alpha}{\sin\gamma\sin\beta},$$

$$\cos\beta^*=\frac{\cos\gamma\cos\alpha-\cos\beta}{\sin\gamma\sin\alpha},$$

$$\cos\gamma^*=\frac{\cos\alpha\cos\beta-\cos\gamma}{\sin\alpha\sin\beta},$$

and the volume of the unit cell in real space is

$$\Omega = abc(1-\cos^2\alpha-\cos^2\beta-\cos^2\gamma+2\cos\alpha\cos\beta\cos\gamma)^{1/2},$$

$$\frac{u}{hb^2c^2\sin^2\alpha+kabc^2(\cos\alpha\cos\beta-\cos\gamma)+lab^2c(\cos\gamma\cos\alpha-\cos\beta)}$$

$$=\frac{v}{habc^2(\cos\alpha\cos\beta-\cos\gamma)+ka^2c^2\sin^2\beta+la^2bc(\cos\gamma\cos\beta-\cos\alpha)}$$

$$=\frac{w}{hab^2c(\cos\gamma\cos\alpha-\cos\beta)+ka^2bc(\cos\gamma\cos\beta-\cos\alpha)+la^2b^2\sin^2\gamma}.$$

The volume of the first Brillouin zone (or the unit cell in reciprocal space) is

$$\Omega^* = \frac{1}{\Omega}.$$

APPENDIX C.2

A FORTRAN program for calculating crystallographic data

Listed below is the FORTRAN source code for calculating the volume of the unit cell in real space, the volume of the first Brillouin zone, interplanar distances, angles between planes, and ratios of *g* vectors. The calculated data are output in a data file, CRYSTAL.DAT. The program has been tested on a VAX computer.

```
C     CRYSTDATA.FOR
C
C     Z.L. Wang
C
C     THIS PROGRAM CALCULATES THE INTERPLANAR DISTANCE,
C     ANGLES BETWEEN PLANES, GRATIO OF RECIPROCAL LATTICE
C     VECTORS AND VOLUME OF UNIT CELL, FOR ANY CRYSTAL SYSTEM.
C
      COMMON H(10),K(10),L(10),DG(10), GRATIO(10,10),THITA(10,10)
      COMMON DU(10), DV(10), DW(10),VOL
      CHARACTER NAME*10 , YON*1, YO*1
      INTEGER H,K,L
      OPEN(UNIT=4,FILE='CRYSTAL.DAT',STATUS='NEW')
 101  A1=90
      B1=90
      C1=90
C
      WRITE(6,100)
 100  FORMAT(/' 1 ---TRICLINIC'/
    1   ' 2 ---MONOCLINIC'/
    2   ' 3 ---ORTHOGONAL'/
    3   ' 4 ---HEXAGONAL '/
    4   ' 5 ---TETRAGONAL'/
    5   ' 6 ---CUBIC    ')
 200  WRITE(6,300)
 300  FORMAT(/' ENTER TYPE OF LATTICE BY NUMBER ?----',$)
      READ(5,400,ERR=200) L1
 400  FORMAT(I1)
 500  WRITE(6,600)
 600  FORMAT(/' ENTER CHEMICAL FORMULA----',$)
      READ(5,800,ERR=500) NAME
 800  FORMAT(A6)
      WRITE(4,900) NAME
 900  FORMAT(/// 10X,'CHEMICAL FORMULA'2X,A6)
      GO TO (10, 20, 30, 40 ,50 ,60), L1
 10   TYPE*,' ENTER PARAMETERS: A,B,C,ALPHA,BETA,GAMMA----'
      ACCEPT * , A,B,C,A1,B1,C1
C
      WRITE(4,120) A,B,C,A1,B1,C1
 120  FORMAT(/ ' TRICLINIC: A=' F10.6,' B='F10.6,' C=' F10.6,
    1   ' ALPHA=' F10.6,' BETA=' F10.6, ' GAMMA=' F10.6)
      GOTO 1000
 20   TYPE *,' ENTER PARAMETERS: A,B,C,BETA----$'
      ACCEPT *, A,B,C,B1
      WRITE(4,130) A,B,C,B1
```

```
 130   FORMAT(/ '  MONOCLINIC: A=' F10.6,'  B=' F10.6,'  C='F10.6,
   1   '  BETA='F10.6)
       GOTO 1000
  30   TYPE *, ' ENTER PARAMETER: A,B,C----$'
       ACCEPT *, A,B,C
       WRITE(4,140) A,B,C
 140   FORMAT(/'  ORTHOGONAL: A=' F10.6,'  B=' F10.6,'  C=' F10.6)
       GOTO 1000
  40   TYPE *,' ENTER PARAMETERS: A,C----$'
       ACCEPT *, A,C
       B=A
       C1=120
       WRITE(4,150) A,C
 150   FORMAT(/ '  HEXAGONAL : A=' F10.6,'  C='F10.6)
       GOTO 1000
  50   TYPE *,'  ENTER PARAMETERS: A,C----$'
       ACCEPT *, A,C
       B=A
       WRITE(4,160) A,C
 160   FORMAT( / '  TETRAGONAL: A='F10.6,'  C=' F10.6 )
       GOTO 1000
  60   TYPE *,'  ENTER PARAMETERS : A----$'
       ACCEPT *, A
       B=A
       C=A
       WRITE(4,180) A
 180   FORMAT(/ '  CUBIC   : A=' F10.6)
1000   CONTINUE
C
       F=3.14159/180.0
       A1=A1*F
       B1=B1*F
       C1=C1*F
C
 190   WRITE(6,210)
 210   FORMAT(/ ' IS THIS A PARTICULAR PROJECTION ? (Y OR N)---$',$)
       READ(5,211,ERR=190) YON
 211   FORMAT(A1)
       IF(YON.NE.'Y') GOTO 2000
       TYPE *,' ENTER , N1,N2,N3----$'
       ACCEPT 220,N1,N2,N3
 220   FORMAT(10I3)
       WRITE(4,260) N1,N2,N3
 260   FORMAT(/4X,' '3I2,'! -- PROJECTION')
2000   CONTINUE
 241   WRITE(6,230)
 230   FORMAT( /' ENTER NUMBER OF PLANES TO BE CALCULATED(<5)',$)
       READ(5,*,ERR=241) NMAX
       DO 240 I=1,NMAX
       TYPE *,'  ENTER H,K,L----$ '
       ACCEPT 2210,H(I),K(I),L(I)
 240   CONTINUE
2210   FORMAT(3I3)
C
       CALL CRYSTAL(A,B,C,A1,B1,C1,NMAX)
C
C Data output
       WRITE(4,721) VOL
 721   FORMAT(2X,'Unit cell volume V = ',E12.5, ' Angstrom**3')
       VOLR=1.0/VOL
       WRITE(4,722) VOLR
 722   FORMAT(2X,'Volume of Brillouin zone V* = ',E12.6,
   A    ' 1/Angstrom**3')
       WRITE(4,723)
```

```
 723   FORMAT(2X,'Interplanar distance and normal direction [uvw]')
       DO 728 I=1,NMAX
       WRITE(4,726)H(I),K(I),L(I),DG(I),DU(I),DV(I),DW(I)
 726   FORMAT(2X,'g = ','(',3I3,')',' dg = ',F8.5,X,'[UVW] = ',
     A    '[',F8.6,X,F8.6,X,F8.6,']',/)
 728   CONTINUE
C
       WRITE(4,729)
 729   FORMAT(//,2X,'Ratio of g and interplanar angles:',/)
       NMA=NMAX-1
       DO 270 I=1,NMAX
       DO 269 J=1,NMAX
       WRITE(4,263) H(I),K(I),L(I),H(J),K(J),L(J),GRATIO(I,J),THITA(I,J)
 263   FORMAT(2X,'g1=(' 3I3,')',2X,'g2=(' 3I3,')',' g2/g1= ',F10.6,X,
     A    'Thita= ',F6.2,/)
 269   CONTINUE
 270   CONTINUE
C
       WRITE(6,280)
 280   FORMAT(/' CALCULATE MORE PLANES ?',$)
       READ(5,211) YO
       IF(YO.EQ.'Y') GOTO 190
       WRITE(6,281)
 281   FORMAT(/' CALCULATION FOR OTHER STRUCTURES ?',$)
       READ(5,211) YO
       IF(YO.EQ.'Y') GOTO 101
       STOP
       END
C
C
       SUBROUTINE CRYSTAL(A,B,C,A1,B1,C1,NMAX)
       COMMON H(10),K(10),L(10),DG(10), GRATIO(10,10), THITA(10,10)
       COMMON DU(10), DV(10), DW(10),VOL
       INTEGER H,K,L
C Volume of the unit cell , A=a, B=b, C=c, A1=Alpha, B1=Beta, C1=Gamma
       VOL=A*B*C*(ABS(1-(COS(A1))**2-(COS(B1))**2-(COS(C1))**2
     A       +2*COS(A1)*COS(B1)*COS(C1)))**0.5
C S11, S22 and S33 are reciprocal space lattice vectors a*, b* and c*
       S11=B**2*C**2*SIN(A1)**2
       S22=(A*C*SIN(B1))**2
       S33=(A*B*SIN(C1))**2
C
       S12=A*B*C**2*(COS(A1)*COS(B1)-COS(C1))
       S23=A**2*B*C*(COS(B1)*COS(C1)-COS(A1))
       S13=A*B**2*C*(COS(C1)*COS(A1)-COS(B1))
C
       DO 10 I=1,NMAX
       R=S11*H(I)**2+S22*K(I)**2+S33*L(I)**2+2*S12*H(I)*K(I)
     A    +2*S23*K(I)*L(I)+2*S13*H(I)*L(I)
C Interplanar distance dg
       DG(I)=VOL/R**0.5
C Direction [u, v, w] perpendicular to (h,k,l)
       DU(I)=H(I)*S11+K(I)*S12+L(I)*S13
       DV(I)=H(I)*S12+K(I)*S22+L(I)*S23
       DW(I)=H(I)*S13+K(I)*S23+L(I)*S33
       SNORM=(DU(I)**2+DV(I)**2+DW(I)**2)**0.5
       IF(SNORM.EQ.0.0) SNORM=1.0
       DU(I)=DU(I)/SNORM
       DV(I)=DV(I)/SNORM
       DW(I)=DW(I)/SNORM
C
 10    CONTINUE
C
       DO 77 I=1,NMAX
       THITA(I,I)=0.0
 77    GRATIO(I,I)=1.0
```

```
C Angles between planes
      DO 20 I=1, NMAX
      DO 30 J=I+1,NMAX
      ARG = S11*H(I)*H(J)+S22*K(I)*K(J)+S33*L(I)*L(J)
     1      +S23*(K(I)*L(J)+K(J)*L(I))+S13*(L(I)*H(J)+L(J)*H(I))
     1      +S12*(H(I)*K(J)+H(J)*K(I))
      X=DG(I)*DG(J)*ARG/VOL**2
      IF(X.EQ.0.0) X=X-0.00001
      PHI=ATAN((1-X*X)**.5/X)
      THITA(I,J)=PHI*180.0/3.14159
      IF(THITA(I,J).LE.0.0) THITA(I,J)=180.0+THITA(I,J)
      GRATIO(I,J)=DG(I)/DG(J)
      THITA(J,I)=THITA(I,J)
      GRATIO(J,I)=1.0/GRATIO(I,J)
   30 CONTINUE
   20 CONTINUE
      DO 66 I=1, NMAX
      DO 66 J=I+1,NMAX
      IF(H(I).EQ.H(J).AND.K(I).EQ.K(J).AND.L(I).EQ.L(J)) THITA(I,J)=0.0
      IF(H(I).EQ.H(J).AND.K(I).EQ.K(J).AND.L(I).EQ.L(J)) THITA(J,I)=0.0
   66 CONTINUE
      RETURN
      END
```

APPENDIX D

Electron diffraction patterns of several types of crystal structures

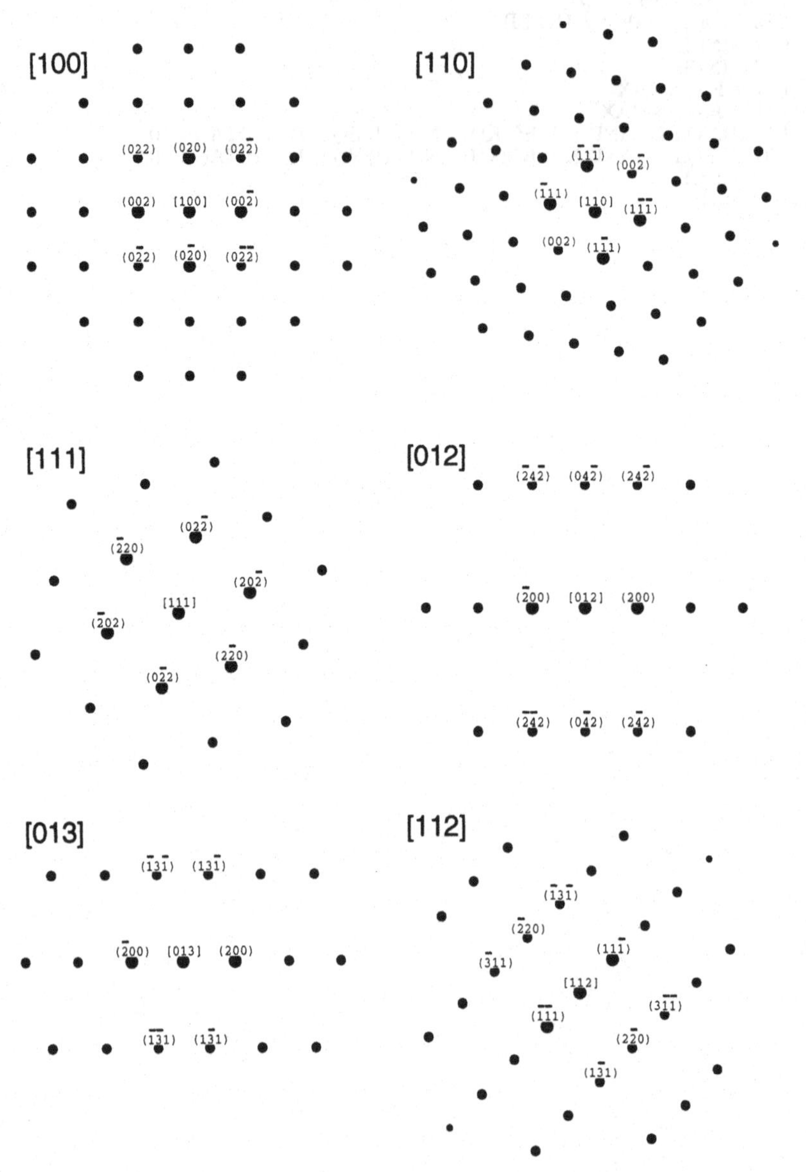

Figure D.1 *Six low-index zone-axis diffraction patterns for face-centered cubic crystals.*

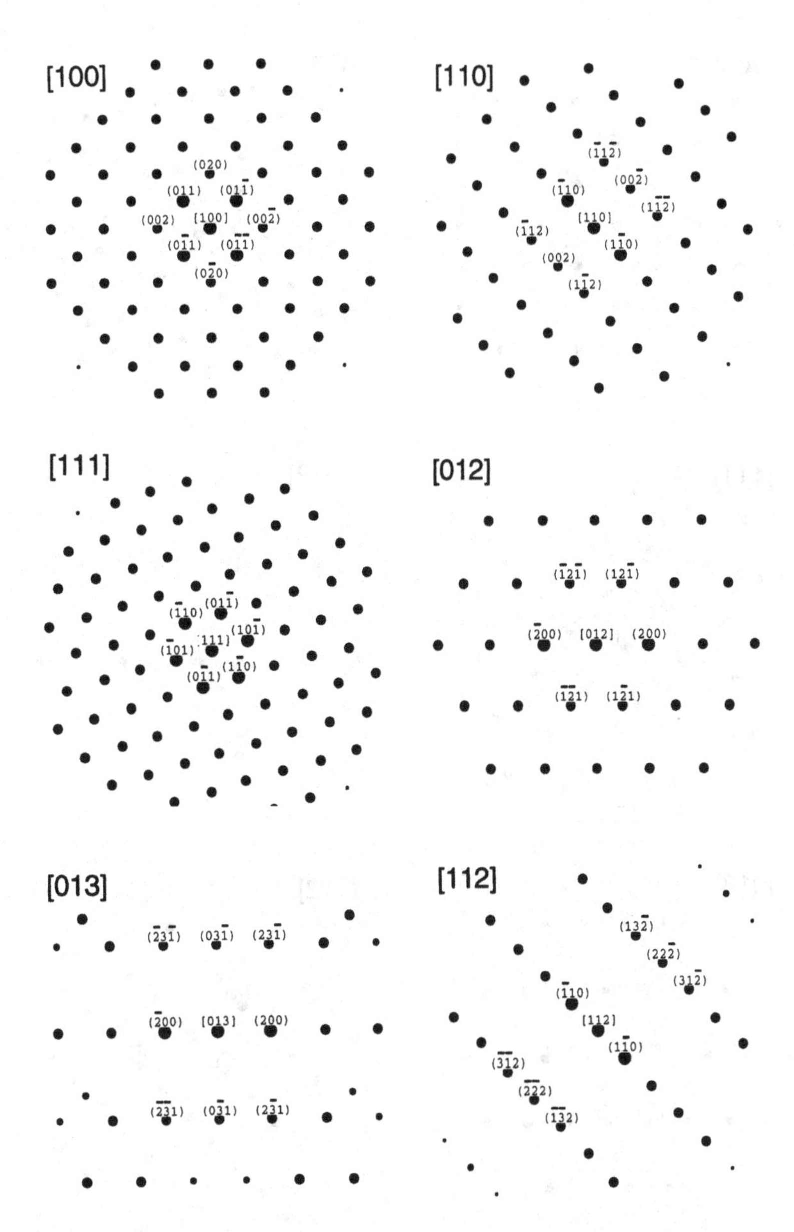

Figure D.2 *Six low-index zone-axis diffraction patterns for body-centered cubic crystals.*

In Figures D.1–D.4 are shown indexed diffraction patterns for commonly encountered structures. These patterns were calculated using the Desktop Microscopist software based on kinematical scattering theory. No double diffraction was included. The beam direction is indicated for each pattern.

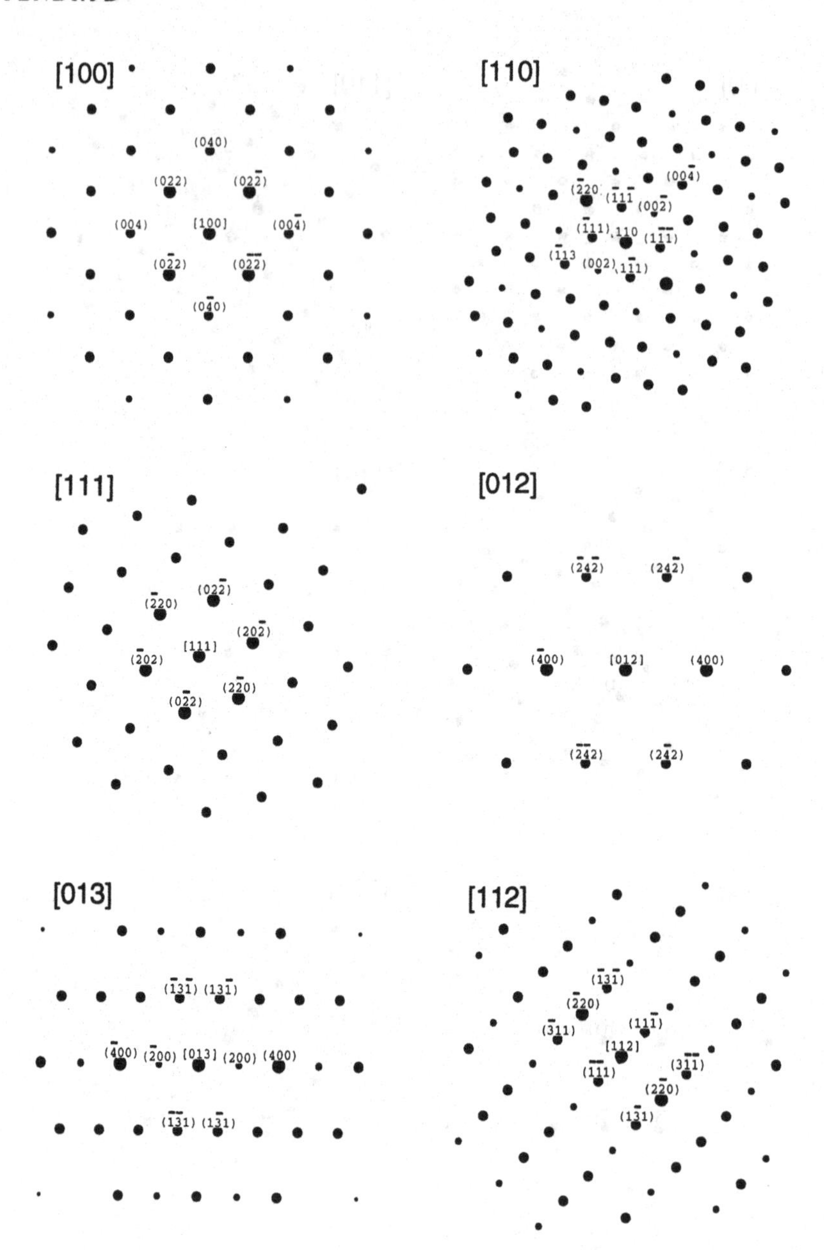

Figure D.3 *Six low-index zone-axis diffraction patterns for crystals with the diamond lattice.*

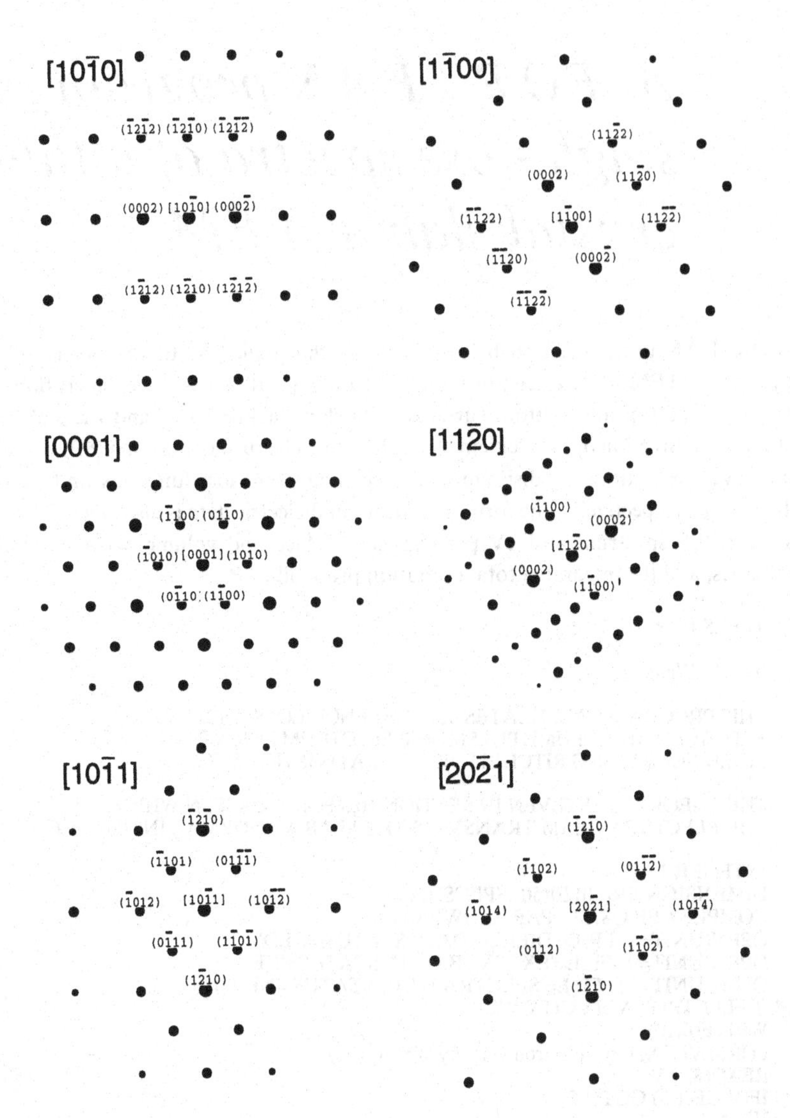

Figure D.4 *Six low-index zone-axis diffraction patterns for the hexagonal lattice with $c/a = 1.633$.*

A FORTRAN program for single-loss spectra of a thin crystal slab in TEM

The FORTRAN program listed below calculates the single-loss EELS spectra of a thin crystal slab in TEM. The calculation is based on Eq. (10.25) with consideration of the excitation of the top and bottom surfaces. The slab thickness is d and the slab normal direction is the direction of the incident electron beam. The surface plasmon (SP) and volume plasmon (VP) excitations are distinguished according to the loss functions Im $[-2/(\varepsilon+1)]$ and Im $(-1/\varepsilon)$, respectively. The format of the input dielectric data is also given. The output gives the EELS spectrum at 1 eV per channel, surface and volume plasmon excitation probabilities, and the integrated total excitation probability P.

```
C       TEELS.FOR
C
C       by Z.L. Wang
C
C       THIS PROGRAM CALCULATES THE VALENCE-LOSS (INCLUDING
C       SURFACE AND VOLUME PLASMONS) SPECTRUM FOR A PARALLEL-
C       SIDED SLAB USING RITCHIE'S NON-RELATIVISTIC THEORY.
C
C       THE THEORY WAS GIVEN IN SECTION 10.5 FOR A CASE IN WHICH
C       THE ELECTRON BEAM TRANSMITS THE SLAB AT NORMAL INCIDENCE.
C
        INTEGER I,J,K
        DIMENSION ER(50),EI(50),SPECSHP(50)
        COMPLEX EE,QSA,UPPART,LOWPART
        OPEN(UNIT=8, FILE='DIELEC.DAT', STATUS='OLD')
        OPEN(UNIT=9, FILE='EXCPROB.DAT', STATUS='NEW')
        OPEN(UNIT=11, FILE='SPECTRA.DAT', STATUS='NEW')
C INPUT ELECTRON VELOCITY V/C
 5      WRITE(6,10)
 10     FORMAT(2X,'Input electron velocity v/c ratio ...')
        READ(5,*) V
        IF(V.GE.1.0) GOTO 5
        VL=1.0
        VB=(V/VL)**2
        WRITE(9,22) V
 22     FORMAT(2X,'Electron velocity v/c = ',F10.8)
        WRITE(11,22) V
C INPUT SPECIMEN THICKNESS
        WRITE(6,30)
 30     FORMAT(2X,'Input specimen thickness (in Angstrom)...')
        READ(5,*) T
        T=ABS(T)
        WRITE(9,32) T
 32     FORMAT(2X,'Specimen thickness d = ',F10.2,' Angstrom')
        WRITE(11,32) T
C INPUT CUT-OFF WAVE VECTOR
        WRITE(6,40)
 40     FORMAT(2X,'Input cut-off wave vector (in 1/Angstrom)...')
        READ(5,*) QC
        WRITE(9,44) QC
```

```
 44     FORMAT(2X,'Cut-off wave vector qc= ',F10.5,' 1/Angstrom')
        WRITE(11,44) QC
        QC=2.0*3.14159*ABS(QC)
C INPUT DIELECTRIC FUNCTION
C       The dielectric function is given in file 8 of name
C       DIELEC.DAT with an energy resolution of 1 eV (= 1 channel)
        READ(8,45) NDP
 45     FORMAT(2X,I2)
        DO 50 I = 1, NDP
 50     READ(8,60) ER(I), EI(I)
 60     FORMAT(2X,F10.7,2X,F10.7)
C
C HEADING FOR OUTPUT
        WRITE(9,64)
 64     FORMAT(2X,'EXCITATION PROBABILITIES:')
        WRITE(11,66)
 66     FORMAT(2X, 'ENERGY-LOSS        EXC. PROB./eV')
        SP=0.0
        VP=0.0
C INTEGRATE OVER ENERGY-LOSS W
C DW is the energy resolution, DW = 1 means 1eV/channel.
        DW=1.0
        DO 400 J = 1, NDP
        W=J
        WW=W**2
C COMPLEX DIELECTRIC FUNCTION
        EE=CMPLX(ER(J), EI(J))
C
        QSA=CMPLX(0.0, 0.0)
C INTEGRATE OVER QY
        NMAX1=4000
        NMAX2=3000
        NMAX=NMAX1+NMAX2
        QYH=0.01
        DO 600 K=1,NMAX
        IF(K.GT.NMAX1) GOTO 80
        QY=(K-1)*QYH/NMAX1+QYH/2.0
        DQY=QYH/NMAX1
        GOTO 85
 80     QY=QYH+(K-NMAX1)*(QC-QYH)/NMAX2+QYH/2.0
        DQY=(QC-QYH)/NMAX2
 85     ALP=(QY**2+WW/(1969.0*V)**2)
        QPP=QY**2/ALP**2
        QYY=QY*T
        IF(QYY.GT.50) QYY=50.0
        UPPART=-DQY*QPP*((1.0-EE)/EE)*(2.0*(EE-1.0)*COS(W*T/(1973.2*V))
   A       +(EE-1.0)*EXP(-QYY)+(1.0-EE**2)*EXP(QYY))
        LOWPART=((EE-1.0)**2*EXP(-QYY)-(EE+1.0)**2*EXP(QYY))
        QSA=QSA+UPPART/LOWPART
 600    CONTINUE
C  SURFACE PLASMON COMPONENT
        SP=AIMAG(QSA)+SP
C VOLUME PLASMON COMPONENT
        VP1=ALOG(1.0+(QC*V*1973.2/W)**2)*AIMAG(-1.0/EE)
        VP=VP1+VP
C VALENCE-LOSS SPECTRUM
        SPECSHP(J)=AIMAG(QSA)+T*VP1/2.0
 400    CONTINUE
        CONS=2.35E-6
C SURFACE PLASMON EXCITATION PROBABILITY
        SP=SP*DW*CONS/V**2
C VOLUME PLASMON EXCITATION PROBABILITY
        VP=VP*T*CONS*DW/V**2
        VP=VP/2.0
C OUTPUT SINGLE-LOSS SPECTRUM
        DO 350 J=1, NDP
```

```
      SPECSHP(J)=SPECSHP(J)*CONS/V**2
      WRITE(11,355)J, SPECSHP(J)
 355  FORMAT(3X,'W=',I2,4X,'dP/dw  =',F20.12)
 350  CONTINUE
C TOTAL EXCITATION PROBABILITY
      TS=VP+SP
      ALAMEDA=T/VP
C OUTPUT SURFACE AND VOLUME PLASMON AND TOTAL EXCITATION
C PROBABILITIES
      WRITE(9,700) SP,VP,TS
 700  FORMAT(2X,'SP P =',F15.10,X,'VP P =',F15.10,X,
    A 'TOTAL P =',F15.10)
      WRITE(9,710) ALAMEDA
 710  FORMAT(8X,'Inelastic mean-free-path =',F12.5,' Angstrom')
      STOP
      END
```

Input data file: DIELEC.DAT

Real and imaginary components of the dielectric function for GaAs. The dielectric function ε is assumed to be independent of q.

The first line is the number of channels. The data are input in a format of FORMAT(2X,F10.7,2X,F10.7) at 1 eV per channel. The energy resolution can be changed by setting the DW value in the program. This data file is also needed for the FORTRAN programs provided in Appendixes E.2–E.4.

```
 22
 -10.180000   8.482000
  14.994      1.636
  16.536     17.567
   9.28      13.828
 -11.51      18.565
  -4.513      6.249
  -2.248      3.908
  -1.251      2.580
  -0.4787     2.047
  -0.1151     1.7785
   0.07577    1.6263
   0.1753     1.4159
   0.2483     1.225
   0.3432     1.0114
   0.6194     0.4715
   0.6544     0.4437
   0.7406     0.3965
   0.8444     0.3332
   0.9392     0.2982
   1.0298     0.2952
   1.093      0.3428
   1.066      0.3087
```

Output for GaAs:

SPECTRA.DAT

```
Electron velocity v/c = 0.54820001
Specimen thickness d =     50.00 Angstrom
Cut-off wave vector qc =  1.00000 1/Angstrom
ENERGY-LOSS         EXC. PROB./eV
 W= 1   dP/dw =     0.001902248827
 W= 2   dP/dw =     0.000086791697
 W= 3   dP/dw =     0.000411188783
```

```
W= 4  dP/dw =   0.000560168293
W= 5  dP/dw =   0.000593677338
W= 6  dP/dw =   0.001533702714
W= 7  dP/dw =   0.001799838967
W= 8  dP/dw =   0.001948193880
W= 9  dP/dw =   0.001966149546
W=10  dP/dw =   0.001972556114
W=11  dP/dw =   0.001963937422
W=12  dP/dw =   0.002053051023
W=13  dP/dw =   0.002160580130
W=14  dP/dw =   0.002278610598
W=15  dP/dw =   0.001816896838
W=16  dP/dw =   0.001638419228
W=17  dP/dw =   0.001283693826
W=18  dP/dw =   0.000923991320
W=19  dP/dw =   0.000690326968
W=20  dP/dw =   0.000584188325
W=21  dP/dw =   0.000536980864
W=22  dP/dw =   0.000556219369
```

EXCPROB.DAT

```
Electron velocity v/c = 0.54820001
Specimen thickness d =    50.00 Angstrom
Cut-off wave vector qc =  1.00000 1/Angstrom
EXCITATION PROBABILITIES:
SP P =  0.0089983502 VP P =  0.0202630609 TOTAL P =  0.0292614121
    Inelastic mean-free-path = 2467.54419 Angstrom
```

It must be pointed out that the calculated inelastic mean-free-path length is smaller than the practically measured one because the dialectric data are supplied only to 22 eV.

APPENDIX E.2

A FORTRAN program for single-loss REELS spectra in RHEED

The theory presented in Section 10.6.3 is applied here to calculate REELS spectra. The electron scattering trajectory is shown in Figure 10.9, in which the electron mean penetration depth into the surface and the mean traveling distance along the surface under resonance conditions are assumed as D_p and L_s, respectively. The outputs are the single-loss REELS spectrum and the integrated scattering probability P.

```
C       REELS.FOR
C
C       by Z.L. Wang
C
C       THIS PROGRAM CALCULATES THE REELS SINGLE-LOSS VALENCE
C       SPECTRUM IN RHEED GEOMETRY USING THE RELATIVISTIC
C       DIELECTRIC RESPONSE THEORY GIVEN IN SECTION 10.6.3
C       THE ELECTRON PENETRATION DEPTH INTO THE SURFACE AND
C       RESONANCE PROPAGATION ALONG THE SURFACE HAVE BEEN
C       INCLUDED.
C
C       THE TOTAL INTEGRATED EXCITATION PROBABILITY P IS ALSO
C       CALCULATED. THIS QUANTITY IS EQUAL TO THE 't/Lameda' IN TEELS.
C
        INTEGER I,J,K
        REAL LS,DP,THITA,NMAX1,NMAX2
        DIMENSION ER(50),EI(50),SPECSHP(50)
        COMPLEX EE,ALP1,QSA,FUNV,FUNC
        OPEN(UNIT=8,FILE='DIELEC.DAT',STATUS='OLD')
        OPEN(UNIT=11, FILE='SPECTRA.DAT', STATUS='NEW')
C INPUT ELECTRON VELOCITY V/C
 5      WRITE(6,10)
 10     FORMAT(2X,'Input electron velocity v/c ratio ...')
        READ(5,*) V
        IF(V.GE.1.0) GOTO 5
        WRITE(11,22) V
 22     FORMAT(2X,'Electron velocity v/c = ',F10.8)
C SPEED OF LIGHT
        VL=1.0
        VB=(V/VL)**2
C BEAM GRAZING INCIDENT ANGLE
        WRITE(6,25)
 25     FORMAT(3X,'Beam incident angle (in mrad)')
        READ(5,*) THITA
        SINT=SIN(ABS(THITA)/1000.0)
        WRITE(11,27) THITA
 27     FORMAT(2X,'Beam incident angle is ',F10.5,' mrad')
C BEAM PENETRATION DEPTH INTO THE SURFACE
        WRITE(6,30)
 30     FORMAT(3X,'Beam penetration depth Dp (in Angstrom)')
        READ(5,*) DP
        DP=ABS(DP)
        WRITE(11,42) DP
 42     FORMAT(2X,'Beam penetration depth is ',F10.4,' Angstrom')
```

```
C BEAM MEAN TRAVELING DISTANCE ALONG THE SURFACE
      WRITE(6,50)
 50   FORMAT(2X,'Input mean traveling distance Ls (in Angstrom)...')
      READ(5,*) LS
      LS=ABS(LS)
      WRITE(11,57) LS
 57   FORMAT(2X,'Beam resonance traveling distance is ',F10.4,' Angstrom')
C INPUT CUT-OFF WAVE VECTOR
      WRITE(6,60)
 60   FORMAT(2X,'Input cut-off wave vector qc (in 1/Angstrom)...')
      READ(5,*) QC
      WRITE(11,61) QC
 61   FORMAT(2X,'Cut-off wave vector qc = ',F10.4,' 1/Angstrom')
      QC=2.0*3.14159*ABS(QC)
      WRITE(6,63)
 63   FORMAT(2X,'Calculation in progress ....')
C INPUT DIELECTRIC FUNCTION
C     The dielectric function is given in file 8 of name DIELEC.DAT with
C     an energy resolution of 1 eV (= 1 channel)
      READ(8,65) NDP
 65   FORMAT(2X,I2)
      DO 70 I = 1, NDP
 70   READ(8,75) ER(I), EI(I)
 75   FORMAT(2X,F10.7,2X,F10.7)
C
C CALCULATE EXCITATION PROB. AS A FUNCTION OF IMPACT DISTANCE
C FOR X < 0  (IN SIDE CRYSTAL) AND X > 0 (IN VACUUM), WHERE X IS
C MEASURED IN ANGSTROM.  HERE WE ASSUME - 30 < X < 30 ANGSTROM.
      TP=0.0
C INTEGRATE OVER ENERGY-LOSS W AT 1eV/Channel
C DW is eV/channel; DW=1 means 1eV/channel.
      DW=1.0
      DO 400 J=1,NDP
      W=J
      WW=W**2
      EE=CMPLX(ER(J), EI(J))
      QSA=CMPLX(0.0, 0.0)
      SP1=0.0
C INTEGRATE OVER QY
      NMAX1=4000
      NMAX2=3000
      NMAX=NMAX1+NMAX2
      QYH=0.01
      DO 300 K=1,NMAX
      IF(K.GT.NMAX1) GOTO 80
      QY=(K-1)*QYH/NMAX1+QYH/2.0
      DQY=QYH/NMAX1
      GOTO 85
 80   QY=QYH+(K-NMAX1)*(QC-QYH)/NMAX2+QYH/2.0
      DQY=(QC-QYH)/NMAX2
 85   ALP1=CSQRT(QY**2+WW*(1.0-VB*EE)/(1973.2*V)**2)
      ALP0=SQRT(QY**2+WW*(1.0-VB)/(1973.2*V)**2)
C
C INSIDE CRYSTAL
      FUNC=((2.0*ALP1**2*(1.0-EE)/(EE*ALP0+ALP1)
     A  -(1.0-EE*VB)*(ALP1-ALP0))/(ALP1*EE*(ALP1+ALP0)))
     A   *((1.0-CEXP(-2.0*ALP1*DP))/(ALP1*SINT)+CEXP(-2.0*ALP1*DP)*LS)
     A   -((1.0-EE*VB)/(ALP1*EE))*(LS+2.0*DP/SINT)
C
C IN VACUUM
 100  FUNV=((2.0*ALP0**2*(EE-1.0)/(EE*ALP0+ALP1)-(1.0-VB)*(ALP0-ALP1))
     A     /(ALP0*(ALP0+ALP1)))*(1.0/(ALP0*SINT))
      QSA=QSA+(FUNC+FUNV)*DQY
 300  CONTINUE
      TP=AIMAG(QSA)+TP
      SPECSHP(J)=AIMAG(QSA)
```

```
 400   CONTINUE
       CONS=2.35E-6
C TOTAL EXCITATION PROBABILITY
       TP=TP*DW*CONS/V**2
C OUTPUT SPECTRUM
       DO 500 J=1,NDP
       RRR=SPECSHP(J)*CONS/V**2
       WRITE(11,450) J, RRR
 450   FORMAT(3X,'W=',I2,3X,'SPECSHP=',F20.12)
 500   CONTINUE
C OUTPUT 't/Lameda'
       WRITE(11,700) TP
 700   FORMAT(2X,'Integrated excitation probability
   A     (t/Lameda) P =',F15.10)
       STOP
       END
```

Output for GaAs:

SPECTRA.DAT

```
Electron velocity v/c = 0.54820001
Beam incident angle is  25.00000 mrad
Beam penetration depth is    5.0000 Angstrom
Beam resonance traveling distance is  500.0000 Angstrom
Cut-off wave vector qc =     1.0000 1/Angstrom
 W= 1   SPECSHP=      0.010349873453
 W= 2   SPECSHP=      0.001384119852
 W= 3   SPECSHP=      0.008757105097
 W= 4   SPECSHP=      0.011791346595
 W= 5   SPECSHP=      0.009134399705
 W= 6   SPECSHP=      0.021467158571
 W= 7   SPECSHP=      0.038230791688
 W= 8   SPECSHP=      0.059711962938
 W= 9   SPECSHP=      0.069641388953
 W=10   SPECSHP=      0.068127840757
 W=11   SPECSHP=      0.064525961876
 W=12   SPECSHP=      0.063530646265
 W=13   SPECSHP=      0.062150344253
 W=14   SPECSHP=      0.059179704636
 W=15   SPECSHP=      0.037474505603
 W=16   SPECSHP=      0.033619437367
 W=17   SPECSHP=      0.026521032676
 W=18   SPECSHP=      0.019368009642
 W=19   SPECSHP=      0.014727880247
 W=20   SPECSHP=      0.012723986991
 W=21   SPECSHP=      0.011853802949
 W=22   SPECSHP=      0.012151909992
Integrated excitation probability  (t/Lameda) P =  0.7164232731
```

A FORTRAN program for single-loss spectra of parallel-to-surface incident beams

Listed below is a FORTRAN source code for calculating the electron inelastic excitation probabilities and single-loss EELS spectra when a 'point' electron probe is positioned at a distance x_0 from the crystal surface, with $x_0 < 0$ in the crystal and $x_0 > 0$ in vacuum. The calculation is based on Eq. (10.46a–d).

```
C       PEELS.FOR
C
C       by Z.L. Wang
C
C       THIS PROGRAM CALCULATES THE EXCITATION PROBABILITY FOR AN
C       ELECTRON WHICH IS TRAVELING AT A DISTANCE X0 FROM AND
C       PARALLEL TO A BULK CRYSTAL SURFACE, USING THE RELATIVISTIC
C       DIELECTRIC RESPONSE THEORY GIVEN IN SECTION 10.6.
C
C       THE SINGLE-LOSS SPECTRUM IS ALSO CALCULATED
C
        INTEGER I,J,K
        DIMENSION ER(50),EI(50),SPECSHP(50)
        COMPLEX EE,ALP1,QSA,FUNV,FUNC
        OPEN(UNIT=8,FILE='DIELEC.DAT',STATUS='OLD')
        OPEN(UNIT=9, FILE='EXCPROB.DAT', STATUS='NEW')
        OPEN(UNIT=11, FILE='SPECTRA.DAT', STATUS='NEW')
C INPUT ELECTRON VELOCITY V/C
 5      WRITE(6,10)
 10     FORMAT(2X,'Input electron velocity v/c ratio ...')
        READ(5,*) V
        IF(V.GE.1.0) GOTO 5
        VL=1.0
        VB=(V/VL)**2
        WRITE(9,22) V
 22     FORMAT(2X,'Electron velocity v/c = ',F10.8)
        WRITE(11,22) V
C BEAM IMPACT DISTANCE AT WHICH THE SPECTRUM IS CALCULATED
 25     WRITE(6,30)
 30     FORMAT(3X,'Beam impact distance x (-30<x<30 Angstrom)')
        READ(5,*) X0
        IF(ABS(X0).GT.30.0) GOTO 25
        WRITE(11,42) X0
 42     FORMAT(2X,'Electron impact distance x = ',F10.6,' Angstrom')
C INPUT SPECIMEN THICKNESS
        WRITE(6,50)
 50     FORMAT(2X,'Input specimen thickness (in Angstrom)...')
        READ(5,*) T
        T=ABS(T)
        WRITE(9,52) T
 52     FORMAT(2X,'Specimen thickness d = ',F10.3,' Angstrom')
        WRITE(11,52) T
```

```
C INPUT CUT-OFF WAVE VECTOR
        WRITE(6,60)
 60     FORMAT(2X,'Input cut-off wave vector qc (in 1/Angstrom)...')
        READ(5,*) QC
        WRITE(9,44) QC
 44     FORMAT(2X,'Cut-off wave vector qc= ',F10.6,' 1/ANGSTROM')
        WRITE(11,44) QC
        QC=2.0*3.14159*ABS(QC)
        WRITE(6,63)
 63     FORMAT(2X,'Calculation in progress ....')
C INPUT DIELECTRIC FUNCTION
C       The dielectric function is given in file 8 of name DIELEC.DAT with
C       an energy resolution of 1 eV (= 1 channel)
        READ(8,65) NDP
 65     FORMAT(2X,I2)
        DO 70 I = 1, NDP
 70     READ(8,75) ER(I), EI(I)
 75     FORMAT(2X,F10.7,2X,F10.7)
        WRITE(9,77)
 77     FORMAT(/,5X,'X',13X,'dP/dz',11X,'VP  dP/dz',8X,'SP  dP/dz',
    A     9X,'Lameda')

C
C CALCULATE EXCITATION PROB. AS A FUNCTION OF IMPACT DISTANCE
C FOR X < 0 (IN SIDE CRYSTAL) AND X > 0 (IN VACUUM), WHERE X IS
C MEASURED IN ANGSTROM.  HERE WE ASSUME - 30 < X < 30 ANGSTROM.
        DO 200 I=1,61
        X=(I-31)
        TP=0.0
        SPP=0.0
C INTEGRATE OVER ENERGY-LOSS W AT 1eV/Channel
C DW is eV/channel; DW=1 means 1eV/channel.
        DW=1.0
        DO 400 J=1,NDP
        W=J
        WW=W**2
        EE=CMPLX(ER(J), EI(J))
        QSA=CMPLX(0.0, 0.0)
        SP1=0.0
C INTEGRATE OVER QY
        NMAX1=4000
        NMAX2=3000
        NMAX=NMAX1+NMAX2
        QYH=0.01
        DO 300 K=1,NMAX
        IF(K.GT.NMAX1) GOTO 80
        QY=(K-1)*QYH/NMAX1+QYH/2.0
        DQY=QYH/NMAX1
        GOTO 85
 80     QY=QYH+(K-NMAX1)*(QC-QYH)/NMAX2+QYH/2.0
        DQY=(QC-QYH)/NMAX2
 85     ALP1=CSQRT(QY**2+WW*(1.0-VB*EE)/(1973.2*V)**2)
        ALP0=SQRT(QY**2+WW*(1.0-VB)/(1973.2*V)**2)
C
        IF(X.GE.0.0) GOTO 100
        FUNC=(2.0*ALP1**2*(1.0-EE)/(EE*ALP0+ALP1)
    A       -(1.0-EE*VB)*(ALP1-ALP0))
    A     *CEXP(-2.0*ALP1*ABS(X))/(ALP1*EE*(ALP1+ALP0))
    A       -(1.0-EE*VB)/(ALP1*EE)
C
        FUNV=(2.0*ALP0**2*(EE-1.0)/(EE*ALP0+ALP1)-(1.0-VB)*(ALP0-ALP1))
    A       *EXP(-2.0*ALP0*ABS(X))/(ALP0*(ALP0+ALP1))
        SP1=SP1+AIMAG(FUNV)*DQY
        QSA=QSA+FUNC*DQY
        GOTO 300
```

```
C
 100    FUNV=(2.0*ALP0**2*(EE-1.0)/(EE*ALP0+ALP1)-(1.0-VB)*(ALP0-ALP1))
  A        *EXP(-2.0*ALP0*ABS(X))/(ALP0*(ALP0+ALP1))
        QSA=QSA+FUNV*DQY
        SP1=SP1+AIMAG(FUNV)*DQY
 300    CONTINUE
        TP=AIMAG(QSA)+TP
C CALCULATE SPECTRUM
        IF(ABS(X-X0).LE.0.8) SPECSHP(J)=AIMAG(QSA)
        SPP=SPP+SP1
 400    CONTINUE
        CONS=2.35E-6
C EXCITATION PROBABILITY (1/Mean-free-path-length)
        TP=TP*DW*CONS/V**2
        ALAMEDA=1.0/TP
C OUTPUT SPECTRUM AT X = X0
        IF(ABS(X-X0).GT.0.8) GOTO 580
        DO 500 J=1,NDP
        RRR=SPECSHP(J)*T*CONS/V**2
        WRITE(11,450)J,RRR
 450    FORMAT(3X,'W=',I2,3X,'SPECSHP=',F20.12)
 500   CONTINUE
C SURFACE PLASMON EXCITATION PROBABILITY
 580    SP=SPP*CONS*DW/V**2
C VOLUME PLASMON EXCITATION PROBABILITY
        VP=TP-SP
C OUTPUT
        WRITE(9,700) X,TP,VP,SP,ALAMEDA
 700    FORMAT(2X,F6.2,5X,3(F10.8,4X),F10.3)
 200   CONTINUE
        STOP
        END
```

Output for GaAs:

SPECTRA.DAT

```
Electron velocity v/c = 0.54820001
Electron impact distance x =   5.000000 Angstrom
Specimen thickness d =   400.000 Angstrom
Cut-off wave vector qc =    1.000000 1/ANGSTROM
 W= 1  SPECSHP=     0.000825980969
 W= 2  SPECSHP=     0.000104130449
 W= 3  SPECSHP=     0.000631376053
 W= 4  SPECSHP=     0.000898271974
 W= 5  SPECSHP=     0.000745139027
 W= 6  SPECSHP=     0.001844869461
 W= 7  SPECSHP=     0.003386186203
 W= 8  SPECSHP=     0.005413890816
 W= 9  SPECSHP=     0.006201905664
 W=10  SPECSHP=     0.005876800977
 W=11  SPECSHP=     0.005413260311
 W=12  SPECSHP=     0.005119679961
 W=13  SPECSHP=     0.004738533404
 W=14  SPECSHP=     0.004088970367
 W=15  SPECSHP=     0.001821006881
 W=16  SPECSHP=     0.001619073213
 W=17  SPECSHP=     0.001299127587
 W=18  SPECSHP=     0.000980766607
 W=19  SPECSHP=     0.000775310386
 W=20  SPECSHP=     0.000699654571
 W=21  SPECSHP=     0.000669916160
 W=22  SPECSHP=     0.000672405586
```

APPENDIX E.3

EXCPROB.DAT

Electron velocity v/c = 0.54820001
Specimen thickness d = 400.000 Angstrom
Cut-off wave vector qc = 1.000000 1/ANGSTROM

X	dP/dz	VP dP/dz	SP dP/dz	Lameda
-30.00	0.00042825	0.00039049	0.00003776	2335.074
-29.00	0.00042838	0.00038922	0.00003917	2334.359
-28.00	0.00042852	0.00038787	0.00004065	2333.618
-27.00	0.00042866	0.00038645	0.00004221	2332.849
-26.00	0.00042881	0.00038495	0.00004386	2332.051
-25.00	0.00042896	0.00038336	0.00004560	2331.222
-24.00	0.00042912	0.00038167	0.00004745	2330.361
-23.00	0.00042928	0.00037988	0.00004940	2329.465
-22.00	0.00042945	0.00037798	0.00005148	2328.533
-21.00	0.00042963	0.00037595	0.00005369	2327.561
-20.00	0.00042982	0.00037378	0.00005605	2326.549
-19.00	0.00043002	0.00037145	0.00005856	2325.492
-18.00	0.00043022	0.00036896	0.00006126	2324.387
-17.00	0.00043044	0.00036627	0.00006417	2323.231
-16.00	0.00043066	0.00036336	0.00006730	2322.019
-15.00	0.00043090	0.00036021	0.00007068	2320.746
-14.00	0.00043114	0.00035678	0.00007436	2319.406
-13.00	0.00043141	0.00035303	0.00007838	2317.991
-12.00	0.00043169	0.00034890	0.00008279	2316.494
-11.00	0.00043198	0.00034433	0.00008766	2314.904
-10.00	0.00043230	0.00033922	0.00009308	2313.208
-9.00	0.00043264	0.00033348	0.00009916	2311.389
-8.00	0.00043301	0.00032693	0.00010608	2309.424
-7.00	0.00043341	0.00031938	0.00011403	2307.281
-6.00	0.00043386	0.00031049	0.00012336	2304.917
-5.00	0.00043436	0.00029979	0.00013457	2302.263
-4.00	0.00043493	0.00028645	0.00014848	2299.208
-3.00	0.00043563	0.00026894	0.00016669	2295.546
-2.00	0.00043653	0.00024381	0.00019272	2290.815
-1.00	0.00043791	0.00020007	0.00023784	2283.567
0.00	0.00044327	0.00000000	0.00044327	2255.941
1.00	0.00023784	0.00000000	0.00023784	4204.422
2.00	0.00019272	0.00000000	0.00019272	5188.962
3.00	0.00016669	0.00000000	0.00016669	5999.260
4.00	0.00014848	0.00000000	0.00014848	6734.828
5.00	0.00013457	0.00000000	0.00013457	7431.317
6.00	0.00012336	0.00000000	0.00012336	8106.139
7.00	0.00011403	0.00000000	0.00011403	8769.326
8.00	0.00010608	0.00000000	0.00010608	9427.271
9.00	0.00009916	0.00000000	0.00009916	10084.388
10.00	0.00009308	0.00000000	0.00009308	10743.891
11.00	0.00008766	0.00000000	0.00008766	11408.221
12.00	0.00008279	0.00000000	0.00008279	12079.325
13.00	0.00007838	0.00000000	0.00007838	12758.789
14.00	0.00007436	0.00000000	0.00007436	13447.942
15.00	0.00007068	0.00000000	0.00007068	14147.938
16.00	0.00006730	0.00000000	0.00006730	14859.780
17.00	0.00006417	0.00000000	0.00006417	15584.381
18.00	0.00006126	0.00000000	0.00006126	16322.554
19.00	0.00005856	0.00000000	0.00005856	17075.090
20.00	0.00005605	0.00000000	0.00005605	17842.666
21.00	0.00005369	0.00000000	0.00005369	18625.990
22.00	0.00005148	0.00000000	0.00005148	19425.697
23.00	0.00004940	0.00000000	0.00004940	20242.404
24.00	0.00004745	0.00000000	0.00004745	21076.729
25.00	0.00004560	0.00000000	0.00004560	21929.248
26.00	0.00004386	0.00000000	0.00004386	22800.553
27.00	0.00004221	0.00000000	0.00004221	23691.221

28.00	0.00004065	0.00000000	0.00004065	24601.809
29.00	0.00003917	0.00000000	0.00003917	25532.885
30.00	0.00003776	0.00000000	0.00003776	26485.008

In Figure E.1 are shown the calculated EELS spectra for a 'point' electron probe which is traveling parallel to the GaAs surface at two impact distances from the surface, $x = 0.5$ nm (in vacuum) and $x = -0.5$ nm (in GaAs). When the electron beam is in the vacuum, only surface losses (5–12 eV) are excited, and the integrated excitation probability is low. Both surface and volume losses are excited when the beam is in the crystal. The peak located at 14 eV is the volume loss.

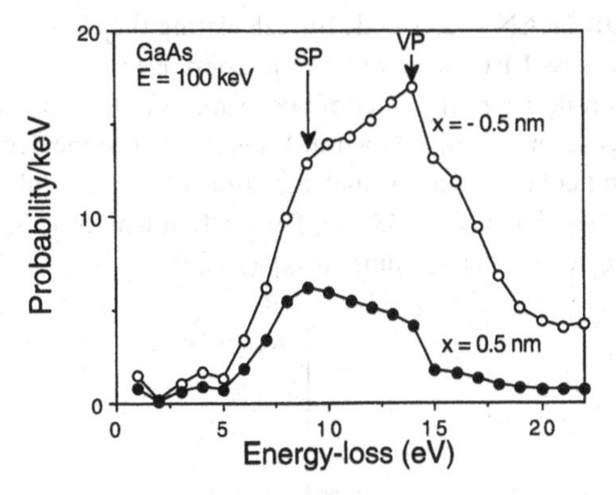

Figure E.1 Calculated single-loss EELS spectra for an incident electron that is traveling parallel to the GaAs surface and at distances $x = -0.5$ nm and $x = 0.5$ nm from the surface.

APPENDIX E.4

A FORTRAN program for single-loss spectra of interface excitation in TEM

Listed below is the FORTRAN source code for calculating the electron inelastic excitation probabilities and single-loss EELS spectra when a 'point' electron probe is positioned at a distance x_0 from and parallel to the interface of two media ($x_0 < 0$ for medium 1 and $x_0 > 0$ for medium 2) in TEM, as schematically shown in Figure E.2. The incident beam direction is assumed to be perpendicular to the normal direction of the crystal slab. The interface excitation is calculated based on Eq. (10.43a–c). The excitation of the top and bottom surfaces is also included in the calculation according to Eq. (10.25).

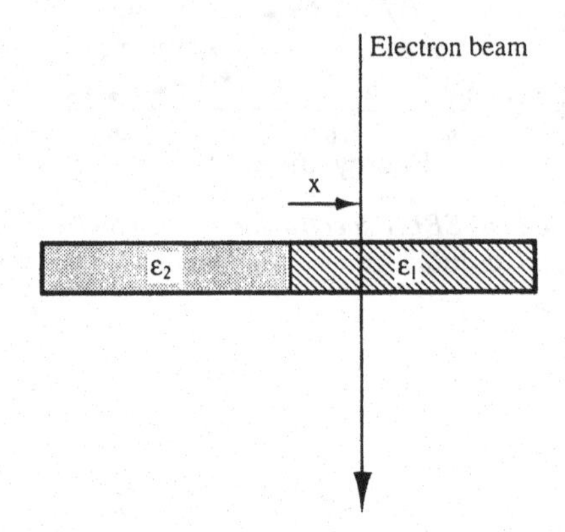

Figure E.2 *A schematic diagram showing the excitation of an interface in TEM.*

```
C       INTERFACE.FOR
C
C       by Z.L. Wang
C
C       THIS PROGRAM CALCULATES THE EXCITATION PROBABILITY FOR AN
C       ELECTRON WHICH IS TRAVELING IN MEDIUM 1 AT A DISTANCE X0 FROM AND
C       PARALLEL TO AN INTERFACE, USING THE RELATIVISTIC
C       DIELECTRIC RESPONSE THEORY GIVEN IN SECTION 10.6.
C
C       THE SINGLE-LOSS SPECTRUM IS ALSO CALCULATED
C
        INTEGER I,J,K
        DIMENSION ER1(50),EI1(50),ER2(50),EI2(50),SPECSHP(50)
        COMPLEX EE1,EE2,ALP1,ALP2,QSA,FUNV,FUNC,UPPART,LOWPART
        OPEN(UNIT=8,FILE='DIELEC1.DAT',STATUS='OLD')
        OPEN(UNIT=10,FILE='DIELEC2.DAT',STATUS='OLD')
        OPEN(UNIT=9, FILE='EXCPROB.DAT', STATUS='NEW')
        OPEN(UNIT=11, FILE='SPECTRA.DAT', STATUS='NEW')
C INPUT ELECTRON VELOCITY V/C
 5      WRITE(6,10)
```

```
 10     FORMAT(2X,'Input electron velocity v/c ratio ...')
        READ(5,*) V
      IF(V.GE.1.0) GOTO 5
        VL=1.0
        VB=(V/VL)**2
        WRITE(9,22) V
 22     FORMAT(2X,'Electron velocity v/c = ',F10.8)
        WRITE(11,22) V
C BEAM IMPACT DISTANCE AT WHICH THE SPECTRUM IS CALCULATED
 25     WRITE(6,30)
 30     FORMAT(3X,'Beam impact distance x (-30<x<30 Angstrom)')
        READ(5,*) X0
        IF(ABS(X0).GT.30.0) GOTO 25
        WRITE(11,42) X0
 42     FORMAT(2X,'Electron impact distance x = ',F10.6,' Angstrom')
C INPUT SPECIMEN THICKNESS
        WRITE(6,50)
 50     FORMAT(2X,'Input specimen thickness (in Angstrom)...')
        READ(5,*) T
        T=ABS(T)
        WRITE(9,52) T
 52     FORMAT(2X,'Specimen thickness d = ',F10.3,' Angstrom')
        WRITE(11,52) T
C INPUT CUT-OFF WAVE VECTOR
      WRITE(6,60)
 60     FORMAT(2X,'Input cut-off wave vector qc (in 1/Angstrom)...')
        READ(5,*) QC
        WRITE(9,44) QC
 44     FORMAT(2X,'Cut-off wave vector qc= ',F10.6,' 1/ANGSTROM')
        WRITE(11,44) QC
        QC=2.0*3.14159*ABS(QC)
      WRITE(6,63)
 63    FORMAT(2X,'Calculation in progress ....')
C INPUT DIELECTRIC FUNCTIONS
C       The dielectric function 1 is given in file 8 of name DIELEC1.DAT with
C       an energy resolution of 1 eV (= 1 channel)
C       The dielectric function 2 is given in file 8 of name DIELEC2.DAT with
C       an energy resolution of 1 eV (= 1 channel)
        READ(8,65) NDP1
      READ(10,65) NDP2
        NDP=MIN(NDP1,NDP2)
 65    FORMAT(2X,I2)
        DO 70 I = 1, NDP
        READ(8,75) ER1(I), EI1(I)
 70     READ(10,75) ER2(I), EI2(I)
 75     FORMAT(2X,F10.7,2X,F10.7)
        WRITE(9,77)
 77     FORMAT(/,5X,'X',8X,'dP/dz',7X,'Lameda')
C
C CALCULATE EXCITATION PROB. AS A FUNCTION OF IMPACT DISTANCE
C FOR X < 0 (IN MEDIUM 2) AND X > 0 (IN MEDIUM 1), WHERE X IS
C MEASURED IN ANGSTROM.  HERE WE ASSUME - 30 < X < 30 ANGSTROM.
        DO 200 I=1,61
        X=(I-31)
        TP=0.0
C INTEGRATE OVER ENERGY-LOSS W AT 1eV/Channel
        DW=1.0
        DO 400 J=1,NDP
        W=J
        WW=W**2
        EE1=CMPLX(ER1(J), EI1(J))
        EE2=CMPLX(ER2(J), EI2(J))
        IF(X.LT.0.0) EE2=CMPLX(ER1(J), EI1(J))
        IF(X.LT.0.0) EE1=CMPLX(ER2(J), EI2(J))
        QSA=CMPLX(0.0, 0.0)
        SP1=0.0
      CP1=0.0
```

```
C INTEGRATE OVER QY
      NMAX1=4000
      NMAX2=3000
      NMAX=NMAX1+NMAX2
      QYH=0.01
      DO 300 K=1,NMAX
      IF(K.GT.NMAX1) GOTO 80
      QY=(K-1)*QYH/NMAX1+QYH/2.0
      DQY=QYH/NMAX1
      GOTO 85
 80   QY=QYH+(K-NMAX1)*(QC-QYH)/NMAX2+QYH/2.0
      DQY=(QC-QYH)/NMAX2
 85   ALP1=CSQRT(QY**2+WW*(1.0-VB*EE1)/(1973.2*V)**2)
      ALP2=CSQRT(QY**2+WW*(1.0-VB*EE2)/(1973.2*V)**2)
C
      FUNV=(2.0*ALP1**2*(EE2-EE1)/(EE1*ALP2+EE2*ALP1)-(1.0-VB*EE1)
   A    *(ALP1-ALP2))*CEXP(-2.0*ALP1*ABS(X))/(EE1*ALP1*(ALP1+ALP2))
   A     -(1.0-VB*EE1)/(ALP1*EE1)
C
C COMPONENT CONTRIBUTED BY THE TOP AND BOTTOM SURFACES OF THE SLAB
      ALP=(QY**2+WW/(1973.2*V)**2)
      QPP=QY**2/ALP**2
      QYY=QY*T
      IF(QYY.GT.50) QYY=50.0
      UPPART=-QPP*((1.0-EE1)/EE1)*(2.0*(EE1-1.0)*COS(W*T/(1973.2*V))
   A       +(EE1-1.0)*EXP(-QYY)+(1.0-EE1**2)*EXP(QYY))
      LOWPART=((EE1-1.0)**2*EXP(-QYY)-(EE1+1.0)**2*EXP(QYY))
      SP1=SP1+AIMAG(UPPART/LOWPART)*DQY
      CP1=CP1+AIMAG(FUNV)*DQY
C
 300  CONTINUE
      TP=CP1*T+SP1+TP
C CALCULATE SPECTRUM
      IF(ABS(X-X0).LE.0.8) SPECSHP(J)=(CP1*T+SP1)
 400  CONTINUE
      CONS=2.35E-6
C EXCITATION PROBABILITY (1/Mean-free-path-length)
      TP=TP*DW*CONS/(V**2)
     TP=TP/T
     ALAMEDA=1.0/TP
C OUTPUT SPECTRUM AT X = X0
      IF(ABS(X-X0).GT.0.8) GOTO 580
      DO 500 J=1,NDP
      RRR=SPECSHP(J)*CONS/V**2
      WRITE(11,450)J,RRR
 450  FORMAT(3X,'W=',I2,3X,'SPECSHP=',F20.12)
 500  CONTINUE
C DATA OUTPUT
 580  WRITE(9,700) X,TP,ALAMEDA
 700  FORMAT(2X,F6.2,5X,F10.8,4X,F10.3)
 200  CONTINUE
      STOP
      END
```

Calculated excitation probability across MgO–GaAs interface.

Excprob.dat:

```
 Electron velocity v/c = 0.77649999
 Specimen thickness d =   100.000 Angstrom
 Cut-off wave vector qc =   1.000000 1/ANGSTROM

    X        dP/dz          Lameda
 -30.00   0.00018677     5354.252
 -29.00   0.00018695     5348.959
```

-28.00	0.00018715	5343.425
-27.00	0.00018735	5337.633
-26.00	0.00018756	5331.563
-25.00	0.00018779	5325.190
-24.00	0.00018802	5318.493
-23.00	0.00018827	5311.440
-22.00	0.00018854	5304.002
-21.00	0.00018882	5296.140
-20.00	0.00018911	5287.814
-19.00	0.00018943	5278.976
-18.00	0.00018977	5269.570
-17.00	0.00019013	5259.534
-16.00	0.00019052	5248.790
-15.00	0.00019094	5237.251
-14.00	0.00019139	5224.805
-13.00	0.00019189	5211.326
-12.00	0.00019243	5196.650
-11.00	0.00019303	5180.575
-10.00	0.00019369	5162.846
-9.00	0.00019443	5143.127
-8.00	0.00019528	5120.965
-7.00	0.00019624	5095.736
-6.00	0.00019737	5066.537
-5.00	0.00019873	5031.979
-4.00	0.00020041	4989.778
-3.00	0.00020260	4935.718
-2.00	0.00020574	4860.557
-1.00	0.00021116	4735.735
0.00	0.00025514	3919.469
1.00	0.00026038	3840.577
2.00	0.00026148	3824.355
3.00	0.00026210	3815.365
4.00	0.00026252	3809.286
5.00	0.00026283	3804.774
6.00	0.00026307	3801.237
7.00	0.00026327	3798.362
8.00	0.00026344	3795.962
9.00	0.00026358	3793.918
10.00	0.00026370	3792.152
11.00	0.00026381	3790.603
12.00	0.00026391	3789.233
13.00	0.00026399	3788.010
14.00	0.00026407	3786.910
15.00	0.00026414	3785.912
16.00	0.00026420	3785.005
17.00	0.00026426	3784.173
18.00	0.00026431	3783.408
19.00	0.00026436	3782.702
20.00	0.00026441	3782.047
21.00	0.00026445	3781.438
22.00	0.00026449	3780.869
23.00	0.00026453	3780.336
24.00	0.00026456	3779.836
25.00	0.00026459	3779.367
26.00	0.00026463	3778.925
27.00	0.00026465	3778.507
28.00	0.00026468	3778.111
29.00	0.00026471	3777.737
30.00	0.00026473	3777.381

Spectra.dat

Electron velocity v/c = 0.77649999
Electron impact distance x = 0.000000 Angstrom
Specimen thickness d = 100.000 Angstrom

```
Cut-off wave vector qc=   1.000000 1/ANGSTROM
 W= 1  SPECSHP=    0.000570254342
 W= 2  SPECSHP=    0.000062943131
 W= 3  SPECSHP=    0.000325356552
 W= 4  SPECSHP=    0.000447751023
 W= 5  SPECSHP=    0.000472897955
 W= 6  SPECSHP=    0.001246014726
 W= 7  SPECSHP=    0.001050096937
 W= 8  SPECSHP=    0.001202774234
 W= 9  SPECSHP=    0.001089790952
 W=10  SPECSHP=    0.001003406127
 W=11  SPECSHP=    0.000965548097
 W=12  SPECSHP=    0.000936761440
 W=13  SPECSHP=    0.000883742701
 W=14  SPECSHP=    0.001628303435
 W=15  SPECSHP=    0.001681482769
 W=16  SPECSHP=    0.001746189781
 W=17  SPECSHP=    0.001789661241
 W=18  SPECSHP=    0.001796069089
 W=19  SPECSHP=    0.001758895465
 W=20  SPECSHP=    0.001678114641
 W=21  SPECSHP=    0.001593395020
 W=22  SPECSHP=    0.001584207057
```

In Figure E.3 is shown the plot of the calculated dP/dz across an MgO–GaAs interface. It is apparent that dP/dz undergoes a sharp variation within a distance smaller than 0.5 nm from the interface. This excitation is approximately localized so that the energy-filtered plasmon-loss electron image may provide some information about the chemical sharpness of the interface.This experiment has been performed by Wang and Shapiro (1995d) using an Al/Ti multilayer specimen.

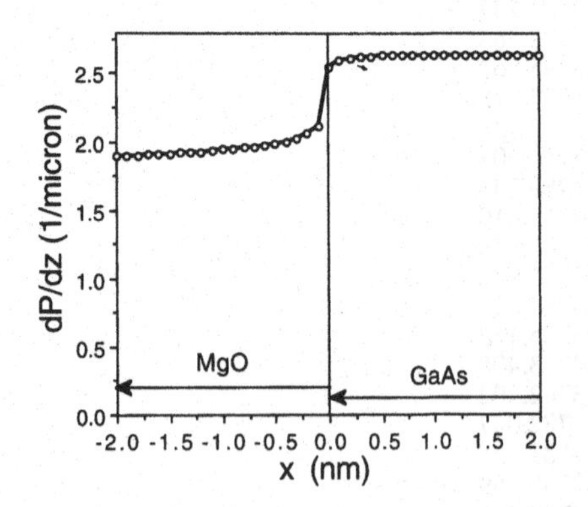

Figure E.3 *The calculated valence-loss excitation probability across the MgO–GaAs interface in TEM. The calculated inelastic mean-free-path length is smaller than the practically measured one because the dielectric data are for energy losses lower than 22 eV.*

APPENDIX F

A bibliography of REM, SREM and REELS

This bibliography lists all papers published on REM, SREM and REELS in the last two decades. The list was based on the bibliography published by Hsu and Peng (1993) and the Science Citation Index. The list includes articles that explicitly deal with the theories, techniques and applications of REM, SREM and REELS for surface studies. Almost exclusively papers published in refereed journals in English are listed.

1995

Wang, Z. L. and Shapiro, A. J. (1995) Studies of $LaAlO_3$ {100} surfaces using RHEED and REM I: twins, steps and dislocations *Surf. Sci.* **328**, 141–58.

Wang, Z. L. and Shapiro, A. J. (1995) Studies of $LaAlO_3$ {100} surfaces using RHEED and REM II 5 × 5 surface reconstruction *Surf. Sci.* **328**, 159–69.

Wang, Z. L. and Shapiro, A. J. (1995) Microstructure and reconstruction of $LaAlO_3$ {100} and {110} surfaces *Proc. Mater. Res. Soc.* **355**, 175–80.

1994

Baba-Kichi, K. Z. (1994) Shadow imaging of surface topography in a conventional scanning electron microscope by forward scattered electrons *Ultramicroscopy* **54**, 1–7.

Horio, Y. and Ichimiya, A. (1994) Origin of phase shift phenomena in RHEED intensity oscillation curves *Ultramicroscopy* **55**, 321–8.

Houchmandzadeh, B. and Misbah, C. (1994) Localized morphological mode during sublimation of a vicinal pyramid *Phys. Rev. Lett.* **73**, 94–7.

Hsu, T. and Cowley, J. M. (1994) Study of twinning with reflection electron microscopy *Ultramicroscopy* **55**, 302–7.

Jung, T. M., Phaneuf, R. J. and Williams, E. D. (1994) Sublimation and phase-transitions on singular and vicinal Si(111) surfaces *Surf. Sci.* **301**, 129–35.

Latyshev, A. V., Krasilnikov, A. B. and Aseev, A. L. (1994) UHV REM study of the anti-band structure on the vicinal Si(111) surface under heating by a direct electric current *Surf. Sci.* **311**, 395–403.

McCoy, J. M. and Maksym, P. A. (1994) Multiple-scattering calculations of step contrast in REM images of the Si(001) surface *Surf. Sci.* **310**, 217–25.

Okamoto, K., Hananoki, R. and Sakiyama, K. (1994) InGaAs epilayers of high In composition grown on GaAs substrates by molecular-beam epitaxy *Japan J. Appl. Phys.* **33**, 28–32.

Omahony, J. D., McGlip, J. F., Flipse, C. F. J., Weightman, P. and Leibsle, F. M. (1994) Nucleation and evolution of the Au-induced 5 × 2 structure on vicinal Si(111) *Phys. Rev.* B **49**, 2527–35.

Pimpinelli, A., Elkinani, I., Karma, A., Misbah, C. and Villain, J. (1994) Step motions on high-temperature vicinal surfaces *J. Phys.: Condensed Matter* **6**, 2661–80.

Pimpinelli, A. and Metois, J. J. (1994) Macrovacancy nucleation on evaporating Si(001) *Phys. Rev. Lett.* **72**, 3566–9.

Porter, S. J., Matthew, J. A. D. and Leggott, R. J. (1994) Inelastic exchange scattering in electron-energy-loss spectroscopy – localized excitations in transition-metal and rare-earth systems *Phys. Rev.* B **50**, 2638–41.

Smith, A. E. and Josefsson, T. W. (1994) The effect of a non-Hermitian crystal potential on the scattering matrix in reflection electron diffraction *Ultramicroscopy* **55**, 247–52.

Spence, J. C. H. (1994) Reflection electron microscopy of buried interfaces in the transmission geometry *Ultramicroscopy* **55**, 293–301.

Suzuki, T. and Nishinaga, T. (1994) Real-time observation and formation mechanism of Ga droplet during molecular-beam epitaxy under excess Ga flux, *J. Cryst. Growth* **142**, 61–7.

Suzuki, T., Tanishiro, Y., Minoda, H., Yagi, K. and Suzuki, M. (1994) REM observations of Si(*Hhk*) surfaces and their vicinal surfaces *Surf. Sci.* **298**, 473–7.

Tsai, F. and Cowley, J. M. (1994) Observation of planar defects by reflection electron-microscopy *Ultramicroscopy* **52**, 400–3.

Uchida, Y., Wang, N. and Zeitler, E. (1994) The alloying effect of Ag on the reconstruction in Au(111) and (100) surfaces observed by REM *Ultramicroscopy* **55**, 308–13.

Vook, R. W. (1994) Transmission and reflection electron-microscopy of electromigration phenomena *Mater. Chem. Phys.* **36**, 199–216.

Wang, L. and Cowley, J. M. (1994) Electron channelling effects at high incident angles in convergent beam reflection high energy electron diffraction *Ultramicroscopy* **55**, 228–40.

Wang, L., Liu, J. and Cowley, J. M. (1994) Studies of single-crystal TiO_2 (001) and (100) surfaces by reflection high-energy electron-diffraction and reflection electron-microscopy *Surf. Sci.* **302**, 141–57.

Williams, E. D. (1994) Surface steps and surface-morphology – understanding macroscopic phenomena from atomic observations *Surf. Sci.* **300**, 502–24.

Wurm, K., Kliese, R., Hong, Y., Rottger, B., Wei, Y., Neddermeyer, H. and Tsong, I. S. T. (1994) Evolution of surface-morphology of Si(100)-(2 × 1) during oxygen-adsorption at elevated temperatures *Phys. Rev.* B **50**, 1567–74.

Yamamoto, T., Akutsu, Y. and Akutsu, N. (1994) Fluctuation of a single-step on the vicinal surface – universal and nonuniversal behaviors *J. Phys. Soc. Japan* **63**, 915–25.

1993

Akita, T., Takeguchi, M. and Shimizu, R. (1993) Observation of reconstructed Pt(100) surface by reflection electron-microscopy *Japan. J. Appl. Phys. Lett.* **32**, L1631–4.

Atwater, H. A., Wong. S. S., Ahn, C. C., Nikzad, S. and Frase, H. N. (1993) Analysis of monolayer films during molecular beam epitaxy by reflection electron energy loss spectroscopy *Surf. Sci.* **298**, 273–83.

Banzhof, H. and Herrmann, K. H. (1993) Reflection electron holography *Ultramicroscopy* **48**, 475–81.

Cowley, J. M. (1993) Electron holography and holographic diffraction for surface studies *Surf. Sci.* **298**, 336–44.

Cowley, J. M. and Liu, J. (1993) Contrast and resolution in REM, SEM and SREM *Surf. Sci.* **298**, 456–67.

Crozier, P. A. and Gajdardziska-Josifovska, M. (1993) Reflection electron energy-loss spectroscopy studies of Ca segregation on MgO (100) surfaces *Ultramicroscopy* **48**, 63–76.

Filicheva, O. D., Davydov, A. A., Belikhmaer, Y. A. and Ivasenko, V. L. (1993) Study of cobalt tetrasulfophthalocyanin applied on γ-Al_2O_3 by the diffusion reflection electron-spectroscopy technique *Zh. Fiz. Khim.* **66**, 3276–80.

Fuwa, K., Yamamuro, K., Yanagida, H. and Shimizu, S. (1993) Development of a scanning reflection electron-microscope for *in-situ* observation of crystal-growth *Nuc. Instrum. Methods Phys. Res.* A **328**, 592–5.

Gajdardziska-Josifovska, M., Crozier, P. A., McCartney, M. R. and Cowley, J. M. (1993) Ca segregation and step modifications on cleaved and annealed MgO(100) surfaces *Surf. Sci.* **284**, 186–99.

Gajdardziska-Josifovska, M., McCartney, M. R. and Smith, D. J. (1993) Reflection electron-microscopy studies of GaP(110) surfaces in UHV-TEM *Surf. Sci.* **287**, 1062–6.

Liu, J. and Cowley, J. M. (1993) Scanning reflection electron-microscopy and associated techniques for surface studies *Ultramicroscopy* **48**, 381–416.

Lo, W. K. and Spence, J. C. H. (1993) Investigation of STM image artifacts by *in-situ* reflection electron-microscopy *Ultramicroscopy* **48**, 433–44.

Marek, T., Strunk, H. P., Bauser, E. and Luo, Y. C. (1993) Networks of growth steps on as-grown GaAs(001) vicinal surfaces *Inst. Phys. Conf. Ser.* **134**, 349–52.

McCoy, J. M. and Maksym, P. A. (1993) Simulation of reflection electron-microscopy images – application to high-resolution imaging of the Si(001)2 × 1 surface *Surf. Sci.* **297**, 113–26.

McCoy, J. M. and Maksym, P. A. (1993) Simulation of high-resolution REM images *Surf. Sci.* **298**, 468–72.

Mondio, G., Neri, F., Curro, G., Patane, S. and Compagnini, G. (1993) The dielectric-constant of TCNQ single-crystals as deduced by reflection electron-energy-loss spectroscopy *J. Mater. Res.* **8**, 2627–33.

Nakahara, H. and Ichikawa, M. (1993) Migration of Ga atoms during Si molecular beam epitaxial growth on a Ga-adsorbed Si(111) surface *Surf. Sci.* **298**, 440–9.

Nikzad, S., Wong, S. S., Ahn, C. C., Smith, A. L. and Atwater, H. A. (1993) *In-situ* reflection electron-energy-loss spectroscopy measurements of low-temperature surface cleaning for Si molecular-beam epitaxy *Appl. Phys. Lett.* **63**, 1414–16.

Osakabe, N., Matsuda, T., Endo, J. and Tonomura, A. (1993) Reflection electron holographic observation of surface displacement field *Ultramicroscopy* **48**, 483–8.

Peng, L. M., Du, A. Y., Jiang, J. and Zhou, J. X. (1993) Reflection electron imaging of semiconductor multilayer materials *Ultramicroscopy* **48**, 453–63.

Sasaki, H., Aoki, M. and Kawarada, H. (1993) Reflection electron-microscope and scanning tunneling microscope observations of CVD diamond (001) surfaces *Diamond Related Mater.* **2**, 1271–6.

Smith, D. J., Gajdardziska-Josifovska, M., Lu, P., McCartney, M. R., Podbrdsky, J., Swann, P. R. and Jones, J. S. (1993) Development and applications of a 300 keV ultrahigh-vacuum high-resolution electron microscope *Ultramicroscopy* **49**, 26–36.

Wang, N., Uchida, Y. and Lehmpfuhl, G. (1993) REM study of Au(100) reconstructed surfaces *Surf. Sci. Lett.* **284**, L419–25.

Wang, Z. L. (1993) Electron reflection, diffraction and imaging of bulk crystal surfaces in TEM and STEM *Rep. Prog. Phys.* **56**, 997–1065.

Wang, Z. L. and Bentley, J. (1993) *In-situ* dynamic processes on cleaved α-alumina bulk crystal-surfaces imaged by reflection electron-microscopy *Ultramicroscopy* **48**, 64–80.

Wang. Z. L. and Bentley, J. (1993) Determination of effective ionization cross-sections for

quantitative surface chemical microanalysis using REELS *Ultramicroscopy* **48**, 465–73.
Yagi, K. (1993) Reflection electron-microscopy – studies of surface-structures and surface dynamic processes *Surf. Sci. Rep.* **17**, 305–62.
Yagi, K., Minoda, H. and Shima, M. (1993) Reflection electron-microscopy study of thin-film growth *Thin Solid Films* **228**, 12–17.
Yamaguchi, H. and Yagi, K. (1993) Current effects on Si(111) surfaces at the phase transition between the 7 × 7 and the 1 × 1 structure. II *Surf. Sci.* **298**, 408–14.

1992

Alfonso, C., Bermond, J. M., Heyraud, J. C. and Metois, J. J. (1992) The meandering of steps and the terrace width distribution on clean Si(111) – an *in situ* experiment using reflection electron-microscopy *Surf. Sci.* **262**, 371–81.
Alfonso, C., Bermond, J. M., Heyraud, J. C. and Metois, J. J. (1992) The meandering of steps and the terrace width distribution on clean Si(111) *Surf. Sci.* **262**, 371–81.
Banzhof, H., Herrmann, K. H. and Lichte, H. (1992) Reflection electron-microscopy and interferometry of atomic steps on gold and platinum single-crystal surfaces *Microsc. Res. Technique* **20**, 450–6.
Claverie, A., Beauvillain, J., Faure, J., Vieu, C. and Jouffrey, B. (1992) Degradation, amorphization, and recrystallization of ion bombarded Si(111) surfaces studied by *in situ* reflection electron-microscopy and reflection high-energy electron-diffraction techniques *Microsc. Res. Technique* **20**, 352–9.
Cowley, J. M. (1992) Resolution limitation in the electron microscopy of surfaces *Ultramicroscopy* **47**, 187–98.
Crozier, P. A. Gajdardziska-Josifovska, M. and Cowley, J. M. (1992) Preparation and characterization of MgO surfaces by reflection electron-microscopy *Microsc. Res. Technique* **20**, 426–38.
Eades, J. A. (1992) Reflection electron microscopy and reflection electron diffraction in the electron microscope, in *Surface Science*, eds. F. A. Ponce and M. Cardona, Springer-Verlag, Berlin, pp. 99–103.
Heinricht, F. and Bostanjoglo, O. (1992) Laser ablation processes imaged by high-speed reflection electron-microscopy *Appl. Surf. Sci.* **54**, 244–54.
Howie, A., Lanzerotti, M. Y. and Wang, Z. L. (1992) Incoherence effects in reflection electron microscopy *Microsc. Microanal. Microstruct.* **3**, 233–41.
Hsu, T. (1992) Technique of reflection electron-microscopy *Microsc. Res. Technique* **20**, 318–32.
Iwanari, S. and Takayanagi, K. (1992) Surfactant epitaxy of Si on Si(111) surface mediated by a Sn layer. 1. Reflection electron-microscope observation of the growth with and without a Sn layer mediating the step flow *J. Cryst. Growth* **119**, 229–40.
Kim, Y. T. and Hsu, T. (1992) Study of α-$Al_2O_3(1\bar{1}02)$ surfaces with the reflection electron-microscopy (REM) technique *Surf. Sci.* **275**, 339–50.
Kleint, C. and Elhalim, S. M. A. (1992) Reflection electron-energy loss spectroscopy of alkali-metals on silicon *Acta Phys. Polonica* A **81**, 47–65.
Krasilnikov, A. B., Latyshev, A. V., Aseev, A. L. and Stenin, S. I. (1992) Monoatomic steps clustering during superstructural transitions on Si(111) surface *J. Cryst. Growth* **116**, 178–84.
Latyshev, A. V., Krasilnikov, A. B. and Aseev, A. L. (1992) Application of ultrahigh-vacuum

reflection electron-microscopy for the study of clean silicon surfaces in sublimation, epitaxy, and phase-transitions *Microsc. Res. Technique* **20**, 341–51.

Latyshev, A. V., Krasilnikov, A. B. and Aseev, A. L. (1992) Direct REM observation of structural processes on clean silicon surfaces during sublimation, phase transition and epitaxy *Appl. Surf. Sci.* **60/61**, 397–404.

Minoda, H., Tanishiro, Y., Yamamoto, N. and Yagi, K. (1992) Growth of Si on Au deposited Si(111) surfaces studied by UHV-REM *Appl. Surf. Sci.* **60/61**, 107–11.

Mondio, G., Neri, F., Patane, S., Arena, A., Marletta, G. and Iacona, F. (1992) Optical-properties from reflection electron-energy loss spectroscopy *Thin Solid Films* **207**, 313–18.

Murooka, K., Tanishiro, Y. and Takayanagi, K. (1992) Dynamic observation of oxygen-induced step movement on the Si(111)7 × 7 surface by high-resolution reflection electron-microscopy *Surf. Sci.* **275**, 26–30.

Nakayama, T., Tanishiro, Y. and Takayanagi, K. (1992) Heterogrowth of Ge on the Si(001)2 × 1 reconstructed surface *Surf. Sci.* **273**, 9–20.

Ndubuisi, G. C., Liu, J. and Cowley, J. M. (1992) Stepped surfaces of sapphire with low Miller indices *Microsc. Res. Technique* **21**, 10–22.

Ndubuisi, G. C., Liu, J. and Cowley, J. M. (1992) Characterization of the annealed (0001) surface of sapphire (α-Al_2O_3) and interaction with silver by reflection electron-microscopy and scanning reflection electron-microscopy *Microsc. Res. Technique* **20**, 439–49.

Nikzad, S., Ahn, C. C. and Atwater, H. A. (1992) Quantitative-analysis of semiconductor alloy composition during growth by reflection-electron energy-loss spectroscopy *J. Vac. Sci. Technol.* B **10**, 762–5.

Osakabe, N. (1992) Observation of surfaces by reflection electron holography *Microsc. Res. Technique* **20**, 457–62.

Peng, L. M., Du, A. Y., Jiang, J. and Zhou, X. C. (1992) Reflection electron-microscopy imaging of GaAs/Al_xGa_{1-x} as multilayer materials *Phil. Mag. Lett.* **66**, 9–17.

Peng, L. M., Gjønnes, K. and Gjønnes, J. (1992) Bloch wave treatment of symmetry and multiple beam cases in reflection high-energy electron-diffraction and reflection electron-microscopy *Microscopy Res. Technique* **20**, 360–70.

Peng, L.-M., Zhou, X. C., Jiang, J. and Du, A. Y. (1992) Application of REM in cross-sectional study of multilayer semiconductor devices *J. Vac. Sci. Technol.* B **10**, 2293–6.

Takayanagi, K. and Tanishiro, Y. (1992) Symmetry and ordering of metal deposits on the Si(111)7 × 7 surface, in *Ordering at Surfaces and Interfaces*, eds. A. Yoshimori, T. Shinjo and H. Watanabe, Springer-Verlag, Berlin, pp. 273–8.

Takeguchi, M., Harada, K. and Shimizu, R. (1992) Study of defects and strains on cleaved GaAs (110) surface by reflection electron-microscopy *J. Electron Microsc.* **41**, 174–8.

Tsai, F. and Cowley, J. M. (1992) Observation of ferroelectric domain boundaries in $BaTiO_3$ single-crystals by reflection electron-microscopy *Ultramicroscopy* **45**, 43–53.

Uchida, Y., Imbihl, R. and Lehmpfuhl, G. (1992) Surface topographic changes on Pt(100) and Pt(111) in the NO + CO reaction: a reflection electron microscopy study *Surf. Sci.* **275**, 253–64.

Wang, N., Uchida, Y. and Lehmpfuhl, G. (1992) Method of directly imaging the reconstruction of Au(111) and Au(100) by reflection electron microscopy *Ultramicroscopy* **45**, 291–8.

Wang, Z. L. (1992) Atomic step structures on cleaved α-alumina ($01\bar{1}2$) surfaces *Surf. Sci.* **271**, 477–92.

Wang, Z. L. and Bentley, J. (1992) Reflection electron energy-loss spectroscopy and imaging for surface studies in transmission electron-microscopes *Microsc. Res. Technique* **20**, 390–405.

Wang, Z. L., Bentley, J., Kenik, E. A., Horton, L. L. and McKee, R. A. (1992) *In-situ* formation of MgO_2 thin films on MgO single crystal surfaces at high temperatures *Surf. Sci.* **273**, 88–108.

Yagi, K., Yamanaka, A. and Homma, I. (1992) Recent studies of surface dynamic processes by reflection electron-microscopy *Microsc. Res. Technique* **20**, 333–40.

Yamaguchi, H., Tanishiro, Y. and Yagi, K. (1992) REM study of surface electromigration of Ge, Au–Cu and Ag on Si(111) surfaces *Appl. Surf. Sci.* **60/61**, 79–84.

Yamanaka, A., Tanishiro, Y. and Yagi, K. (1992) Surface electromigration of Au on Si(111) studied by REM, in *Ordering at Surfaces in Interfaces*, eds. A. Yoshimori, T. Shinjo and H. Watanabe, Springer-Verlag, Berlin, pp. 215–26.

Yamanaka, A., Tanishiro, Y. and Yagi, K. (1992) Surface electromigration of Au on Si(001) studied by REM *Surf. Sci.* **264**, 55–64.

Yubero, F. and Tougaard, S. (1992) Model for quantitative-analysis of reflection-electron-energy-loss spectra *Phys. Rev.* **B 46**, 2486–97.

1991

Aseev, A. L. Latyshev, A. V. and Krasilnikov, A. B. (1991) Direct observation of monoatomic step behaviour in MBE on Si by reflection electron microscopy *J. Cryst. Growth* **115**, 393–7.

Gajdardziska-Josifovska, M., Crozier, P. A. and Cowley, J. M. (1991) A $\sqrt{3}\times\sqrt{3}R30°$ reconstruction on annealed (111) surfaces of MgO *Surf. Sci. Lett.* **248**, L259–64.

Gajdardziska-Josifovska, M., Crozier, P. A. and Cowley, J. M. (1991) Characterization of (100) and (111) surfaces of MgO by reflection electron microscopy, in *The Structure of Surfaces III*, eds. S. Y. Tong, M. A. Van Hove, K. Takayanagi and X. D. Xie, Springer-Verlag, Berlin, pp. 660–4.

Homma, I., Tanishiro, Y. and Yagi, K. (1991) REM and TEM studies of 2D Au–Cu alloy adsorbates on a Si(111) surface *Surf. Sci.* **242**, 81–9.

Homma, I., Tanishiro, Y. and Yagi, K. (1991) Structures formed by co-deposition of two metals on Si(111)7 × 7 surfaces studied by reflection electron microscopy, in *The Structure of Surfaces III*, eds. S. Y. Tong, M. A. Van Hove, K. Takayanagi and X. D. Xie, Springer-Verlag, Berlin, pp. 610–14.

Hsu, T. (1991) Study of surface defects with the reflected electrons, in *Defects in Materials*, eds. P. D. Bristowe, J. E. Epperson, J. E. Griffith and Z. Liliental-Weber, Materials Research Society, New York, pp. 629–35.

Hsu, T. and Kim, Y. T. (1991) Reconstruction of the α-$Al_2O_3(11\bar{2}0)$ surfaces *Surf. Sci. Lett.* **243**, L63–6.

Hsu, T. and Kim, Y. T. (1991) Structure of α-$Al_2O_3(11\bar{2}0)$ surface: facets and reconstruction *Surf. Sci.* **258**, 119–30.

Hsu, T., Wang, T. Y. and Stringfellow, G. B. (1991) Study of atomic structure of OMVPE grown GaInAs/InP quantum wells *J. Microsc.* **163**, 275–86.

Iwatsuki, M., Murooka, K., Kitamura, S.-I., Takayanagi, K. and Harada, Y. (1991) A scanning tunneling microscope (STM) for conventional transmission electron microscopy (TEM) *J. Electron Microsc.* **40**, 48–53.

Kim, Y. T. and Hsu, T. (1991) A reflection electron microscopy (REM) study of α-$Al_2O_3(0001)$ surfaces *Surf. Sci.* **258**, 131–46.

Kondo, Y., Ohi, K., Ishibashi, Y., Hirano, H., Harada, Y., Takayanagi, K., Tanishiro, Y.,

Kobayashi, K. and Yagi, K. (1991) Design and development of an ultrahigh vacuum high-resolution transmission electron microscope *Ultramicroscopy* **35**, 111–18.
Kondo, Y., Yagi, K., Kobayashi, K., Kobayashi, H., Yanaka, Y., Kise, K. and Ohkawa, T. (1991) Design features of an ultrahigh-vacuum electron microscope for REM-PEEM studies of surfaces *Ultramicroscopy* **36**, 142–7.
Latyshev, A. V. and Krasilnikov, A. B. (1991) *In situ* REM study of silicon surface during MBE processes *Mater. Sci. Forum* **69**, 159–62.
Latyshev, A. V., Krasilnikov, A. B., Aseev, A. L. and Stenin, S. I. (1991) REM study of clean Si(111) surface reconstruction during the (7×7)–(1×1) phase transition *Surf. Sci.* **254**, 90–6.
Lehmpfuhl, G., Ichimiya, A. and Nakahara, N. (1991) Interpretation of RHEED oscillations during MBE growth *Surf. Sci. Lett.* **245**, L159–62.
Li, Z. Q., Li, G. H., Qin, Y., Shen, H. and Wu, X. J. (1991) Fractography of Bi-permeated copper bicrystal *Scripta Metall. Mater.* **25**, 367–70.
Litvin, L. V., Krasilnikov, A. B. and Latyshev, A. V. (1991) Transformations of stepped Si(001) surface structure induced by specimen heating current *Surf. Sci. Lett.* **244**, L121–4.
Peng, L.-M. and Czernuszka, J. T. (1991) REM observations of dislocations and dislocation motion in zinc oxide *Phil. Mag.* A **64**, 533–41.
Peng, L.-M. and Czernuszka, J. T. (1991) Studies on the etching and annealing behaviour of α-$Al_2O_3(\bar{1}102)$ surface by reflection electron microscopy *Surf. Sci.* **243**, 210–18.
Peng, L.-M., Du, A. Y., Jiang, J. and Zhou, X. C. (1991) A combined REM and WTEM study of $GaAs/Al_xGa_{1-x}As$ multilayer structures *Phil. Mag. Lett.* **64**, 261–7.
Shima, M., Tanishiro, Y., Kobayashi, K. and Yagi, K. (1991) UHV-REM study of homoepitaxial growth of Si *J. Cryst. Growth* **115**, 359–64.
Tanishiro, Y., Fukuyama, M. and Yagi, K. (1991) REM and RHEED studies of lead adsorption on silicon(111) surfaces, in *The Structure of Surfaces III*, eds. S. Y. Tong, M. A. Van Hove, K. Takayanagi and X. D. Xie, Springer-Verlag, Berlin, pp. 623–7.
Tanishiro, Y., Takayanagi, K. and Yagi, K. (1991) Density of silicon atoms in the Si(111) $\sqrt{3} \times \sqrt{3}$–Ag structure studied by *in situ* UHV reflection electron microscopy *Surf. Sci. Lett.* **258**, L687–90.
Wang, Z. L. (1991) Reflection electron imaging and spectroscopy studies of aluminium metal reduction at α-aluminium oxide $(01\bar{1}2)$ surfaces *J. Microsc.* **163**, 261–74.
Wang. Z. L. (1991) Imaging friction tracks at natural polished diamond surfaces using reflection electron microscopy *J. Electron Microsc. Technique* **17**, 231–40.
Wang, Z. L. and Bentley, J. (1991) Imaging and spectroscopy of α-alumina, diamond, Ni and Fe bulk crystal surfaces *Ultramicroscopy* **37**, 103–15.
Wang, Z. L. and Bentley, J. (1991) *In-situ* observations of high temperature surface processes on α-alumina bulk crystals, in *Advances in Surfaces and Thin Film Diffraction*, eds. T. C. Huang, P. I. Cohen and D. J. Eaglesham, Materials Research Society, New York, pp. 155–60.
Wang, Z. L. and Bentley, J. (1991) Z-contrast imaging of bulk crystal surfaces in scanning reflection electron microscopy *Ultramicroscopy* **37**, 39–49.
Wang, Z. L., Feng, Z. and Field, J. E. (1991) Imaging friction tracks on diamond surfaces using reflection electron microscopy (REM) *Phil. Mag.* A **63**, 1275–89.
Yagi, K., Yamanaka, A., Sato, H., Shima, M., Ohse, H., Ozawa, S.-I. and Tanishiro, Y. (1991) UHV-TEM-REM studies of Si(111) surfaces *Prog. Theor. Phys.* Supplement 106, 303–14.
Yamanaka, A., Ohse, N., Kahata, H. and Yagi, K. (1991) Current effects on clean Si(111) and

(001) surfaces studied by reflection electron microscopy, in *The Structure of Surfaces III*, eds. S. Y. Tong, M. A. Van Hove, K. Takayanagi and X. D. Xie, Springer-Verlag, Berlin, pp. 502–6.

Yamanaka, A. and Yagi, K. (1991) Surface electromigration of In and Cu on Si(111) surfaces studied by REM *Surf. Sci.* **242**, 181–90.

1990

Banzhof, H. and Herrmann, K. H. (1990) Comparison of surface step images in reflection electron microscopy and scanning reflection electron microscopy *Ultramicroscopy* **33**, 23–6.

Bostanjoglo, O. and Heinricht, F. (1990) A reflection electron microscope for imaging of fast phase transitions on surfaces *Rev. Sci. Instrum.* **61**, 1223–9.

Hsu, T. and Kim, Y. T. (1990) Surface structure of the α-Al_2O_3 single crystals, indexing the facets with computer simulation *Ultramicroscopy* **32**, 103–12.

Latyshev, A. V., Aseev, A. L., Krasilnikov, A. B. and Stenin, S. I. (1990) Reflection electron microscopy study of structural transformations on a clean silicon surface in sublimation, phase transition and homoepitaxy *Surf. Sci.* **227**, 24–34.

Lehmpfuhl, G. and Uchida, Y. (1990) Observation of surface crystallography by reflection electron microscopy *Surf. Sci.* **235**, 295–306.

Li, Z.-Q., Li, Q.-H., Qin, Y. and Shen, H. (1990) Characterization of the fracture surface of bi-permeated copper bicrystal by reflection electron microscopy *Phil. Mag. Lett.* **62**, 125–30.

Li, Z.-Q., Shen, H. and Qin, Y. (1990) Growth steps on Pt(110) *Chinese Phys. Lett.* **7**, 245–7.

Peng, L.-M. (1990) Illumination of crystal surface in the electron microscope under RHEED and REM geometry *Ultramicroscopy* **32**, 169–76.

Spence, J. C. H., Lo, W. and Kuwabara, M. (1990) Observation of the graphite surface by reflection electron microscope during STM operation *Ultramicroscopy* **33**, 69–82.

Tanishiro, Y., Yagi, K. and Takayanagi, K. (1990) Gold adsorption processes on Si(111)7×7 studied by *in-situ* reflection electron microscopy *Surf. Sci.* **234**, 37–42.

Uchida, Y. and Lehmpfuhl, G. (1990) Estimation of ad-vacancy formation energy on the Pt(111) surface by using reflection electron microscopy *Surf. Sci.* **243**, 193–8.

Uchida, Y., Lehmpfuhl, G. and Imbihl, R. (1990) Reflection electron microscopy of the catalytic etching of Pt single-crystal spheres in $CO + O_2$ *Surf. Sci.* **234**, 27–36.

Wang, Z. L. and Howie, A. (1990) Electron beam radiation damage at α-alumina($01\bar{1}1$) surfaces of different atomic terminations *Surf. Sci.* **226**, 293–306.

Wang, Z. L. and Spence, J. C. H. (1990) Magnetic contrast in reflection electron microscopy *Surf. Sci.* **234**, 98–107.

Yao, N. and Cowley, J. M. (1990) Electron diffraction conditions and surface imaging in reflection electron microscopy *Ultramicroscopy* **33**, 237–54.

1989

Bleloch, A. L., Howie, A., Milne, R. H. and Walls, M. G. (1989) Elastic and inelastic scattering effects in reflection electron microscopy *Ultramicroscopy* **29**, 175–182.

Buffat, P. A., Ganiere, J. D., Stadelmann, P. (1989) Transmission and reflection electron microscopy on cleaved edges of III–V multilayered structures, in *Evaluation of Advanced Semiconductor Materials by Electron Microscopy*, ed. D. Cherns, Plenum Press, New York, pp. 319–34.

Buffat, P. A., Ganiere, J. D., Stadelmann, P. (1989) High resolution observation and image simulation on cleaved wedges of III–V semiconductors *Mater. Res. Soc. Symp. Proc.* **139**, 111–16.

Cowley, J. M. (1989) Imaging and analysis of surfaces with high spatial resolution *J. Vac. Sci. Technol.* A **7**, 2823–8.

Faist, J., Ganiere, J.-D., Buffat, P., Sampson, S. and Reinhart, F.-K. (1989) Characterization of GaAs/$(GaAs)_n(AlAs)_m$ surface-emitting laser structures through reflectivity and high-resolution electron microscopy measurements *J. Appl. Phys.* **66**, 1023–32.

Heinricht, F. (1989) Reflection electron microscopy of ultra fast surface processes *Phys. Stat. Sol*, (a) **116**, 145–52.

Hsu, T. and Lehmpfuhl, G. (1989) Streaking of reflection high energy electron diffraction spots as a result of refraction on a curved specimen surface *Ultramicroscopy* **27**, 359–66.

Inoue, N. and Yagi, K. (1989) *In situ* observation by ultrahigh vacuum reflection electron microscopy of terrace formation processes on (100) silicon surfaces during annealing *Appl. Phys. Lett.* **55**, 1400–2.

Kagiyama, K., Tanishiro, Y. and Takayanagi, K. (1989) Reconstructions and phase transitions of Ge on the Si(111)7 × 7 surface, I. Structural changes *Surf. Sci.* **222**, 38–46.

Kahata, H. and Yagi, K. (1989) Preferential diffusion of vacancies perpendicular to the dimers on Si(001)2 × 1 surfaces studied by UHV REM *Japan. J. Appl. Phys.* **28**, L1042–4.

Kahata, H. and Yagi, K. (1989) The effect of surface anisotropy of Si(001)2 × 1 on hollow formation in the initial stage of oxidation as studied by reflection electron microscopy *Surf. Sci.* **220**, 131–6.

Kahata, H. and Yagi, K. (1989) REM observation on conversion between single-domain surfaces of Si(001)2 × 1 and 1 × 2 induced by specimen heating current *Japan. J. Appl. Phys.* **28**, L858–61.

Koike, H., Kobayashi, K., Ozawa, S.-I. and Yagi, K. (1989) High-resolution reflection electron microscopy of Si(111)7 × 7 surface using a high-voltage electron microscope *Japan. J. Appl. Phys.* **28**, 861–5.

Kuwabara, M., Lo, W. and Spence, J. C. H. (1989) Reflection electron microscope imaging of an operating scanning tunneling microscope *J. Vac. Sci. Technol.* **7**, 2745.

Latyshev, A. V., Aseev, A. L., Krasilnikov, A. B. and Stenin, S. I. (1989) Transformations on clean Si(111) stepped surface during sublimation *Surf. Sci.* **213**, 157–69.

Latyshev, A. V., Aseev, A. L., Krasilnikov, A. B. and Stenin, S. I. (1989) Initial stages of silicon homoepitaxy studied by *in situ* reflection electron microscopy *Phys. Stat. Sol.* (a) **113**, 421–30.

Milne, R. H. (1989) STEM imaging of monatomic surface steps and emerging dislocations *J. Microsc.* **153**, 22

Ohse, N. and Yagi, K. (1989) Reflection electron microscope study of hydrogen atom adsorption on Si(111)7 × 7 surfaces *Surf. Sci.* **217**, L430–4.

Osakabe, N., Endo, J., Matsuda, T., Fukuhara, A. and Tonomura, A. (1989) Observation of surface undulation due to single atomic shear of a dislocation by reflection electron holography *Phys. Rev. Lett.* **62**, 2969–72.

Peng, L.-M. (1989) REM observation of electron-beam-induced reactions on GaAs(110) surface *Ultramicroscopy* **27**, 423–6.

Peng, L.-M. (1989) The importance of the surface illumination in reflection electron microscopy, in *EMAG-MICRO 89*, eds. P. J. Goodhew and H. Y. Elder, Institute of Physics, Bristol, Vol. 1 pp. 9–12.

Peng, L.-M. and Cowley, J. M. (1989) Thermal diffuse scattering and REM image-contrast preservation *Ultramicroscopy* **29**, 168–74.

Peng, L.-M., Cowley, J. M. and Hsu, T. (1989) Reflection electron imaging of free surface and surface/dislocation interactions *Ultramicroscopy* **29**, 135–46.

Takayanagi, K. (1989) Surface reconstructions and dynamics: UHV electron microscopy and diffraction studies *Int. J. Mod. Phys.* B **3**, 509–19.

Takayanagi, K., Tanishiro, Y., Ishitaba, T. and Akiyama, K. (1989) *In-situ* UHV electron microscope study of metal on silicon surfaces *Appl. Surf. Sci.* **41–42**, 337–41.

Takayanagi, K., Tanishiro, Y., Murooba, K. and Mitome, M. (1989) High resolution UHV electron microscopy on surfaces and heteroepitaxy *Mater. Res. Soc. Symp. Proc.* **139**, 59–66.

Tanishiro, Y., and Takayanagi, K. (1989) Dynamic observation of gold adsorption on Si(111)7 × 7 surface by high resolution reflection electron microscopy *Ultramicroscopy* **31**, 20–8.

Wang, Z. L. and Howie, A. (1989) Electron beam radiation damages of α-alumina $(01\bar{1}1)$ surfaces with different atomic terminations *Surf. Sci.* **226**, 293–306.

Wang. Z. L. and Howie, A. (1989) REM studies of surface cleavage, friction and damage processes, in *Proc. EMAG-MICRO*, Institute of Physics, Bristol, Vol. 1, pp. 17–20.

Yagi, K. (1989) Electron microscopy of surface structure *Adv. Optical Electron Microsc.* **11**, 57–100.

Yamamoto, N. (1989) Cross-sectional reflection electron microscopy of a GaAs–AlGaAs single quantum well structure *Japan. J. Appl. Phys.* **28**, L2147–9.

Yamanaka, A. and Yagi, K. (1989) Surface electromigration of metal atoms on Si(111) surfaces studied by UHV reflection electron microscopy *Ultramicroscopy* **29**, 161–7.

Yao, N., Wang, Z. L. and Cowley, J. M. (1989) REM and REELS identifications of atomic terminations at α-alumina $(01\bar{1}1)$ surface *Surf. Sci.* **208**, 533–49.

1988

Akiyama, K., Takayanagi, K. and Tanishiro, Y. (1988) UHV electron microscopy and diffraction analyses of the $\sqrt{3} \times 3$ structure formed by Pd on Si(111)7 × 7 *Surf. Sci.* **205**, 177–86.

Banzhof, H., Hermann, K. H. and Lichte, H. (1988) Reflexion electron interferometry of single crystal surfaces, in *EUREM 88*, Institute of Physics, Bristol, Vol. 1, pp. 263–4.

Cowley, J. M. (1988) Reflection electron microscopy in TEM and STEM instruments, in *Reflection High-Energy Electron Diffraction and Reflection Electron Imaging of Surfaces*, eds. P. K. Larsen and P. J. Dobson, NATO ASI Series B: Phys. Vol. 188, Plenum Press, New York, pp. 261–84.

Cowley, J. M. (1988) Reflection electron microscopy, in *Surface and Interface Characterization by Electron Optical Methods*, eds. A. Howie and U. Valdrè, Plenum Press, New York, pp. 127–58.

Cowley, J. M. (1988) High resolution electron microscopy of the solid-vacuum interface *J.*

Vac. Sci. Technol. A **6**, 1–4.

Hsu, T. and Peng, L. M. (1988) Contrast of surface steps and dislocations under resonance, non-resonance, Bragg, and non-Bragg conditions, in *Reflection High-Energy Electron Diffraction and Reflection Electron Imaging of Surfaces*, eds. P. K. Larsen and P. J. Dobson, NATO ASI Series B: Phys. Vol. 188, Plenum Press, New York, pp. 329–42.

Ichikawa, M. and Doi, T. (1988) Microprobe reflection high-energy electron diffraction, in *Reflection High-Energy Electron Diffraction and Reflection Electron Imaging of Surfaces*, eds. P. K. Larsen and P. J. Dobson, NATO ASI Series B: Phys. Vol. 188, Plenum Press, New York, pp. 343–70.

Ikarashi, N., Kobayashi, K., Koike, H., Hasegawa, H. and Yagi, K. (1988) Profile and plan-view imaging of reconstructed surface structures of gold *Ultramicroscopy* **26**, 195–204.

Kambe, K. (1988) Anomalous reflected images of objects on crystal surfaces in reflection electron microscopy *Ultramicroscopy* **25**, 259–64.

Latyshev, A. V., Aseev, A. L. and Stenin, S. I. (1988) Anomalous behaviour of monoatomic steps during structural transition (1 × 1)–(7 × 7) on atomic clean silicon surface (111) *JETP Lett.* **47**, 448–50.

Latyshev, A. V., Krasilnikov, A. B., Aseev, A. L. and Stenin, S. I. (1988) Effect of electric current on the ratio of the areas with (2 × 1) and (1 × 2) domains on clean (001) silicon surface during sublimation *JETP Lett.* **48**, 529–32.

Lehmpfuhl, G. and Uchida, Y. (1988) Electron diffraction conditions for surface imaging *Ultramicroscopy* **26**, 177–88.

Milne, R. H. (1988) Reflection microscopy in a scanning transmission electron microscope, in *Reflection High-Energy Electron Diffraction and Reflection Electron Imaging of Surfaces*, eds. P. K. Larsen and P. J. Dobson, NATO ASI Series B: Phys. Vol. 188, Plenum Press, New York, pp. 317–28.

Milne, R. H., Howie, A. and Walls, M. G. (1988) Reflection imaging using a scanning transmission electron microscope, in *Proc. EUREM*, Institute of Physics, Bristol, Vol. 1, p. 261.

Ogawa, S., Tanishiro, Y. and Yagi, K. (1988) *In situ* studies of fast atom bombardment and annealing processes by reflection electron microscopy *Nuc. Instrum. Methods Phys. Res.* B **33**, 474–8.

Osakabe, N., Matsuda, T., Endo, J. and Tonomura, A. (1988) Observation of atomic steps by reflection electron holography *Japan. J. Appl. Phys.* **27**, L1772–4.

Ou, H.-J. and Cowley, J. M. (1988) The surface reaction of Pd/MgO studied by scanning reflection electron microscopy *Phys. Stat. Sol.* **107**, 719–29.

Peng, L.-M. and Cowley, J. M. (1988) Surface resonance effects and beam convergence in REM *Ultramicroscopy* **26**, 161–8.

Peng, L.-M. and Cowley, J. M. (1988) A multislice approach to the RHEED and REM calculation *Surf. Sci.* **199**, 609–22.

Spence, J. C. H. (1988) A scanning tunneling microscopy in a side-entry holder for reflection electron microscopy in the Philips EM 400 *Ultramicroscopy* **25**, 165–70.

Twomey, T., Uchida, Y., Lehmpfuhl, G. and Kolb, D. M. (1988) The initial stages of electrochemically induced facetting of platinum: an electron microscopic investigation *Z. Physik. Chemie (NF)* **160**, 1–10.

Uchida, Y. (1988) Application of reflection electron microscopy for surface science (observation of cleaned crystal surfaces of Si, Pt, Au and Ag), in *Reflection High-Energy Electron Diffraction and Reflection Electron Imaging of Surfaces*, eds. P. K. Larsen and P. J. Dobson,

NATO ASI Series B: Phys. Vol. 188, Plenum Press, New York, pp. 302–16.

Wang, Z. L. (1988) Dynamical investigation of electron channeling process in a stepped crystal surface in reflection electron microscopy *Ultramicroscopy* **24**, 371–86.

Wang, Z. L. (1988) Reflection high resolution analytical electron microscopy: a technique for studying crystal surfaces *J. Electron Microsc. Techniques* **10**, 35–43.

Wang, Z. L. (1988) REM and REELS characterization of MgO (100) surfaces – a technique for studying the cleaving mechanisms and reactions of crystal surfaces *Mater. Lett.* **6**, 105–11.

Wang, Z. L. (1988) Surface structures and beam induced metallic reduction in InP *Mater. Lett.* **7**, 40–43.

Wang, Z. L. (1988) Dynamical investigations of electrons channelling on a stepped crystal surfaces in the REM geometry *Ultramicroscopy* **24**, 371–86.

Yagi, K., Ogawa, S. and Tanishiro, Y. (1988) Reflection electron microscopy with use of CTEM: studies of Au growth on Pt(111), in *Reflection High-Energy Electron Diffraction and Reflection Electron Imaging of Surfaces*, eds. P. K. Larsen and P. J. Dobson, NATO ASI Series B: Phys. Vol. 188, Plenum Press, New York, pp. 285–302.

1987

Aseev, A. L., Latyshev, A. V., Krasilnikov, A. B. and Stenin, S. I. (1987) Reflection electron microscopy study of the structure of atomic clean silicon surface, in *Defects in Crystal*, ed. E. Mizera, World Scientific, Singapore, pp. 231–7.

Buffat, P. A., Stadelmann, P., Ganiere, J. D., Martin, D. and Reinhart, F. K. (1987) HREM and REM observations of multiquantum well structures (AlGaAs/GaAs), in *Proc. Microscopy of Semiconducting Materials Conf.*, Institute of Physics, Bristol, pp. 207–12.

Canullo, J., Uchida, Y., Lehmpfuhl, G., Twomey, T. and Kolb, D. M. (1987) An electron microscopic investigation of the electrochemical facetting of platinum *Surf. Sci.* **188**, 350–63.

Cowley, J. M. (1987) High resolution imaging and diffraction studies of crystal surfaces *J. Electron Microsc.* **36**, 72–81.

Howie, A., Milne, R. H. and Walls, M. G. (1987) Dielectric excitations at surfaces and interfaces, *Proc. EMAG*, Institute of Physics, Bristol, pp. 327–32.

Hsu, T. (1987) Calibration of focusing steps and image rotation of electron microscopes *J. Electron Microsc. Techniques* **5**, 75–80.

Hsu, T., Cowley, J. M., Peng, L.-M. and Ou, H.-J. (1987) Reflection electron microscopy for the study of surfaces *J. Microsc.* **146**, 7–27.

Hsu, T. and Peng, L.-M. (1987) Experimental studies of atomic step contrast in reflection electron microscopy (REM) *Ultramicroscopy* **22**, 217–24.

Inoue, N., Tanishiro, Y. and Yagi, K. (1987) UHV-REM study of changes in the step structures of clean (100) silicon surfaces by annealing *Japan. J. Appl. Phys.* **26**, L293–5.

McKernan, S., De Cooman, B. C., Conner, J. R., Summerfelt, S. R. and Carter, C. B. (1987) Electron microscope imaging of III–V compound superlattices, in *Proc. Microscopy of Semiconducting Materials Conf.*, Institute of Physics, Bristol, pp. 201–6.

Milne, R. H. and Fan, T. W. (1987) Electron beam modifications of InP surfaces *J. Microsc.* **147**, 75.

Nakayama, T., Tanishiro, Y. and Takayanagi, K. (1987) Monolayer and bilayer high steps on

Si(001)2 × 1 vicinal surface *Japan. J. Appl. Phys.* **26**, L1186–8.

Ogawa, S., Tanishiro, Y., Takayanagi, K. and Yagi, K. (1987) Reflection electron microscope study of Pt(111) surfaces *J. Vac. Sci. Technol.* A **5**, 1735–8.

Ou, H.-J. and Cowley, J. M. (1987) SREM of MgO crystal surface structure and *in-situ*-deposited metallic particles on MgO surface *Ultramicroscopy* **22**, 207–16.

Ou, H.-J. and Cowley, J. M. (1987) Study of freshly deposited metallic particles on MgO crystal surfaces by scanning reflection electron microscopy *Ultramicroscopy* **23**, 263–70.

Peng, L.-M. and Cowley, J. M. (1987) Diffraction contrast in reflection electron microscopy I: screw dislocation *Micron Microsc. Acta* **18**, 171–8.

Peng, L. M., Cowley, J. M. and Hsu, T. (1987) Diffraction contrast in reflection electron microscopy II: surface steps and dislocations under the surface *Micron Microsc. Acta* **18**, 179–86.

Shimizu, N. and Muto, S. (1987) Reflection electron microscopic observation of high-temperature growth GaAs surfaces of molecular beam epitaxy *Appl. Phys. Lett.* **51**, 743–5.

Shimizu, N., Tanishiro, Y., Takayanagi, K. and Yagi, K. (1987) On the vacancy formation and diffusion on the Si(111)7 × 7 surfaces under exposures of low oxygen pressure studied by *in situ* reflection electron microscopy *Surf. Sci.* **191**, 28–44.

Takayanagi, K., Tanishiro, Y., Kobayashi, K., Akiyama, K. and Yagi, K. (1987) Surface structures observed by high-resolution UHV electron microscopy at atomic level *Japan. J. Appl. Phys.* **26**, L957–60.

Uchida, Y. and Lehmpfuhl, G. (1987) Observation of double contours of monoatomic steps on single crystal surfaces in reflection electron microscopy *Ultramicroscopy* **23**, 53–9.

Uchida, Y. and Lehmpfuhl, G. (1987) Reflection electron microscopic observation of crystal surfaces on an Si cylindrical specimen *Surf. Sci.* **188**, 364–77.

Yagi, K. (1987) Reflection electron microscopy *J. Appl. Cryst.* **20**, 147–60.

Zinke-Allmang, M., Feldman, L. C. and Nakahara, S. (1987) Role of Ostwald ripening in islanding processes *Appl. Phys. Lett.* **51**, 975–7.

1986

Claverie, A., Faure, J., Vieu, C., Beauvillain, J. and Jouffrey, B. (1986) RHEED and REM study of Si(111) surface degradation under Ar bombardment *J. Physique* **47**, 1805–12.

Cowley, J. M. (1986) Electron microscopy and surface structure *Prog. Surf. Sci.* **21**, 209–50.

Honda, K., Ohsawa, A. and Toyokura, N. (1986) Silicon surface roughness structural observation by reflection electron microscopy *Appl. Phys. Lett.* **48**, 779–81.

Howie, A. (1986) The revolution in the electron microscopy of surfaces *Proc. R. Microsc. Soc.* **21**, 141–4.

Hsu, T. and Nutt, S. R. (1986) High resolution imaging of as-grown sapphire surfaces, in *Materials Problem Solving with the Transmission Electron Microscope*, eds. L. W. Hobbs, K. H. Westmacott and D. B. Williams, Materials Research Society, New York, pp. 387–94.

Lehmpfuhl, G. and Dowell, W. C. T. (1986) Convergent-beam reflection high energy electron diffraction (RHEED) observations from a Si(111) surface *Acta Cryst.* A **42**, 569–77.

Marks, L. D. (1986) High resolution electron microscopy of surfaces, in *Structure and Dynamics of Surfaces I*, eds. W. Schommer and P. von Blanckenhagen, Springer-Verlag, Berlin, pp. 71–109.

Peng, L.-M. and Cowley, J. M. (1986) Dynamical diffraction calculation for RHEED and

REM *Acta Cryst.* A **42**, 545–52.

Smith, D. J. (1986) High-resolution electron microscopy in surface science, in *Chemistry and Physics of Solid Surfaces VI*, eds. R. Vanselow and R. Howe, Springer-Verlag, Berlin, pp. 413–33.

Smith, D. J. (1986) Atomic imaging of surfaces by electron microscopy *Surf. Sci.* **178**, 462–74.

Tanishiro, Y., Takayanagi, K. and Yagi, K. (1986) Observation of lattice fringes of the Si(111)-7 × 7 structure by reflection electron microscopy *J. Microsc.* **142**, 211–21.

Venables, J. A., Smith, D. J. and Cowley, J. M. (1986) HREM, STEM, REM, STEM – and STM *Surf. Sci.* **181**, 235–49.

1985

Beauvillain, J., Claverie, A. and Jouffrey, B. (1985) Double tilt heating specimen holder for surface image by reflection electron microscopy using an EM 300 Philips microscope *Rev. Sci. Instrum.* **56**, 418–20.

Cowley, J. M. and Peng, L.-M. (1985) The image contrast of surface steps in reflection electron microscopy *Ultramicroscopy* **16**, 59–69.

DeCooman, B. C., Kuester, K.-H. and Carter, C. B. (1985) Cross-sectional reflection microscopy of III–V compound epilayers *J. Electron Microsc. Techniques* **2**, 533–46.

Hsu, T. (1985) Superlattice image of cleaved $GaAs/Al_{0.39}Ga_{0.61}As$ by reflection electron microscopy (REM) *J. Vac. Sci. Technol.* B **3**, 1035–6.

Hsu, T. and Cowley, J. M. (1985) Reflection electron microscopy studies of crystal lattice terminations at surfaces, in *The Structure of Surfaces*, eds. M. A. Van Hove and S. Y. Tong, Springer-Verlag, Berlin, pp. 55–9.

Hsu, T. and Cowley, J. M. (1985) Surface characterization by reflection electron microscopy (REM), in *Advanced Photon and Particle Techniques for the Characterization of Defects in Solids*, eds. J. B. Roberto, R. W. Carpenter and M. C. Wittel, Materials Research Society, New York, pp. 121–7.

Shimizu, N., Tanishiro, Y., Kobayashi, K., Tagayanki, K. and Yagi, K. (1985) Reflection electron microscope study of the initial stages of oxidation of Si (111)-7 × 7 surfaces *Ultramicroscopy* **18**, 453–62.

Susnitzky, D. W., Kouh Simpson, Y., De Cooman, B. C. and Carter, C. B. (1985) *The Structure of Surface Steps on Low-Index Planes of Oxides*, Materials Research Society Symposium Proceedings, Materials Research Society, New York, Vol. 60, pp. 219–26.

1984

Cowley, J. M. (1984) Reflection electron microscopy and diffraction from crystal surfaces, in *Mater. Res. Soc. Proc.* **31**, 177–88.

Cowley, J. M. (1984) The use of scanning transmission electron microscopes to study surfaces and small particles, in *Catalytic Materials: Relationship Between Structure and Reactivity*, eds. T. E. Whyte, R. A. D. Betta, E. G. Derouane and R. T. K. Baker, American Chemical Society, Washington, DC, pp. 353–66.

Cowley, J. M. and Neumann, K. D. (1984) The alignment of gold particles on MgO crystal faces *Surf. Sci.* **145**, 301–12.

De Cooman, B. C., Kuester, K.-H., Wicks, G., Carter, C. B. and Hsu, T. (1984) Reflection electron microscopy of MBE-grown epilayers *Phil. Mag.* A **50**, 849–56.

Howie, A. and Milne, R. H. (1984) Energy loss spectra and reflection images from surfaces *J. Microsc.* **136**, 279–85.

Hsu, T., Ijima, S. and Cowley, J. M. (1984) Atomic and other structures of cleaved GaAs(110) surfaces *Surf. Sci.* **137**, 551–69.

Milne, R. H. and Howie, A. (1984) Electron microscopy of copper oxidation *Phil. Mag.* A **49**, 665–73.

Uchida, Y., Jaeger, J. and Lehmpfuhl, G. (1984) Direct imaging of atomic steps in reflection electron microscopy *Ultramicroscopy* **13**, 325–8.

Uchida, Y., Lehmpfuhl, G. and Jaeger, J. (1984) Observation of surface treatments on single crystals by reflection electron microscopy *Ultramicroscopy* **15**, 119–30.

Yamamoto, N. and Muto, S. (1984) Direct observation of $Al_xGa_{1-x}As/GaAs$ superlattices *Japan. J. Appl. Phys.* **23**, L806–8.

1983

Beauvillian, J., Claverie, A. and Jouffrey, B. (1983) Observation en microscopie électronique de la croissance d'oxides sur une surface de silicium *J. Cryst. Growth* **64**, 549–57.

Cowley, J. M. (1983) The STEM approach to the imaging of surfaces and small particles *J. Microsc.* **129**, 253–61.

Cowley, J. M. (1983) Microdiffraction in a STEM instrument and application to surface structures *Scanning Electron Microsc.* **1**, 51–60.

Cowley, J. M. and Kang, Z.-C. (1983) STEM imaging and analysis of surfaces *Ultramicroscopy* **11**, 131–40.

Howie, A. (1983) Surface reactions and excitations *Ultramicroscopy* **11**, 141–8.

Hsu, T. and Cowley, J. M. (1983) Reflection electron microscopy (REM) of fcc metals *Ultramicroscopy* **11**, 239–50.

Tanishiro, Y., Takayanagi, K. and Yagi, K. (1983) On the phase transition between the (7 × 7) and (1 × 1) structure of silicon (111) surface studied by reflection electron microscopy *Ultramicroscopy* **11**, 95–102.

Yamamoto, N. and Spence, J. C. H. (1983) Surface imaging of III–V semiconductors by reflection electron microscopy and inner potential measurements *Thin Solid Films* **104**, 43–55.

1982

Cowley, J. M. (1982) The accomplishments and prospects of high resolution imaging methods *Ultramicroscopy* **8**, 1–12.

Cowley, J. M. (1982) The STEM approach to the imaging of surfaces and small particles *J. Microsc.* **129**, 253–61.

Yagi, K. (1982) Surface studies by ultra-high-vacuum transmission and reflection electron microscopy, in *Scanning Electron Microscopy*, ed. O. Johari, SEM Inc., Chicago, Vol. 4, pp. 1421–8.

Yagi, K., Takayanagi, K. and Honjo, G. (1982) *In-situ* UHV electron microscopy of surfaces, in *Crystals, Growth, Properties and Applications*, Springer-Verlag, Berlin, Vol. 7, pp. 47–74.

1981

Osakabe, N., Tanishiro, Y., Yagi, K. and Honjo, G. (1981) Image contrast of dislocations and atomic steps on (111) silicon surface in reflection electron microscopy *Surf. Sci.* **10**, 424–42.

Osakabe, N., Tanishiro, Y., Yagi, K. and Honjo, G. (1981) Direct observation of the phase transition between the (7 × 7) and (1 × 1) structures of clean (111) silicon surfaces *Surf. Sci.* **109**, 353–66.

Turner, P. S. and Cowley, J. M. (1981) STEM and CTEM observations of interference between Laue- and Bragg-diffracted electrons in images of polyhedral crystals *Ultramicroscopy* **6**, 125–38.

Venables, J. A. (1981) Electron microscopy of surfaces *Ultramicroscopy* **7**, 81–98.

1980

Osakabe, N., Tanishiro, Y., Yagi, K. and Honjo, G. (1980) Reflection electron microscopy of clean and gold deposited (111) silicon surfaces *Surf. Sci.* **97**, 393–408.

Osakabe, N., Yagi, K. and Honjo, G. (1980) Reflection electron microscope observations of dislocations and surface structure phase transition on clean (111) silicon surfaces *Japan. J. Appl. Phys.* **19**, L309–12.

1979

Hembree, G. G., Cowley, J. M. and Otooni, M. A. (1979) The oxidation of copper studied by electron scattering techniques *Oxidation Metals* **13**, 331–51.

1977

Shuman, H. (1977) Bragg diffraction imaging of defects at crystal surfaces *Ultramicroscopy* **2**, 361–9.

1976

Hojlund Nielsen, P. E. and Cowley, J. M. (1976) Surface imaging using diffracted electrons *Surf. Sci.* **54**, 340–54.

Russel, G. J. and Wood, J. (1976) A specimen holder for reflection electron microscopy using a JEM 120 or 7A electron microscope *J. Phys.* E **9**, 98–9.

1975

Cowley, J. M. and Hojlund Nielsen, P. E. (1975) Magnification variations in reflection electron microscopy using diffracted beams *Ultramicroscopy* **1**, 145–50.

References

Aharonov, Y. and Bohm, D. (1959) *Phys. Rev.* **115** 485.

Alfonso, C., Bermond, J. M., Heyraud, J. C. and Metois, J. J. (1992) *Surf. Sci.* **262** 371.

Anstis, G. R. (1989) in *Computer Simulation of Electron Microscope Diffraction and Images*, eds. Krakow, W. and O'Keefe, M., The Minerals, Metals & Materials Society, Washington D.C., pp. 229.

Anstis, G. A. (1994) in *Proc. 13th Int. Cong. on Electron Microscopy*, eds. Jouffrey, B. and Colliex, C., Les Editions de Physique, Paris, p. 1027.

Anstis, G. R. and Gan, X. S. (1992) *Scanning Microsc. Suppl.* **6** 185.

Anstis, G. R. and Gan, X. S. (1994) *Surf. Sci.* **314** L919.

Ashley, J. C. and Ferrell, T. L. (1976) *Phys. Rev.* B **14** 3277.

Ashley, J. C., Ferrell, T. L. and Ritchie, R. H. (1974) *Phys. Rev.* **B 10** 554.

Atwater, H. A. and Ahn, C. C. (1991) *Appl. Phys. Lett.* **58** 269.

Atwater, H. A., Ahn, C. C., Nikzad, S. and Crozier, P. A. (1991) *Microbeam Analysis* 150.

Atwater, H. A., Wong, S. S., Ahn, C. C., Nikzad, S. and Frase, H. N. (1993) *Surf. Sci.* **198** 273.

Banzhof, H. and Herrmann, K. H. (1993) *Ultramicroscopy* **48** 475.

Banzhof, H., Herrmann, K. H. and Lichte, H. (1992) *Mater. Res. Technique* **20** 450.

Bauer, E. (1994) *Rep. Prog. Phys.* **57** 895.

Bauer, E., Mundschau, M., Swiech, W. and Telieps, W. (1989) *Ultramicroscopy* **31** 49.

Beeby, J. L. (1993) *Surf. Sci.* **298** 307.

Berger, S. D., Salisbury, I. G., Milne, R. H., Imeson, D. and Humphreys, C. J. (1987) *Phil. Mag.* B **55** 3410.

Berthelet, T., Kingery, W. D. and Van der Sande, J. B. (1976) *Ceram. Int.* **2** 62.

Bethe, H. (1928) *Ann. Phys.* **87** 55.

Binnig, G., Quate, C. F. and Gerber, C. (1986) *Phys. Rev. Lett.* **56** 930.

Binnig G., Rohrer, H., Gerber, C. and Weibel, E. (1983) *Phys. Rev. Lett.* **50** 120.

Björkman, G., Lundqvist, B. I. and Sjölander, A. (1967) *Phys. Rev.* **159** 551.

Bleloch, A. L., Howie, A., Milne, R. H. and Walls, M. G. (1987) in *Reflection High-Energy Electron Diffraction and Reflection Electron Imaging of Surfaces*, eds. Larsen, P. K. and Dobson, P. J., Plenum Press, New York, p. 77.

Bleloch, A. L., Howie A. and Milne, R. H. (1989b) *Ultramicroscopy* **31** 99.

Bleloch, A. L., Howie, A., Milne, R. H. and Walls, M. G. (1989a) *Ultramicroscopy* **29** 175.

Bohm, D. (1953) *Phys. Rev.* **92** 609.

Booker, G. R. (1970) in *Modern Diffraction and Imaging Techniques in Materials Science*, eds. Amelinckx, S. *et al.*, North-Holland, Amsterdam, pp. 57–68.

Born, M. (1942a) *Rep. Prog. Phys.* **9** 294.

Born, M. (1942b) *Proc. R. Soc. Lond.* A **180** 397.

Bostanjoglo, O. and Heinricht, F. (1987) *J. Phys.* E **20** 1491.

Brewer, L. and Searcy, A. W. (1951) *J. Am. Chem. Soc.* **73** 5307.

Brüesch, P. (1982) *Phonons: Theory and Experiments I – Lattice Dynamics and Models of Interatomic Force*, Springer-Verlag, Berlin.

Buseck, P. R., Cowley, J. M. and Eyring, L. (eds.) (1989) *High-Resolution Transmission Electron Microscopy and Associated Techniques*, Oxford University Press, New York.

Canullo, J., Uchida, Y., Lehmpfuhl, G., Twomey, T. and Kolb, D. M. (1987) in *Reflection High-Energy Electron Diffraction and Reflection Electron Imaging of Surfaces*, eds. Larsen, P. K. and Dobson, P. J., Plenum Press, New York, p. 350.

Castaing, R. and Henry, L. (1962) *C. R. Acad. Sci. Paris* B **255** 76.

Chadi, D. J. (1978) *J. Vac. Sci. Technol.* **5** 631.

Claverie, A., Beauvillain, J., Faure, J., Vieu, C. and Jouffrey, B. (1992) *Microsc. Res. Technique* **20** 352.

Colbourn, E. A., Kendrick, J. and Mackrodt, W. C. (1983) *Surf. Sci.* **126** 550.

Colella, R. (1972) *Acta Cryst.* A **28** 11.

Cowley, J. M. (1981) *Diffraction Physics*, 2nd edn, North-Holland, New York.

Cowley, J. M. (1986) *Prog. Surf. Sci.* **21** 209.

Cowley, J. M. (1987) in *Reflection High-Energy Electron Diffraction and Reflection Electron Imaging of Surfaces*, eds. Larsen, P. K. and Dobson, P. J., Plenum Press, New York, p. 261.

Cowley, J. M. (1988) in *High Resolution Transmission Electron Microscopy and Associated Techniques*, eds. Buseck, P., Cowley, J. and Eyring, L., Oxford University Press, New York, ch. 1.

Cowley, J. M. and Hojlund Nielsen, P. E. (1975) *Ultramicroscopy* **1** 145.

Cowley, J. M. and Liu, J. (1993) *Surf. Sci.* **298** 456.

Cowley, J. M. and Moodie, A. F. (1957) *Acta Cryst.* **10** 609.

Cowley, J. M. and Peng, L. M. (1985) *Ultramicroscopy* **16** 59.

Crozier, P. A. and Gajdardziska-Josifovska, M. (1993) *Ultramicroscopy* **48** 63.

Crozier, P. A., Gajdardziska-Josifovska, M. and Cowley, J. M. (1992) *Microsc. Res. Technique* **20** 426.

Dobson, P. J. (1987) in *Surface and Interface Characterization by Electron Optical Methods*, eds. Howie, A. and Valdrè, U., Plenum Press, New York, p. 159.

Dobson, P. J., Joyce, B. A., Neave, J. H. and Zhang, J. (1987) in *Surface and Interface Characterization by Electron Optical Methods*, eds. Howie, A. and Valdrè, U., Plenum Press, New York, p. 185.

Doyle, P. A. and Turner, P. S. (1968) *Acta Cryst.* A **24** 390.

Drucker, J., Krishnamurthy, M. and Hembree, G. (1991) *Ultramicroscopy* **35** 323.

Dudarev, S., Peng, L. M. and Ryazanov, M. I. (1991) *Acta Cryst.* A **47** 170.

Dudarev, S. L., Peng, L. M. and Whelan, M. J. (1993) *Phys. Rev.* B **48** 13408.

Dudarev, S. and Whelan, M. J. (1994) *Surf. Sci.* **310** 373.

Echenique, P. M., Flores, F. and Ritchie, R. H. (1990) *Solid State Phys.* **43** 229.

Edington, J. W. (1976) *Practical Transmission Electron Microscopy*, Van Nostrand Reinhold Co., New York.

Egerton, R. F. (1986) *Electron Energy-Loss Spectroscopy in the Electron Microscope*, Plenum Press, New York.

Egerton, R. F. (1989) *Ultramicroscopy* **28** 215.

Endo, A. (1993) *Surf. Sci.* **297** 71.

Endo, A. and Ino, S. (1993a) *Surf. Sci.* **293** 165.

Endo, A. and Ino, S. (1993b) *Japan J. Appl. Phys.* **32** 4718.

Ferrell, R. (1956) *Phys. Rev.* **101** 554.
Ferrell, R. (1957) *Phys. Rev.* **107** 450.
Ferrell, T. L. and Echenique, P. M. (1985) *Phys. Rev. Lett.* **55** 1526.
Fert, C. and Saport, R. (1952) *C. R. Acad. Sci. Paris* **235** 1490.
Floquet, N. and Dufour, L. C. (1983) *Surf. Sci.* **126** 543.
Fujimoto, F. and Komaki, K. (1968) *J. Phys. Soc. Japan* **25** 1679.
Fujiwara, K. (1961) *J. Phys. Soc. Japan* **16** 2226.

Gabor, D. (1949) *Proc. R. Soc. Lond.* A **197** 454.
Gajdardziska-Josifovska, M. (1994) *Microsc. Soc. Am. Bulletin* **24** 507.
Gajdardziska-Josifovska, M. and Cowley, J. M. (1991) *Acta Cryst.* A **47** 74.
Gajdardziska-Josifovska, M., Crozier, P. A. and Cowley, J. M. (1991) *Surf. Sci. Lett.* **248** L259.
Gajdardziska-Josifovska, M., McCartney, M. R. and Smith, D. J. (1993) *Surf. Sci.* **287** 1062.
Gajdardziska-Josifovska, M. and Smith, D. J. (1994) *Proc. 13th Int. Cong. on Electron Microscopy*, eds. Jouffrey, B. and Colliex, C., Les Editions de Physique, Paris, p. 1067.
Garcia-Molina, R., Gras-Marti A., Howie, A. and Ritchie, R. H. (1985) *J. Phys. C. Solid State Phys.* **18** 5335.
Gevers, R. and David, M. (1982) *Phys. Stat. Sol.* **113** 665.
Gjønnes, J. and Moodie, A. F. (1965) *Acta Cryst.* A **19** 65.

Halliday, J. S. and Newman, R. C. (1960) *Br. J. Appl. Phys.* **11** 158.
Harris, J. J., Joyce, B. A. and Dobson, P. J. (1981) *Surf. Sci.* **103** L90.
Hasegawa, S., Ino, S., Yamamoto, Y. and Daimon, H. (1985) *Japan J. Appl. Phys.* **24** L387.
Hayakawa, K. and Miyake, S. (1974) *Acta Cryst.* A **30** 374.
Heinricht, F. (1989) *Phys. Stat. Sol.* **116** 145.
Heinricht, F. and Bostanjoglo, O. (1992) *Appl. Surf. Sci.* **54** 244.
Hirsch, P. B., Howie, A., Nicholson, R. B., Pashley, D. W. and Whelan, M. J. (1977) *Electron Microscopy of Thin Crystals*, Krieger, New York.
Hojlund Nielsen, P. E. and Cowley, J. M. (1976) *Surf. Sci.* **54** 340.
Homma, Y., Suzuki, M. and Tomita, M. (1993b) *Appl. Phys. Lett.* **62** 3276.
Homma, Y., Tanishiro, Y. and Yagi, K. (1991a) *Surf. Sci.* **242** 81.
Homma, Y., Tomita, M. and Hayashi, T. (1991b) *Surf. Sci.* **258** 147.
Homma, Y., Tomita, M. and Hayashi, T. (1993a) *Ultramicroscopy* **52** 187.
Howie, A. (1962) *J. Phys. Soc. Japan* **17** suppl. B-H, 122.
Howie, A. (1983) *Ultramicroscopy* **11** 141.
Howie, A., Lanzerotti, M. Y. and Wang, Z. L. (1992) *Microsc. Microanal. Microstruct.* **3** 233.
Howie, A. and Milne, R. H. (1984) *J. Microsc.* **136** 279.
Howie, A. and Milne, R. H. (1985) *Ultramicroscopy* **18** 427.
Howie, A., Milne, R. H. and Walls, M. G. (1985) *Inst. Phys. Conf. Ser.* **78** 117.
Hsu, T. (1983) *Ultramicroscopy* **11** 167.
Hsu, T. (1985) *J. Vac. Sci. Technol.* B **3** 1035.
Hsu, T. (1992) *Microsc. Res. Technique* **20** 318.
Hsu, T. (1994) in *Proc. 52nd Annual Meeting of the Microscopy Society of America*, eds. Bailey, G. W. and Garratt-Rheed, A. J. San Francisco Press, San Francisco, p. 808.
Hsu, T. and Cowley, J. M. (1983) *Ultramicroscopy* **11** 239.
Hsu, T. and Cowley, J. M. (1985) in *The Structure of Surfaces*, eds. Van Hove, M. A. and Tong, S. Y., Springer-Verlag, New York, p. 55.

Hsu, T. and Cowley, J. M. (1994) *Ultramicroscopy* **55** 302.
Hsu, T., Cowley, J. M., Peng, L. M. and Ou, H. J. (1987) *J. Microsc.* **146** 17.
Hsu, T., Iijima, S. and Cowley, J. M. (1984) *Surf. Sci.* **137** 551.
Hsu, T. and Kim, Y. (1990) *Ultramicroscopy* **32** 103.
Hsu, T. and Kim, Y. (1991) *Surf. Sci.* **258** 119.
Hsu, T. and Nutt, S. R. (1986) *Mater. Res. Soc. Proc.* **62** 387.
Hsu, T. and Peng, L. M. (1987a) in *Reflection High-Energy Electron Diffraction and Reflection Electron Imaging of Surfaces*, eds. Larsen, P. K. and Dobson, P. J., Plenum Press, New York, p. 329.
Hsu, T. and Peng, L. M. (1987b) *Ultramicroscopy* **22** 217.
Hsu, T. and Peng, L. M. (1993) *Ultramicroscopy* **48** 489.
Hsu, T., Petrich, G. S. and Cohen, P. I. (1990) *Proc. Int. Cong. Electron Microscopy*, eds. Peachey, L. D. and Williams, D. B., San Francisco Press, San Francisco, Vol. **4** p. 626.
Hsu, T., Wang, T. Y. and Stringfellow, G. B. (1991) *J. Microsc.* **163** 275.
Huang, Y., Gajdardziska-Josifovska, M. and Cowley, J. M. (1994) in *Proc. 52nd Annual Meeting of the Microscopy Society of America*, eds. Bailey, G. W. and Garratt-Rheed, A. J., San Francisco Press, San Francisco, p. 804.
Humphreys, C. J. (1979) *Rep. Prog. Phys.* **42** 1825.
Humphreys, C. J., Salisbury, I. G., Berger, S. D., Timsit, R. S. and Mochel, M. E. (1985) *Inst. Phys. Conf. Ser.* **78** 1.

Ibach, H. and Mills, D. A. (1982) *Electron Energy Loss Spectroscopy*, Academic Press, London.
Ichikawa, M. and Doi, T. (1987) in *Reflection High-Energy Electron Diffraction and Reflection Electron Imaging of Surfaces*, eds. Larsen, P. K. and Dobson, P. J., Plenum Press, New York, p. 343.
Ichikawa, M. and Doi, T. (1990) *Vacuum* **41** 933.
Ichikawa, M., Doi, T., Ichihashi, M. and Hayakawa, M. (1985) *Japan. J. Appl. Phys.* **24** L387.
Ichimiya, A. (1983) *Japan. J. Appl. Phys.* **22** 176.
Ichimiya, A. (1993) *Ultramicroscopy* **48** 425.
Ichimiya, A., Kambe, K. and Lehmpfuhl, G. (1980) *J. Phys. Soc. Japan* **49** 684.
Imeson, D., Milne, R. H., Berger, S. D. and McMullan, D. (1985) *Ultramicroscopy* **17** 243.
Ino, S. (1977) *J. Appl. Phys.* **16** 891.
Ino, S. (1987) in *Reflection High-Energy Electron Diffraction and Reflection Electron Imaging of Surfaces*, eds. Larsen, P. K. and Dobson, P. J., Plenum Press, New York, p. 3.
Ino, S. (1993) *Microsc. Soc. Am. Bulletin* **23** 109.
Ino, S., Ichikawa, T., and Okada, S. (1980) *Japan J. Appl. Phys.* **19** 1451.
Ino, S. and Yamanaka, T. (1993) *Surf. Sci.* **298** 432.
Ino, S., Yamanaka, T. and Ito, S. (1993) *Surf. Sci.* **283** 319.
Inokuti, M. (1971) *Rev. Mod. Phys.* **43** 297.
Ishizuka, K. (1982) *Acta Cryst.* A **38** 773.
Ishizuka, K., Takayanagi, K., Tanishiro, Y. and Yagi, K. (1986) *Proc. 11th Int. Conf. on Electron Microscopy* Japanese Society of Electron Microscopy, Kyoto, Vol. **2** p. 1347.
Ishizuka, K. and Uyeda, N. (1977) *Acta Cryst.* A **33** 740.
Isu, T., Watanabe, A., Hata, M. and Katayama, Y. (1990) *J. Cryst. Growth* **100** 433.
Iwanari, S. and Takayanagi, K. (1991) *Japan. J. Appl. Phys.* **30** L1978.
Iwatsuki, M., Murooka, K., Kitamura, S., Takayanagi, K. and Harada, Y. (1991) *J. Electron Microsc.* **40** 48.

Jackson, A. G. (1991) *Handbook of Crystallography for Electron Microscopists and Others*, Springer-Verlag, New York.
Johnson, D. E. and Spence, J. C. H. (1974) *J. Phys. D: Appl. Phys.* 7 771.
Joyce, B. A., Neave, J. H., Zhang, J. and Dobson, P. J. (1987) in *Reflection High-Energy Electron Diffraction and Reflection Electron Imaging of Surfaces*, eds. Larsen, P. K. and Dobson, P. J., Plenum Press, New York, p. 397.

Kahata, H. and Yagi, K. (1989a) *Japan. J. Appl. Phys.* **28** L858.
Kahata, H. and Yagi, K. (1989b) *Surf. Sci.* **220** 131.
Kahata, H. and Yagi, K. (1989c) *Japan. J. Appl. Phys.* **28** L1042.
Kajiyama, K., Takayanagi, K., Tanishiro, Y. and Yagi, K. (1986) in *Proc. 11th Int. Conf. on Electron Microscopy* Japanese Society of Electron Microscopy, Kyoto, Vol. 2, p. 1341.
Kambe, K. (1988) *Ultramicroscopy* **25** 259.
Kang, Z. C. (1984) *Chinese Phys.* **4** 18.
Kelly, P. M., Jostons, A., Blake, R. G. and Napier, J. G. (1975) *Phys. Stat. Sol* a **31** 771.
Kikuchi, S. and Nakagawa, S. (1933) *Sci. Pap. Inst. Phys. Chem. Res.* **21** 256.
Kim, H. S. and Shenin, S. S. (1982) *Phys. Stat. Sol.* b **109** 807.
Kim, Y. and Hsu, T. (1991) *Surf. Sci.* **258** 131.
Kim, Y. and Hsu, T. (1992) *Surf. Sci.* **275** 339.
Kingery, W. D., Miramura, T., Van der Sande J. B. and Hall, E. L. (1979) *J. Mater. Sci.* **14** 1766.
Kliewer, K. L. and Fuchs, R. (1974) in *Aspects of the Study of Surfaces*, eds. I. Prigogine and S. A. Rice, Wiley (UK), **27** 355.
Knotek, M. L. and Feibelman, P. J. (1978) *Phys. Rev. Lett.* **40** 964.
Kohl, H. (1983) *Ultramicroscopy* **11** 53.
Kohl, H. and Rose, H. (1985) *Adv. Electronics Electron Phys.* **65** 173.
Kohra, K., Moliere, K., Nakano, S. and Ariyama, M. (1962) *J. Phys. Soc. Japan* **17** Suppl. B II, 82.
Koike, H., Kobayashi, K., Ozawa, S. and Yagi, K. (1989) *Japan. J. Appl. Phys.* **28** 861.
Kondo, Y., Yagi, K., Kobayashi, K., Kobayashi, H., Yanaka, Y., Kise K. and Ohkawa, T. (1991) *Ultramicroscopy* **36** 142.
Konopinski, E. J. (1981) *Electromagnetic Fields and Relativistic Particles*, McGraw-Hill Book Company, New York, section 10.5.
Korte, U. and Meyer-Ehmsen (1992a) *Surf. Sci.* **271** 616.
Korte, U. and Meyer-Ehmsen (1992b) *Surf. Sci.* **277** 109.
Korte, U. and Meyer-Ehmsen (1993a) *Surf. Sci.* **298** 299.
Korte, U. and Meyer-Ehmsen (1993b) *Phys. Rev.* B **48** 8345.
Krasilnikov, A. B., Latyshev, A. V., Aseev, A. L. and Stenin, S. I. (1992) *J. Cryst. Growth* **116** 178.
Krivanek, O. L., Gubbens, A. J. and Dellby, N. (1991) *Microsc. Microanal. Microstruct.* **2** 315.
Kronberg, M. L. (1957) *Acta Met.* **5** 508.

Lagally, M. G., Savage, D. E. and Tringides, D. L. (1987) in *Reflection High-Energy Electron Diffraction and Reflection Electron Imaging of Surfaces*, eds. Larsen, P. K. and Dobson, P. J., Plenum Press, New York, p. 139.
Landau, L. (1944) *J. Phys. USSR* **8** 201.

Latyshev, A. V., Aseev, A. L., Krasilnikov, A. B. and Stenin, S. I. (1989a) *Phys. Stat. Sol.* **113** 421.
Latyshev, A. V., Aseev, A. L., Krasilnikov, A. B. and Stenin, S. I. (1989b) *Surf. Sci.* **213** 157.
Latyshev, A. V., Krasilnikov, A. B. and Aseev, A. L. (1992) *Microsc. Res. Technique* **20** 341.
Latyshev, A. V., Krasilnikov, A. B. and Aseev, A. L. (1994) *Surf. Sci.* **311** 395.
Latyshev, A. V., Krasilnikov, A. B., Sokolov, L. B. and Stenin, S. I. (1988) *Pis'ma Zh. Eksp. Teor. Fiz.* **48** 484 (in Russian).
Latyshev, A. V., Krasilnikov, A. B., Sokolov, L. V. and Stenin, S. I. (1991) *Surf. Sci.* **254** 90.
Lee, T. C., Yen, M. Y., Chen, P. and Madhukar, A. (1986) *J. Vac. Sci. Technol.* A **4** 884.
Lee, V. C. and Wong, H. S. (1978) *J. Phys. Soc. Japan* **45** 895.
Lehmpfuhl, G. and Dowell, W. C. T. (1986) *Acta Cryst.* A **42** 569.
Lehmpfuhl, G. and Uchida, Y. (1988) *Ultramicroscopy* **26** 177.
Lehmpfuhl, G. and Uchida, Y. (1990) *Surf. Sci.* **235** 295.
Lewis, B. F., Lee, T. C., Grunthaner, F. J., Madhukar, A., Fernandez, R. and Maserjian, J. (1984) *J. Vac. Sci. Technol.* B **2** 419.
Li, Z. Q., Li, A. H., Qin, Y. and Shen, H. (1990) *Phil Mag. Lett.* **62** 125.
Lichte, H. (1986) *Ultramicroscopy* **20** 293.
Litvin, L. V., Krasilnikov, A. B. and Latyshev, A. V. (1991) *Surf. Sci. Lett.* **244** L121.
Liu, J. and Cowley, J. M. (1988) *Scanning Microsc.* **2** 65.
Liu, J. and Cowley, J. M. (1989) *Proc. 47th Annual Meeting of EMSA*, San Francisco Press, San Francisco, p. 542.
Liu, J. and Cowley, J. M. (1991) *Ultramicroscopy* **37** 50.
Liu, J. and Cowley, J. M. (1993) *Ultramicroscopy* **48** 381.
Liu, J., Crozier, P. A. and Cowley, J. M. (1990) in *Proc. XIIth Int. Cong. for Electron Microscopy*, eds. Peachey, L. D. and Williams D. B., San Francisco Press, San Francisco, Vol. 1, p. 334.
Liu, J., Hembree, G. G., Spinnler, G. E. and Venables, J. A. (1992a) *Surf. Sci.* **262** L111.
Liu, J., Wang, L. and Cowley, J. M. (1992b) *Surf. Sci. Lett.* **268** L293.
Lo, W. K. and Spence, J. C. H. (1993) *Ultramicroscopy* **48** 433.
Locquet, J. P. and Machler, E. (1994) *Mater. Res. Soc. Bulletin* **39** 39.
Lu, P., Liu, J. and Cowley, J. M. (1991) *Acta Cryst.* A **47** 317.
Lu, P. and Smith, D. J. (1991) *Surf. Sci.* **254** 119.
Lynch, D. F. and Moodie, A. F. (1972) *Surf. Sci.* **32** 422.
Lynch, D. F. and Smith, A. E. (1983) *Phys. Stat. Sol.* **119** 355.

Ma, Y. (1991) *Acta Cryst.* A **47** 137.
Ma, Y., Lordi, S., Flynn, C. P. and Eades, J. A. (1994) *Surf. Sci.* **302** 241.
Ma, Y., Lordi, S., Larsen, P. K. and Eades, J. A. (1993) *Surf. Sci.* **289** 47.
Ma, Y. and Marks, L. D. (1989) *Acta Cryst.* A **45** 174.
Ma, Y. and Marks, L. D. (1990) *Acta Cryst.* A **46** 594.
Ma, Y. and Marks, L. D. (1992) *Microsc. Res. Technique* **20** 371.
Mahan, J. E., Geib, K. M., Robinson, G. Y. and Long, R. G. (1990) *J. Vac. Sci. Technol.* A **8** 3692.
Maksym, P. A. and Beeby, J. L. (1981) *Surf. Sci.* **110** 423.
Marks, L. D. and Ma, Y. (1988) *Acta Cryst.* A **44** 392.
Marten, H. (1987) in *Reflection High-Energy Electron Diffraction and Reflection Electron Imaging of Surfaces*, eds. Larsen , P. K. and Dobson, P. J., Plenum Press, New York, p. 108.

Marten, H. and Meyer-Ehmsen, G. (1985) *Surf. Sci.* **151** 570.
Marten, H. and Meyer-Ehmsen, G. (1988) *Acta Cryst.* A **44** 853.
Maxwell-Garnett (1904) *Phil. Trans. R. Soc.* **203** 385.
Maxwell-Garnett (1906) *Phil. Trans. R. Soc.* **205** 237.
McCoy, J. M. and Maksym, P. A. (1993) *Surf. Sci.* **297** 113.
McCoy, J. M. and Maksym, P. A. (1994) *Surf. Sci.* **310** 217.
McRae, E. G. (1979) *Rev. Mod. Phys.* **51** 541.
Menter, J. W. (1953) *J. Photo. Sci.* **1** 12.
Meyer-Ehmsen, G. (1987) in *Reflection High-Energy Electron Diffraction and Reflection Electron Imaging of Surfaces*, eds. Larsen, P. K. and Dobson, P. J., Plenum Press, New York, p. 99.
Meyer-Ehmsen, G. (1989) *Surf. Sci.* **219** 177.
Milne, R. H. (1987) in *Reflection High-Energy Electron Diffraction and Reflection Electron Imaging of Surfaces*, eds. Larsen, P. K. and Dobson, P. J., Plenum Press, New York, p. 317.
Milne, R. H. (1989) *Ultramicroscopy* **27** 433.
Milne, R. H. (1990) *Surf. Sci.* **232** 17.
Mitra, S. S. and Massa, N. E. (1982) in *Handbook on Semiconductors*, Vol. 1, eds. Mos, T. S. and Paul, W., North Holland, Amsterdam, p. 81.
Miyake, S. (1937) *Sci. Pap. Inst. Phys. Chem. Res. Tokyo* **31** 161.
Miyake, S. and Hayakawa, K. (1970) *Acta Cryst.* A **26** 60.
Moon, A. R. (1972) *Z. Naturforsch.* a **27** 390.
Murooka, K, Tanishiro, Y. and Takayanagi, K. (1992) *Surf. Sci.* **275** 26.

Nakada, T., Ikeda, T., Yata, M. and Osaka, T. (1989) *Surf. Sci.* **222** L825.
Nakahara, H. and Ichikawa, M. (1993) *Surf. Sci.* **298** 440.
Nakahara, H., Ichikawa, M. and Stoyanov, S. (1993) *Ultramicroscopy* **48** 417.
Nakayama, H., Nishino, T., Ueda, K., Hoshi, T., Taguchi, M. and Oiwa, R. (1993) *J. Vac. Soc. Japan.* **36** 869.
Nakayama, H., Nichino, T., Ueda, K., Takeno, S. and Fujita, H. (1991) *Surf. Sci.* **39** 329.
Nakayama, H., Takenaka, T., Maeda, H., Fujita, H. and Ueda, K. (1990) *J. Electronic Mater.* **19** 801.
Ndubuisi, G. C., Liu, J. and Cowley, J. M. (1992a) *Microsc. Res. Technique* **20** 439.
Ndubuisi, G. C., Liu, J. and Cowley, J. M. (1992b) *Microsc. Res. Technique* **21** 10.
Neave, J. H., Joyce, B. A., Dobson, P. J. and Norton, N. (1983) *Appl. Phys.* A **31** 1.
Nikzad, S., Ahn, C. C. and Atwater, H. A. (1992) *J. Vac. Sci. Technol.* B **10** 762.
Nozières, P. and Pine, D. (1958) *Phys. Rev.* **109** 741.

Ogawa, S., Tanishiro, Y., Takayanagi, K. and Yagi, K. (1986) in *Proc. 11th Int. Conf. on Electron Microscopy* Japanese Society of Electron Microscopy, Kyoto, Vol. 2, p. 1351.
Ogawa, S., Tanishiro, Y., Takayanagi, K. and Yagi, K. (1987) *J. Vac. Sci. Technol.* A **5** 1735.
Ogawa, S., Tanishiro, K. and Yagi, K. (1988) *Nucl. Instrum. Methods Phys. Res.* B **33** 474.
Ohkawa, T., Kobayashi, K., Kise, K., Shidahara, Y., Yagi, K. and Kondo, Y. (1994) *Proc. 13th Int. Cong. on Electron Microscopy*, eds. Jouffrey, B. and Colliex, C., Les Editions de Physique, Paris, p. 1035.
Ohse, N. and Yagi, K. (1989) *Surf. Sci.* **217** L430.
O'Keefee, M. and Spence, J. C. H. (1994) *Acta Cryst.* A **50** 33.
Osakabe, N. (1992) *Microsc. Res. Technique* **20** 457.

Osakabe, N., Endo, J., Matsuda, T., Tonomura, A. and Fukahara, A. (1989) *Phys. Rev. Lett.* **62** 2969.

Osakabe, N., Matsuda, T., Endo, J. and Tonomura, A. (1993) *Ultramicroscopy* **48** 483.

Osakabe, N., Tanishito, Y. and Yagi, K. (1981a) *Surf. Sci.* **102** 424.

Osakabe, N., Tanishiro, Y., Yagi, K. and Honjo, G. (1980) *Surf. Sci.* **97** 393.

Osakabe, N., Tanishiro, Y., Yagi, K. and Honjo, G. (1981b) *Surf. Sci.* **109** 353.

Ou, H. J. and Cowley, J. M. (1987) *Ultramicroscopy* **22**, 207.

Peng, L. M. (1995) *Adv. Electronics Electron Phys.* **90**, 205.

Peng, L. M. and Cowley, J. M. (1986) *Acta Cryst.* A **42** 545.

Peng, L. M. and Cowley, J. M. (1987) *J. Electron Microsc. Technique* **6** 43.

Peng, L. M. and Cowley, J. M. (1988a) *Acta Cryst.* A **44** 1.

Peng, L. M. and Cowley, J. M. (1988b) *Surf. Sci.* **199** 609.

Peng, L. M. and Cowley, J. M. (1988c) *Surf. Sci.* **201** 559.

Peng, L. M., Cowley, J. M. and Hsu, T. (1987) *Micron Microsc. Acta* **18** 171.

Peng, L. M., Cowley, J. M. and Hsu, T. (1989) *Ultramicroscopy* **29** 135.

Peng, L. M., Cowley, J. M. and Yao, T. (1988) *Ultramicroscopy* **29** 189.

Peng, L. M. and Czernuszka, J. T. (1991a) *Surf. Sci.* **243** 210.

Peng, L. M. and Czernuszka, J. T. (1991b) *Phil. Mag.* A **64** 533.

Pukite, P. R., Lent, C. S. and Cohen, P. I. (1985) *Surf. Sci.* **161** 39.

Qian, W., Spence, J. C. H. and Zuo, J. M. (1993) *Acta Cryst.* A **49** 436.

Raether, H. (1980) *Excitation of Plasmons and Interband Transitions by Electrons*, Springer-Verlag, New York.

Reimer, L., Bakenfelder, A., Fromm, I., Rennekamp, R. and Moss-Messemer, M. (1990) *EMSA Bulletin* **20** 73.

Reimer, L., Fromm, I. and Rennekamp, R. (1988) *Ultramicroscopy* **24** 339.

Rez, P., Weng, X. and Ma, H. (1991) *Microsc. Microanal. Microstruct.* **2** 143.

Ritchie, R. H. (1957) *Phys. Rev.* **106** 874.

Ritchie, R. H. (1981) *Phil. Mag.* A **44** 931.

Ritchie, R. H. and Howie, A. (1988) *Phil. Mag.* A **58** 753.

Ross, F. M. and Gibson, J. M. (1992) *Phys. Rev. Lett.* **68** 1782.

Ross, F. M., Gibson, J. M. and Twesten, R. D. (1994) *Surf. Sci.* **310** 243.

Ruska, E. (1933) *Z. Phys.* **83** 492.

Sakamoto, T., Funabashi, H., Ohta, K., Nakagawa, T., Kawai, N. J. and Kojima, T. (1984) *Japan. J. Appl. Phys.* **23** L657.

Sayers, D. E., Stern, E. A. and Lytle, F. W. (1971) *Phys. Rev. Lett.* **27** 1204.

Schattschneider, P. and Jouffrey, B. (1995) in *Energy Filtering Transmission Electron Microscopy*, Springer Series in Optical Science Vol. 71, Springer Verlag, Berlin, p. 151.

Scherzer, O. (1949) *J. Appl. Phys.* **20** 20.

Schwoebel, R. N. and Shipsey, E. J. (1966) *J. Appl. Phys.* **37** 3682.

Senoussi, S., Henry, L. and Castaing, R. (1971) *J. Microsc.* **11** 19.

Sewell, P. B. and Cohen, M. (1967) *Appl. Phys. Lett.* **11** 298.

Shimizu, N. and Muto, S. (1987) *Appl. Phys. Lett.* **51** 743.

Shimizu, N., Tanishiro, Y., Kobayashi, K., Takayanagi, K. and Yagi, K. (1985) *Ultramicroscopy* **18** 453.

Shuman, H. (1977) *Ultramicroscopy* **2** 361.

Shuman, H., Chang, C. F. and Somlyo, A. P. (1986) *Ultramicroscopy* **19** 121.
Sinha, S. K. (1973) *Crit. Rev. Solid State Sci.* **3** 273.
Smith, A. E., Lehmpfuhl, G. and Uchida, Y. (1992) *Ultramicroscopy* **41** 367.
Smith, A. E. and Lynch, D. F. (1988) *Acta Cryst.* A **44** 780.
Smith, D. J. (1987) in *Chemistry and Physics of Solid Surfaces VI.*, eds. Vanselow, R. and Howe, R., Springer-Verlag, New York, ch. 15.
Spence, J. C. H. (1988a) *Experimental High-Resolution Electron Microscopy*, 2nd edition, Oxford University Press, New York, p. 128.
Spence, J. C. H. (1988b) *Ultramicroscopy* **25** 165.
Spence, J. C. H. (1993) in *Electron Diffraction Techniques*, Vol. II, ed. Cowley, J. M., International Union of Crystallography and Oxford University Press, Oxford.
Spence, J. C. H. and Kim, Y. (1987) in *Reflection High-Energy Electron Diffraction and Reflection Electron Imaging of Surfaces*, eds. Larsen, P. K. and Dobson, P. J., Plenum Press, New York, p. 117.
Spence, J. C. H., Lo, W. and Kuwabara, M. (1990) *Ultramicroscopy* **33** 69.
Spence, J. C. H. and Mayer, J. (1991) in *Proc. 49th Annual Meeting of the Electron Microscopy Society of America*, ed. Bailey, G. W., San Francisco Press, San Francisco, p. 616.
Spence, J. C. H. and Zuo, J. M. (1992) *Electron Microdiffraction*, Plenum Press, New York.
Stern, E. A. (1974) *Phys. Rev.* B **10** 3027.
Stratton, J. S. (1941) *Electromagnetic Theory* McGraw-Hill, New York, pp. 28 and 573.
Susnitzky, D. W., Simpson, Y. K., De Cooman, B. C. and Carter, C. B. (1986) *Mater. Res. Soc. Symp. Proc.* **60** 219.
Suzuki, T., Tanishiro, Y., Minoda, H. and Yagi, K. (1994) *Proc. 13th Int. Cong. on Electron Microscopy*, eds. Jouffrey, B. and Colliex, C., Les Editions de Physique, Paris, p. 1033.
Swiech, W., Bauer, E. and Mundschau, M. (1991) *Surf. Sci.* **253** 283.

Tabor, D. (1979) in *The Properties of Diamond*, ed. Field, J. E., Academic Press, London, ch. 10.
Takagi, S. (1958a) *J. Phys. Soc. Japan* **13** 278.
Takagi, S. (1958b) *J. Phys. Soc. Japan* **13** 287.
Takayanagi, K., Tanishiro, Y., Takahashi, S. and Takahashi, M. (1985) *Surf. Sci.* **164** 367.
Takeguchi, M., Harada, K. and Shimizu, R. (1992) *J. Electron Microsc.* **41** 174.
Tanishiro, Y., Fukuyama, M. and Yagi, K. (1991b) in *The Structure of Surfaces III*, eds. Tong, S. Y., Van Hove, M. A., Takayanagi, K. and Xie, X. D., Spring-Verlag, Berlin, p. 623.
Tanishiro, Y. and Takayanagi, K. (1989) *Ultramicroscopy* **31** 20.
Tanishiro, Y., Takayanagi, K., Kanamori, H., Kobayashi, K. and Yagi, K. (1981) *Acta Cryst. Suppl.* A **37** C300.
Tanishiro, Y., Takayanagi, K. and Yagi, K. (1986) *J. Microsc.* **11** 95.
Tanishiro, Y., Takayanagi, K. and Yagi, K. (1991a) *Surf. Sci. Lett.* **258** L687.
Tanishiro, Y., Yagi, K. and Takayanagi, K. (1990) *Surf. Sci.* **234** 37.
Thornton, A. G. and Wilks, J. (1976) *J. Phys. D: Appl. Phys.* **9** 27.
Tonomura, A. (1987) *Rev. Mod. Phys.* **59** 639.
Tonomura, A. (1990) *Phys. Today*, April, p. 22.
Tougaard, S. and Kraaer, J. (1991) *Phys. Rev.* B **43** 1651.
Tsai, F. and Cowley, J. M. (1992) *Ultramicroscopy* **45** 43.

Uchida, Y. (1987) in *Reflection High-Energy Electron Diffraction and Reflection Electron Imaging of Surfaces*, eds. Larsen, P. K. and Dobson, P. J., Plenum Press, New York, p. 303.
Uchida, Y., Imbihl, R. and Lehmpfuhl (1992a) *Surf. Sci.* **275** 253.
Uchida, Y., Jäger, J. and Lehmpfuhl, G. (1984b) *Ultramicroscopy* **55** 308.
Uchida, Y. and Lehmpfuhl, G. (1987) *Ultramicroscopy* **23** 53.
Uchida, Y. and Lehmpfuhl, G. (1991) *Surf. Sci.* **243** 27.
Uchida, Y., Lehmpfuhl, G. and Imbihl, R. (1990) *Surf. Sci.* **234** 193.
Uchida, Y., Lehmpfuhl, G. and Jäger, J. (1984a) *Ultramicroscopy* **15** 119.
Uchida, Y., Wang, N. and Zeitler, E. (1994) *Ultramicroscopy* **55** 308.
Uchida, Y., Weinberg, G. and Lehmpfuhl, G. (1992b) *Microsc. Res. Technique* **20** 406.
Ueda, K., Nakayama, H., Sakinet, M. and Fujita, H. (1991) *Vacuum* **42** 547.

Van Dyck, D. (1985) *Adv. Electronics Electron Phys.* **65** 295.
Van Hove, H. M., Lent, C. S., Pukite, P. R. and Cohen, P. I. (1983) *J. Vac. Sci. Technol.* **B 1** 741.
Völkl, E. and Lichte, H. (1990) *Ultramicroscopy* **32** 177.
Vook, R. W. (1994) *Mater. Chem. Phys.* **36**, 199.

Walls, M. G. and Howie, A. (1989) *Ultramicroscopy* **28** 40.
Wang, L., Liu, J. and Cowley, J. M. (1994) *Surf. Sci.* **302** 141.
Wang, N., Uchida, Y. and Lehmpfuhl, G. (1992) *Ultramicroscopy* **45** 291.
Wang, N., Uchida, Y. and Lehmpfuhl, G. (1993) *Surf. Sci. Lett.* **284** L419.
Wang, Z. L. (1988a) *Ultramicroscopy* **24** 371.
Wang, Z. L. (1988b) *Mater. Lett.* **7** 40.
Wang, Z. L. (1988c) *Mater. Lett.* **6** 105.
Wang, Z. L. (1988d) *Ultramicroscopy* **26** 321.
Wang, Z. L. (1988e) *Micron Microsc. Acta* **19** 201.
Wang, Z. L. (1988f) *J. Electron Microsc. Technique* **10** 35.
Wang, Z. L. (1989a) *Surf. Sci.* **214** 44.
Wang, Z. L. (1989b) *Phil Mag.* **B 60** 617.
Wang, Z. L. (1989c) *Surf. Sci.* **215** 201; *ibid.* **215** 217.
Wang, Z. L. (1989d) *J. Electron Microsc. Technique* **14** 13.
Wang, Z. L. (1991) *J. Microsc.* **163** 261.
Wang, Z. L. (1992a) *Phil. Mag.* **B 65** 559.
Wang, Z. L. (1992b) *Surf. Sci.* **271** 477.
Wang, Z. L. (1993) *Rep. Prog. Phys.* **56** 997.
Wang, Z. L. (1995) *Elastic and Inelastic Scattering in Electron Diffraction and Imaging*, Plenum Press, New York.
Wang, Z. L. (1996) *Micron*, in press.
Wang, Z. L. and Bentley, J. (1991a) *Ultramicroscopy* **37** 39.
Wang, Z. L. and Bentley, J. (1991b) *Microsc. Microanal. Microstruct.* **2** 301.
Wang, Z. L. and Bentley, J. (1991c) *Ultramicroscopy* **37** 103.
Wang, Z. L. and Bentley, J. (1991d) *Mater. Res. Soc. Symp. Proc.* **208** 155.
Wang, Z. L. and Bentley, J. (1992) *Microsc. Res. Technique* **20** 390.
Wang, Z. L. and Bentley, J. (1993a) *Ultramicroscopy* **48** 64.
Wang, Z. L. and Bentley, J. (1993b) *Ultramicroscopy* **48** 465.
Wang, Z. L., Bentley, J. Clausing, R. E., Heatherly, L. and Horton, L. L. (1993) *Proc. 51st Annual Meeting of the Microscopy Society of America*, eds. Bailey, G. W. and Rieder, C. L., San Francisco Press, San Fancisco, p. 1006.

Wang, Z. L., Bentley, J., Horton, L. L. and McKee, R. A. (1992) *Surf. Sci.* **273** 88.
Wang, Z. L. and Cowley, J. M. (1987a) *Ultramicroscopy* **21** 77.
Wang, Z. L. and Cowley, J. M. (1987b) *Ultramicroscopy* **21** 335.
Wang, Z. L. and Cowley, J. M. (1988a) *J. Microsc. Spectrosc. Electron.* **13** 189.
Wang, Z. L. and Cowley, J. M. (1988b) *Surf. Sci.* **193** 501.
Wang, Z. L. and Egerton, R. F. (1989) *Surf. Sci.* **205** 25.
Wang, Z. L., Feng, Z. and Field J. E. (1991) *Phil. Mag.* A **63** 1275.
Wang, Z. L. and Howie, A. (1990) *Surf. Sci.* **226** 293.
Wang, Z. L., Liu, J. and Cowley, J. M. (1989b) *Acta Cryst.* A **45** 325.
Wang, Z. L., Liu, J. and Cowley, J. M. (1989c) *Surf. Sci.* **216** 528.
Wang, Z. L., Liu, J., Lu, P. and Cowley, J. M. (1989a) *Ultramicroscopy* **27** 101.
Wang, Z. L. and Lu, P. (1988) *Ultramicroscopy* **26** 217.
Wang, Z. L., Lu, P. and Cowley, J. M. (1987) *Ultramicroscopy* **23** 205.
Wang, Z. L. and Rez, P. (1987) in *Proc. 45th Annual Meeting of the Electron Microscopy Society of America*, ed. Baily, G., San Francisco Press, San Francisco, p. 120.
Wang, Z. L. and Shapiro, A. J. (1995a) *Surf. Sci.* **328** 141.
Wang, Z. L. and Shapiro, A. J. (1995b) *Surf. Sci.* **328** 159.
Wang, Z. L. and Shapiro, A. J. (1995c) in *Evolution of Thin-Film and Surface Structure and Morphology*, eds. Demczyk, B. G., Garfunkel, E., Clemens, B. M., Williams, E. D. and Cuomo, J. J., Materials Research Society, Pittsburgh.
Wang, Z. L. and Shapiro, A. J. (1995d) *Ultramicroscopy*, submitted.
Wang, Z. L. and Spence, J. C. H. (1990) *Surf. Sci.* **234** 98.
Warren, B. E. (1990) *X-ray Diffraction*, Dover Publication, Inc., New York.
Watanabe, M. (1957) *J. Phys. Soc. Japan* **12** 874.
Whelan, M. J. (1965a) *J. Appl. Phys.* **36** 2099.
Whelan, M. J. (1965b) *J. Appl. Phys.* **36** 2103.
Whelan, M. J. (1976) *J. Phys.* C **9** L195.
Wilks, E. M. and Wilks, J. (1972) *J. Phys. D: Appl. Phys.* **5** 1902.
Williams, B. G., Sparrow, T. G. and Egerton, R. F. (1984) *Proc. R. Soc. Lond.* A **393** 409.
Woll, E. J. and Kohn, W. (1962) *Phys. Rev.* **126** 1693.
Wong, S. S., Atwater, H. A. and Ahn, C. C. (1993) *Microbeam Analysis* 290.
Wood, E. A. (1964) *J. Appl. Phys.* **4** 1305.
Woodruff, D. P. and Delchar, T. A. (1994) *Modern Techniques of Surface Science*, 2nd Edition, Cambridge University Press, Cambridge.

Yagi, K. (1987) *J. Appl. Cryst.* **20** 147.
Yagi, K. (1993a) *Surf. Sci. Rep.* **17** 305.
Yagi, K. (1993b) in *Electron Diffraction Techniques*, ed. Cowley, J. M., International Union of Crystallography and Oxford University Press, Oxford, Vol. 260.
Yagi, K., Ogawa, S. and Tanishiro, Y. (1987) in *Reflection High-Energy Electron Diffraction and Reflection Electron Imaging of Surfaces*, eds. Larsen , P. K. and Dobson, P. J., Plenum Press, New York, p. 285.
Yagi, K., Osakabe, N., Tanishiro, Y. and Honjo, G. (1980) *Proc. 4th ICSS (Cannes)* Vol. 2, p. 1007.
Yagi, K., Yamanaka, A. and Homma, I. (1992) *Microsc. Res. Technique* **20** 333.
Yamaguchi, H. and Yagi, K. (1993) *Appl. Phys. Lett.* **55** 622.
Yamamoto, N. and Muto, S. (1984) *Japan. J. Appl. Phys.* **23** L806.
Yamamoto, N. and Spence, J. C. H. (1983) *Thin Solid Films* **104** 43.

Yamamoto, Y. (1993) *Surf. Sci.* **281** 253.
Yamanaka, A., Ohse, H. and Yagi, K. (1990) *Proc. XIIth Int. Cong. for Electron Microscopy (Seattle)*, San Francisco Press, San Francisco, p. 306.
Yamanaka, A., Tanishiro, Y. and Yagi, K. (1992) *Surf. Sci.* **264** 55.
Yamanaka, A. and Yagi, K. (1991) *Surf. Sci.* **242** 181.
Yamanaka, A., Yagi, K. and Yasunaga, H. (1989) *Ultramicroscopy* **29** 161.
Yamanaka, T., Endo, A. and Ino, S. (1993) *Surf. Sci.* **294** 53.
Yao, N. and Cowley, J. M. (1989) *Ultramicroscopy* **31** 149.
Yao, N. and Cowley, J. M. (1990) *Ultramicroscopy* **33** 237.
Yao, N. and Cowley, J. M. (1992) *Microsc. Res. Technique* **20** 413.
Yao, N., Wang, Z. L. and Cowley, J. M. (1989) *Surf. Sci.* **208** 533.
Yasunaga, H. and Natori, A. (1992) *Surf. Sci. Rep.* **15** 205.
Yen, M. Y., Lee, T. C., Chen, P. and Madhukar, A. (1986) *J. Vac. Sci. Technol.* B **4** 590.
Yoshioka, H. (1957) *J. Phys. Soc. Japan* **12** 618.

Zachariasen, W. H. (1945) *Theory of X-ray Diffraction in Crystals*, John Wiley & Sons, New York.
Zhao, T. C., Poon, H. C. and Tong, S. Y. (1988) *Phys. Rev.* B **38** 1172.
Zuo, J. M. and Liu, J. (1992) *Surf. Sci.* **271** 253.

Materials index

Subject index